Ligand Design in Metal Chemistry

Ligand Design in Metal Chemistry

Reactivity and Catalysis

Edited by

Mark Stradiotto
Department of Chemistry, Dalhousie University
Canada

Rylan J. Lundgren
Department of Chemistry, University of Alberta
Canada

WILEY

This edition first published 2016

Registered Office
John Wiley & Sons, Ltd, The Atrium, Southern Gate, Chichester, West Sussex, PO19 8SQ, United Kingdom

For details of our global editorial offices, for customer services and for information about how to apply for permission to reuse the copyright material in this book please see our website at www.wiley.com.

Library of Congress Cataloging-in-Publication Data

Names: Stradiotto, Mark, author. | Lundgren, Rylan, author.
Title: Ligand design in metal chemistry : reactivity and catalysis /
[edited by] Mark Stradiotto, Rylan Lundgren.
Description: Chichester, UK ; Hoboken, NJ : John Wiley & Sons, 2016. |
Includes bibliographical references and index.
Identifiers: LCCN 2016023026 | ISBN 9781118839836 (cloth) | ISBN 9781118839812 (epub)
Subjects: LCSH: Ligands. | Organometallic compounds–Reactivity. | Homogeneous catalysis.
Classification: LCC QD474 .L54 2016 | DDC 546/.3–dc23
LC record available at https://lccn.loc.gov/2016023026

A catalogue record for this book is available from the British Library.

Set in 10.5/12.5pt Times by SPi Global, Pondicherry, India

Printed in the UK

Contents

List of Contributors xii
Foreword by Stephen L. Buchwald xiv
Foreword by David Milstein xvi
Preface xvii

1 Key Concepts in Ligand Design: An Introduction **1**
Rylan J. Lundgren and Mark Stradiotto
1.1 Introduction 1
1.2 Covalent bond classification and elementary bonding concepts 2
1.3 Reactive versus ancillary ligands 4
1.4 Strong- and weak-field ligands 4
1.5 *Trans* effect 6
1.6 Tolman electronic parameter 6
1.7 Pearson acid base concept 8
1.8 Multidenticity, ligand bite angle, and hemilability 8
1.9 Quantifying ligand steric properties 10
1.10 Cooperative and redox non-innocent ligands 12
1.11 Conclusion 12
References 13

2 Catalyst Structure and *Cis–Trans* Selectivity in Ruthenium-based Olefin Metathesis **15**
Brendan L. Quigley and Robert H. Grubbs
2.1 Introduction 15
2.2 Metathesis reactions and mechanism 17
2.2.1 Types of metathesis reactions 17
2.2.2 Mechanism of Ru-catalyzed olefin metathesis 19
2.2.3 Metallacycle geometry 19
2.2.4 Influencing syn–anti *preference of metallacycles* 22
2.3 Catalyst structure and *E*/*Z* selectivity 24
2.3.1 Trends in key catalysts 24
2.3.2 Catalysts with unsymmetrical NHCs 26
2.3.3 Catalysts with alternative NHC ligands 29
2.3.4 Variation of the anionic ligands 31

2.4 Z-selective Ru-based metathesis catalysts 33
2.4.1 Thiophenolate-based Z-selective catalysts 33
2.4.2 Dithiolate-based Z-selective catalysts 34
2.5 Cyclometallated Z-selective metathesis catalysts 36
2.5.1 Initial discovery 36
2.5.2 Model for selectivity 37
2.5.3 Variation of the anionic ligand 38
2.5.4 Variation of the aryl group 40
2.5.5 Variation of the cyclometallated NHC substituent 41
2.5.6 Reactivity of cyclometallated Z-selective catalysts 42
2.6 Conclusions and future outlook 42
References 43

3 Ligands for Iridium-catalyzed Asymmetric Hydrogenation of Challenging Substrates **46**
Marc-André Müller and Andreas Pfaltz
3.1 Asymmetric hydrogenation 46
3.2 Iridium catalysts based on heterobidentate ligands 49
3.3 Mechanistic studies and derivation of a model for the enantioselective step 57
3.4 Conclusion 63
References 64

4 Spiro Ligands for Asymmetric Catalysis **66**
Shou-Fei Zhu and Qi-Lin Zhou
4.1 Development of chiral spiro ligands 66
4.2 Asymmetric hydrogenation 73
4.2.1 Rh-catalyzed hydrogenation of enamides 73
4.2.2 Rh- or Ir-catalyzed hydrogenation of enamines 73
4.2.3 Ir-catalyzed hydrogenation of α,β-unsaturated carboxylic acids 75
4.2.4 Ir-catalyzed hydrogenation of olefins directed by the carboxy group 78
4.2.5 Ir-catalyzed hydrogenation of conjugate ketones 79
4.2.6 Ir-catalyzed hydrogenation of ketones 80
4.2.7 Ru-catalyzed hydrogenation of racemic 2-substituted aldehydes via dynamic kinetic resolution 81
4.2.8 Ru-catalyzed hydrogenation of racemic 2-substituted ketones via DKR 82
4.2.9 Ir-catalyzed hydrogenation of imines 84
4.3 Carbon–carbon bond-forming reactions 85
4.3.1 Ni-catalyzed hydrovinylation of olefins 85
4.3.2 Rh-catalyzed hydroacylation 85
4.3.3 Rh-catalyzed arylation of carbonyl compounds and imines 86
4.3.4 Pd-catalyzed umpolung allylation reactions of aldehydes, ketones, and imines 87

4.3.5 *Ni-catalyzed three-component coupling reaction* 87
4.3.6 *Au-catalyzed Mannich reactions of azlactones* 89
4.3.7 *Rh-catalyzed hydrosilylation/cyclization reaction* 89
4.3.8 *Au-catalyzed [2 + 2] cycloaddition* 90
4.3.9 *Au-catalyzed cyclopropanation* 91
4.3.10 *Pd-catalyzed Heck reactions* 91
4.4 Carbon–heteroatom bond-forming reactions 91
4.4.1 *Cu-catalyzed N—H bond insertion reactions* 91
4.4.2 *Cu-, Fe-, or Pd-catalzyed O—H insertion reactions* 93
4.4.3 *Cu-catalyzed S—H, Si—H and B—H insertion reactions* 95
4.4.4 *Pd-catalyzed allylic amination* 95
4.4.5 *Pd-catalyzed allylic cyclization reactions with allenes* 97
4.4.6 *Pd-catalyzed alkene carboamination reactions* 98
4.5 Conclusion 98
References 98

5 Application of Sterically Demanding Phosphine Ligands in Palladium-Catalyzed Cross-Coupling leading to C(sp^2)—E Bond Formation (E = NH_2, OH, and F) 104
Mark Stradiotto and Rylan J. Lundgren
5.1 Introduction 104
5.1.1 *General mechanistic overview and ancillary ligand design considerations* 105
5.1.2 *Reactivity challenges* 107
5.2 Palladium-catalyzed selective monoarylation of ammonia 108
5.2.1 *Initial development* 109
5.2.2 *Applications in heterocycle synthesis* 110
5.2.3 *Application of Buchwald palladacycles and imidazole-derived monophosphines* 112
5.2.4 *Heterobidentate κ^2-P,N ligands: chemoselectivity and room temperature reactions* 115
5.2.5 *Summary* 117
5.3 Palladium-catalyzed selective hydroxylation of (hetero)aryl halides 117
5.3.1 *Initial development* 118
5.3.2 *Application of alternative ligand classes* 120
5.3.3 *Summary* 122
5.4 Palladium-catalyzed nucleophilic fluorination of (hetero)aryl (pseudo)halides 123
5.4.1 *Development of palladium-catalyzed C(*sp^2*)—F coupling employing (hetero)aryl triflates* 124
5.4.2 *Discovery of biaryl monophosphine ancillary ligand modification* 125
5.4.3 *Extending reactivity to (hetero)aryl bromides and iodides* 127
5.4.4 *Summary* 128
5.5 Conclusions and outlook 129
Acknowledgments 130
References 131

6 Pd-*N*-Heterocyclic Carbene Complexes in Cross-Coupling Applications **134**
Jennifer Lyn Farmer, Matthew Pompeo, and Michael G. Organ
6.1 Introduction 134
6.2 *N*-heterocyclic carbenes as ligands for catalysis 135
6.3 The relationship between *N*-heterocyclic carbene structure and reactivity 136
6.3.1 Steric parameters of NHC ligands 136
6.3.2 Electronic parameters of NHC ligands 138
6.3.3 Tuning the electronic properties of NHC ligands 139
6.4 Cross-coupling reactions leading to C–C bonds that proceed through transmetalation 140
6.5 Kumada–Tamao–Corriu 141
6.6 Suzuki–Miyaura 148
*6.6.1 The formation of tetra-*ortho*-substituted (hetero)biaryl compounds* 149
6.6.2 Enantioselective Suzuki–Miyaura coupling 153
6.6.3 Formation of sp^3–sp^3 or sp^2–sp^3 bonds 156
6.6.4 The formation of (poly)heteroaryl compounds 158
6.7 Negishi coupling 163
6.7.1 Mechanistic studies: investigating the role of additives and the nature of the active transmetalating species 166
6.7.2 Selective cross-coupling of secondary organozinc reagents 168
6.8 Conclusion 170
References 171

7 Redox Non-innocent Ligands: Reactivity and Catalysis **176**
Bas de Bruin, Pauline Gualco, and Nanda D. Paul
7.1 Introduction 176
7.2 Strategy I. Redox non-innocent ligands used to modify the Lewis acid–base properties of the metal 179
7.3 Strategy II. Redox non-innocent ligands as electron reservoirs 181
7.4 Strategy III. Cooperative ligand-centered reactivity based on redox active ligands 192
7.5 Strategy IV. Cooperative substrate-centered radical-type reactivity based on redox non-innocent substrates 195
7.6 Conclusion 200
References 201

8 Ligands for Iron-based Homogeneous Catalysts for the Asymmetric Hydrogenation of Ketones and Imines **205**
Demyan E. Prokopchuk, Samantha A. M. Smith, and Robert H. Morris
8.1 Introduction: from ligands for ruthenium to ligands for iron 205
8.1.1 Ligand design elements in precious metal homogeneous catalysts for asymmetric direct hydrogenation and asymmetric transfer hydrogenation 205

8.1.2 Effective ligands for iron-catalyzed ketone and imine reduction 212
8.1.3 Ligand design elements for iron catalysts 213
8.2 First generation iron catalysts with symmetrical [6.5.6]-P-N-N-P ligands 216
8.2.1 Synthetic routes to ADH and ATH iron catalysts 217
8.2.2 Catalyst properties and mechanism of reaction 218
8.3 Second generation iron catalysts with symmetrical [5.5.5]-P-N-N-P ligands 220
8.3.1 Synthesis of second generation ATH catalysts 220
8.3.2 Asymmetric transfer hydrogenation catalytic properties and mechanism 222
8.3.3 Substrate scope 226
8.4 Third generation iron catalysts with unsymmetrical [5.5.5]-P-NH-N-P′ ligands 227
8.4.1 Synthesis of bis(tridentate)iron complexes and P-NH-NH_2 ligands 227
8.4.2 Template-assisted synthesis of iron P-NH-N-P′ complexes 228
8.4.3 Selected catalytic properties 229
8.4.4 Mechanism 230
8.5 Conclusions 231
Acknowledgments 232
References 232

9 Ambiphilic Ligands: Unusual Coordination and Reactivity Arising from Lewis Acid Moieties **237**
Ghenwa Bouhadir and Didier Bourissou
9.1 Introduction 237
9.2 Design and structure of ambiphilic ligands 238
9.3 Coordination of ambiphilic ligands 242
9.3.1 Complexes featuring a pendant Lewis acid 242
9.3.2 Bridging coordination involving M → Lewis acid interactions 243
9.3.3 Bridging coordination of M—X bonds 248
9.3.4 Ionization of M—X bonds 250
9.4 Reactivity of metallic complexes deriving from ambiphilic ligands 251
9.4.1 Lewis acid enhancement effect in Si—Si and C—C coupling reactions 251
9.4.2 Hydrogenation, hydrogen transfer and hydrosilylation reactions assisted by boranes 255
9.4.3 Activation/functionalization of N_2 and CO 262
9.5 Conclusions and outlook 264
References 266

10 Ligand Design in Enantioselective Ring-opening Polymerization of Lactide **270**
Kimberly M. Osten, Dinesh C. Aluthge, and Parisa Mehrkhodavandi
10.1 Introduction 270
10.1.1 Tacticity in PLA 271
10.1.2 Metal catalysts for the ROP of lactide 272
10.1.3 Ligand design in the enantioselective polymerization of racemic lactide 274
10.2 Indium and zinc complexes bearing chiral diaminophenolate ligands 292
10.2.1 Zinc catalysts supported by chiral diaminophenolate ligands 292
10.2.2 The first indium catalyst for lactide polymerization 294
10.2.3 Polymerization of cyclic esters with first generation catalyst 295
10.2.4 Ligand modifications 296
10.3 Dinuclear indium complexes bearing chiral salen-type ligands 297
10.3.1 Chiral indium salen complexes 297
10.3.2 Polymerization studies 297
10.4 Conclusions and future directions 301
References 302

11 Modern Applications of Trispyrazolylborate Ligands in Coinage Metal Catalysis **308**
Ana Caballero, M. Mar Díaz-Requejo, Manuel R. Fructos, Juan Urbano, and Pedro J. Pérez
11.1 Introduction 308
11.2 Trispyrazolylborate ligands: main features 310
11.3 Catalytic systems based on Tp^xML complexes (M = Cu, Ag) 311
11.3.1 Carbene addition reactions 312
11.3.2 Carbene insertion reactions 314
11.3.3 Nitrene addition reactions 319
11.3.4 Nitrene insertion reactions 321
11.3.5 Oxo transfer reactions 322
11.3.6 Atom transfer radical reactions 324
11.4 Conclusions 326
Acknowledgments 326
References 327

12 Ligand Design in Modern Lanthanide Chemistry **330**
David P. Mills and Stephen T. Liddle
12.1 Introduction and scope of the review 330
12.2 C-donor ligands 333
12.2.1 Silylalkyls 333
12.2.2 Terphenyls 335
12.2.3 Substituted cyclopentadienyls 336
12.2.4 Constrained geometry cyclopentadienyls 338
12.2.5 Benzene complexes 340

12.2.6 Zerovalent arenes 342
12.2.7 Tethered N-*heterocyclic carbenes* 343
12.3 N-donor ligands 344
12.3.1 Hexamethyldisilazide 344
12.3.2 Substituted trispyrazolylborates 347
12.3.3 Silyl-substituted triamidoamine, $[N(CH_2CH_2NSiMe_2Bu^t)_3]^{3-}$ 348
12.3.4 NacNac, $\{N(Dipp)C(Me)CHC(Me)N(Dipp)\}^-$ 349
12.4 P-donor ligands 349
12.4.1 Phospholides 349
12.5 Multiple bonds 350
12.5.1 $Ln{=}CR_2$ 350
12.5.2 $Ln{=}NR$ 354
12.5.3 $Ln{=}O$ 355
12.6 Conclusions 356
Notes 357
References 357

13 Tight Bite Angle *N,O*-Chelates. Amidates, Ureates and Beyond 364
Scott A. Ryken, Philippa R. Payne, and Laurel L. Schafer
13.1 Introduction 364
13.1.1 N,O-*Proligands* 366
13.1.2 Preparing metal complexes 367
13.2 Applications in reactivity and catalysis 377
13.2.1 Polymerizations 377
13.2.2 Hydrofunctionalization 385
13.3 Conclusions 400
References 401

Index 406

List of Contributors

Dinesh C. Aluthge
The University of British Columbia, Canada

Ghenwa Bouhadir
CNRS, Université Paul Sabatier, France

Didier Bourissou
CNRS, Université Paul Sabatier, France

Bas de Bruin
University of Amsterdam, The Netherlands

Ana Caballero
Universidad de Huelva, Spain

M. Mar Díaz-Requejo
Universidad de Huelva, Spain

Jennifer Lyn Farmer
York University, Canada

Manuel R. Fructos
Universidad de Huelva, Spain

Robert H. Grubbs
California Institute of Technology, USA

Pauline Gualco
University of Amsterdam, The Netherlands

Stephen T. Liddle
The University of Manchester, UK

Rylan J. Lundgren
University of Alberta, Canada

Parisa Mehrkhodavandi
The University of British Columbia, Canada

David P. Mills
The University of Manchester, UK

Robert H. Morris
University of Toronto, Canada

Marc-André Müller
University of Basel, Switzerland

Michael G. Organ
York University, Canada

Kimberly M. Osten
The University of British Columbia, Canada

Nanda D. Paul
Indian Institute of Engineering Science and Technology, India

Philippa R. Payne
The University of British Columbia, Canada

Pedro J. Pérez
Universidad de Huelva, Spain

Andreas Pfaltz
University of Basel, Switzerland

Matthew Pompeo
York University, Canada

Demyan E. Prokopchuk
University of Toronto, Canada

Brendan L. Quigley
California Institute of Technology, USA

Scott A. Ryken
The University of British Columbia, Canada

Laurel L. Schafer
The University of British Columbia, Canada

Samantha A. M. Smith
University of Toronto, Canada

Mark Stradiotto
Dalhousie University, Canada

Juan Urbano
Universidad de Huelva, Spain

Qi-Lin Zhou
Nankai University, China

Shou-Fei Zhu
Nankai University, China

Foreword

Ligands have the ability to dramatically affect the way that metal complexes react.

In the context of this book, their ability to enhance the reactivity and/or selectivity in the transformation of small molecules is at the heart of the matter. In recent years there has been a growing emphasis on developing an understanding of how structural features of ligands play out in the catalytic transformations in which they are employed. In our work at MIT (described in part by Stradiotto and Lundgren in Chapter 5), we have found that the use of very bulky (steric), electron-rich (electronic) ligands can be particularly effective in palladium-catalyzed carbon–heteroatom bond-forming reactions. We have systematically examined how the change in ligand structure impacts the observed catalytic activity. In addition to the obvious effects of size and the arrangement of substituents, issues such as how coordination number affects the stability and reactivity of the catalytically active intermediates must be taken into account. Most of the basic strategies that we have relied upon were built on the fundamental research conducted by legions of chemists over the years. It is this continued, combined effort, that ultimately leads to successful outcomes.

This book describes the efforts of organic, inorganic and organometallic chemists to apply old principles and develop new ones in an incredible set of contexts. Those with experience in the field realize that good ligands for metals in one area of the periodic table often cannot be used when moving to the right or left. This has led to the need to find different creative solutions to, for example, develop catalysts for hydroamination reactions using group 4 metals rather than for the use of group 8 metals for asymmetric hydrogenation. The many exciting chapters in this book lay out how this has been achieved. Included are some of the most important and topical areas of research in organometallic chemistry. From the perspective of organic synthesis, olefin metathesis, asymmetric hydrogenation and palladium-catalyzed reactions have become some of the most widely used transformations in both the fine chemical industry and academia. The use of metals other than palladium, rhodium, iridium and ruthenium is of growing interest and chapters describing the use of iron catalysts for asymmetric hydrogenation and coinage metals for a variety of reactions are illustrative of this. The chemistry of early transition metal and lanthanide complexes which possess intriguing reactivity and with very different ligands than, for example, with palladium or rhodium is nicely described in two chapters. Finally, two chapters describe "less conventional" types of ligands: non-innocent ligands and ambiphilic ligands. The first of these describes a situation where the ligand may change structure or have some sort of secondary function (e.g., recognition). The second reflects ligands that combine donor and acceptor capabilities.

Overall, this book provides a broad overview of both many areas in which ligands hold sway and the means by which they accomplish this. I am certain it will serve as a great resource for students and practitioners in the field alike.

Stephen L. Buchwald
Department of Chemistry
Massachusetts Institute of Technology
USA

Foreword

These are great times for catalysis research. It is widely recognized that catalysis is of key importance in addressing the central societal needs of sustainability, including sustainable chemical synthesis, energy, and the environment. Aided by the current knowledge base in the field, and by advanced computational methods, much progress has already been made in catalytic design aimed at these goals.

The Editors of this book, Professors Mark Stradiotto and Rylan Lundgren, are to be commended for assembling an impressive book of excellent chapters, covering key aspects of the important and timely field of ligand design, which is of course essential to the development of selective and efficient reactions catalyzed by transition metal complexes.

Historically, the development of the fundamentals of ligand design has been largely driven by industrial needs. For example, some of the basic concepts, such as the Tolman ligand cone angle, and the Tolman electronic parameter, described by the Editors in the first chapter of this book, were postulated by Chad Tolman at DuPont Central Research in conjunction with the development of the industrially very important nickel-catalyzed process of butadiene hydrocyanation to adiponitrile *en route* to nylon 6,6, pioneered by Bill Drinkard. The success of this ligand design approach has led to further long-term intensive research on organometallic ligand design, as I had the privilege to personally experience in both industry and academia.

Several useful new families of ligands have evolved in the last few decades. Among those, NHC-type and pincer-type ligands have become quite popular and influential in organometallic chemistry and homogeneous catalysis. A particularly fascinating aspect for me is the ability of pincer-type complexes to effectively function by metal–ligand cooperation, in which both the ligand and the metal are involved in bond breaking and making. This has resulted in recent developments of various environmentally benign synthetic reactions, as well as findings relevant to sustainable energy.

I believe that the reported key concepts of ligand design and the catalytic reactions based on them, covered in this book by leading groups in this field, will capture the imagination of practitioners and students in this exciting field, and will likely lead to further exciting developments in catalysis.

David Milstein
The Kimmel Centre for Molecular Design
Department of Organic Chemistry
The Weizmann Institute of Science
Israel

Preface

Synthetic inorganic/organometallic chemistry represents a burgeoning field of study, in which the discovery of fundamentally new bonding motifs and stoichiometric reactivity can in turn underpin the practical development of catalytic substrate transformations on bench-top and industrial scales. The design and application of ancillary ligands to modify the reactivity properties of metal complexes has figured and continues to figure directly in enabling such advances. A number of important ancillary ligand design strategies have emerged that have served to advance the state-of-the-art across a range of reaction classes.

In recognizing the difficulty associated with comprehensively documenting all aspects of ancillary ligand design within a single, accessible monograph, we opted instead to assemble a diverse collection of cutting-edge chapters from international leaders in synthetic inorganic/organometallic chemistry and homogeneous catalysis that highlight the breadth and depth of modern ancillary ligand design. In some cases, we have directed the reader to allied texts that may be informative.

We envision that this book will be of particular interest to academic and industry practitioners working in the field of ancillary ligand design. Furthermore, given the significant impact of ancillary ligand design in transition metal catalysis, this text is also likely to be informative to scientists in the fields of synthetic organic chemistry, medicinal chemistry, polymer science, materials chemistry, and beyond. The relatively short "readable" chapters, each featuring a brief historical account followed by more advanced aspects of modern ancillary ligand design, renders this text well-suited to students in advanced undergraduate and graduate chemistry programs, as well as related short courses.

The book is organized into thirteen chapters, with Chapter 1 providing a brief overview of some of the key concepts and terminology that are employed within the ensuing chapters. Chapter 2 covers aspects of ancillary ligand design related to selectivity in ruthenium-catalyzed olefin metathesis. Chapter 3 describes the design of ancillary ligands for use in the iridium-catalyzed asymmetric hydrogenation of challenging unsaturated substrates, while Chapter 4 details the development of chiral spirocyclic ligands for such applications and beyond. Chapters 5 and 6 describe the development of sterically demanding phosphine and *N*-heterocyclic carbene ancillary ligands, respectively, for use in addressing challenges in palladium-catalyzed cross-coupling chemistry. Redox non-innocent ancillary ligands are the focus of Chapter 7, while Chapters 8 and 9 deal with divergent facets of metal–ligand cooperative reactivity. Ancillary ligand design related to the enantioselective ring-opening polymerization of lactide is the focus of Chapter 10, while the application of trispyrazolylborate ancillary ligands in advancing coinage-metal chemistry is presented in Chapter 11. Chapter 12 details ancillary ligand strategies employed in lanthanide chemistry. Finally, Chapter 13 is focused on the development of tight bite angle *N,O*-chelates and their application in supporting catalytically active early metal complexes.

Our goal is that the collective insights provided by these diverse chapters will serve to educate experts and novice readers alike, so as to inspire future advances in the field.

Mark Stradiotto and Rylan J. Lundgren
Halifax, Nova Scotia, Canada, and Edmonton, Alberta, Canada

1

Key Concepts in Ligand Design: An Introduction

Rylan J. Lundgren[1] **and Mark Stradiotto**[2]

[1] *Department of Chemistry, University of Alberta, Edmonton, Alberta, Canada T6G 2G2*
[2] *Department of Chemistry, Dalhousie University, 6274 Coburg Road, PO Box 15000, Halifax, Nova Scotia, Canada B3H 4R2*

1.1 Introduction

Organic or main-group molecules and ions that bind to metal centers to generate coordination complexes are referred to as ligands. Metal–ligand bonding interactions that arise upon coordination of a ligand to a metal serve both to modulate the electronic properties of the metal, and to influence the steric environment of the metal coordination sphere, thereby allowing for some control over the structure and reactivity of metal complexes. Thus, the fields of transition metal and organometallic chemistry, as well as homogeneous metal catalysis, have been greatly enriched by the design and study of new ligand motifs. An understanding of how ligands influence the structural and reactivity properties of metal species has allowed for the discovery of new and improved metal-catalyzed reactions that are exploited widely in the synthesis of a broad spectrum of molecules (e.g., pharmaceuticals) and materials (e.g., polymers). Moreover, such an understanding has enabled chemists to isolate and interrogate reactive intermediates of relevance to important biological or industrial processes, and to uncover fundamentally new modes of bonding between metal centers and

Ligand Design in Metal Chemistry: Reactivity and Catalysis, First Edition. Edited by Mark Stradiotto and Rylan J. Lundgren.

organic or main-group compounds. This chapter is meant to serve as a brief overview of what the authors believe are some of the important basic concepts when considering how ligands can alter the behavior of soluble metal complexes with respect to chemical reactivity and catalysis. General overviews of ligand structure, bonding, and nomenclature can be found in most introductory inorganic or organometallic textbooks, as can historical aspects of the importance of ligands in the development of these fields. We direct the reader to such resources for a more thorough treatment of the subject.[1]

1.2 Covalent bond classification and elementary bonding concepts

In most simple cases, ligands act as Lewis bases, donating electron density to Lewis acidic metal centers. A prevailing method to classify the number and type of interactions between a metal and ligand, the Covalent Bond Classification, has been formulated by Green and Parkin (Figure 1).[2] Using this formalism, neutral two electron donor fragments are described as L-type ligands. The metal–ligand bond can be considered a dative interaction, whereby the valence of the metal is not changed upon ligand coordination. For simplicity, formal atom charges on the donor (ligand) and acceptor (metal) atom are invariably not depicted in chemical structures featuring such L-type interactions. Examples of L-type ligands include many classical Lewis bases, such as amines and phosphines. Single electron donors (or alternatively described, anionic two electron donors), such as halides, alkoxides, or carbon-based aryl or alkyl groups, are described as X-type ligands. The metal–ligand bond can be considered a covalent bond whereby one electron comes from both the metal and the ligand, raising the valence of the metal by one upon ligand coordination. Certain molecules can bind to metals in a fashion such that they accept, rather than donate, two electrons and are classified as Z-type ligands. This type of dative interaction formally increases the valence state of the metal by two. The most common Z-type ligands feature B or Al acceptor atoms.

Ligands can bind to metals via one or more points of attachment, and/or can engage simultaneously in multiple bonding interactions with a metal center, via combinations

M←L	M—X	M→Z
L ligand 2-electron donor	**X** ligand 1-electron donor (1-electron from M)	**Z** ligand 0-electron donor (2-electrons from M)
M—PR_3	M—OR	M—BR_3
M—NR_3	M—aryl	M—AlR_3

Figure 1 Classification and examples of L, X, and Z ligands according to the Covalent Bond Classification method

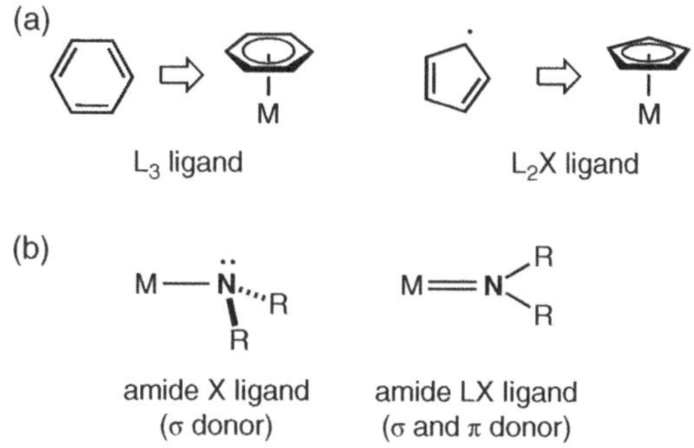

Figure 2 *(a) Examples of ligands which bind to metals via multiple L- or LX-type interactions. (b) Examples of metal–amide single (X) and double (LX) bonding*

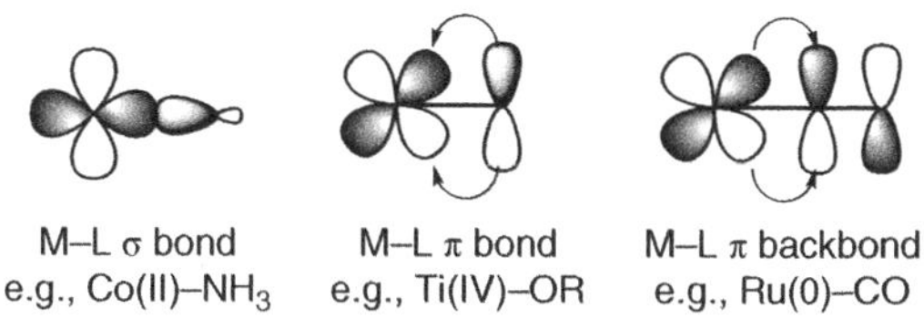

Figure 3 *Simplified schematic of metal–ligand σ and π bonding, as well as π backbonding*

of L-, X- and Z-type interactions. The type and strength of the metal–ligand bonding involved will depend on the metal and oxidation state, among other factors. Prototypical examples of such bonding scenarios include arene–metal structures, where the three double bonds of the aromatic act as electron pair donors (an L_3-type ligand), as well as the cyclopentadienyl group, an L_2X-type ligand (Figure 2a). Simultaneous LX-type bonding can also arise to generate formal M=L double bonds, as is prevalent in many amide and alkoxide complexes (Figure 2b). The classification of these ligands as X-type or LX-type ligands is usually evidenced by the crystallographically determined bond angles about the donor atom, in addition to the observed M–L interatomic distance.

From an elementary molecular orbital perspective, filled ligand orbitals, such as lone pairs, donate to metals to form metal–ligand σ bonds while generating an accompanying empty metal–ligand σ* orbital. Ligands can also donate electron density from orbitals of π symmetry. In instances where the metal has empty dπ orbitals, for example d^0 metals such as Ti^{4+}, the bond between the metal and the π-donor ligand can be particularly strong. Ligands possessing empty p orbitals or π* orbitals can act as π acids, accepting electron density from filled metal d orbitals of appropriate energy and symmetry (Figure 3). This type of π backbonding renders the metal center more electrophilic and strengthens the metal–ligand interaction. The combination of σ- and π-bonding interactions will dictate the overall M–L bond strength, as well as the reactivity properties of the M–L fragment.

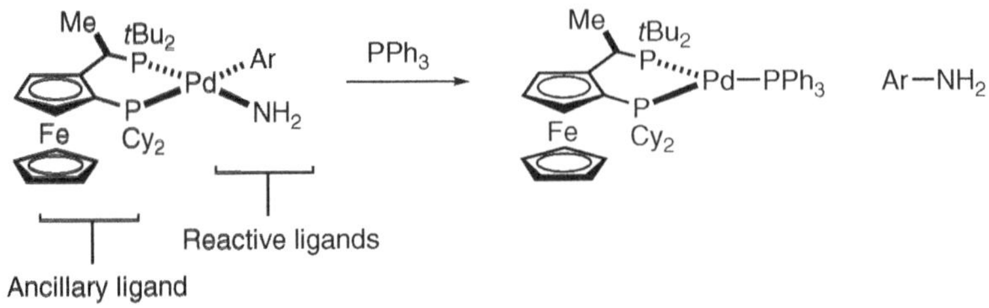

Figure 4 *An example of a metal complex with ancillary and reactive ligands*

1.3 Reactive versus ancillary ligands

When considering the behavior of ligands coordinated to a metal center, two general classifications arise. Reactive ligands, when bound to a metal, undergo chemical change, which can include irreversible chemical transformations or dissociation from the metal. Prototypical examples of reactive ligands include hydride, aryl or alkyl groups. Ancillary ligands are defined as supporting ligands that can modulate the reactivity of a metal center, but do not themselves undergo irreversible transformations (Figure 4). The contents of this book deal generally with ancillary ligand design aimed at modulating the behavior of reactive ligands in reaction chemistry and catalysis. Undesired ancillary ligand reactivity, such as oxidation or cyclometallation, is a common cause of metal complex decomposition or deactivation during catalysis. It should be noted that depending on the reaction setting, a coordinated ligand could behave in a reactive or ancillary manner; CO and olefins serve as examples of such ligands. Non-innocent and cooperative ligands,[3] discussed in more detail below, operate between these definitions.

1.4 Strong- and weak-field ligands

Ligands have a large influence over the electronic configuration (or spin state), as well as the geometry, of transition metal complexes. Moreover, the ability of ligands to act as π donors or π acceptors can alter the relative energies of the d orbitals on the ligated metal center. Ligands that are π-accepting, such as CO, CN^- or imine-type donors such as bipyridines, cause a large splitting in the energies of the d orbitals in a ligand field. For example, in ideal octahedral complexes the large energy difference between t_{2g} orbitals (d_{xz}, d_{yz}, d_{xy}) and e_g orbitals ($d_{x^2-y^2}$, d_{z^2}) causes metals of certain d-electron counts to adopt low-spin configurations, as in $Fe(CN)_6^{4-}$. Conversely, π-donating ligands, such as halides or alkoxides reduce the energy difference of the t_{2g} and e_g orbitals and promote high-spin configurations as in $Fe(H_2O)_6^{2+}$ (Figure 5).[1c] Similar trends occur for metals in other coordination geometries, such as tetrahedral or trigonal bipyramidal structures. The ability of ligands to act as donors or acceptors to induce changes in d-orbital energies (especially for octahedral complexes) can be easily assessed by use of spectroscopic methods, thus giving rise to the

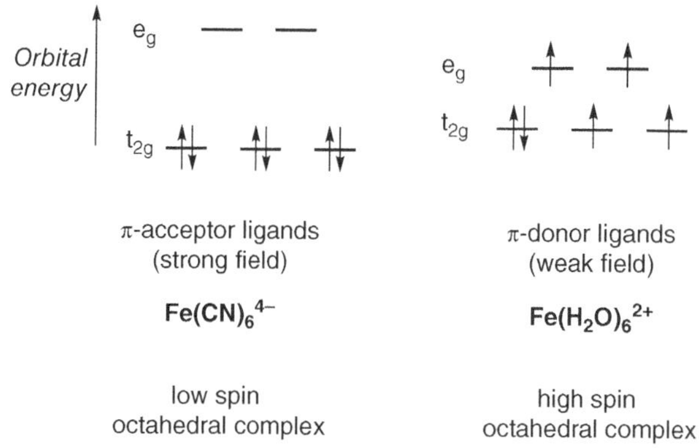

Figure 5 *Influence of weak-field and strong-field ligands on the spin state of two prototypical octahedral d⁶ metal complexes*

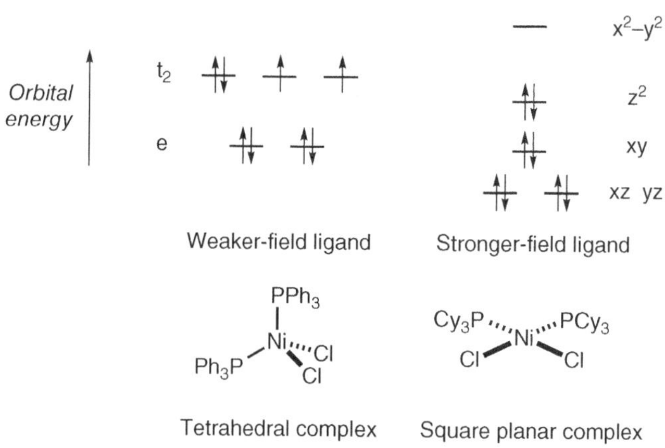

Figure 6 *Coordination geometry controlled by ligand field strength in four-coordinate Ni(II) complexes*

spectrochemical series, which ranks ligand π-bonding strength indirectly by measuring the octahedral e_g/t_{2g} energy gap.

Ligand field strength can also affect the geometry of transition metal complexes. An illustrative example is that of four coordinate d^8 complexes. Binding to weak-field ligands promotes the formation of tetrahedral complexes, for example $NiCl_4^{2-}$ or $NiCl_2(PPh_3)_2$, whereas strong-field ligands promote the formation of square planar complexes, such as $Ni(CN)_4^{2-}$ or $NiCl_2(PCy_3)_2$ (Figure 6). A similar phenomenon is observed with d^6 Fe(II) complexes, where strong field phosphine ligands can promote square planar geometries over the typically observed tetrahedral arrangement.[4] While strong-field or weak-field ligands generally influence coordination geometry to much lesser extent with second- or third-row transition metals [most Rh(I), Ir(I), Pd(II), and Pt(II) complexes are square planar], they can influence the relative d-orbital energies, thus altering the ordering of the metal-based molecular orbitals within derived coordination complexes.[5]

1.5 *Trans* effect

Ligand coordination can influence a metal ion so as to alter the kinetics of ligand substitution and the bond strengths of the donor groups located at the *cis* or *trans* positions. This topic has been described in detail elsewhere.[1b] The kinetic *trans* effect observed for square planar d^8 complexes is illustrative. In these cases, ligands that are good π acceptors or strong σ donors can increase the rate of associative ligand substitution at the *trans* position by several orders of magnitude. Upon the formation of the trigonal bipyramidal structure by incoming ligand association, strong π-acceptor ligands (such as olefins) bind favourably to the more π-basic equatorial sites and labilize the other equatorial positions (Figure 7). By contrast, strong σ donors, for example silyl or alkyl groups, weaken the *trans* M—L bonds in square planar species by overlapping with the same metal orbitals as those involved in bonding with the *trans* L group.

1.6 Tolman electronic parameter

The ability to measure and predict ligand donor (or acceptor) strength is an important tool in ligand design. Lone-pair basicity can be determined by pKa measurements of the corresponding conjugate acid, but as most metals are softer Lewis acids than a proton, these values can be misleading. The overall donor strength of a ligand when bonding with soft transition metals can be determined more accurately by measuring carbonyl stretching frequencies of ligated $M(CO)_n$ species, as originally described by Tolman's study of $Ni(CO)_3L$ species (Tolman electronic parameter, TEP).[6] In such complexes a reduction in the carbonyl stretching frequency wavenumber correlates to a metal center being made more electron rich via ligand (L) donation. Select TEP values for representative phosphine and carbene ligands are provided in Figure 8. More

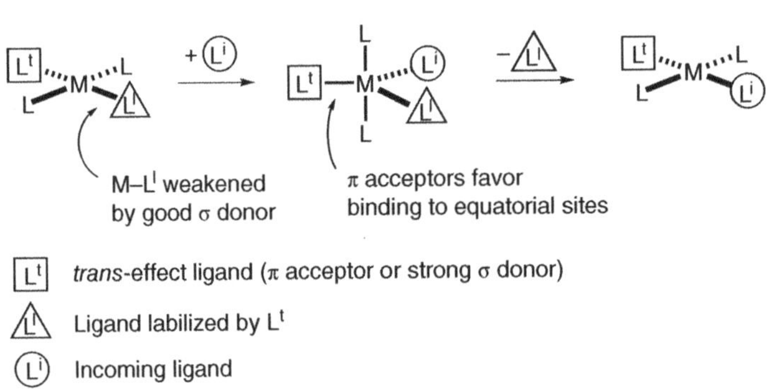

Figure 7 Overview of the trans effect for square planar complexes

Ligand	TEP (cm^{-1})
$P(C_6F_5)_3$	2091
$P(OPh)_3$	2085
$P(OBu)_3$	2077
PPh_3	2069
$P(NMe_2)_3$	2062
PEt_3	2062
$P(iPr)_3$	2059
$P(tBu)_3$	2056
IPr	2024
CAAC	2020

Increasing donor strength

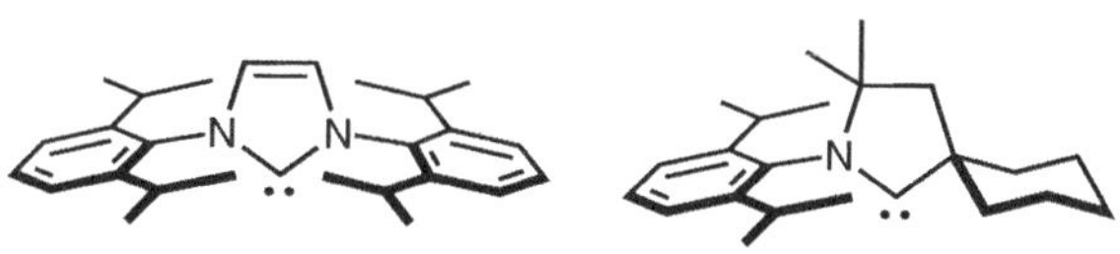

IPr CAAC

Figure 8 *Selected TEP values for phosphines and carbenes*

comprehensive data on a vast range of ligands are available in the literature, including values that have been obtained through computational analysis with other metals.[6,7] Given the toxicity of $Ni(CO)_4$, it is more common to benchmark ligand donicity experimentally with carbonyl stretching frequencies of $Ir(CO)_2ClL$ complexes, which Crabtree has correlated to the TEP.[8]

From the large collection of TEP data on ligand donor ability, a few generalizations can be made with regard to ligand structure and metal bonding. Trialkylphosphines are stronger donors than aryl phosphines. The donicity of aryl phosphines can be modulated by the introduction of P–*aryl*-group substituents, thus allowing for some control over the electron-richness of the ligated metal. Many *N*-heterocyclic carbenes are very strong donors, even stronger than bulky trialkylphosphines. Nitrogen-based ligands are generally poorer σ donors, especially when binding to low oxidation state late transition metals. Pyridines, imines, and related *N*-heterocyclic donors (such as oxazolines) are good π-acceptor ligands and can be used to enhance metal electrophilicity. These N-ligand frameworks have most commonly been exploited with success in combination with first-row transition metals (e.g., Fe, Ni, and Cu) or metals in relatively high oxidation states (e.g., Pt^{4+}). In all cases, the donor ability and nature of the metal–ligand interaction will depend highly on the transition metal, oxidation state, and other connected ligands.

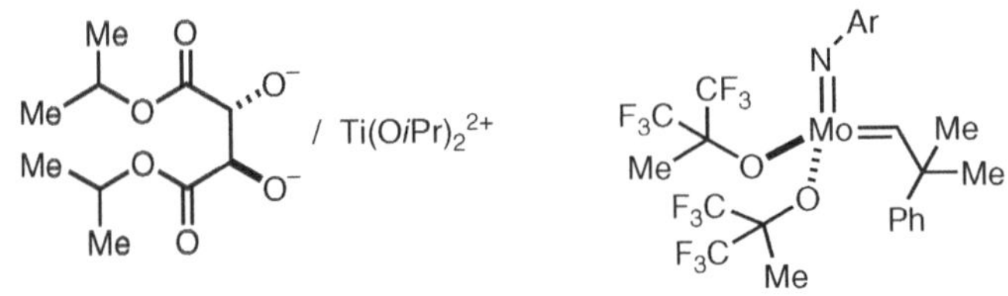

Sharpless' epoxidation catalyst   Schrock's olefin metathesis catalyst

Figure 9 *Representative early/mid transition metal catalysts featuring hard metal/ligand interactions*

Dow's LP Oxo Catalyst   High-Activity Cross-Coupling Catalyst

Figure 10 *Representative late transition metal catalysts featuring soft metal/ligand interactions*

1.7 Pearson acid base concept

Consideration of hard and soft Lewis acid/base properties (Pearson acid base concept) can provide an intuitive concept to estimate metal–ligand bond strength on the basis of electrostatics and orbital overlap.[9] Hard transition metals, such as high oxidation state complexes of the Group 3, 4, or 5 metals, as well as lanthanides, form strong bonds to hard Lewis basic ligands such as those featuring N- and O-donor atoms. Particularly noteworthy examples of catalysts demonstrating this trend include Sharpless' Ti catalysts for epoxidation[10] and Schrock's catalysts for olefin metathesis[11] (Figure 9). Conversely, larger and more polarizable transition metals in relatively low oxidation states bind strongly to softer donors, such as phosphines and carbenes. This helps explain the domination of these ligand classes for platinum-group metal catalyzed reactions such as Rh-catalyzed hydroformylation[12] and Pd-catalyzed cross-coupling (Figure 10).[13]

1.8 Multidenticity, ligand bite angle, and hemilability

Species that bind by more than one point of attachment to a metal center are described as chelating or polydentate ligands. The increased favorability of polydentate ligand binding to a given metal, compared with that of similar monodentate ligands, is referred to as the chelate effect. Chelating ligands are ubiquitous in transition metal bond activation and catalysis, as they can provide increased stability and a higher degree of control over the coordination environment of a metal compared with analogous monodentate ligands.

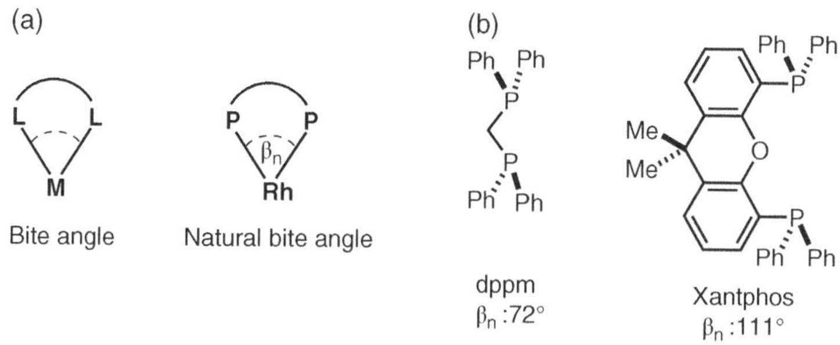

Figure 11 *Bite angle and natural bite angle of bidentate ligands, with representative examples of small (dppm) and large (Xantphos) bite angle bisphosphine ligands*

Bite angle is a parameter that is used to describe the angle between two donor atoms and the metal (i.e., the L–M–L angle; Figure 11a). Geometric constraints imposed by the backbone of chelating ligands can restrict how close the donor groups can be to a metal; as such, chelating ligands can bind with bite angles that are much larger or smaller than the geometrical ideal, thereby influencing the ground state or transition state of metal complexes undergoing a chemical reaction. The term natural bite angle (β_n) is used to describe the preferred chelation angle determined by the ligand backbone constraints, and is obtained using molecular mechanics calculations employing a standard Rh—P bond length of 231.5 pm, rather than being determined experimentally.[14] For example, bis(diphenylphosphino)methane (dppm) binds to five-coordinate Rh(I) with a β_n of 72°, whereas the wide bite angle bisphosphine Xantphos binds at an angle of 111°. Both observed structures represent a major deviation from the ideal bond angle of 90° between axial and equatorial positions in trigonal pyramidal structures, or *cis* positions in square planar structures (Figure 11b). A noteworthy example of the impact ligand bite angle has on metal-centered reactivity is seen in Rh-catalyzed hydroformylation reactions, in which increasing the ligand bite angle provides favorable reactivity.[14,15] The flexibility range of chelating ligands, defined as the range of bite angles adoptable within 3 kcal/mol of strain energy, is a useful term to judge the degree of allowable distortion chelating ligands can undergo upon metal binding. The widespread utility of Xantphos-type ligands in homogeneous catalyst highlights the importance of bite angle tuning in ligand design.

Conceptually similar "constrained geometry" catalysts feature distorted cyclopentadienyl/σ-donor ligands and have found widespread use in early transition metal catalysis, most notably polymerization.[16] In these cases, the bite angle of the cyclopentadienyl (Cp)–metal–amido is compressed by ~25–30° compared to unstrained metallocenes (Figure 12).[17]

A particularly important class of chelating ligands called pincers feature three adjacent, often coplanar, donors. Pincer complexes tend to display high stability due to their tight binding and rigid structure. Metal complexes supported by pincer ligands have been demonstrated to exhibit a wide range of reactivity in bond activation and catalysis.[17,18] A commonly employed pincer ligand motif features two phosphorus

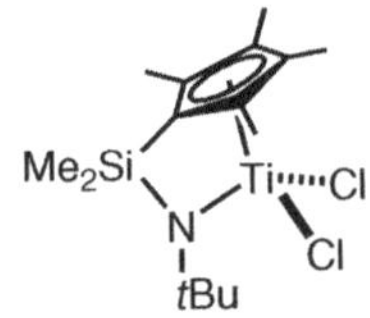

Figure 12 Constrained geometry catalyst precursor featuring Cp–metal–amido bond angle compression

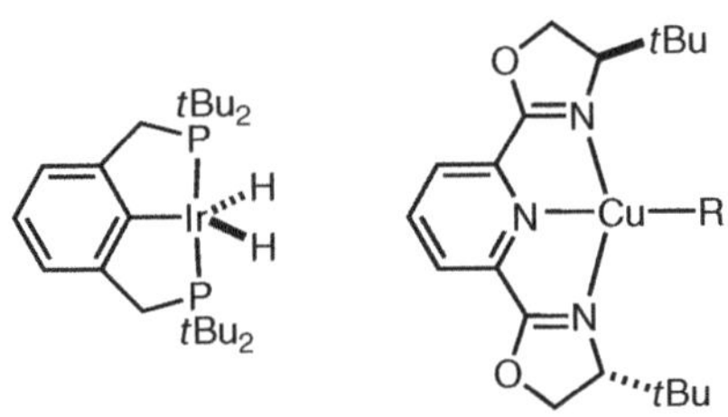

Figure 13 Structure of an Ir-PCP pincer complex and of a Cu-PyBox complex

L donors flanking a central cyclometalled aryl carbon group (i.e., an X donor), referred to as a PCP pincer (Figure 13), although an extremely broad range of pincer ligands with B, C, N, O, Si and P donor groups have been reported. Planar, tridentate ligands have also found widespread utility in metal-mediated asymmetric catalysis. For example, pyridine-centered bisoxazolines, termed PyBox ligands, have found utility in many Cu-, Ni-, and Ru-catalyzed reactions thanks to their tight chelation and tunable C_2 symmetry.[19] The utility of pincer complexes has been reviewed extensively and we direct the reader to a recent monograph.[20]

Flexible polydentate ligands featuring a combination of strong and weak donor groups (often a mix of hard and soft donors) can often undergo facile coordination/decoordination events in the presence of reactants or under catalytic conditions. This dynamic property is called hemilability. The design of hemilabile ligands offers the opportunity to employ relatively stable coordinatively saturated complexes, which in turn provide access to highly reactive low-coordinate metal species in the presence of substrate molecules. The remarkable effects of ligand hemilability can be found in both foundational and modern organometallic chemistry and catalysis. For example, Shaw demonstrated that a phosphine ligand containing a hemilabile methoxy group dramatically enhanced the rates of oxidation addition to Ir(I) complexes (Figure 14).[21]

1.9 Quantifying ligand steric properties

The space a ligand occupies around the metal center, or steric bulk, is an extremely important parameter for modulating the reactivity and stability of complexes. Ligand sterics are most often modified by changing the substituent groups of the donor atom.

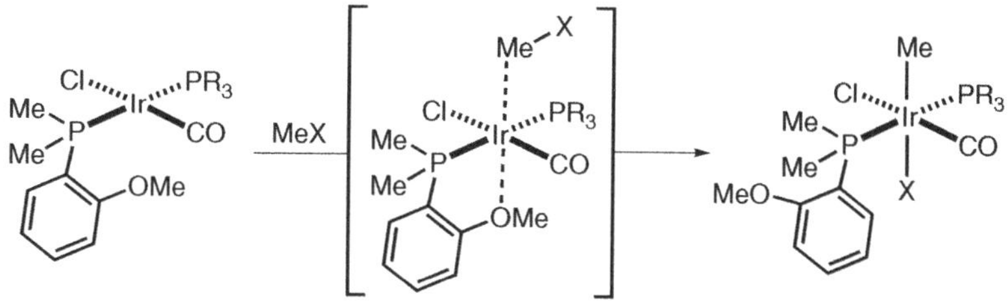

Figure 14 *Hemilabile oxygen donor promotes oxidative addition of MeX by Ir(I) complexes over 100 times faster than with the analogous 4-OMe-substituted ligand*

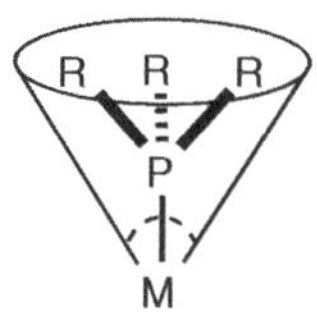

Ligand	cone angle
PF_3	104°
PMe_3	118°
PEt_3	132°
PPh_3	145°
$P(tBu)_3$	182°
PAr_3	212°

Ar:

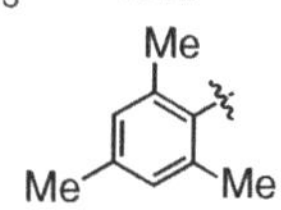

Figure 15 *Schematic depiction of ligand cone angle for a generic phosphine ligand and representative Tolman cone angle values*

Ligand cone-angle provides a metric for determining the space a ligand occupies from the view of a metal center (Figure 15). As the name implies, it is the solid angle formed between the metal center and the outer edge of the ligand hydrogen atoms. The Tolman cone angle is based on standard Ni—P bond distances of 2.28 Å. This calculation does not account for flexibility (ligands rarely form perfect cones), thus alternative descriptors have been developed, such as White's solid angle.[22] Related data are also available for bidentate ligands.[23]

To better describe the steric environment conferred on a metal by non-conical ligands, particularly *N*-heterocyclic carbene ligands, the concept of percent buried volume ($\%V_{bur}$) was introduced by Nolan and Cavallo.[24] This term is defined as the percent of the total volume of a sphere, with a radius of 3.50 Å, occupied by a ligand at a given metal—ligand bond distance (Figure 16). These values were initially determined on the basis of X-ray crystallographic data. Computational software is now available to calculate $\%V_{bur}$ for new ligands and a large collection of values has been assembled by Nolan and co-workers. $\%V_{bur}$ values are typically standardized and computed according to the SambVca platform developed by Cavallo and co-workers, whereby the sphere radius is set at 3.50 Å and metal–ligand bond lengths of 2.00 and 2.28 Å are employed.[25] To place a few values in context, relatively small phosphine ligands,

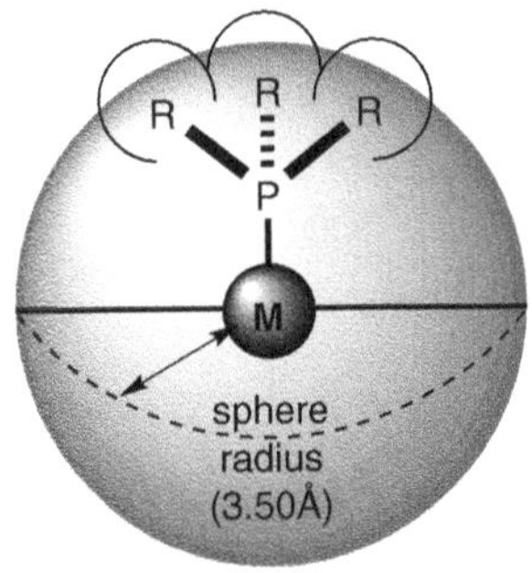

Figure 16 Schematic depiction of percent buried volume ($\%V_{bur}$)

such as PMe_3, will have $\%V_{bur}$ of approximately 25%, whereas very sterically demanding ligands will have values >40%, for example $P(tBu)_3$ $\%V_{bur}=42\%$, $PMes_3$ $\%V_{bur}=53\%$. Finally, for chiral, polydentate ligands and those ligands that are not tripodal, ligand substituent size can be derived from the Sterimol parameter, as recently demonstrated by Sigman.[26]

1.10 Cooperative and redox non-innocent ligands

Some ligands can actively participate in substrate activation in concert with the bound metal. These ligands are termed cooperative ligands.[3] The synergistic nature of the metal–ligand interaction is usually imparted by the ability of the ligand to act as a proton shuttle (via protonation or deprotonation) or electron shuttle (by electron transfer to or from the metal center), although ligands that promote reactivity by other effects, such as hydrogen bonding or electrostatic interactions may be considered cooperative as well. Noyori's Ru-based catalysts featuring diamine ligands are classical examples of cooperative metal–ligand systems, whereby the ligand acts as a proton shuttle and basic site in the bifunctional activation of H_2 (either directly or from an H_2 donor molecule) for carbonyl reduction (Figure 17a).[27] Ligands that can reversibly change their oxidation state under conditions that do not result in metal oxidation state change are termed redox non-innocent ligands.[28] The ability of ligands to act as single-electron donors and acceptors is a well-established facet of bioinorganic chemistry, and has been exploited in the design of ligands to enable catalysis. For example, bis(imino)pyridine ligands can exist in several stable oxidation states and can confer unique reactivity to Fe complexes, for example the 2+2 cycloaddition of olefins (Figure 17b).[29]

1.11 Conclusion

The remarkable repertoire of unique chemical reactivity that has been established to date in transition metal chemistry can be attributed in large part to the action of ancillary ligands. Although the *a priori* design of ancillary ligands as a means of enabling

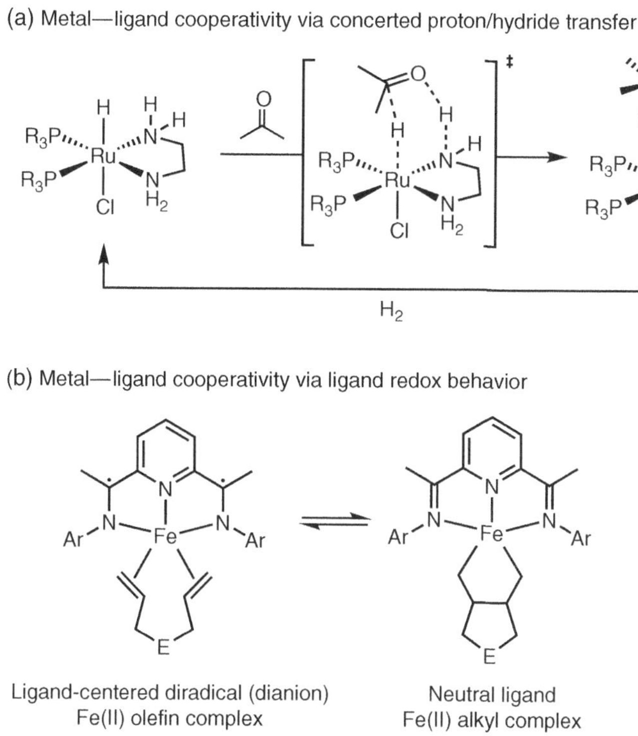

Figure 17 *Representative examples of ligand participation in substrate activation*

the targeted application of transition metal complexes in bond activation and homogeneous catalysis has yet to be realized reliably in practice, important concepts relating to ligand structure and bonding have emerged over the past decades, some of which were highlighted herein. Consideration of these design criteria can greatly assist in guiding the development of ancillary ligand architectures in the quest for new transition metal reactivity.

References

[1] (a) J. Hartwig, *Organotransition Metal Chemistry: From Bonding to Catalysis*, University Science Books, Sausalito, CA, **2010**; (b) R. H. Crabtree, *The Organometallic Chemistry of the Transition Metals*, *Vol. 4*, John Wiley & Sons, Inc., Hoboken, NJ, **2005**; (c) G. L. Miessler, P. J. Fischer, D. A. Tarr, *Inorganic Chemistry*, 5th edn, Pearson, Upper Saddle River, NJ, **2014**.

[2] M. L. H. Green, G. Parkin, *J. Chem. Ed.* **2014**, *91*, 807–816.

[3] M. Trincado, H. Grutzmacher, in *Cooperative Catalysis* (ed. R. Peters), Wiley-VCH, Weinheim, **2015**, pp. 67–110.

[4] E. J. Hawrelak, W. H. Bernskoetter, E. Lobkovsky, G. T. Yee, E. Bill, P. J. Chirik, *Inorg. Chem.* **2005**, *44*, 3103–3111.

[5] J. Börgel, M. G. Campbell, T. Ritter, *J. Chem. Educ.* **2016**, *93*, 118–121.

[6] C. A. Tolman, *Chem. Rev.* **1977**, *77*, 313–348.
[7] (a) D. G. Gusev, *Organometallics* **2009**, *28*, 763–770; (b) D. S. Coll, A. B. Vidal, J. A. Rodriguez, E. Ocando-Mavarez, R. Anez, A. Sierraalta, *Inorg. Chim. Acta* **2015**, *436*, 163–168.
[8] A. R. Chianese, X. W. Li, M. C. Janzen, J. W. Faller, R. H. Crabtree, *Organometallics* **2003**, *22*, 1663–1667.
[9] R. G. Pearson, *J. Chem. Ed.* **1987**, *64*, 561–567.
[10] M. G. Finn, K. B. Sharpless, *J. Am. Chem. Soc.* **1991**, *113*, 113–126.
[11] R. R. Schrock, *Angew. Chem. Int. Ed.* **2006**, *45*, 3748–3759.
[12] R. Tudor, M. Ashley, *Platin. Met. Rev.* **2007**, *51*, 116–126.
[13] A. Deangelis, T. J. Colacot, in *RSC Catalysis Series, Vol. 2015* (ed. T. J. Colacot), The Royal Society of Chemistry, Cambridge, **2015**, pp. 20–90.
[14] P. van Leeuwen, P. C. J. Kamer, J. N. H. Reek, P. Dierkes, *Chem. Rev.* **2000**, *100*, 2741–2769.
[15] P. van Leeuwen, P. C. J. Kamer, J. N. H. Reek, *Pure Appl. Chem.* **1999**, *71*, 1443–1452.
[16] H. Braunschweig, F. M. Breitling, *Coord. Chem. Rev.* **2006**, *250*, 2691–2720.
[17] M. E. van der Boom, D. Milstein, *Chem. Rev.* **2003**, *103*, 1759–1792.
[18] C. Gunanathan, D. Milstein, *Chem. Rev.* **2014**, *114*, 12024–12087.
[19] R. Rasappan, D. Laventine, O. Reiser, *Coord. Chem. Rev.* **2008**, *252*, 702–714.
[20] K. J. Szabo, O. F. Wendt, *Pincer and Pincer-Type Complexes: Applications in Organic Synthesis and Catalysis*, Wiley-VCH, Weinheim, **2014**.
[21] E. M. Miller, B. L. Shaw, *J. Chem. Soc. Dalton Trans.* **1974**, 480–485.
[22] D. White, B. C. Taverner, P. G. L. Leach, N. J. Coville, *J. Comput. Chem.* **1993**, *14*, 1042–1049.
[23] T. Niksch, H. Goerls, W. Weigand, *Eur. J. Inorg. Chem.* **2010**, 95–105.
[24] (a) H. Clavier, S. P. Nolan, *Chem. Commun.* **2010**, *46*, 841–861; (b) A. C. Hillier, W. J. Sommer, B. S. Yong, J. L. Petersen, L. Cavallo, S. P. Nolan, *Organometallics*, **2003**, *22*, 4322–4326.
[25] A. Poater, B. Cosenza, A. Correa, S. Giudice, F. Ragone, V. Scarano, L. Cavallo, *Eur. J. Inorg. Chem.* **2009**, 1759–1766.
[26] K. C. Harper, E. N. Bess, M. S. Sigman, *Nature Chem.* **2012**, *4*, 366–374.
[27] R. Noyori, T. Ohkuma, *Angew. Chem. Int. Ed.* **2001**, *40*, 40–73.
[28] V. Lyaskovskyy, B. de Bruin, *ACS Catal.* **2012**, *2*, 270–279.
[29] P. J. Chirik, K. Wieghardt, *Science* **2010**, *327*, 794–795.

2

Catalyst Structure and *Cis–Trans* Selectivity in Ruthenium-based Olefin Metathesis

Brendan L. Quigley and Robert H. Grubbs

Arnold and Mabel Beckman Laboratories of Chemical Synthesis, Division of Chemistry and Chemical Engineering, California Institute of Technology, Pasadena, CA 91125, USA

2.1 Introduction

In the 1950s and 1960s, the field of olefin metathesis began to emerge from the realization that the unexpected olefinic products observed in a variety of industrially relevant transformations catalyzed by transition metals (TMs) were all derived from a common mechanistic pathway [1]. Initial developments in the field were largely based on heterogeneous or ill-defined catalysts and it was through careful mechanistic study of these systems that TM carbenes were identified as the potential catalysts responsible for these alkene products [2]. The drive to further understand this reactivity prompted the development of well-defined carbene complexes that could catalyze homogeneous olefin metathesis with comparable efficiency to the early heterogeneous systems [3].

A wide variety of early-TM carbenes, particularly those of molybdenum and tungsten, were found to be efficient olefin metathesis catalysts and could be tuned for various applications through careful ligand selection [3b, 4]. However, they typically were sensitive to air and water and had poor tolerance to particular functional groups, such as

Ligand Design in Metal Chemistry: Reactivity and Catalysis, First Edition. Edited by Mark Stradiotto and Rylan J. Lundgren.

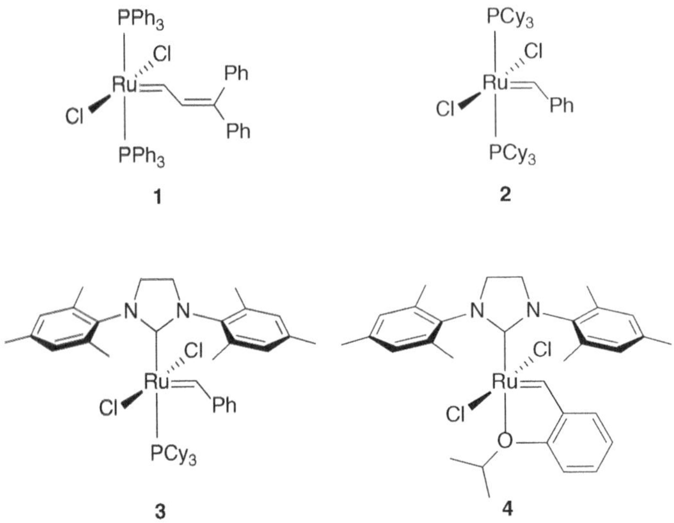

Figure 1 *Key Ru-based metathesis catalysts*

acids and alcohols. While ill-defined Ruthenium metathesis catalysts were known and had demonstrated stability to air, moisture and protic solvents [5], well-defined Ru carbenes that were proficient at metathesis remained elusive for many years. In 1991, the discovery of stable Ru alkylidene **1** (Figure 1) which could effect olefin metathesis was therefore of great interest and inspired significant research efforts [6].

Similar to the development of Mo and W catalysts, several key advances in Ru-based metathesis catalysts were achieved through tuning of the ligand environment [7]. Catalyst **2** incorporated more electron-donating tricyclohexylphosphine ligands in place of the original triphenylphosphine ligands and a benzylidene in place of the vinylidene ligand [8]. These changes both resulted in a significant increase in activity and a new synthetic route allowed preparation of **2** on a large scale. Subsequent substitution of one phosphine by an *N*-heterocyclic carbene (NHC) ligand resulted in the significantly more active catalyst **3**, which could catalyze challenging metathesis reactions previously confined to early-TM catalysts [9]. It was later found that introduction of a chelating isopropoxy substituent on the benzylidene, as in catalyst **4**, significantly enhanced the stability of these Ru-based catalysts [10]. These key developments in activity and stability have inspired the development of a vast number of further Ru-based catalysts tailored for specific applications. Today, catalysts **2–4** are commercially available and have been widely employed in a variety of olefin metathesis transformations carried out from academic to industrial scales, in applications as diverse as biomimetic materials [11], olefin upgrading [12], and fine chemical synthesis [13].

Initially, there were two limiting factors in the utility of metathesis: selective formation of a desired olefinic product from the mixture of possible olefins and formation of that product as a single *cis* or *trans* stereoisomer. A model for olefin reactivity has allowed for selective formation of a single product in a number of cases but achieving high selectivity for one olefin isomer has remained more challenging [14]. This is typically achieved by exploiting a preference for the thermodynamically favored isomer, which is generally (but not always) the *trans* olefin product. Achieving levels of selectivity either above the thermodynamic preference or opposite to it requires imposing a kinetic preference for the chosen isomer. Understanding these kinetic and thermodynamic aspects at play in metathesis reactions and how catalyst structure influences them is key to achieving the desired stereochemical outcome. The first major success in this area was reported in 2009 when the Schrock and Hoveyda groups demonstrated that Mo- and W-based monoaryloxide pyrrolide catalysts could selectively form *cis*-olefin products [15]. Outstanding progress has been made in the use of these *Z*-selective Mo- and W-based metathesis catalysts in synthesis and parallel developments in the discovery of further *Z*-selective Mo and W catalysts have been made. These developments have been reviewed elsewhere [4e,16].

This chapter commences with an introduction to the different types of metathesis reactions and introduces the model for selective formation of a single product in cross metathesis (CM) reactions. The CM reaction mechanism is then discussed with particular focus on the geometry of key intermediates that influence product stereochemistry and the role of ligands in influencing this selectivity. Other factors that can alter the stereochemical outcome are then addressed, including non-productive metathesis pathways and secondary reactions of the products. These principles will then be illustrated through general strategies for achieving *cis*-selective metathesis catalysts.

In order to give perspective on the kinetic versus thermodynamic balance, the *cis–trans* selectivity of some commonly utilized Ru metathesis catalysts is presented. A number of catalysts with modified ligands that result in a distinct stereochemical preference are then compared with these original catalysts and their reactivity discussed. Finally, the successful implementation of ligand-driven selectivity has led to three families of Ru-based metathesis catalysts that can perform *Z*-selective metathesis. For each of these catalyst families, a model for the origin of *Z*-selectivity, the role of ligands in influencing stereochemistry and trends in their reactivity are examined.

2.2 Metathesis reactions and mechanism

2.2.1 Types of metathesis reactions

Olefin metathesis reactions can be categorized based on the types of olefins undergoing reaction and the types of olefinic products formed (Scheme 1). Of these reactions, initial attention was largely devoted to ring-closing metathesis (RCM) and ring-opening metathesis polymerization (ROMP), which both have a strong driving force toward product formation. In the case of RCM, there is an entropic driving force due to the production of two molecules from one (the product and ethylene), whereas ROMP relies on release of ring-strain in the monomer to drive the reaction. In contrast, exploration of CM reactivity was slower, as there are a number of inherent challenges, particularly with regard to selectivity. CM lacks an inherent driving force for product

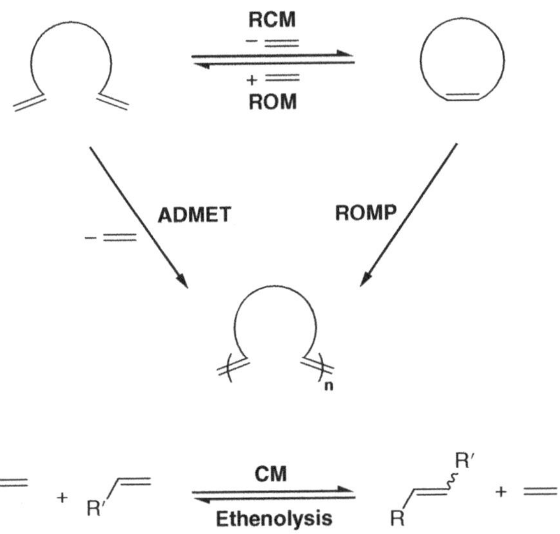

Scheme 1 *Types of olefin metathesis reactions. ADMET = acyclic diene metathesis*

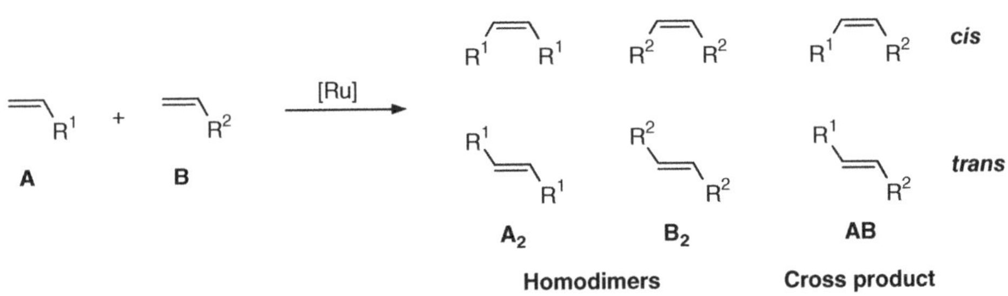

Scheme 2 *Products formed in CM of two terminal olefins*

formation and so can be reversible under conditions where ethylene is not efficiently removed from the reaction mixture.

Typically, if two distinct terminal olefins (**A** + **B**, Scheme 2) undergo CM, multiple products result: the homodimers of **A** and **B** are formed ($\mathbf{A}_2$ + $\mathbf{B}_2$) in addition to the desired cross product (**AB**). In addition, each of these products is typically formed as a mixture of the *cis* and *trans* isomers, generating six different products.

A general model for selectivity in CM has been proposed in which olefins are classified based on their relative rate of homodimer formation and consumption in CM reactions with a given catalyst [14]. There are four olefin categories of decreasing reactivity: Type I olefins, which homodimerize readily and whose homodimers can readily undergo further reaction; Type II olefins, which homodimerize slowly and whose homodimers are slow to undergo further reactions; Type III olefins, which do not homodimerize and only form cross products with Type I or II olefins; and Type IV olefins, which are spectators to metathesis. This differentiation in reactivity is based on steric factors, electronic factors or both, with increased steric bulk proximal to the double bond and electron-withdrawing substituents generally leading to reduced reactivity.

If two terminal olefins of Type I are used, a statistical distribution of products results. Hence, if a 1:1 ratio of **A**:**B** is used, then 25% of each homodimer is produced in addition to 50% of the desired cross product. The amount of the cross product can be increased by introducing an excess of one terminal olefin, for example using a 4:1 ratio of **A**:**B** increases the potential yield of cross product (**AB**) to 80%. When the two olefins of different types are used, they can undergo selective CM, favoring formation of the cross product over formation of homodimers, leading to yields greater than the statistical product distribution. The "Type" is specific for a given catalyst and therefore through judicious catalyst selection, metathesis can be rendered selective for the desired cross product in a wide variety of cases. In general, less active catalysts show greater differentiation of more similar olefins (e.g., allyl pinacol boronate and the bulkier styrene are Type I and Type II, respectively, for catalyst **2**, while for more active catalyst **3**, they are both Type I).

In a wide variety of cases, application of this model allows the desired cross product to be generated in high yield. In order to generate the product as a single stereoisomer, an understanding of the relationship between catalyst structure in Ru-based catalysts and *cis–trans* selectivity is necessary. This requires a closer examination of the olefin metathesis mechanism.

2.2.2 Mechanism of Ru-catalyzed olefin metathesis

Over the past 40 years, there has been a vast accumulation of experimental [17] and computational [18] evidence in support of the general olefin metathesis mechanism, as originally proposed by Hérisson and Chauvin. Here, we will illustrate the mechanism for CM of two terminal olefins with a phosphine-based catalyst (Scheme 3). Initial phosphine dissociation forms a 14-electron complex **I**, which then binds an equivalent of terminal olefin in a η^2 fashion (**II**). Subsequent [2+2] cycloaddition generates the intermediate 1,2-disubstituted metallacyclobutane **III**, which can then undergo a cycloreversion (**IV**) to release an equivalent of internal olefin and generate ruthenium methylidene **V**. The methylidene then binds another equivalent of terminal olefin (**VI**) and cycloaddition generates a 1-substituted metallacyclobutane **VII**. Cycloreversion in the opposite sense to formation (**VIII**) and release of ethylene regenerates the 14-electron alkylidene **I**, which can then re-enter the catalytic cycle. Non-productive metathesis can occur through isomer **II′**, which leads to formation of 1,3-metallacycle **IX**, resulting in alkylidene exchange. While these general steps are broadly accepted, the geometry of the intermediates (in particular, **III**), which has serious implications for product stereochemistry, has remained contentious.

2.2.3 Metallacycle geometry

Two distinct pathways have been proposed for the formation of the metallacyclobutane that differ in the orientation of the metallacycle with respect to the other ligands around Ru (Figure 2). In the "bottom-bound" pathway, metallacycle formation takes place with an olefin bound *trans* to the NHC, leading to a metallacycle on the opposite face to the NHC and the two anionic ligands (X) being *trans* to each other. Alternatively, in the "side-bound" pathway, metallacycle formation takes place with an olefin bound *cis*

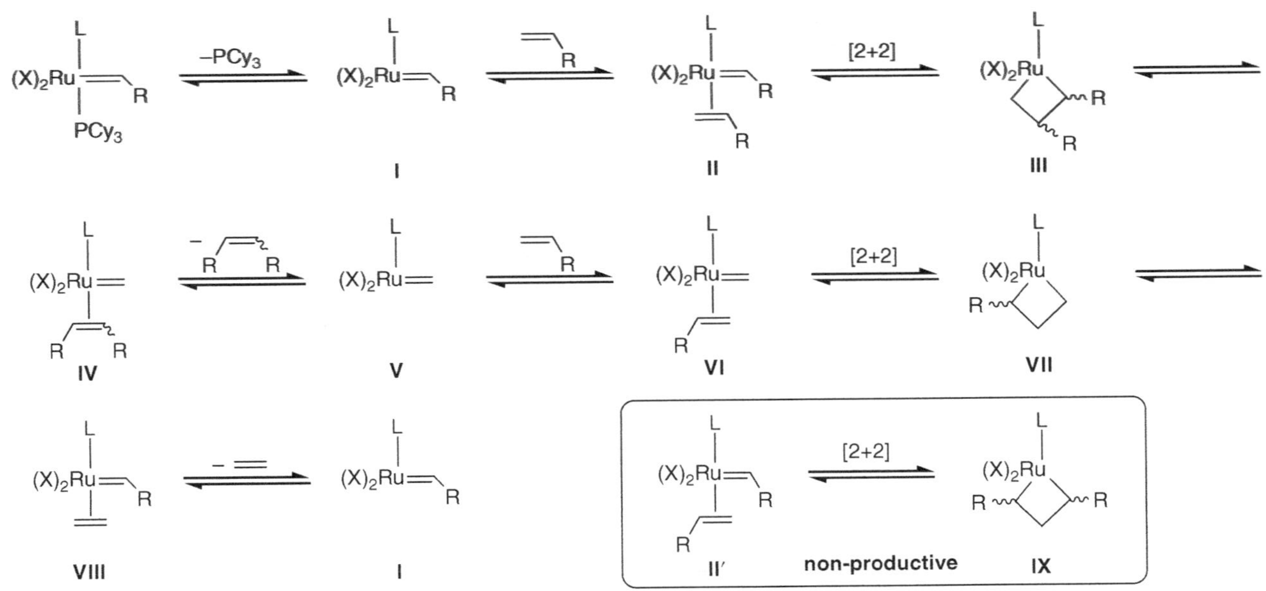

Scheme 3 Mechanism for productive CM of terminal olefins

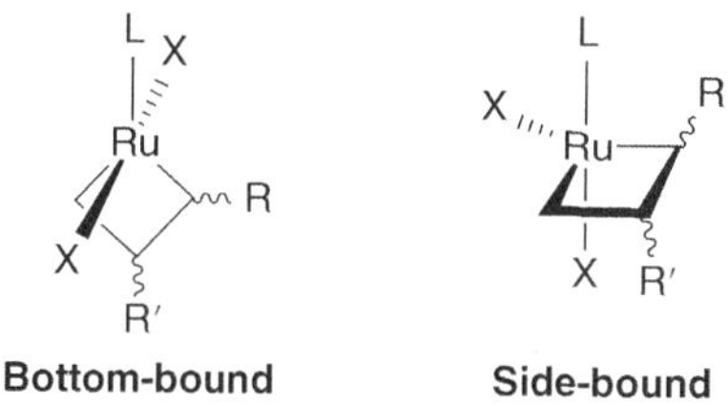

Figure 2 *Possible orientations for ruthenacyclobutanes*

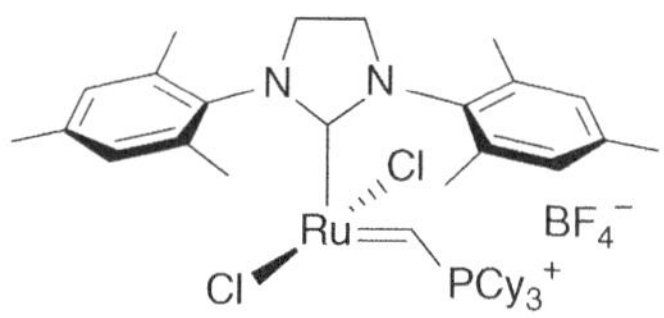

5

Figure 3 *Phosphonium alkylidene catalyst* **5**

to the NHC and the resulting metallacycle is oriented in a perpendicular plane to the NHC and the two anionic ligands are *cis* to each other. While there was indirect computational [18] and experimental [19] evidence in support of both side-bound and bottom-bound ruthenacycles, the first direct evidence arose in 2004, when Romero and Piers reported that ruthenacycles could be generated at low temperatures and studied using NMR spectroscopy, via the "pre-initiated" catalyst **5** (Figure 3) [20].

Using this methodology, a small number of metallacyclobutanes relevant to CM and RCM have been generated and their structures studied [20, 21]. Of particular relevance to CM, metallacycles have been accessed by reaction of **5** with ethylene, propene, 1-butene and 1-hexene. In all cases, NMR spectroscopic evidence supported a C_{2v}-symmetric metallacycle, which corresponds to a bottom-bound geometry, and no evidence for side-bound metallacycles has been observed in these experiments. An unsymmetrical NHC ligand bearing *N*-2,4,6-trimethylphenyl (Mes) and *N*′-2,6-diisopropylphenyl (DIPP) substituents was used to elucidate which of the two possible metallacycle orientations proposed for ethylene-derived bottom-bound metallacycles was present (Figure 4) [21a]. NMR spectroscopic analysis revealed the presence of **A**, in which the bottom-bound metallacycle lies coincident with the span of the NHC, by its diastereotopic α-methylene groups, which result from the asymmetry of the NHC. In addition to ethylene-derived unsubstituted metallacycles, 1-monosubstituted and 1,3-disubstituted metallacycles have been observed in these experiments. As discussed earlier, 1,3-disubstituted metallacycles result in non-productive metathesis, generating terminal olefins via olefin–alkylidene exchange. 1,2-Disubstituted metallacycles of terminal olefins, which lead to productive metathesis, have not been observed to date.

In both side-bound and bottom-bound metallacycle orientations, disubstituted metallacycles can adopt one of two stereochemical configurations: one where the substituents are *syn* to each other, which would lead to formation of the *Z*-alkene; and

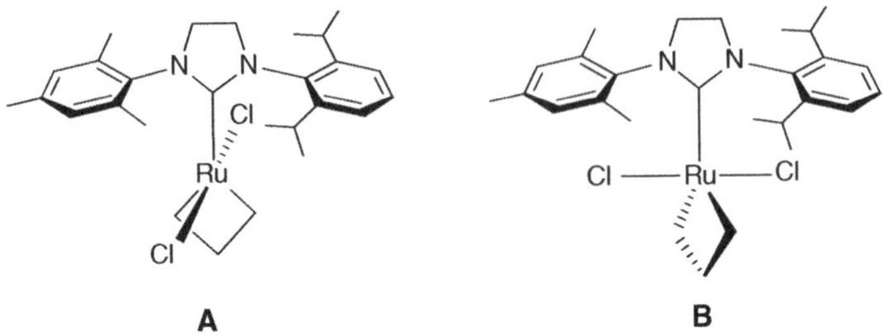

Figure 4 Possible orientations of the bottom-bound metallacycle relative to the NHC ligand

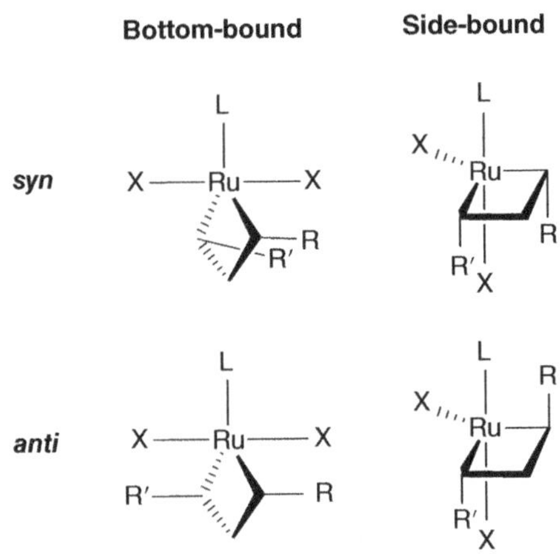

Figure 5 Syn and anti geometries of bottom- and side-bound metallacycles

one where the substituents are *anti* to each other, which would lead to formation of the *E*-alkene (Figure 5). Both *syn*- and *anti*- bottom-bound 1,3-disubstituted metallacycles were observed by ^{1}H-NMR spectroscopy, with the *anti*-metallacycle being two to three times more abundant than the *syn*-metallacycle in all cases [21e]. The two stereoisomers were found to be interconverting and a mechanism invoking cycloreversion, alkylidene rotation and cycloaddition was proposed. This implies that alkylidene dynamics as well as metallacycle stability plays a role in the stereochemical outcome of olefin metathesis reactions.

2.2.4 Influencing *syn–anti* preference of metallacycles

It is expected that ligand-induced asymmetry with respect to the two faces of the metallacycle will have the most discernible influence on the *syn–anti* preference of disubstituted metallacycles. These effects will only be evident, however, if ligand rotations with respect to the metallacycle plane are slower than either productive

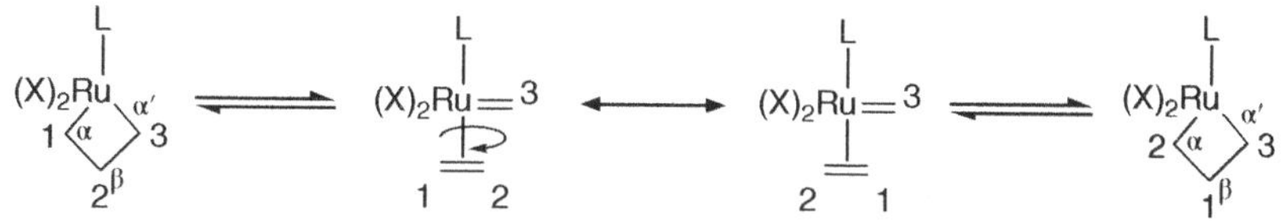

Scheme 4 *Proposed mechanism for exchange of α and β protons*

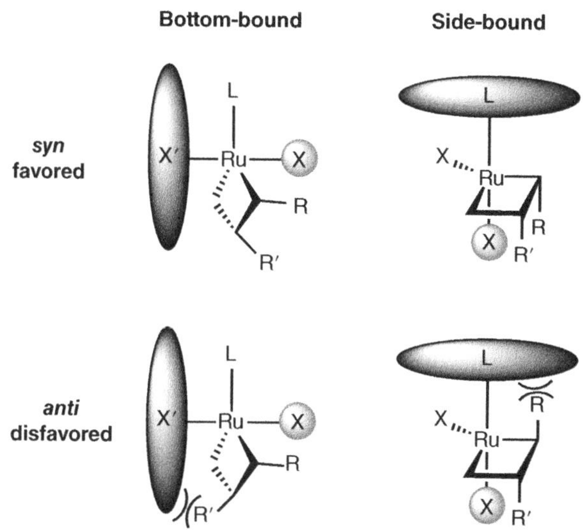

Figure 6 *Proposed models for favoring syn-metallacycles*

cycloreversion or interconversion of the *syn* and *anti* isomers. In ethylene-derived metallacycles, rapid exchange of the α and β protons was observed via non-productive cycloreversion/cycloaddition pathways (Scheme 4) [21e]. Such processes can result in rapid inversion of stereochemistry at Ru, thereby compromising its ability to retain stereochemical integrity of the metallacycle, potentially resulting in the formation of the thermodynamically favored product.

Since Ru-based catalysts generally yield *E*-dominant mixtures of olefin products, the development of catalysts that are selective for the *Z*-olefin has been a major focus of research. If the steric environment above and below the plane of the metallacycle can be differentiated, by introduction of steric bulk on one face, then this may impose a preference for both metallacycle substituents to occupy the opposite face, resulting in a *syn* orientation (Figure 6). In the case of a bottom-bound metallacycle, the ligands above and below the plane of the metallacycle are the two anionic ligands. By increasing the size of one of these ligands, it might be expected that substituents would prefer to occupy the opposite face, leading to *Z*-olefin products. In the case of the side-bound pathway, the ligands above and below the metallacycle are one L-type and one X-type ligand. Here, by increasing the bulk of the L-type ligand relative to the X-type ligand, preference for the *syn*-metallacycle may be imposed.

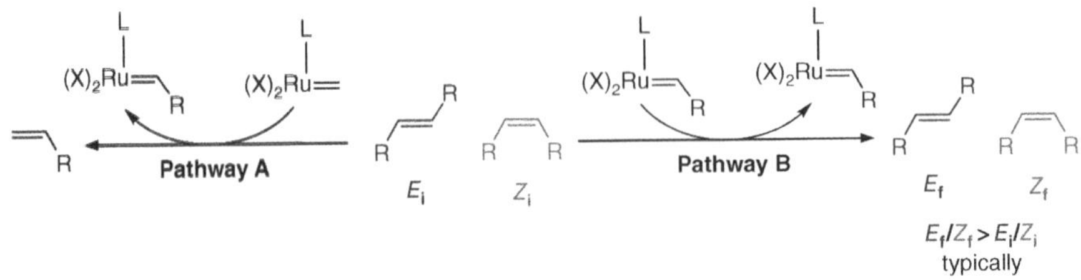

Scheme 5 *Secondary metathesis processes in CM. (See insert for color/color representation of this figure)*

While control of these parameters may lead to initial formation of the product with high stereoselectivity, the stereochemical integrity of the product can be compromised by secondary metathesis events, where the initially formed products undergo subsequent reaction. For CM of two terminal olefins, there are two major secondary metathesis processes to consider. In the first pathway (**Pathway A**, Scheme 5), ethenolysis of the product can take place to regenerate the starting materials. In the second (**Pathway B**), internal olefins are interconverted via trisubstituted metallacycles to generate products with opposite stereochemistry. In general, secondary metathesis processes result in an increase in the *E*-olefin content of the mixture due to two factors: the Z-olefin is typically less thermodynamically stable than the *E*-olefin and is also kinetically more reactive. At low conversions, the *E*/*Z* ratio is largely governed by primary metathesis and hence reflects the inherent (or kinetic) selectivity of the catalyst. At high conversion, secondary metathesis is expected to govern the *E*/*Z* ratio, reflecting the relative kinetic and thermodynamic stability of the products. Hence, in order to achieve *Z*-selectivity, a catalyst must both have an inherent preference for the *syn*-metallacycle and either exhibit minimal secondary metathesis or the secondary metathesis must also be highly selective for increasing the *cis*-isomer content.

2.3 Catalyst structure and *E*/*Z* selectivity

2.3.1 Trends in key catalysts

In order to directly compare the efficiency of catalysts in olefin metathesis, a standard set of reactions was established [22]. Among the reactions included was the CM of allylbenzene and *cis*-1,4-diacetoxy-2-butene (CDAB), which display similar reactivity, both being Type I olefins (Figure 7). For a number of widely utilized Ru-based metathesis catalysts, the rate of the reaction and the *E*/*Z* ratio of the cross product over time were monitored. Two major trends were observed with regard to selectivity of the product **6**. First, while the rate of conversion differed between catalysts, the *E*/*Z* ratio with respect to conversion was consistent within phosphine-substituted first-generation catalysts (**2** and **7**) and within NHC-substituted second-generation catalysts (**3**, **4**, **8** and **9**). Secondly, whereas first-generation catalysts maintained a relatively constant *E*/*Z* ratio of ~5 over the course of the reaction, second-generation catalysts had a lower initial *E*/*Z* ratio of ~3

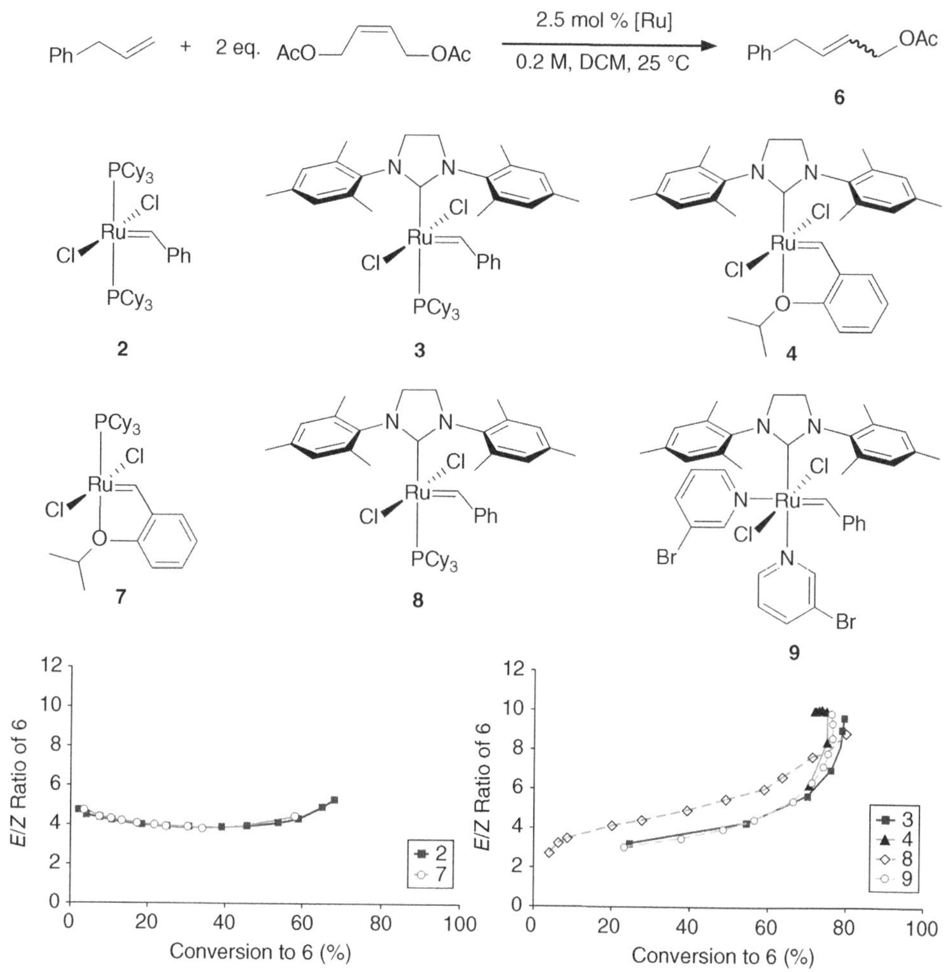

Figure 7 *Cross metathesis of allylbenzene and CDAB with common metathesis catalysts. Adapted from [22].*

but this greatly increased at higher conversions, leading to a final *E*/*Z* ratio of ~10. The increase in the *E*/*Z* ratio at high conversion for second-generation catalysts was attributed to secondary metathesis events. No increase in *E*/*Z* ratio at high conversions was observed with first-generation catalysts, which are significantly less active and as a result exhibit negligible secondary metathesis.

If one considers the bottom-bound metallacycle orientation observed in previous studies (**A**, Figure 4), kinetic selectivity for the *E*-isomer can be rationalized on the basis of minimizing steric interactions both between the two metallacycle substituents and between the substituents and the chloride ligands. The difference in the initial *E*/*Z* ratio between the first- and the second-generation catalysts is less well understood. A potential explanation lies between the different distribution of steric bulk in the tricyclohexylphosphine ligand and the NHC ligand. This was invoked to rationalize the faster initiation rate of first-generation catalysts over second-generation catalysts, where increased effective steric bulk of the phosphine ligand over the NHC would facilitate faster dissociation of the phosphine *trans* to it [17c]. A similar argument could be used to explain influence of these ligands on the metallacycle, which is also located *trans* to the ligand. Where catalysts with various dissociating ligands (**2** and **7** or **3**, **4** and **9**) were used, the *E*/*Z* selectivity was generally similar, as is expected since exchange with one of the terminal olefins must occur before CM can take place. At low conversion, the only observable difference based on catalyst structure was that the unsaturated NHC catalyst **8** displayed a slightly higher *E*/*Z* ratio than the corresponding saturated catalyst **3** at low conversion, although the same value was reached at high conversion.

These experiments demonstrate the challenge in achieving a catalyst that is selective for the *Z*-olefin. First, the catalyst must have a kinetic selectivity for the *Z*-olefin, potentially dictated by steric bulk of the ligands. All of the abovementioned catalysts are kinetically *E*-selective to a greater or lesser extent. Hence, the design of a catalyst environment that can impose a steric or electronic preference for *syn*- over *anti*-metallacycles needs to be achieved. Secondly, secondary metathesis needs to be prevented or else this will lead to an increase in the more thermodynamically stable olefin, which is the *E*-olefin in the vast majority of cases. For NHC-containing catalysts, this has a greater effect on the final *E*/*Z* ratio than the initial kinetic selectivity, but selectivity is ultimately limited by the inherent catalyst preference.

2.3.2 Catalysts with unsymmetrical NHCs

The modularity of the ligand framework in Ru-based metathesis catalysts has provided access to a wide variety of catalysts, allowing systematic variation of steric and electronic parameters [23]. While these experimental studies [17, 22, 24] and parallel computational investigations [18, 25] have provided much insight into structure–activity relationships in Ru-based metathesis catalysts, definitive relationships between catalyst structure and *E*/*Z* selectivity have been harder to elucidate. In general, significant variation of the NHC ligand was required to induce a substantial difference in catalyst selectivity, although this did not always occur. In particular, a number of catalysts with unsymmetrical NHC-type ligands (e.g., NHCs with very different *N* and *N′* substituents) have demonstrated differences in *E*/*Z* selectivity. Examples of such catalysts will be presented, which either have differences in kinetic selectivity compared with other second-generation catalysts or reduced rates of secondary metathesis. Where data are available, the reactivity of these catalysts in CM of allylbenzene and CDAB to form olefin **6** will be discussed in order to allow ready comparison between catalysts.

Catalysts **10–15** (Figure 8), which possess unsymmetrical NHC ligands bearing one *N*-mesityl ligand and various *N′*-fluorophenyl substituents display distinct behavior from their symmetrical analogs [26]. In a variety of standard metathesis reactions, catalytic efficiencies were found to be generally comparable with that of **3** and **4**, with the relative performance of these catalysts being substrate-dependent. No clear correlation between the electronic character of the NHC and activity was observed, including in the CM of allylbenzene and CDAB. In this reaction, **10–15** displayed similar *E*/*Z* selectivity to their *N*,*N′*-dimesityl analogs at conversions below 50%. However, they

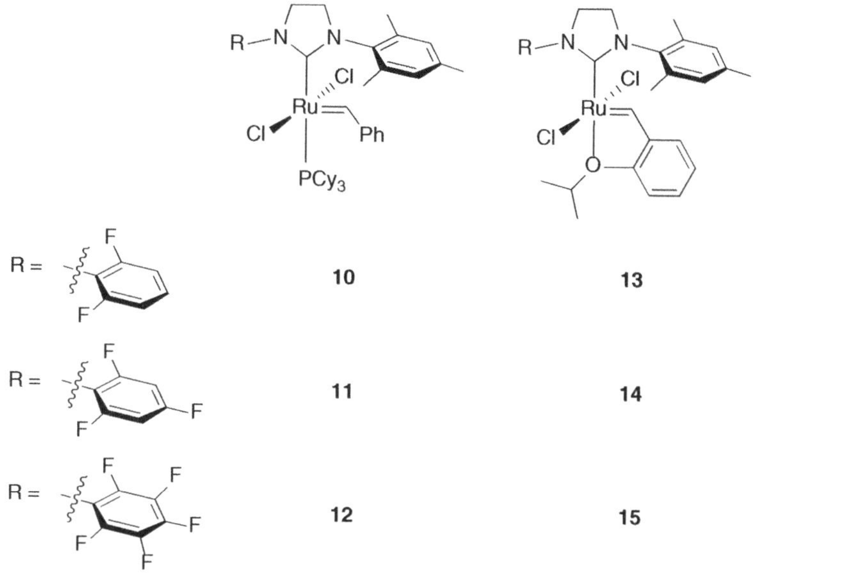

Figure 8 *Metathesis catalysts bearing N-mesityl, N′-fluorophenyl NHCs*

exhibited significantly lower *E*/*Z* ratios at higher conversions, giving *E*/*Z* ratios of ~5–6 at 80% conversion compared with ~10 for catalyst **3**, likely indicating slower secondary metathesis processes.

More drastic variation of the two NHC substituents was achieved through the use of *N*-aryl, *N′*-alkyl NHCs, which led to a greater impact on catalytic behavior (Scheme 6). Blechert and co-workers prepared analogs of **3** and **4** bearing *N*-mesityl, *N′*-Me NHCs (**16** and **17**) [27]. In both catalysts, the mesityl group is preferentially oriented over the benzylidene both in the solid state and in solution. Consistent with the expected increase in σ-donation ability of the NHC, shorter Ru—C(NHC) and Ru=C bonds were observed in the solid state and the benzylidene proton signal was shifted upfield in the ^{1}H-NMR spectrum. In CM reactions of allylbenzene and CDAB, the phosphine-substituted catalyst **16** generated product **6** with an *E*/*Z* ratio of 3, whereas under identical conditions **17** gave an *E*/*Z* ratio of 6, identical to that reported for **3** and **4**. More interestingly, catalyst **16** could generate allylic-substituted cross product **18** with an *E*/*Z* selectivity of only 3, whereas **17**, **3** and **4** all delivered the *E*-product exclusively (>20 *E*/*Z*), though quantitative yields were obtained in all cases. The difference in selectivity observed for **16** and **17** is difficult to reconcile with the mechanism proposed for catalysts bearing *N*,*N′*-bisaryl NHCs, where the dissociating ligand demonstrates no influence on *cis–trans* selectivity. Interesting behavior was also observed in the CM of allylbenzene with acrylonitrile. In contrast to most substrates, CM of acrylonitrile with terminal olefins shows a slight preference for the *Z*-isomer with catalysts **3** and **4** (~0.6–0.8 *E*/*Z*). While both **16** and **17** gave low yields of product **19** under identical conditions, they showed a complete reversal in selectivity, generating predominantly the *E*-isomer in *E*/*Z* ratios of 2.4 and 1.8, respectively.

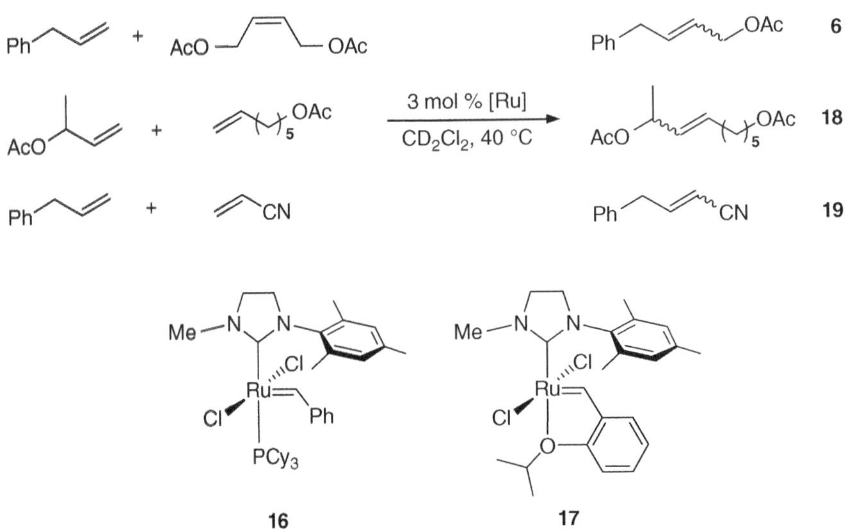

Scheme 6 Cross metathesis reactions with catalysts bearing N-Mes, N′-Me NHCs

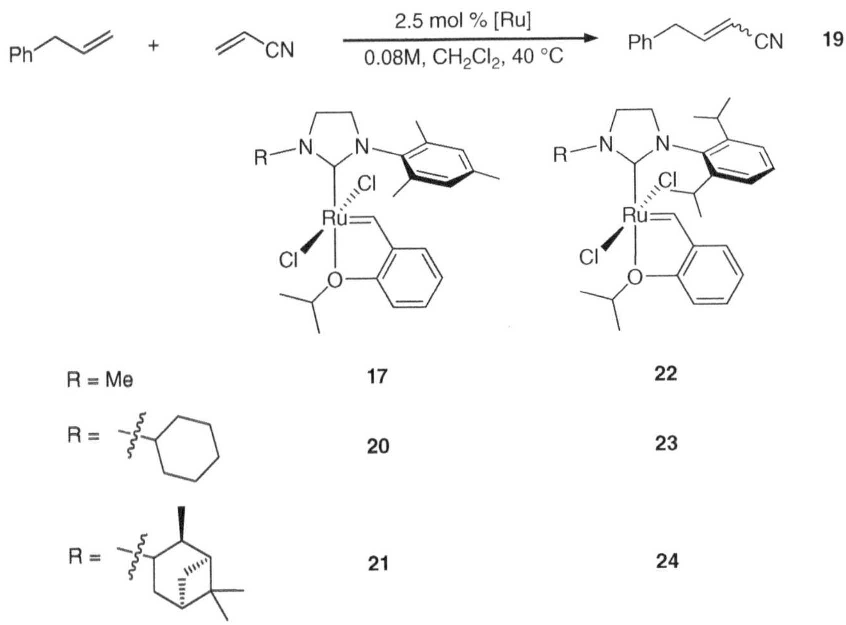

Scheme 7 Cross metathesis of allylbenzene and acrylonitrile with catalysts bearing N-aryl, N′-alkyl NHCs

In related studies, Ledoux, Verpoort and co-workers prepared several *N*-aryl, *N*′-alkyl NHC derivatives **20–24**, where the steric bulk of both substituents was varied (Scheme 7) [28]. Consistent with **17**, these catalysts were generally found to be less active than **4** in CM of acrylonitrile and allylbenzene, with *N*-DIPP-substituted complexes showing lower activity than *N*-Mes counterparts. In the *N*-Mes-substituted catalysts, increasing the steric bulk of the *N′*-alkyl substituent correlated with both decreased activity and decreased *E*-selectivity, with **21** giving a comparable *E*/*Z* ratio to **4**. Interestingly, for *N*-DIPP-substituted catalysts, higher *E*/*Z* ratio was again correlated with decreased reactivity, although the lowest activity (and highest *E*/*Z* ratio) was observed with *N′*-cyclohexyl catalyst **23**, which gave an *E*/*Z* ratio of 3.2 at low conversion. In all cases, overall conversion was low (<30%) and so *E*/*Z* ratios are unlikely to reflect the propensity for secondary metathesis but rather reflect a difference in the inherent catalyst selectivity.

2.3.3 Catalysts with alternative NHC ligands

Imidazol-2-ylidines and imidazolin-2-ylidenes are the most commonly employed phosphine surrogates in Ru-based metathesis catalysts but a variety of other related carbenes have also been explored. Catalysts **25–29** containing *N*-aryl-thiazol-2-ylidenes with varying steric bulk of the aryl substituent have been prepared (Figure 9) [29].

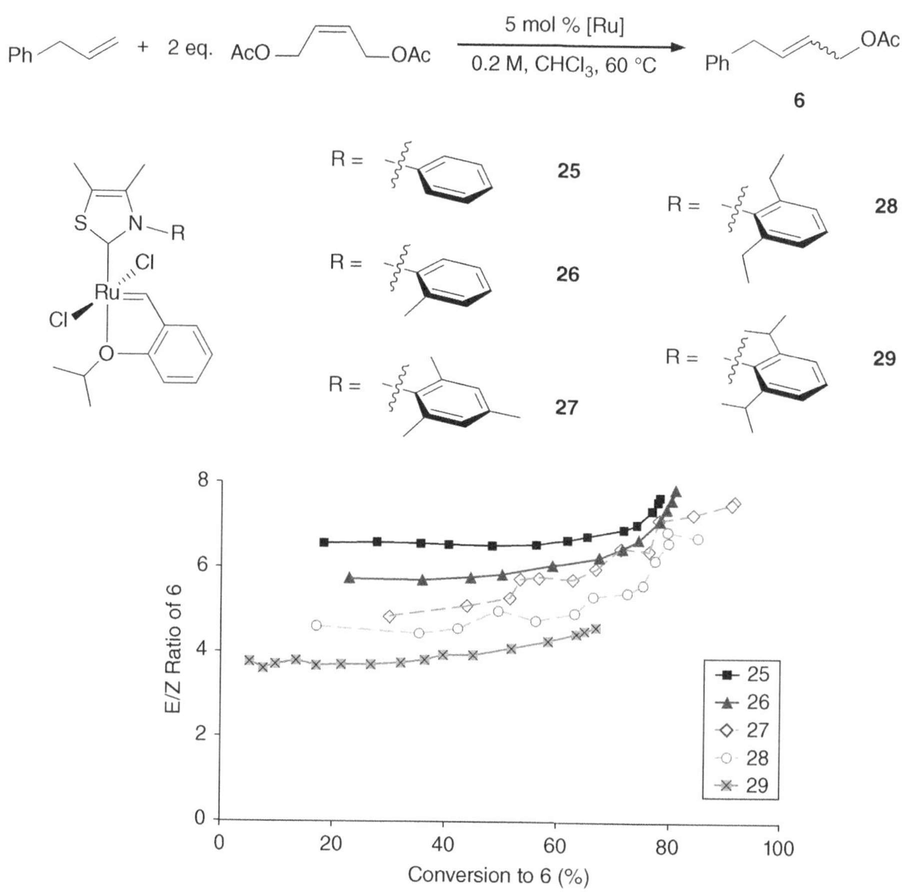

Figure 9 *Cross metathesis of allylbenzene and CDAB with thiazolium-based catalysts*

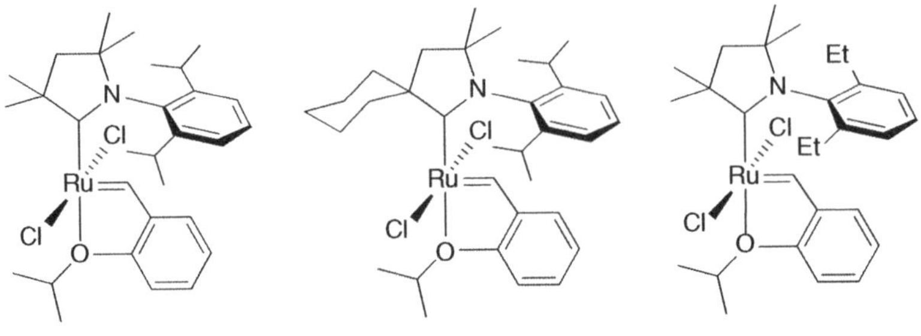

30 **31** **32**

Figure 10 *Cyclic (alkyl)(amino)carbene-based Ru metathesis catalysts*

Similar to catalysts **3** and **4**, the *E*/*Z* ratio for catalysts **25–29** remains relatively constant at conversions below 60%, but is higher, ranging from 4 to 6.5 across the series. Notably, this is one of the few cases where there is a close correlation between increased steric bulk of the aryl group (going from **25** to **29**) and a change in the *E*/*Z* ratio (here, a decrease). At higher conversions, an increase in the *E*/*Z* ratio to ~7–8 is observed, indicating secondary metathesis processes are in play. Interestingly, in the RCM of a macrocyclic 14-membered lactone, all of the thiazol-2-ylidene-containing catalysts performed similarly, giving an *E*/*Z* ratio of ~3 at low conversion and ~6 at high conversion, which was similar to that observed with **2**. In contrast, **4** gave an initial *E*/*Z* ratio of ~7 and a final value of ~10.

Catalysts bearing cyclic (alkyl)(amino)carbenes (CAACs) **30–32** (Figure 10) were shown to generate cross product **6** in the standard CM reaction with significantly lower *E*/*Z* selectivity than standard catalysts [30]. *E*/*Z* ratios of 1.5–2.5 were observed at conversions <60% and had minimally increased to ~3 at 70% conversion. This contrasts with **3** and **4** where initial *E*/*Z* ratios of ~3–4 increase to ~6 by 70% conversion. While catalyst **32** reached 60% conversion in 1 h at 25 °C, **30** and **31** took 32 and 45 h, respectively. This is indicative of the *cis–trans* selectivity being largely independent of catalyst activity and rather being influenced by metallacycle preference.

Catalysts **33** and **34** contain acyclic diaminocarbene ligands, which have wider N—C—N angles and are stronger σ-donors than their cyclic analogs [31]. While catalyst **33** showed similar *cis–trans* selectivity to catalyst **4** in CM, the bulkier catalyst **34** showed distinct behavior (Figure 11). In CM reactions of allylbenzene and CDAB, an *E*/*Z* ratio of ~1 was maintained until 80% conversion and a final *E*/*Z* value of 1.9 was observed at a maximum 86% conversion to the cross product. Allowing a further 24 h of reaction time resulted in a comparably small increase to 3, indicating that secondary metathesis processes are slow for this catalyst. To the best of our knowledge, this was the best result for generation of the *Z*-product in this standard CM reaction at the time. In the more challenging CM reaction of acrylonitrile and allylbenzene, both catalysts gave poor conversions to product **19**: <45% at 24 h. Catalyst **34** demonstrated a minor preference for the *E*-isomer (*E*/*Z* of 1.2) at ~30% conversion (3 h) but only a modest increase in conversion was observed at longer reaction times and preference for the *E*-isomer was eroded.

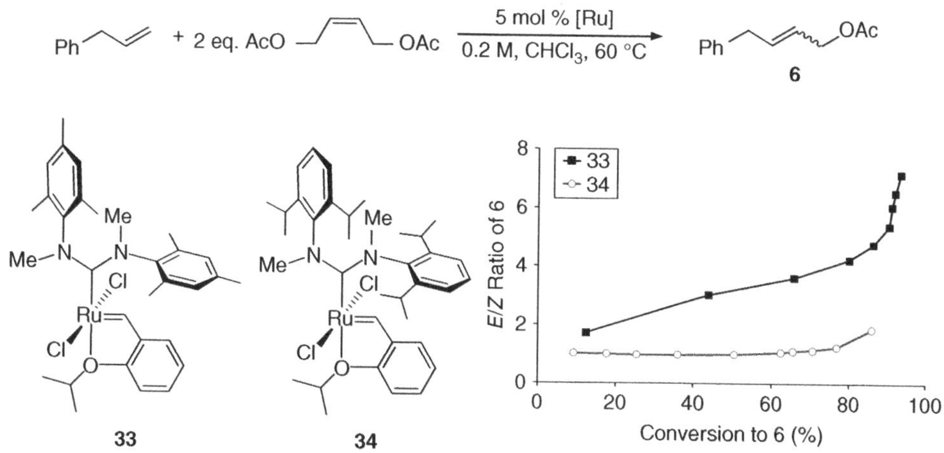

Figure 11 *Cross metathesis of allylbenzene and CDAB with acyclic diaminocarbene-based catalysts*

These examples illustrate that variation of the NHC ligand can lead to significant changes in *E*/*Z* selectivity but these differences can be highly sensitive to subtle changes in sterics and electronics. Hence, it has been difficult to generalize the modulation of *E*/*Z* selectivity across wider series of catalysts with the result that definitive structure–activity relationships have remained elusive. Until recently, no alterations of the NHC ligand led to selective production of the *Z*-isomer in high yields. Additionally, variation of the NHC ligand typically involves multistep synthesis and is not always efficient.

2.3.4 Variation of the anionic ligands

In contrast to altering the NHC ligand, variation of the anionic ligands offers an alternative approach to modulate the *cis*–*trans* selectivity that can be achieved readily by anion metathesis with commercially available catalysts. This approach has been demonstrated to influence polymer *cis* content in alternating ROMP using a Ru catalyst bearing a chelating phosphine-phenoxide ligand [32]. It was thought that a similar approach might be able to affect selectivity in CM and so catalysts bearing one sulfonate (**35**–**40**) or phosphate (**41** and **42**) ligand and one chloride ligand were prepared by anion metathesis using the corresponding silver salts (Figure 12) [33]. Of the various complexes prepared, the combination of the bulky H_2IDIPP NHC and mesitylsulfonate ligand in **39** was found to give the lowest *E*/*Z* ratio of 2.7 at high conversion (78%) in CM of allylbenzene and CDAB. While its *bis*-mesitylsulfonate analog was found to be significantly more *Z*-selective, its activity was low enough to necessitate high catalyst loadings and longer reaction times in order to reach modest conversions (*E*/*Z* ratio of 1.1 and 34% conversion at 24 h). While only moderate success was achieved in this case, this design inspired further exploration of anionic ligands as a route to modulate *E*/*Z* selectivity.

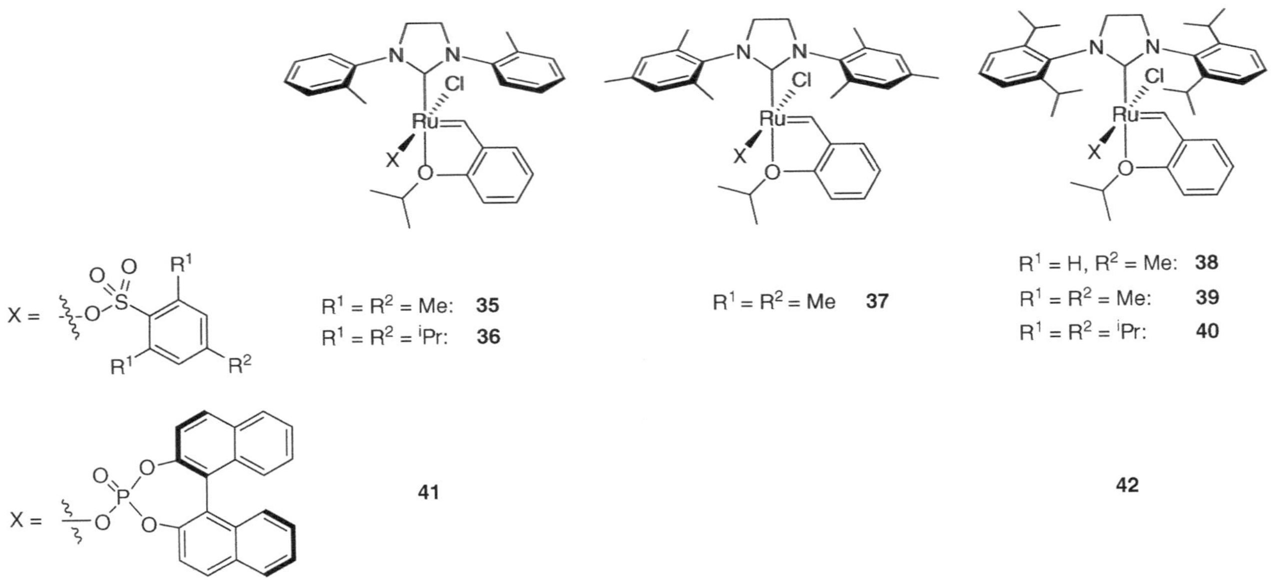

Figure 12 *Monosulfonate and monophosphate metathesis catalysts*

2.4 *Z*-selective Ru-based metathesis catalysts

2.4.1 Thiophenolate-based *Z*-selective catalysts

Catalyst **43** (Figure 13) developed by Jensen and co-workers represents the first successful implementation of employing two anionic ligands with different steric profiles to achieve high *Z*-selectivity [34]. Here, a small chloride ligand and a bulky thiophenolate ligand (2,4,6-triphenylbenzene thiolate) fulfill these criteria and the resulting catalyst was shown to impart good *Z*-selectivity in a variety of CM applications. Notably, this catalyst was readily prepared from commercially available **4** in high yield. Catalyst **43** emerged as the most promising candidate from density functional theory (DFT) calculations used to establish a general set of ligand requirements necessary to impart high *Z*-selectivity [35]. The essential features were determined to be a large X-type ligand that makes a Ru—X—E (e.g., ∠Ru—S—C in **43**) bond angle of >90° and <120°, and a small or planar X-type ligand, which is preferably electron-withdrawing. In their initially optimized catalyst, these qualities are fulfilled by the bulky thiophenolate (∠Ru—S—C = 113°) and chloride. Notably phenolate analogs which have Ru—O—C bond angles of ~130° were determined to be poor candidates for *Z*-selective catalysts.

In these catalysts, olefin approach occurs from the bottom-face of the catalyst (i.e., *anti* to the NHC) leading to a bottom-bound metallacyclobutane (Figure 14). DFT calculations using allylbenzene as a model substrate indicate that the transition state

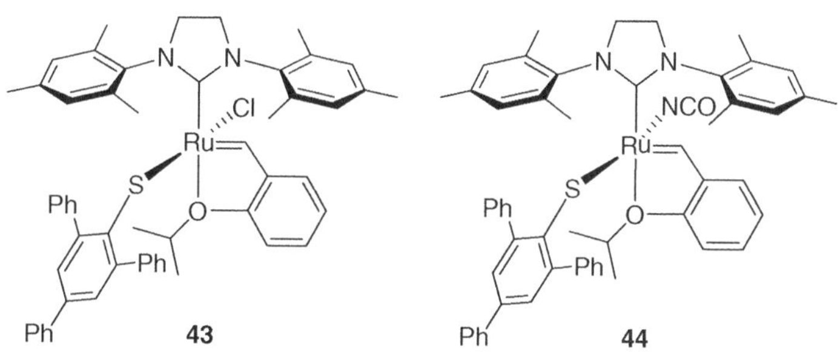

43 **44**

Figure 13 *Thiophenolate-based Z-selective catalysts*

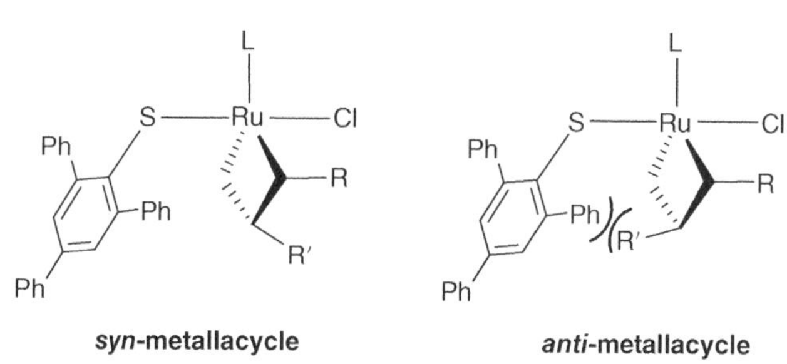

Figure 14 *Model for selectivity in thiophenolate-based Z-selective catalysts*

for cycloreversion leading to product formation is the highest energy barrier in the reaction and thus governs the overall selectivity of the process. In the lowest energy transition state, both substituents of the collapsing metallacycle are oriented away from both the NHC and bulky thiophenolate ligand in a *syn* fashion, which results in formation of the Z-olefin product. Consistent with the steric bulk of the thiophenolate being important for Z-selectivity, a variant containing the less bulky 2,4,6-trimethylbenzene thiolate ligand led to significantly reduced Z-selectivity.

While catalyst **43** demonstrated high Z-selectivity at low conversions in a variety of CM reactions, Z-selectivity was compromised at high conversion. This change in selectivity was attributed to in-situ formation of dichloro catalyst **4**, which is both highly active and *E*-selective. Suspecting formation of **4** was facilitated by the lability of the remaining chloride ligand, DFT calculations were employed to identify a suitable alternative ligand with reduced lability [36]. Isocyanate-containing catalyst **44** was selected and prepared in two steps via formation of the bis-isocyanate catalyst and subsequent monosubstitution with the thiophenolate. The solid-state structure of **44** demonstrated a reduced Ru—S—C bond angle of 110° and slightly reduced Ru–S bond length, bringing the bulky thiophenolate closer to the metallacycle. Additionally, a slight reduction in the Ru—C(NHC) and Ru—O bond lengths was observed. Consistent with tightening of the steric environment around the metallacycle, DFT calculations predicted that catalyst **44** would show improved Z-selectivity over **4**.

When catalyst **44** was tested in various CM reactions under an Ar atmosphere, it was found to be less active but more Z-selective than its progenitor **43**, although significant reductions in Z-selectivity were again observed with prolonged reaction times. For both catalysts, extensive olefin-migration isomerization was observed for allylbenzene, which can be indicative of catalyst decomposition to hydride-containing species. However, **44** was found to be relatively tolerant to both acids and air in comparison with **43** and when CM reactions of unpurified substrates were conducted under air, significant improvements were observed for catalyst **44**. While activity was still low, both olefin-migration isomerization of starting materials and *cis–trans* isomerization of products were inhibited. This is suggestive of the species causing isomerization having a low stability to oxygen relative to the catalyst itself. Neat homodimerization of allylbenzene could now be achieved with **44** in 52% isolated yield and 80% Z-selectivity, with 58% overall conversion of starting material. A variety of other homodimerization reactions were carried out and moderate yields (48–69%) were obtained for hydrocarbon substrates with good Z-selectivities (67–80%). Heteroatom-containing substrates, such as *N*-allylaniline led to significantly reduced yields (5–19%) but Z-selectivities were still generally high (65–91%). To date, this is the only catalyst demonstrated to effect Z-selective CM with unpurified substrates and under air.

2.4.2 Dithiolate-based Z-selective catalysts

A second design strategy based on choice of anionic ligand has proved successful in achieving Z-selective Ru-based metathesis catalysts by the Hoveyda group [37]. In contrast to the previous strategy, the selectivity of these catalysts relies on the formation

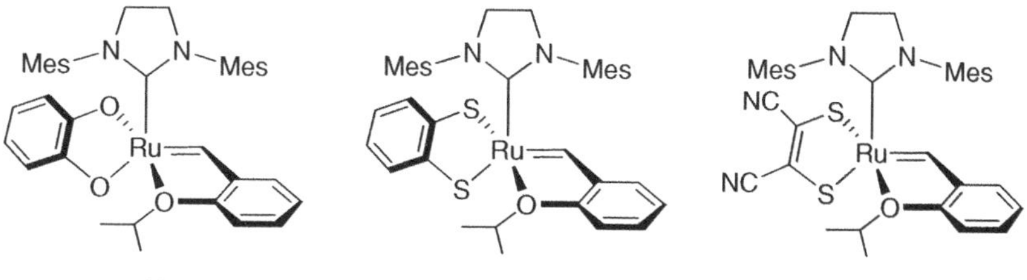

Figure 15 *Catechol- and dithiolate-based Z-selective catalysts*

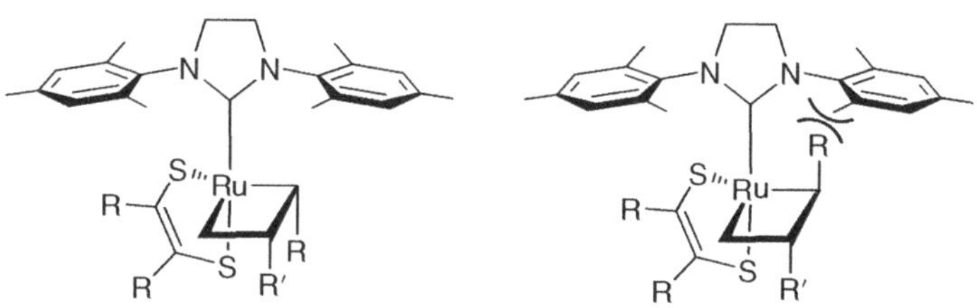

Figure 16 *Model for selectivity in dithiolate catalysts*

of side-bound metallacycles. In order to enforce preferential formation of the side-bound metallacycle, a chelating *bis*-anionic ligand is used to restrict the geometry of the complex. Catalysts **45–47** (Figure 15) were readily prepared from catalyst **4** and an excess of the corresponding sodium salt. In these catalysts, steric differentiation of the two metallacycle faces is thought to rely on the difference in bulk of the "large" NHC ligand, which occupies the top face, and the low steric requirement the bidentate ligand imposes on the bottom face (Figure 16). As a result, the substituents of the metallacycle are expected to preferentially orient downward away from the NHC, giving a *syn*-metallacycle and, hence, Z-olefin products.

Solid-state structures of catecholate **45** and dithiocatecholate **46** revealed that the isopropoxy group was oriented *syn* to the NHC and there was significant deviation from linearity between the NHC and "*trans*" heteroatom (149 and 143°, respectively, for **45** and **46**). This was attributed to minimization of donor–donor interactions between the two ligands. To date, the reactivity of these complexes has only been reported in ROMP and ring-opening/cross metathesis (ROCM) reactions, which are both energetically driven by release of ring strain. While the catecholate catalyst **45** delivered only moderate Z-selectivities, both sulfur-containing catalysts (**46** and **47**) demonstrated efficient ROMP of norbornene and 1,5-cyclooctadiene, giving >98% Z-selectivity under various conditions. Turnover numbers of over 40 000 were observed for ROMP of norbornene with dithiolene catalyst **47**. In addition, **46** was found to effect ROCM of norbornenes, cyclobutenes and cyclopropenes with various cross partners, including enol ethers, heterocyclic alkenes and allylic alcohols [37, 38].

Z-selectivities in all cases were high (>88% Z) and for many substrates the Z-product was formed essentially exclusively (>98% Z).

2.5 Cyclometallated Z-selective metathesis catalysts

2.5.1 Initial discovery

While variation of the anionic ligands has led to notable successes in Z-selective Ru-based olefin metathesis, a different class of catalysts with a structurally unique architecture were the first Z-selective Ru-based catalysts to be reported and these catalysts have enjoyed the widest variety of Z-selective metathesis applications to date. This series of catalysts arose from a serendipitous discovery, which occurred during an attempt to prepare monopivalate catalyst **48** [39]. Computational evidence had suggested that **48** (Figure 17) would show improved Z-selectivity over the monophosphonate and monosulfonate catalysts previously explored in our laboratories (e.g., **35–42**), which had only achieved a moderate decrease in the *E*/*Z* ratio. Attempts to prepare **48** analogously to the previous catalysts, using a slight excess of silver pivalate, led to a mixture of products. When two equivalents (or greater) of silver pivalate was used, the major product could be isolated cleanly, in which one *ortho*-Me of the *N*-mesityl NHC ligand was found to have undergone a carboxylate-driven C—H activation to generate stable cyclometallated product **49** [40]. Previously, when C—H activation had occurred in other Ru metathesis catalysts, it was typically followed by insertion of the Ru—C bond into the carbene, leading to metathesis-inactive species.

The reactivity of catalyst **49** was then explored and, while less active than **4** in various standard metathesis reactions, it showed a promising *E*/*Z* ratio in CM of allylbenzene and CDAB of 1.4 (41% Z) with conversion approaching 60%. Two further analogs with increased steric bulk of the non-chelating NHC substituent, **50** and **51**, were targeted via similar reaction conditions (Figure 18). Catalyst **50**, bearing a bulkier *N*-DIPP group, could be prepared and gave improved selectivity in CM of allylbenzene and CDAB (51% Z at 60% conversion). In attempting to prepare catalyst **51**, which contains an *N*-adamantyl group, under similar reaction conditions, no formation of the desired product was observed. Instead, C—H activation occurred exclusively at the

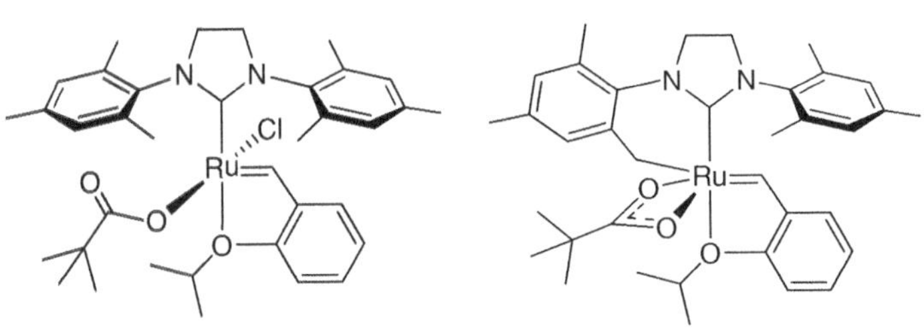

48 **49**

Figure 17 Targeted monopivalate Z-selective catalyst **48** and unexpected product **49**

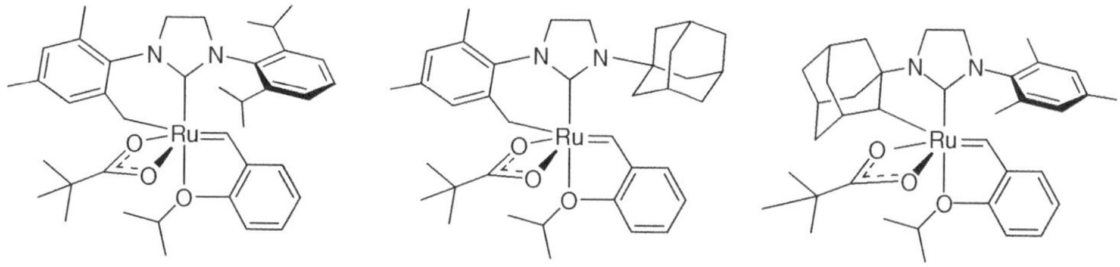

50 51 52

Figure 18 *Targeted mesityl-activated catalysts* ***50*** *and* ***51*** *and adamantyl-activated catalyst* ***52***

β methylene-carbon of the adamantyl NHC substituent, leading to product **52**, which contains a new five-membered metallacycle. While adamantyl-chelated catalyst **52** demonstrated significantly lower activity at room temperature than **49** or **50**, appreciable conversion could be obtained at higher temperatures. Catalyst **52** was tested in CM of allylbenzene and CDAB and, for the first time, high *Z*-selectivity was achieved in the CM of allylbenzene and CDAB, where the desired product was formed in up to 90% *Z*-selectivity (corresponding to an *E*/*Z* ratio of 0.12).

2.5.2 Model for selectivity

A general model for selectivity in these catalysts bearing cyclometallated NHCs has been established through experimental observations and computational analysis [39, 41]. In this model, preferential formation of side-bound metallacycles occurs over bottom-bound metallacycles. DFT calculations based on catalyst **52** have elucidated two contributing factors for this preference. First, in the bottom-bound metallacycle (and the transition states for its formation and collapse), there are significant steric clashes between the chelating adamantyl group and the metallacyclobutane, whereas these are absent in the case of the side-bound metallacycle. Secondly, in the bottom-bound pathway, both the NHC and alkylidene π* orbitals are located in the same plane and hence align with the same Ru d orbital (Figure 19a), leading to weaker π back-donation and hence having a destabilizing effect. This is alleviated in the side-bound pathway where the NHC and alkylidene π* orbitals are perpendicular to each other and hence align with two distinct Ru d orbitals allowing for more efficient π back-donation, which has an overall stabilizing effect. In the side-bound geometry, the chelating NHC substituent forces the non-chelating NHC substituent (in the case of **52**, a mesityl group) to be held directly over the metallacyclobutane (Figure 19b). The anionic ligand located on the bottom face of the metallacyclobutane can orient away from the metallacycle, leading to minimal steric influence. Hence, in the *syn*-metallacycle the two substituents can point downwards with minimal steric clash, whereas in the *anti*-metallacycle, there is a significant steric clash between one substituent and the *N*-mesityl group. Of note in these calculations is the flexible coordination mode of the carboxylate ligand, either mono- or bidentate, which allowed for increased stabilization of key transitions states and intermediates.

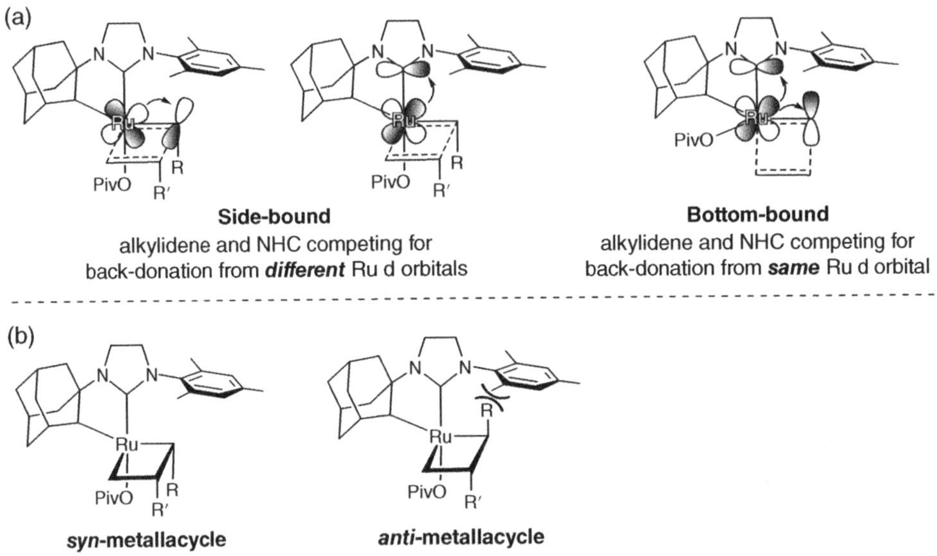

Figure 19 *(a) Electronic factors stabilizing side-bound over bottom-bound metallacycle. (b) Model for Z-selectivity in cyclometallated catalysts*

Further insights into the relationships between structure and catalyst activity and selectivity have been achieved through variation of the three major components: the anionic ligand; the non-chelating NHC substituent; and the nature of the chelating substituent.

2.5.3 Variation of the anionic ligand

Systematic variation of the steric bulk of the carboxylate has been undertaken in the case of catalysts bearing the six-membered chelating mesityl group (Figure 20) [42]. Catalysts **53–55** were compared with original catalyst **49** in the CM reaction of allylbenzene and CDAB. Here, increased bulk of the carboxylate led to a reduction in Z-selectivity, presumably due to increased steric clash of the phenyl groups with the downward-facing metallacyclobutane substituents. Replacement of the bidentate pivalate ligand by a monodentate chloride ligand as in catalyst **56** led to a significant loss of Z-selectivity and preferential formation of the *E*-isomer (81% *E*) was observed, which is more similar to ratios typically observed in the case of bottom-bound metallacycles. In the case of less-active **52** which contains the five-membered metallacycle, replacement of the pivalate with a monodentate iodide (**57**) or phenolate (**58**) ligand (Figure 21) completely attenuated metathesis activity of the resulting catalysts for homodimerization of allylbenzene, both catalysts instead effecting significant olefin-migration isomerization [43]. However, **57** was found to be a useful precursor for the preparation of a variety of catalysts with alternative bidentate ligands (**59–61**).

While these catalysts could effect metathesis with similar levels of Z-selectivity, they had lower levels of metathesis activity and caused significant amounts of olefin-migration

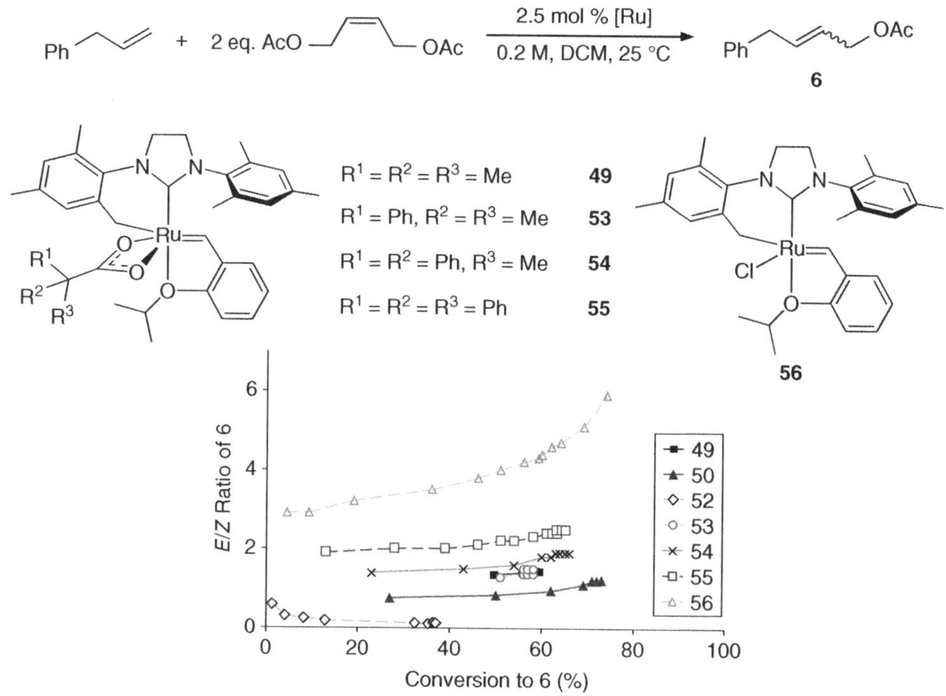

Figure 20 Mesityl-activated catalysts bearing different anionic ligands and their E/Z ratio vs. conversion for CM of allylbenzene and CDAB

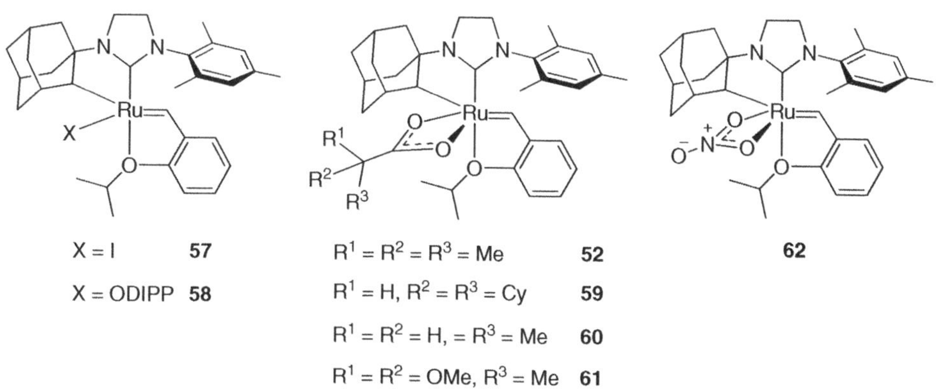

Figure 21 Adamantyl-activated catalysts with varying anionic ligands

isomerization. A significant advance came about in the form of catalyst **62** where the pivalate ligand was replaced by a nitrate ligand. Catalyst **62** demonstrated higher metathesis activity than **52** for a range of substrates, while maintaining high Z-selectivity and high metathesis to olefin-migration ratio (Scheme 8). Crystallographic analysis revealed analogous bidentate coordination of the nitrate ligand and only subtle

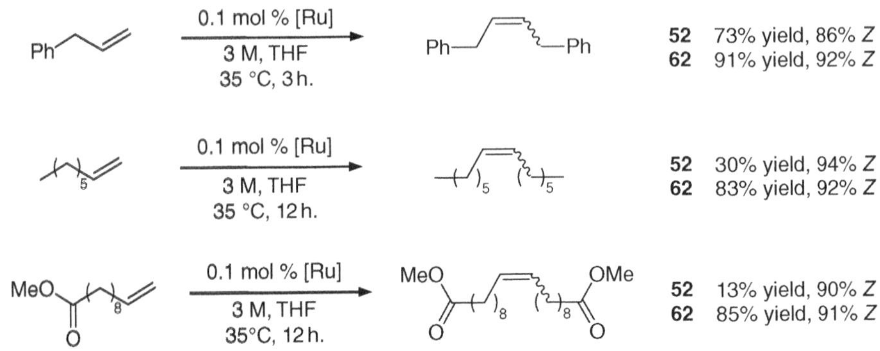

Scheme 8 *Comparison of pivalate-substituted catalyst* **62**...

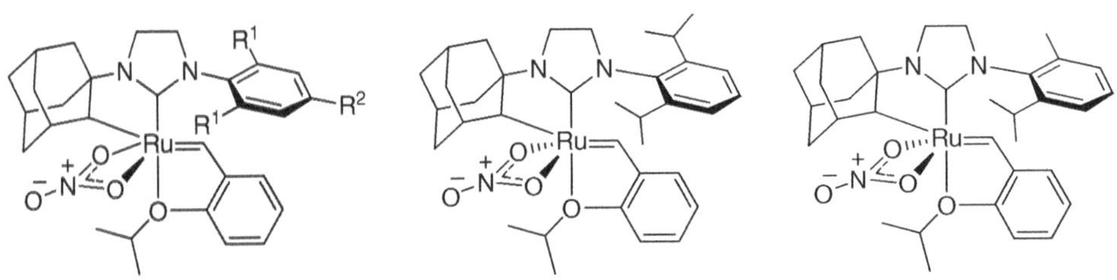

66

67

R^1 = Me, R^2 = OMe **63**
R^1 = Me, R^2 = Cl **64**
R^1 = Et, R^2 = Me **65**

Figure 22 *Variation of the N-aryl group in adamantyl-activated catalysts*

deviations in bond lengths and angles was observed. Largely similar behavior of the two ligands was also supported by DFT calculations, where the main difference was that the small size of the nitrate ligand resulted in more stable bidentate coordination over the reaction pathway [44]. Experimentally, catalyst **62** showed greater stability both in the presence of oxygen and at high temperatures when compared with pivalate-containing catalyst **52**.

2.5.4 Variation of the aryl group

Subtle alterations to sterics and electronics of the aryl group as in catalysts **63–65** (Figure 22) were generally well tolerated, but only led to minor differences in overall metathesis activity and *Z*-selectivity [43]. Catalyst **66**, which contains an *N*-DIPP substituent, was an obvious target, given the improved *Z*-selectivity observed in the case of the six-membered chelate (**50**). However, initial attempts to prepare **66** or other catalysts with significant variation of electronic and steric parameters were hampered by the instability of the products under the reaction conditions, which used silver pivalate.

While formation of the desired complexes could be observed by ^{1}H-NMR spectroscopy, they typically underwent rapid decomposition to form multiple species. It was later shown that using sodium pivalate (in place of silver pivalate) offered a milder method for salt metathesis and C—H activation to form cyclometallated catalysts [45]. These conditions allowed the preparation of a number of catalysts, including **66**, which could not be obtained using silver pivalate. As expected based on the model where *Z*-selectivity is governed by interactions of metallacycle substituents with the *N*-aryl group, catalyst **66** with the bulkier *N*-DIPP group showed improved *Z*-selectivity over its *N*-Mes analog **62** and comparable activity in a variety of CM reactions. Further exploration of reaction conditions revealed that *N*-DIPP-substituted catalyst **66** was active even at extremely low catalyst loadings (0.01 mol%) and achieved turnover numbers of >7000 in the homodimerization of allylbenzene. While catalyst **67** demonstrated comparable *Z*-selectivity with **66**, it had lower metathesis activity. Consistent with previous studies of related catalysts, NHC aryl substituents with *ortho*-C—H bonds were not tolerated under C—H-activation conditions, as C—H activation was followed by rapid insertion into the benzylidene moiety [46].

2.5.5 Variation of the cyclometallated NHC substituent

As previously noted, there were very significant differences in both activity and *Z*-selectivity observed between **52**, which has a five-membered adamantyl-based chelate, and **49**, with a six-membered mesityl-based chelate. Though many attempts at further variation of the chelating group have been made, these changes are less well tolerated and efforts have only yielded a small number of new stable catalysts (Figure 23). Hence, it is difficult to draw further conclusions about the role of the chelate in the selectivity and activity of these catalysts. Interestingly, catalyst **68**, which contains an *N*-3,5-dimethyladamantyl-substituted NHC, had similar *Z*-selectivity to its adamantyl analog (**52**) but was noticeably less active and required significantly higher catalyst loadings to achieve comparable yield [45]. Catalyst **69**, which contains a chelating *tert*-butyl group, could be isolated but was found to have limited stability at room temperature and was a poor catalyst in CM applications [47]. Attempts to prepare related catalysts **70** and **71** were not successful.

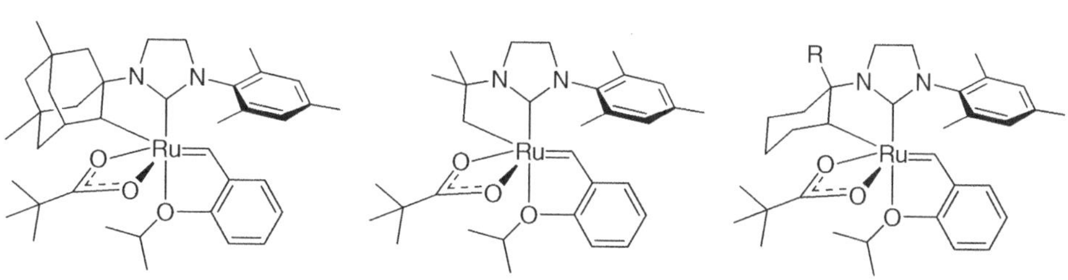

68 **69** R = H **70** R = Me **71**

Figure 23 *Variation of the cyclometallated NHC substituent*

2.5.6 Reactivity of cyclometallated Z-selective catalysts

Of the catalysts prepared, **62** and later **66** emerged as having an appropriate balance of activity and Z-selectivity and have been employed in a variety of metathesis reactions. CM with catalyst **62** provided access to a variety of Lepidopteran pheromones in good yield and high Z-selectivity (76–88%) [48]. Additionally, a variety of macrocycles were prepared by RCM in moderate to good yields and generally high Z-selectivity (65–94%) [49]. When more-selective catalyst **66** was later employed in a sample of these reactions, Z-selectivity was uniformly high (>95%), although in some cases reduced yields were observed [45]. In addition to these reactions, **62** was employed in Z-selective ROMP [50] and a resolved enantiopure version of **62** achieved high Z-selectivity and high enantiomeric excess in a variety of asymmetric ROCM applications [51].

The high barrier for the formation of *anti*-metallacycles with catalysts **62** and **66** has been exploited to allow orthogonal reactivity of terminal- or Z-olefins in the presence of *E*-olefins. Catalyst **62** has demonstrated Z-selective ethenolysis of *E*/*Z* mixtures of olefins, allowing isolation of the pure *E*-olefin with high stereopurity (>95% *E*) [44a, 49]. In addition, **66** demonstrated Z-selective CM of substrates containing both a terminal olefin and an internal *E*-olefin to yield *E*,*Z*-dienes with >95% Z-selectivity [52]. While the Z-selectivity observed in the abovementioned cases is impressive, all CM examples were confined to unhindered, Type I olefins. Some categories of allylic-substituted olefins, such as vinyl acetals and vinyl epoxides, fall into the Type II category for catalyst **66**, which can effect Z-selective CM of these substrates with Type I olefins in good yield and high Z-selectivity (89 to >95% *Z*) [53]. In contrast, other classes of allylic-substituted olefins demonstrated low or negligible reactivity under identical conditions. Hence, in order to expand the scope of reactivity to these and more challenging substrates, such as 1,1-disubstituted olefins, further improvements in catalyst activity are required.

2.6 Conclusions and future outlook

To date, three distinct strategies have been utilized to impose a preference for *syn*-metallacycles in Ru-based olefin metathesis catalysts, resulting in three families of Z-selective catalysts. Promising initial reactivity has been observed with both thiophenolate- and dithiolate-based catalysts and both frameworks offer many opportunities for further tuning of activity and Z-selectivity. The cyclometallated catalysts have been further developed and have demonstrated high activity and Z-selectivity for a wide variety of substrates. However, in all cases, further improvements will be necessary to achieve Z-selective metathesis across the broad substrate scope demonstrated by previous generations of Ru-based catalysts.

While these families of catalyst demonstrate a high kinetic preference for the Z-olefin, catalysts that demonstrate a comparably high kinetic preference for the *E*-olefin have yet to be achieved. The generation of *E*-olefins instead relies on thermodynamic control and so selectivity is substrate-dependent. While some insights into the structure-selectivity relationships in Ru-based metathesis catalysts have been gained, the design of catalysts that lead to complementary selectivity for *trans*-metallacyclobutanes has not been realized to date and represents an outstanding challenge in the field of olefin metathesis.

References

[1] (a) N. Calderon, H. Y. Chen, K. W. Scott, *Tetrahedron Lett.* **1967**, *8*, 3327–3329; (b) N. Calderon, *Acc. Chem. Res.* **1972**, *5*, 127–132.
[2] P. Jean-Louis Hérisson, Y. Chauvin, *Makromol. Chem.* **1971**, *141*, 161–176.
[3] (a) R. Schrock, S. Rocklage, J. Wengrovius, G. Rupprecht, J. Fellmann, *J. Mol. Catal.* **1980**, *8*, 73–83; (b) J. S. Murdzek, R. R. Schrock, *Organometallics* **1987**, *6*, 1373–1374; (c) C. Copéret, F. Lefebvre, J.-M. Basset, in *Handbook of Metathesis, Vol. 1*, 1st edn (ed. R. H. Grubbs), Wiley-VCH, Weinheim, **2003**, pp. 33–46.
[4] (a) R. R. Schrock, R. T. DePue, J. Feldman, C. J. Schaverien, J. C. Dewan, A. H. Liu, *J. Am. Chem. Soc.* **1988**, *110*, 1423–1435; (b) R. R. Schrock, *Tetrahedron* **1999**, *55*, 8141–8153; (c) R. R. Schrock, J. S. Murdzek, G. C. Bazan, J. Robbins, M. DiMare, M. O'Regan, *J. Am. Chem. Soc.* **1990**, *112*, 3875–3886; (d) R. R. Schrock, *Acc. Chem. Res.* **1990**, *23*, 158–165; (e) R. R. Schrock, *Chem. Rev.* **2009**, *109*, 3211–3226.
[5] (a) R. E. Rinehart, H. P. Smith, *J. Polym. Sci., Part B: Polym. Lett.* **1965**, *3*, 1049–1052; (b) F. W. Michelotti, W. P. Keaveney, *J. Polym. Sci., Part A: Gen. Pap.* **1965**, *3*, 895–905.
[6] S. T. Nguyen, L. K. Johnson, R. H. Grubbs, J. W. Ziller, *J. Am. Chem. Soc.* **1992**, *114*, 3974–3975.
[7] T. M. Trnka, R. H. Grubbs, *Acc. Chem. Res.* **2000**, *34*, 18–29.
[8] P. Schwab, M. B. France, J. W. Ziller, R. H. Grubbs, *Angew. Chem. Int. Ed. Engl.* **1995**, *34*, 2039–2041.
[9] (a) M. Scholl, S. Ding, C. W. Lee, R. H. Grubbs, *Org. Lett.* **1999**, *1*, 953–956; (b) J. Huang, E. D. Stevens, S. P. Nolan, J. L. Petersen, *J. Am. Chem. Soc.* **1999**, *121*, 2674–2678; (c) T. Weskamp, F. J. Kohl, W. A. Herrmann, *J. Organomet. Chem.* **1999**, *582*, 362–365.
[10] S. B. Garber, J. S. Kingsbury, B. L. Gray, A. H. Hoveyda, *J. Am. Chem. Soc.* **2000**, *122*, 8168–8179.
[11] S. Sutthasupa, M. Shiotsuki, F. Sanda, *Polym. J.* **2010**, *42*, 905–915.
[12] J. C. Mol, *J. Mol. Catal. A: Chem.* **2004**, *213*, 39–45.
[13] J. Prunet, *Curr. Top. Med. Chem.* **2005**, *5*, 1559-1577.
[14] A. K. Chatterjee, T.-L. Choi, D. P. Sanders, R. H. Grubbs, *J. Am. Chem. Soc.* **2003**, *125*, 11360–11370.
[15] (a) I. Ibrahem, M. Yu, R. R. Schrock, A. H. Hoveyda, *J. Am. Chem. Soc.* **2009**, *131*, 3844–3845; (b) M. M. Flook, A. J. Jiang, R. R. Schrock, A. H. Hoveyda, *J. Am. Chem. Soc.* **2009**, *131*, 7962–7963; (c) A. J. Jiang, Y. Zhao, R. R. Schrock, A. H. Hoveyda, *J. Am. Chem. Soc.* **2009**, *131*, 16630–16631.
[16] (a) A. H. Hoveyda, *J. Org. Chem.* **2014**, *79*, 4763–4792; (b) S. Shahane, C. Bruneau, C. Fischmeister, *Chem. Cat. Chem.* **2013**, *5*, 3436–3459; (c) C. Deraedt, M. d'Halluin, D. Astruc, *Eur. J. Inorg. Chem.* **2013**, *2013*, 4881–4908.
[17] (a) E. L. Dias, S. T. Nguyen, R. H. Grubbs, *J. Am. Chem. Soc.* **1997**, *119*, 3887–3897; (b) M. S. Sanford, M. Ulman, R. H. Grubbs, *J. Am. Chem. Soc.* **2001**, *123*, 749–750; (c) M. S. Sanford, J. A. Love, R. H. Grubbs, *J. Am. Chem. Soc.* **2001**, *123*, 6543–6554.

[18] (a) L. Cavallo, *J. Am. Chem. Soc.* **2002**, *124*, 8965–8973; (b) C. Adlhart, P. Chen, *J. Am. Chem. Soc.* **2004**, *126*, 3496–3510; (c) B. F. Straub, *Angew. Chem. Int. Ed.* **2005**, *44*, 5974–5978.
[19] (a) J. A. Tallarico, P. J. Bonitatebus, M. L. Snapper, *J. Am. Chem. Soc.* **1997**, *119*, 7157–7158; (b) T. M. Trnka, M. W. Day, R. H. Grubbs, *Organometallics* **2001**, *20*, 3845–3847; (c) D. R. Anderson, D. D. Hickstein, D. J. O'Leary, R. H. Grubbs, *J. Am. Chem. Soc.* **2006**, *128*, 8386–8387; (d) D. R. Anderson, D. J. O'Leary, R. H. Grubbs, *Chem. Eur. J.* **2008**, *14*, 7536–7544; (e) I. C. Stewart, D. Benitez, D. J. O'Leary, E. Tkatchouk, M. W. Day, W. A. Goddard, R. H. Grubbs, *J. Am. Chem. Soc.* **2009**, *131*, 1931–1938.
[20] P. E. Romero, W. E. Piers, *J. Am. Chem. Soc.* **2005**, *127*, 5032–5033.
[21] (a) A. G. Wenzel, R. H. Grubbs, *J. Am. Chem. Soc.* **2006**, *128*, 16048–16049; (b) P. E. Romero, W. E. Piers, *J. Am. Chem. Soc.* **2007**, *129*, 1698–1704; (c) E. F. van der Eide, P. E. Romero, W. E. Piers, *J. Am. Chem. Soc.* **2008**, *130*, 4485–4491; (d) F. van der EideEdwin, W. E. Piers, *Nat Chem* **2010**, *2*, 571–576; (e) A. G. Wenzel, G. Blake, D. G. VanderVelde, R. H. Grubbs, *J. Am. Chem. Soc.* **2011**, *133*, 6429–6439.
[22] T. Ritter, A. Hejl, A. G. Wenzel, T. W. Funk, R. H. Grubbs, *Organometallics* **2006**, *25*, 5740–5745.
[23] G. C. Vougioukalakis, R. H. Grubbs, *Chem. Rev.* **2010**, *110*, 1746–1787.
[24] M. S. Sanford, L. M. Henling, M. W. Day, R. H. Grubbs, *Angew. Chem. Int. Ed.* **2000**, *39*, 3451–3453.
[25] N. Bahri-Laleh, R. Credendino, L. Cavallo, *Beilstein J. Org. Chem.* **2011**, *7*, 40–45.
[26] G. C. Vougioukalakis, R. H. Grubbs, *Organometallics* **2007**, *26*, 2469–2472.
[27] K. Vehlow, S. Maechling, S. Blechert, *Organometallics* **2005**, *25*, 25–28.
[28] N. Ledoux, A. Linden, B. Allaert, H. V. Mierde, F. Verpoort, *Adv. Synth. Catal.* **2007**, *349*, 1692–1700.
[29] (a) G. C. Vougioukalakis, R. H. Grubbs, *J. Am. Chem. Soc.* **2008**, *130*, 2234–2245; (b) G. C. Vougioukalakis, R. H. Grubbs, *Chem. Eur. J.* **2008**, *14*, 7545–7556.
[30] D. R. Anderson, T. Ung, G. Mkrtumyan, G. Bertrand, R. H. Grubbs, Y. Schrodi, *Organometallics* **2008**, *27*, 563–566.
[31] E. L. Rosen, D. H. Sung, Z. Chen, V. M. Lynch, C. W. Bielawski, *Organometallics* **2009**, *29*, 250–256.
[32] S. Torker, A. Müller, P. Chen, *Angew. Chem. Int. Ed.* **2010**, *49*, 3762–3766.
[33] P. Teo, R. H. Grubbs, *Organometallics* **2010**, *29*, 6045–6050.
[34] G. Occhipinti, F. R. Hansen, K. W. Törnroos, V. R. Jensen, *J. Am. Chem. Soc.* **2013**, *135*, 3331–3334.
[35] V. R. Jensen, G. Occhipinti, F. R. Hansen, *Patent WO 2012/032131 A1*, **2012**.
[36] G. Occhipinti, V. Koudriavtsev, K. W. Tornroos, V. R. Jensen, *Dalton Trans.* **2014**, *43*, 11106–11117.
[37] R. K. M. Khan, S. Torker, A. H. Hoveyda, *J. Am. Chem. Soc.* **2013**, *135*, 10258–10261.
[38] M. J. Koh, R. K. M. Khan, S. Torker, A. H. Hoveyda, *Angew. Chem. Int. Ed.* **2014**, *53*, 1968–1972.
[39] K. Endo, R. H. Grubbs, *J. Am. Chem. Soc.* **2011**, *133*, 8525–8527.
[40] J. S. Cannon, L. Zou, P. Liu, Y. Lan, D. J. O'Leary, K. N. Houk, R. H. Grubbs, *J. Am. Chem. Soc.* **2014**, *136*, 6733–6743.
[41] (a) P. Liu, X. Xu, X. Dong, B. K. Keitz, M. B. Herbert, R. H. Grubbs, K. N. Houk, *J. Am. Chem. Soc.* **2012**, *134*, 1464–1467; (b) Y. Dang, Z.-X. Wang, X. Wang, *Organometallics* **2012**, *31*, 7222–7234.

[42] K. Endo, M. B. Herbert, R. H. Grubbs, *Organometallics* **2013**, *32*, 5128–5135.
[43] B. K. Keitz, K. Endo, P. R. Patel, M. B. Herbert, R. H. Grubbs, *J. Am. Chem. Soc.* **2012**, *134*, 693–699.
[44] (a) H. Miyazaki, M. B. Herbert, P. Liu, X. Dong, X. Xu, B. K. Keitz, T. Ung, G. Mkrtumyan, K. N. Houk, R. H. Grubbs, *J. Am. Chem. Soc.* **2013**, *135*, 5848–5858; (b) Y. Dang, Z.-X. Wang, X. Wang, *Organometallics* **2012**, *31*, 8654–8657.
[45] L. E. Rosebrugh, M. B. Herbert, V. M. Marx, B. K. Keitz, R. H. Grubbs, *J. Am. Chem. Soc.* **2013**, *135*, 1276–1279.
[46] M. B. Herbert, Y. Lan, B. K. Keitz, P. Liu, K. Endo, M. W. Day, K. N. Houk, R. H. Grubbs, *J. Am. Chem. Soc.* **2012**, *134*, 7861–7866.
[47] L. E. Rosebrugh, V. M. Marx, B. K. Keitz, R. H. Grubbs, *J. Am. Chem. Soc.* **2013**, *135*, 10032–10035.
[48] M. B. Herbert, V. M. Marx, R. L. Pederson, R. H. Grubbs, *Angew. Chem. Int. Ed.* **2013**, *52*, 310–314.
[49] V. M. Marx, M. B. Herbert, B. K. Keitz, R. H. Grubbs, *J. Am. Chem. Soc.* **2013**, *135*, 94–97.
[50] B. K. Keitz, A. Fedorov, R. H. Grubbs, *J. Am. Chem. Soc.* **2012**, *134*, 2040–2043.
[51] (a) J. Hartung, R. H. Grubbs, *J. Am. Chem. Soc.* **2013**, *135*, 10183–10185; (b) J. Hartung, R. H. Grubbs, *Angew. Chem. Int. Ed.* **2014**, *53*, 3885–3888.
[52] J. S. Cannon, R. H. Grubbs, *Angew. Chem. Int. Ed.* **2013**, *52*, 9001–9004.
[53] B. L. Quigley, R. H. Grubbs, *Chem. Sci.* **2014**, *5*, 501–506.

3

Ligands for Iridium-catalyzed Asymmetric Hydrogenation of Challenging Substrates

Marc-André Müller and Andreas Pfaltz

Department of Chemistry, University of Basel, St. Johanns-Ring 19, 4056 Basel, Switzerland

3.1 Asymmetric hydrogenation

Enantioselective homogeneous hydrogenation catalyzed by chiral transition metal complexes is one of the most well established transformations in asymmetric synthesis [1]. Excellent enantioselectivities have been achieved in the hydrogenation of a wide range of substrates, often with very low catalyst loadings. High reliability, mild reaction conditions, and perfect atom economy are further attractive attributes of this method. In particular complexes based on Ru or Rh have found broad application in industrial processes [1] and the impact of these catalysts has been recognized by the Nobel Prize awarded to Ryoji Noyori and William S. Knowles in 2001 [2].

In this chapter, we focus on a more recent direction in asymmetric hydrogenation, the development of Ir catalysts based on heterobidentate ligands, which have considerably enhanced the substrate scope for this transformation. In particular, the major ligand classes and the underlying design principles as well as the special features and selected applications of these catalysts are discussed. We also dwell on recent mechanistic

Ligand Design in Metal Chemistry: Reactivity and Catalysis, First Edition. Edited by Mark Stradiotto and Rylan J. Lundgren.

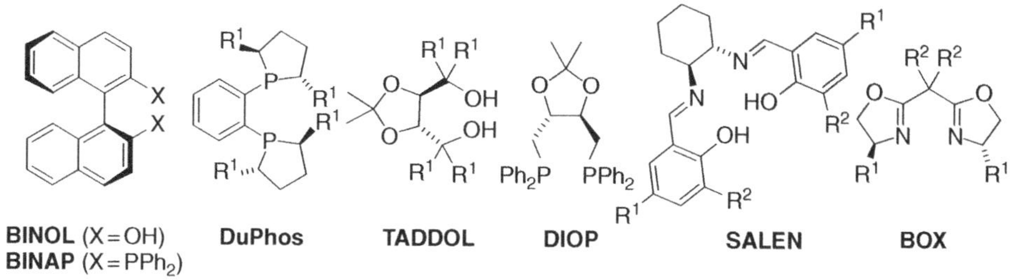

Figure 1 *Selection of widely used C_2-symmetric ligands*

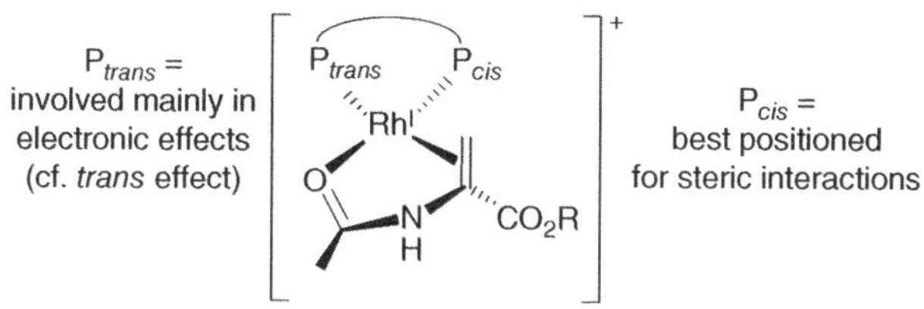

Figure 2 *Intermediate in the Rh-catalyzed asymmetric hydrogenation used to explain Achiwa's "Respective Control Concept"*

insights that have led to a qualitative model for the enantioselective step, which may serve as a useful guide in the search for new ligands.

Progress in asymmetric hydrogenation was driven largely by the design and discovery of new chiral ligands. The initial breakthrough came with the introduction of C_2-symmetric bidentate diphosphine ligands [3], based on the idea that C_2 symmetry reduces the number of undesired competing isomeric transition states that lower the enantioselectivity, compared with reactions with nonsymmetric ligand complexes. Furthermore, the lower number of possible catalytic intermediates and the reduced complexity of NMR spectra of C_2-symmetric ligand complexes facilitate mechanistic studies and a rationalization of the observed enantioselectivities [3]. As a consequence, C_2-symmetric ligands dominated the field of asymmetric catalysis for a long time (Figure 1).

However, despite the benefits of C_2 symmetry, there are arguments that for certain reactions, including asymmetric hydrogenation, nonsymmetric ligands should be the better choice. The "Respective Control Concept" introduced by Achiwa is a good example [4]. In the structurally well characterized catalytic intermediates of the hydrogenation of acetamido-acrylic acid derivatives, for example, the two P atoms of a bidentate ligand occupy different positions with respect to the substrate (Figure 2). The phosphino group oriented *cis* to the coordinated C=C bond is closer to the prochiral unit of the substrate and, therefore, was postulated to be responsible for enantiocontrol through steric effects. The P atom in *trans* position to the C=C bond is better positioned for electronic interaction with the substrate and hence was assumed to mainly affect the reaction rate. As a consequence, the steric and electronic

Table 1 *Asymmetric hydrogenation of dimethyl itaconate (**1**) using ligands with different phosphine units*

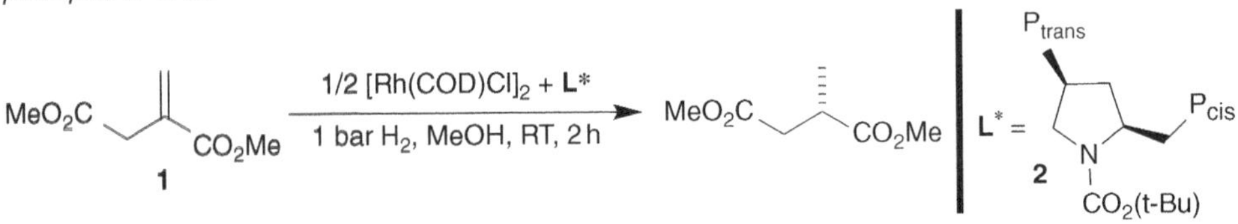

Entry	P_{cis}	P_{trans}	Catalyst loading (mol%)	Conv. (%)	*ee* (%)
1	PPh_2	PPh_2	0.1	36	5
2	P(*p*-$Me_2NPh)_2$	PPh_2	0.1	55	68
3	PPh_2	P(*p*-$Me_2NPh)_2$	0.1	>99	93

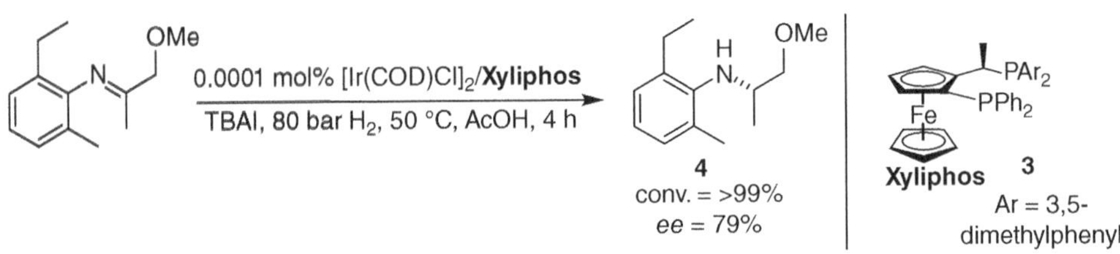

Scheme 1 *Asymmetric hydrogenation as key step in the production of (S)-metolachlor*

properties of each of the two phosphine units should be tuned individually, in order to obtain an optimal ligand for a given substrate.

To validate this concept, various ligands with two different phosphine units were tested in the asymmetric hydrogenations and compared with analogous ligands with two identical phosphine units. In the hydrogenation of a carbonyl compound both the enantioselectivity and conversion could be improved by replacing one PPh_2 group of DIOP by a dicyclohexylphosphine unit. The results obtained in the hydrogenation of dimethyl itaconate (**1**) as well demonstrated that switching the positions of the *cis* and *trans* phosphine units of the ligand scaffold **2** has a strong effect on the enantiomeric excess (*ee*) and conversion (Table 1).

Although these results emphasize the potential advantages of electronically and sterically unsymmetrical ligands, this concept does not guarantee improved catalyst performance. If isomeric complexes are formed, in which the two phosphine units have switched positions, all efforts to optimize these groups individually may be futile. However, in the absence of such complications catalyst performance can be impressive as in the example of the industrial production of (*S*)-metolachlor (Scheme 1) [5]. Individual structural optimization of the two phosphine units led to an Ir catalyst based on the Xyliphos ligand (**3**) that catalyzed the hydrogenation of the C=N bond with more than 10^6 turnovers and a turnover frequency of 10^5 per hour producing the (*S*)-metolachlor precursor **4** in 79% *ee*.

3.2 Iridium catalysts based on heterobidentate ligands

Many efficient Rh- and Ru-based catalysts are known that allow highly enantioselective hydrogenation of a wide variety of functionalized olefins. However, their application range is limited to alkenes bearing a coordinating group near the C=C bond, dehydroamino acid derivatives or allylic alcohols being typical substrates for these catalysts. On the other hand, olefins lacking coordinating groups normally show poor enantioselectivity and low reactivity.

A key step towards hydrogenation catalysts with broader substrate scope was the development of mixed N,P ligands. Such ligands represent an even more effective way of desymmetrization than diphosphines, based on the different electronic properties of the two coordinating heteroatoms, a "soft" P atom with σ-donor and π-acceptor properties and a "hard" N atom with mainly σ-donor character (Figure 3).

The concept of electronic desymmetrization led to the phosphinooxazoline (**5**) (PHOX) ligands, which were originally designed and successfully used for the enantiocontrol of Pd-catalyzed allylic substitution [6]. Subsequently, many further applications of these versatile, readily accessible ligands were found [7]. Inspired by the Crabtree catalyst (**6**) [8], which in contrast to Rh and Ru complexes shows high reactivity in the hydrogenation of unfunctionalized C=C bonds, Ir complexes such as **7** derived from PHOX ligands were evaluated as chiral analogs of **6** and found to give encouraging results in the asymmetric hydrogenation of olefins lacking coordinating groups (Scheme 2a and b) [9].

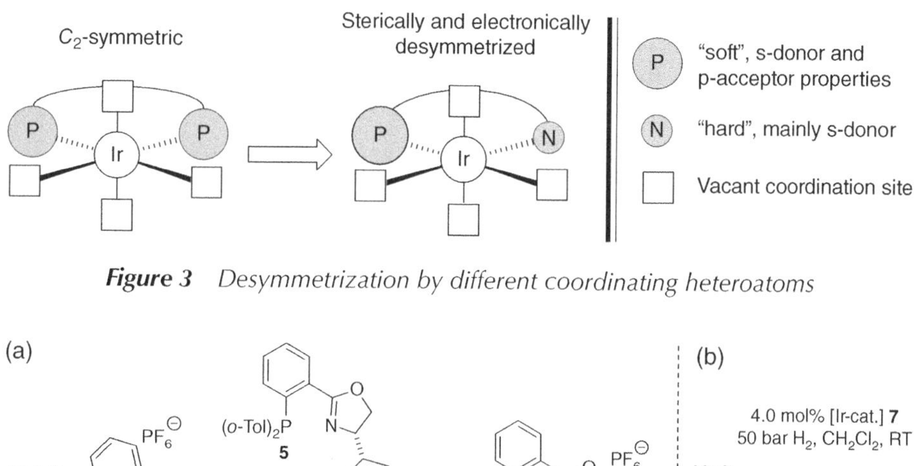

Figure 3 *Desymmetrization by different coordinating heteroatoms*

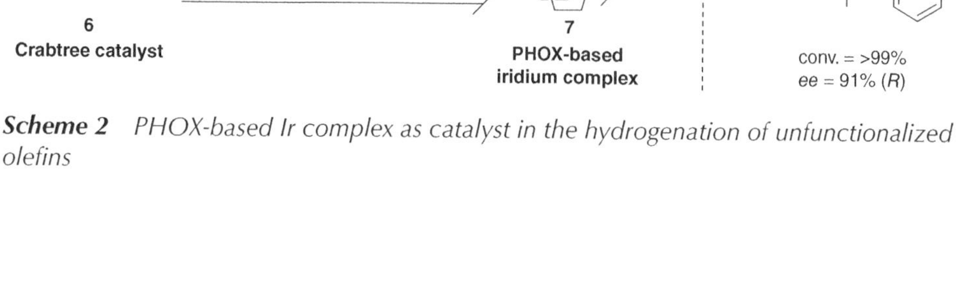

Scheme 2 *PHOX-based Ir complex as catalyst in the hydrogenation of unfunctionalized olefins*

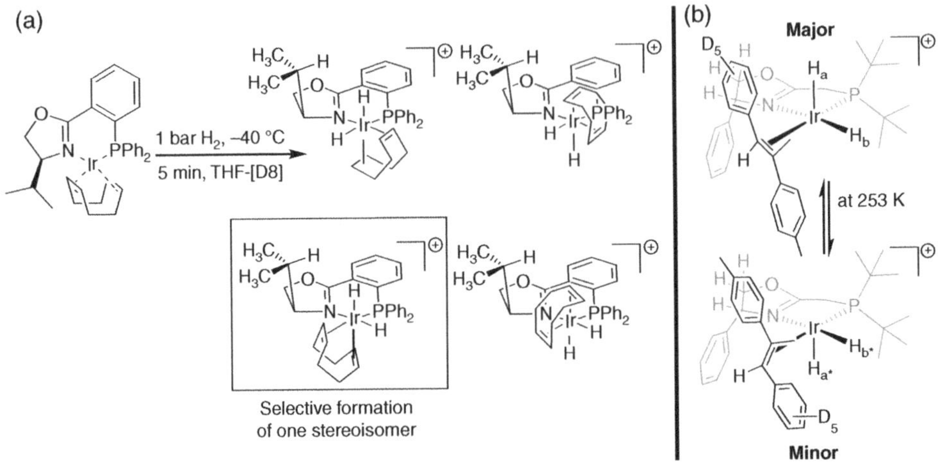

Scheme 3 *Selective formation of specific iridium dihydride species due to electronic and steric differentiation*

^{1}H-NMR experiments conducted at low temperature demonstrated that only one of the four possible stereoisomers was formed upon oxidative addition of dihydrogen, due to electronic and steric differentiation of the N,P ligand (Scheme 3a) [10]. Recently it became even possible to characterize dihydride intermediates with a coordinated alkene, representing the resting state of the catalyst (Scheme 3b) [11]. These findings showed that electronic discrimination of the N,P ligand results in the oxidative addition of hydrogen exclusively *trans* to the Ir—N bond, which is electronically favored [12], whereas the olefin is bound *trans* to the Ir—P bond in agreement with computational studies [13].

The counterion as well was found to strongly influence catalyst performance. Initial experiments with Ir-PHOX complexes gave high enantioselectivity and full conversion, but only at high catalyst loadings of 4 mol% (Scheme 2b) [9]. Lower catalyst loadings resulted in decreased conversion due to catalyst deactivation [14] with concomitant formation of an inactive trinuclear iridium hydride cluster **8** (Scheme 4) [15], analogous to the deactivation products observed with the Crabtree catalyst **6** [16].

While all attempts to avoid formation of the hydride cluster or to convert it back to a catalytically active species failed, the solution to the deactivation problem turned out to be as simple as surprising. Exchange of the counterion from PF_6 to tetrakis[3,5-bis(trifluoromethyl)phenyl]borate (BAr_F) strongly increased conversion (Scheme 5). Using catalyst **9** with BAr_F as the counterion, several thousand turnovers could be obtained in the hydrogenation of *trans*-methyl stilbene **10** under optimized conditions.

Kinetic studies conducted with Ir complexes bearing different counterions provided a plausible explanation of this unusual anion effect [17]. Catalysts with PF_6 as counterion showed first-order rate dependence on the olefin concentration, whereas for catalysts containing BAr_F as counterion the rate dependence was close to zero order, implying a much faster reaction of the olefin with the catalyst in this case.

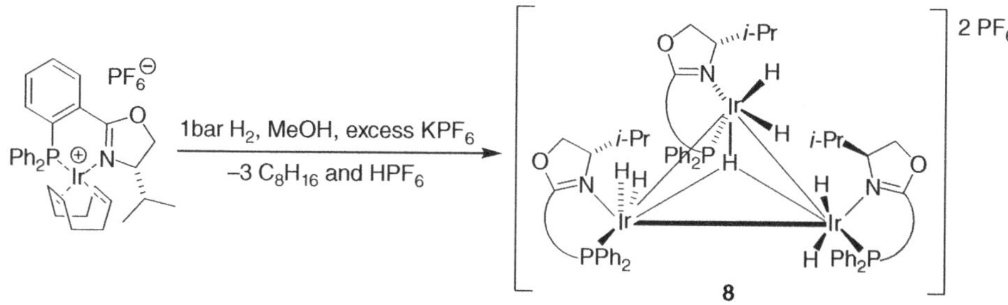

Scheme 4 *Formation of trinuclear iridium hydride cluster as main deactivation pathway of catalysts*

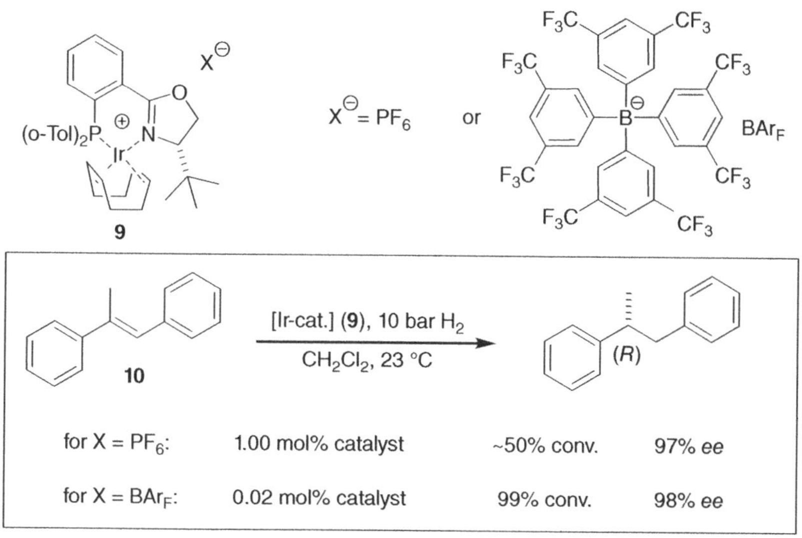

Scheme 5 *Enantioselective hydrogenation of trans-methyl stilbene* ***10*** *with PHOX-based Ir catalyst bearing different counterions*

This rate difference may be explained by the stronger coordination ability of the PF_6 counterion or the formation of a tighter ion pair compared with BAr_F, hampering coordination of the olefin to the Ir center. As a result, the reaction with the olefin becomes rate-determining. In contrast, the extremely weakly coordinating BAr_F ion does not impede olefin coordination and therefore the catalyst remains saturated with substrate. As a consequence, migratory insertion is much faster than the deactivation reaction in the case of the BAr_F salt and, therefore, the catalyst has a much longer life time than the PF_6 analog, which undergoes migratory insertion and deactivation at similar rates.

An additional advantage of BAr_F-based Ir complexes is their stability against moisture and oxygen and hence they can be handled in air, whereas reactions with PF_6-based complexes require set-up under inert gas atmosphere. Furthermore, it is

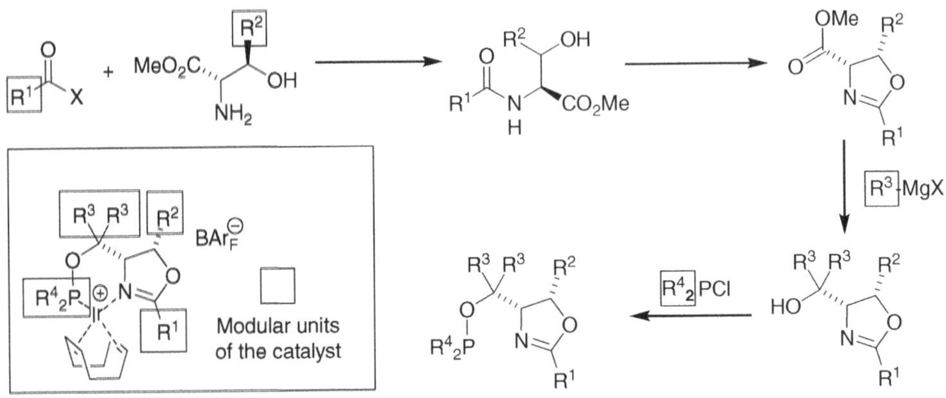

Scheme 6 *Modular synthesis of SerPHOX and ThrePHOX ligands*

possible to purify Ir-BAr_F complexes by column chromatography on silica gel if necessary, which facilitates purification significantly.

After the first successful applications of chiral Ir N,P ligand complexes, many new ligand classes were explored, in order to expand the substrate scope. Because of the limited mechanistic insights at that time, most of the new ligands were found by chance, intuition or systematic screening. To facilitate structural optimization of a ligand, modular structures that allow individual variation of different parts are clearly desirable. Examples of structures that ideally fulfill this criterion are the SerPHOX and ThrePHOX ligands derived from the amino acids serine or threonine [18]. Overall this ligand class contains four structural elements, which can be modified in a four-step reaction sequence, employing different carboxylic acid derivatives, chlorophosphines and Grignard reagents (Scheme 6).

The high modularity of this ligand class allowed preparation of a diverse library of chiral Ir complexes. The effectiveness of this approach was amply demonstrated by using the three isomeric olefins **11**, **12** and **13**, which provide after hydrogenation the same product. In this case, different catalysts performed best for each substrate and high *ee* values between 94% and >99% were obtained (Scheme 7). The fact that these catalysts showed high selectivity in the hydrogenation of many different substrates such as **14**, **15** and **16** highlights the importance of a modular catalyst structure [18b, 19].

Although reasonably high enantioselectivity for substrate **13** (94% *ee*) was obtained, in general terminal olefins remained challenging substrates in terms of enantioselectivity compared with trisubstituted ones. The difficulty arises from the fact that in this case discrimination of the two enantiofaces of the C=C bond depends on the ability of the catalyst to differentiate between the two geminal substituents, which often are of similar size. In contrast, enantiocontrol in the hydrogenation of a trisubstituted C=C bond relies on the more pronounced difference in size between the H atom and an aryl or alkyl group at the less substituted olefinic C atom. Furthermore, terminal olefins with α H atoms next to the double bond are known to undergo a competitive isomerization

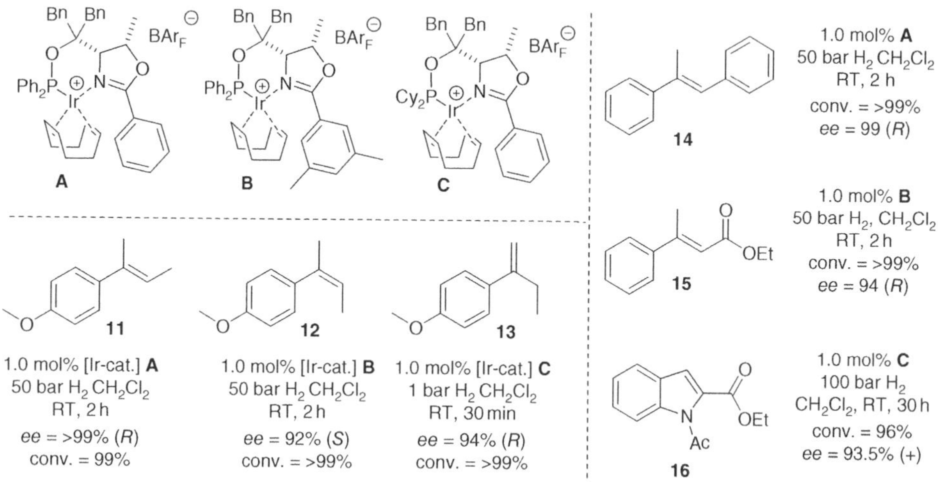

Scheme 7 *Different ThrePHOX-based Ir catalysts achieving high selectivity in the hydrogenation of different substrates*

under hydrogenation conditions to the thermodynamically more stable trisubstituted isomers [20], which can produce products with opposite configuration. In order to overcome these issues, a library containing 96 different phosphite-oxazoline ligands was prepared. By systematic variation of different residues on the ligand scaffold of catalyst **17**, axial chiral biarylphosphites were found to be most effective in the hydrogenation of 1,1-disubstituted alkenes [20, 21]. These ligands enabled the hydrogenation of a wide range of different 1,1-disubstituted alkenes with excellent enantioselectivity (Scheme 8a). In addition, the presence of neighboring groups such as hydroxyl, acetoxy, silyl or heteroaryl was tolerated as well and *ee* values between 91% and 99% were obtained. Furthermore, catalyst **18** showed good to excellent results in the hydrogenation of sterically and electronically diverse terminal biaryl alkenes. For example, the terminal olefin consisting of a phenyl and an *ortho*-tolyl substituent on the double bond was reduced with up to >99% *ee*. In the case of electronically different aryl substituents such as *para*-trifluoromethylphenyl and a *para*-methoxyphenyl a promising *ee* value of 65% was obtained (Scheme 8b).

Although the complexes discussed so far proved to be efficient catalysts for the hydrogenation of trisubstituted and terminal olefins, they all performed poorly in the hydrogenation of tetrasubstituted olefins. This limitation may be explained by the increased steric demand of tetrasubstituted olefins compared with trisubstituted ones, impeding formation of a catalyst–substrate complex. Although tetrasubstituted alkenes have been successfully hydrogenated by using chiral zirconocene complexes, long reaction times, high pressures and high catalyst loadings are drawbacks of these systems [22]. Eventually, more efficient and more practical catalysts for tetrasubstituted alkenes were found based on easily accessible phosphinomethyl-oxazoline ligands that had been originally reported for allylic substitution reactions [6a]. The corresponding

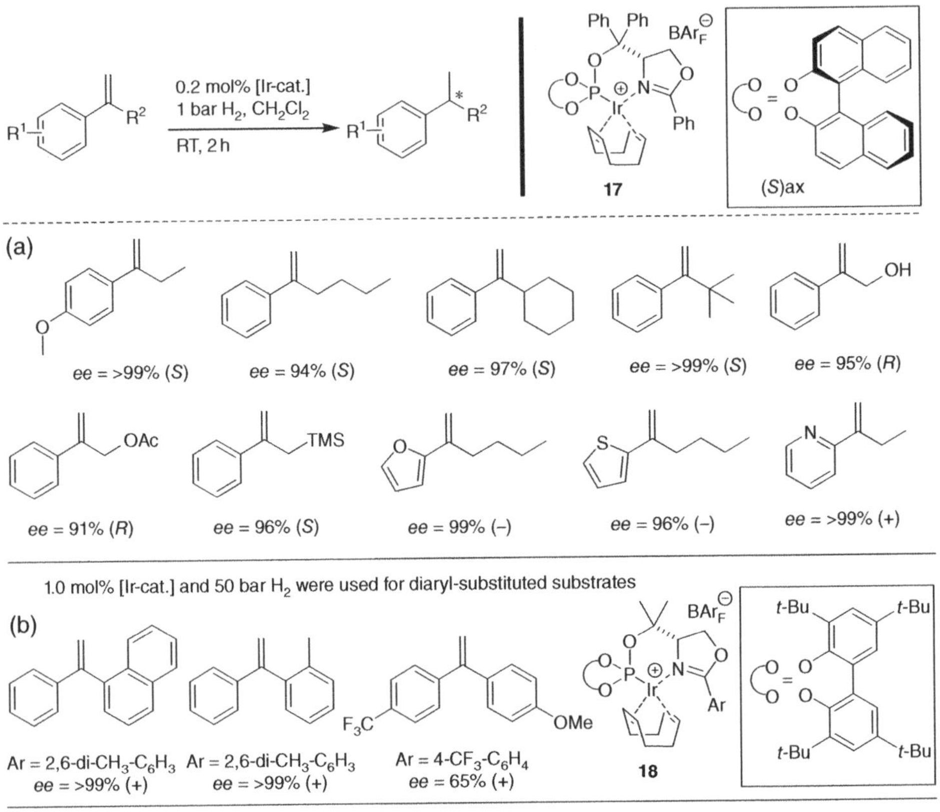

Scheme 8 *Hydrogenation of terminal olefins with oxazoline-based catalysts bearing an axial chiral biarylphosphite moiety. Full conversion was observed for all entries*

Ir complexes gave high enantioselectivities for a range of tetrasubstituted arylalkenes (Scheme 9) [23]. For substrates with at least one methyl substituent essentially full conversion was obtained, whereas more sterically demanding substrates showed reduced reactivity.

The higher reactivity of these catalysts may be explained by the more open coordination sphere of a five-membered chelate complex compared with an analogous six-membered chelate complex formed with a typical PHOX ligand. As illustrated by the two crystal structures in Figure 4, substituents at the oxazoline ring and at the P atom of the six-membered chelate PHOX complex hinder the approach of a substrate to the Ir center, whereas the five-membered chelate analog allows easier access of the substrate.

Another notable development are ligands forming larger chelate rings. Especially the SIPHOX-based Ir catalysts developed by Zhou and co-workers [24], bearing a rigid axially chiral spirobiindane backbone, have considerably extended the scope of asymmetric hydrogenation. These catalysts showed remarkable selectivities and activities in the hydrogenation of a wide variety of different unsaturated carboxylic acids. Furthermore, the optimal reaction conditions with methanol as solvent and a base as

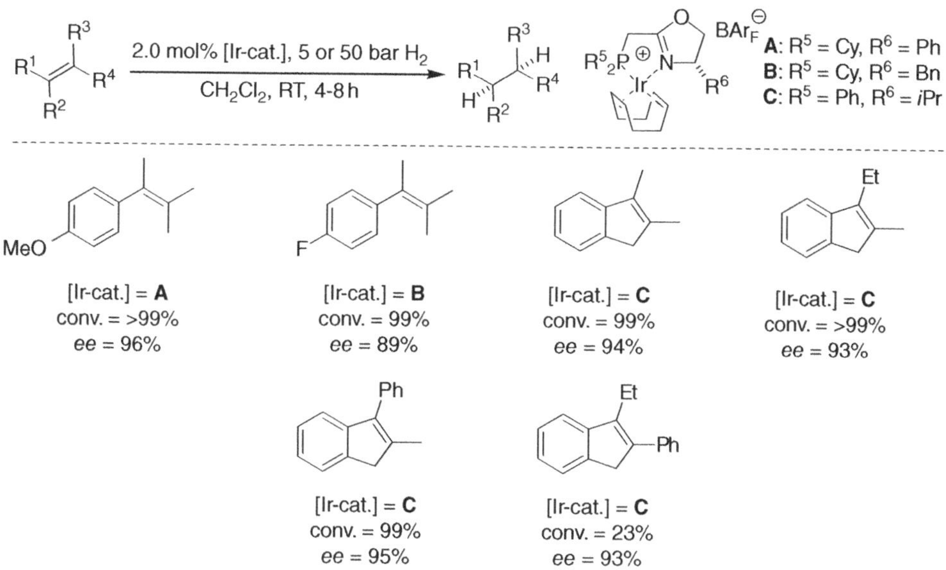

Scheme 9 *Phosphanyl oxazoline-derived Ir catalysts in the hydrogenation of tetra-substituted unfunctionalized olefins*

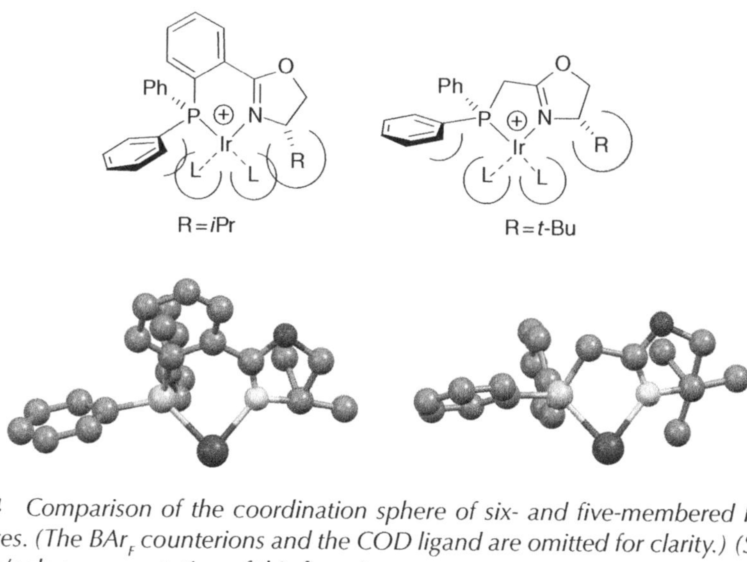

Figure 4 *Comparison of the coordination sphere of six- and five-membered Ir chelate complexes. (The BAr_F counterions and the COD ligand are omitted for clarity.) (See insert for color/color representation of this figure)*

additive are rather unique for Ir-based catalysts. In general, asymmetric Ir-based catalysts require weakly coordinating solvents such as CH_2Cl_2 or toluene [25], while coordinating solvents or additives, such as methanol or triethylamine are known to significantly reduce catalyst reactivity [26]. As shown in Scheme 10, SIPHOX-based

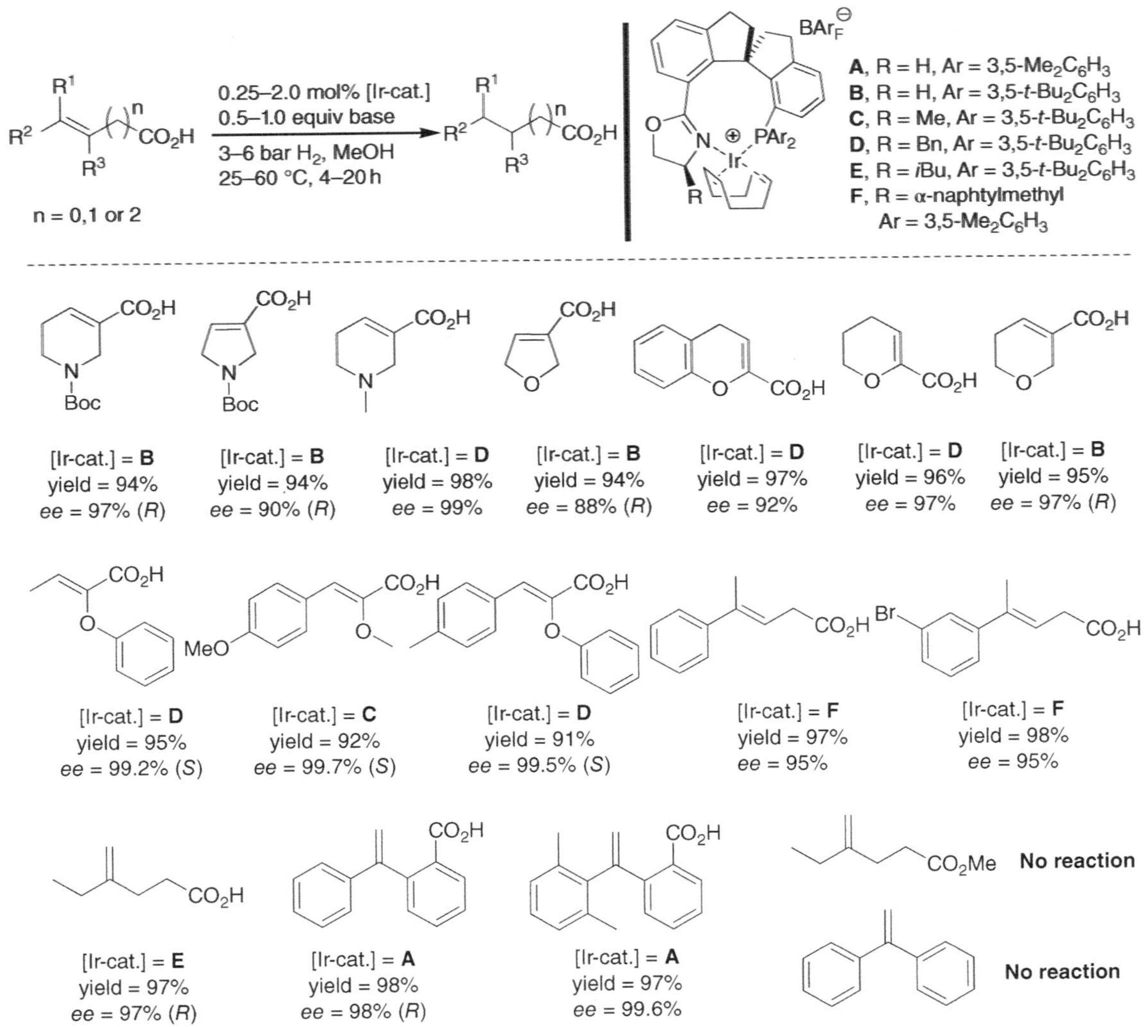

Scheme 10 *Iridium SIPHOX-based catalysts in the asymmetric hydrogenation of unsaturated carboxylic acids*

catalysts perform very well with unsaturated heterocyclic acids [27], α-aryloxy and α-alkoxy α,β-unsaturated carboxylic acids [28], β,γ-unsaturated carboxylic acids [29] or terminal double bonds with a carboxyl group in the β-position [30]. The presence of a free carboxyl proved to be crucial, as analogous carboxylic esters showed no reactivity in the hydrogenation with these catalysts.

While most Ir catalysts developed so far are based on N,P ligands, C,N ligands have been successfully used as well. Exchange of the phosphine unit of a PHOX type ligand by a N-heterocyclic carbene moiety resulted in iridium complexes with different electronic properties at the metal center, which turned out to be advantageous for acid-labile substrates [31]. Iridium hydrides that are formed as intermediates in the catalytic cycle are known to display Brønsted acidity. Density functional theory (DFT) calculations of putative iridium(V) hydride intermediates predicted that Ir N,P ligand complexes form hydrides being up to 7.6 pK_a units more acidic than the corresponding C,N intermediates (Scheme 11a) [32]. These calculations were supported by experimental observations in the hydrogenation of acid sensitive substrates [33]. For example the silyl ether **19** was converted to the desired product **20** in high yield using the catalyst derived from C,N ligand **21**, whereas the corresponding N,P complex based on ligand **22** produced significant amounts of the desilylated hydrogenation product **23** (Scheme 11b) [34].

3.3 Mechanistic studies and derivation of a model for the enantioselective step

Despite the extensive literature on Ir-catalyzed asymmetric hydrogenations and the remarkable progress on ligand development in recent years, experimental data concerning the mechanism and the structures of the catalytic intermediates are still scarce. Based on DFT calculations Andersson and co-workers proposed a catalytic cycle going through iridium(III) and iridium(V) hydrides as intermediates (Scheme 12b, left side) [35]. Computational studies by Burgess and Hall as well supported a similar Ir(III)/Ir(V) cycle (Scheme 12b, right side) for the C,N ligand complex shown in Scheme 11a [35b]. On the other hand gas phase experiments, based on electrospray ionization tandem mass spectrometry, were consistent with an Ir(I)/Ir(III) cycle [36], analogous to the well-established mechanism for Rh-catalyzed hydrogenation (Scheme 12a) [37]. However, because these experiments were conducted in the gas phase, they do not rule out a Ir(III)/Ir(V) cycle in solution. Only recently, experimental evidence for an Ir(III)/Ir(V) cycle was obtained in solution. $[Ir(III)(H)_2(alkene)(L)]^+$ intermediates, which were characterized by NMR spectroscopy at low temperature (see Scheme 3), were found to require addition of H_2 to induce migratory insertion of the alkene upon warming. In the absence of H_2 only dissociation of the coordinated alkene was observed, speaking against an Ir(I)/Ir(III) pathway [11].

So far computational studies had only limited influence on the development of new catalysts and were mainly used to explain results after the fact. Given the high fluxionality and multifaceted aggregation behavior of iridium hydride species and the many

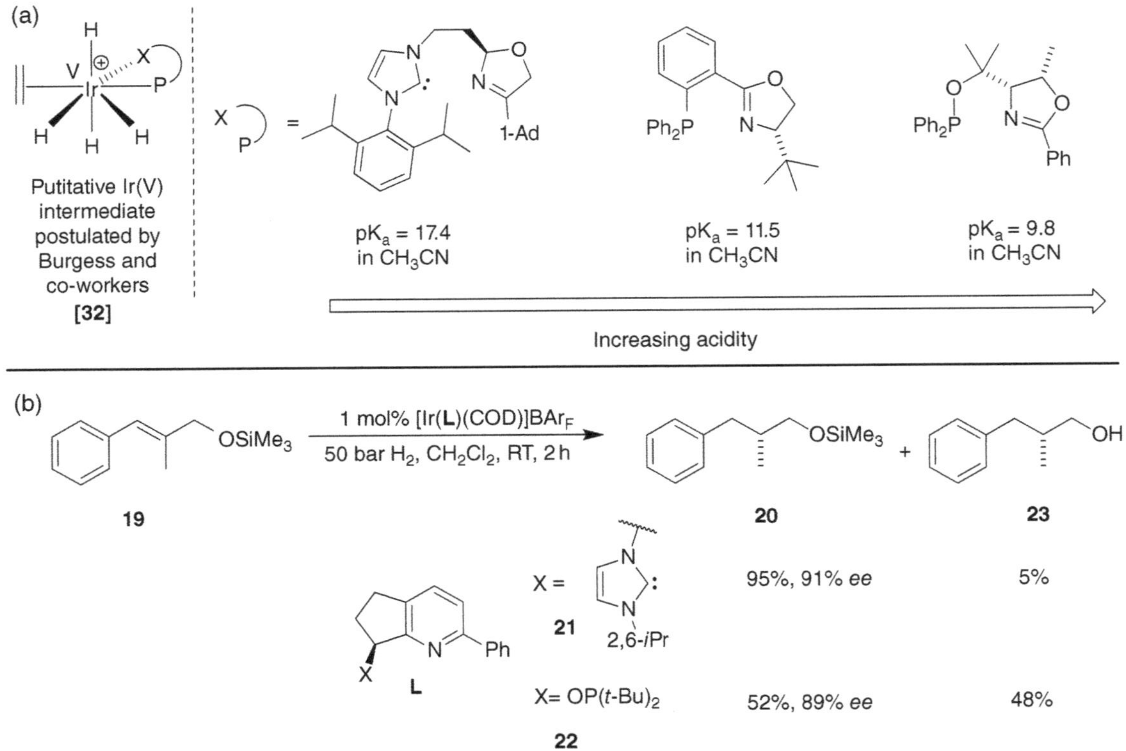

Scheme 11 *Iridium C,N ligand-based complexes as superior catalysts for substrates containing acid-labile groups*

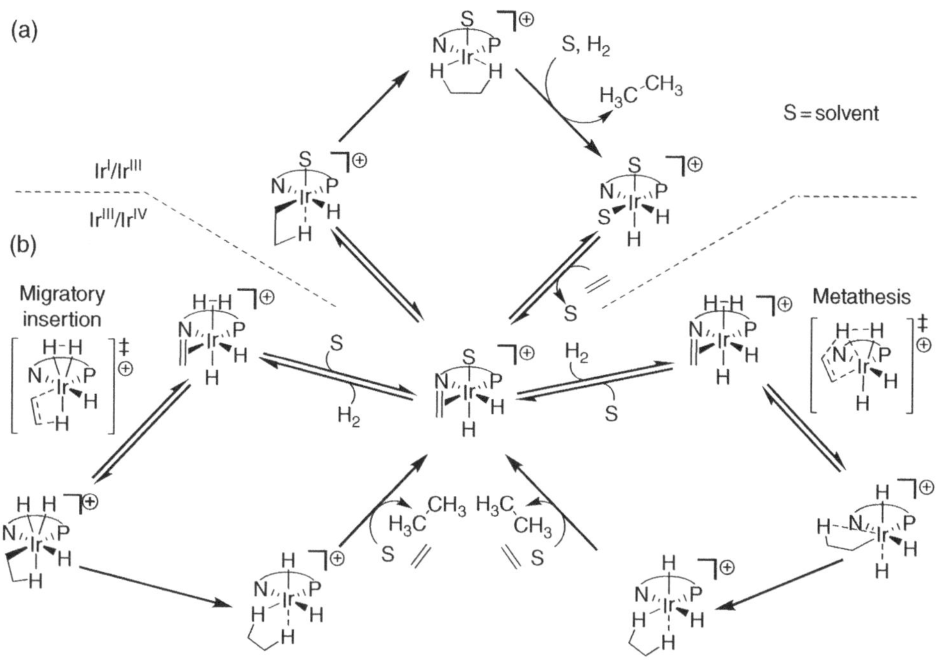

Scheme 12 *Different catalytic cycles proposed for Ir-catalyzed asymmetric hydrogenation*

stereoisomers that can be formed when a heterobidentate ligand complex reacts with an alkene and H_2, it will remain a significant challenge if not an impossible task to design new ligands in a truly rational manner. Furthermore, it cannot be ruled out that different mechanisms may operate (even in parallel) depending on the catalyst, substrate, and the reaction conditions. For example, substrates with an additional coordinating group should disfavor coordination of a second hydrogen molecule that would be required for the formation of an Ir(V) intermediate.

Nevertheless, based on DFT calculations a useful qualitative model was proposed that predicts to which enantioface H_2 is preferentially added (Scheme 13a) [38]. The model, which is also consistent with the experimentally determined structure of substrate–catalyst intermediates (see Scheme 3), displays the steric situation near the coordinated alkene in a schematic manner. As shown in a 3D model of an Ir-PHOX complex (Scheme 13b), the R substituent at the stereogenic center strongly shields the lower right quadrant. Therefore, the alkene, which is bound *trans* to the P atom with the C=C bond axis orthogonal to the coordination plane, reacts preferentially through a transition state with the smallest substituent, the H atom, positioned in the sterically hindered quadrant. The substituents on the P atom are too remote to exert strong steric interactions with the substrate, but can still influence the enantioselectivity through electronic effects, interaction with axial ligands, and/or influencing the geometry of the ligand backbone.

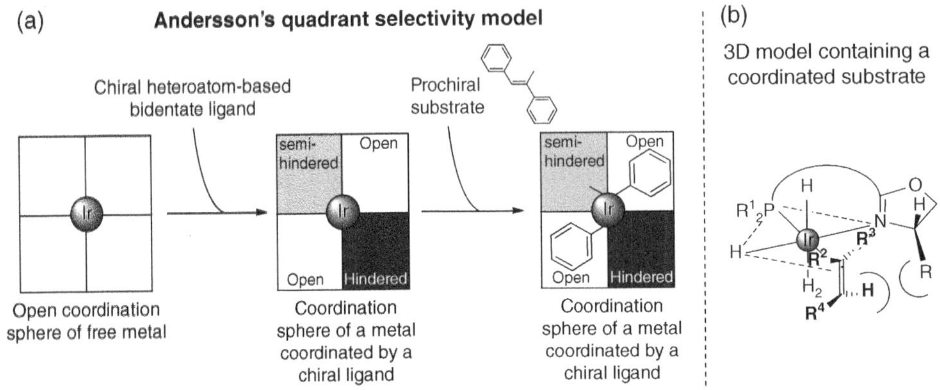

Scheme 13 *Andersson's quadrant model developed to rationalize the enantioselectivity*

Although the model is consistent with the observed absolute configuration induced in many hydrogenations with a wide range of catalysts, it has to be used with caution. In substrates with a strongly polarized C=C bond such as in α,β-unsaturated carboxylic esters, electronic effects can override steric interactions resulting in a break-down of the model [39]. Nevertheless, the quadrant or 3D models shown in Scheme 13 provide useful guidelines for improving the catalyst structure for a specific substrate, taking into account that it is mainly the N part of the ligand that sterically interacts with the alkene and hence should strongly affect the outcome of the reaction (for examples, see Figure 7c).

Despite the remarkable success with oxazoline-based N,P ligands, there were still important substrate classes left that provided unsatisfactory results. Therefore, the search for new ligands with different steric and electronic properties went on. With the idea to mimic the coordination sphere of the Crabtree catalyst **6** more closely, pyridine-based complexes such as **24** and **25** were evaluated (Figure 5a) [40]. Encouraged by the promising results provided by these ligands, further structural variations were carried out that led to the bicyclic conformationally more rigid pyridine-phosphinite ligands **26** and **27** [41]. The pyridine-based ligands as well are highly modular and a range of derivatives containing five-, six-, and seven-membered carbocyclic rings, various phosphinite units and different substituents at the 2-position of the pyridine ring were prepared and tested with a wide range of substrates [41]. Two derivatives, **26** with a di-*tert*-butyl phosphinite moiety, a five-membered carbocyclic ring in the back-bone, and a phenyl group at the 2-position of the pyridine ring, and the di-*ortho*-tolyl-phosphinite analog **27** with a six-membered carbocyclic ring, emerged as the most efficient, most versatile ligands, which provided excellent results in the hydrogenation of several new substrate classes such as purely alkyl-substituted alkenes or furans, for which no suitable catalysts were known before [41, 42].

The hydrogenation of γ-tocotrienyl acetate, a vitamin E precursor [43], is a striking example demonstrating the potential of these catalysts. With complex **27** all three double bonds in the side chain were reduced with very high enantio- and diastereoselectivity to give the natural (*R*,*R*,*R*)-isomer of γ-tocotrienyl acetate almost exclusively. Complexes

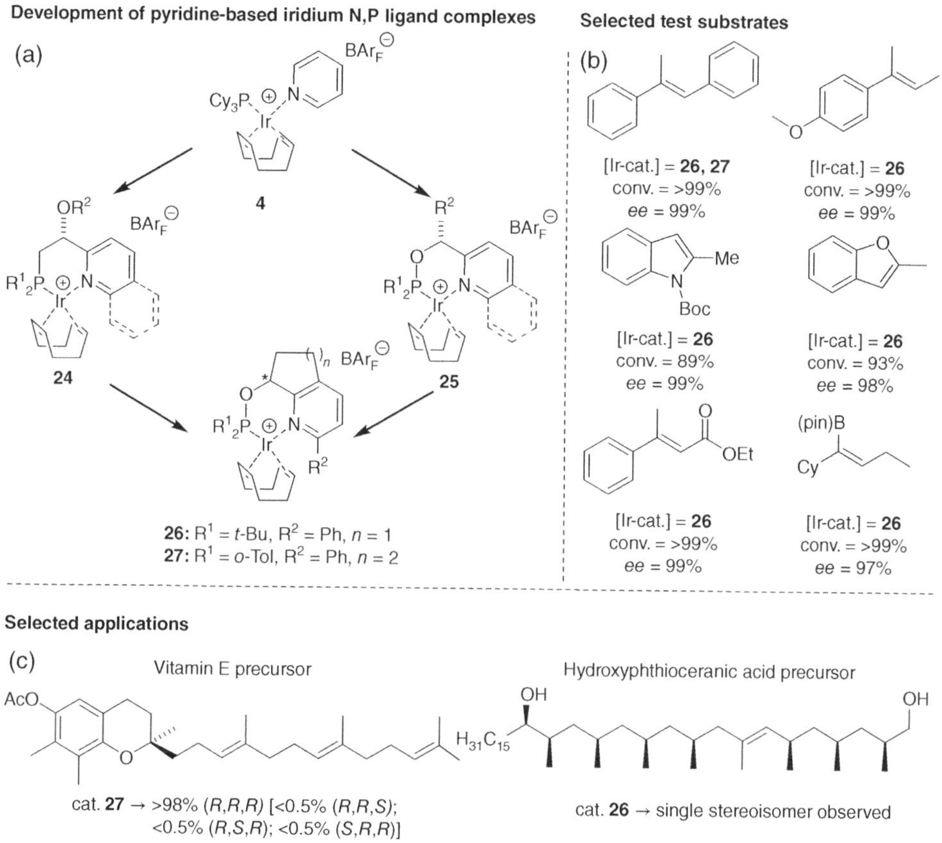

Figure 5 *(a) Development of pyridine-based catalysts; (b) hydrogenation of selected model substrates; and (c) selected application in total synthesis*

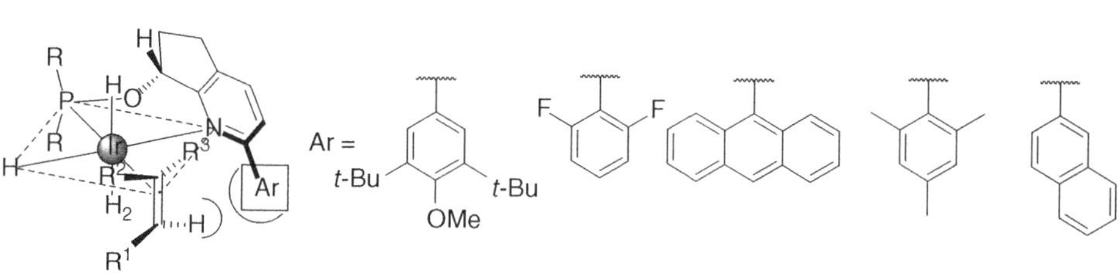

Figure 6 *Variation of the aryl substituent of catalyst **26***

such as **26** and **27** have also been successfully used in the total synthesis of natural products [44], for example in the catalyst-controlled diastereoselective hydrogenation of a hydroxyphthioceranic acid precursor (Figure 5c [44a]).

Based on crystal structures of Ir complexes with bicyclic pyridine-phosphinite ligands of this type, the structure shown in Figure 6 was postulated for the catalyst–substrate

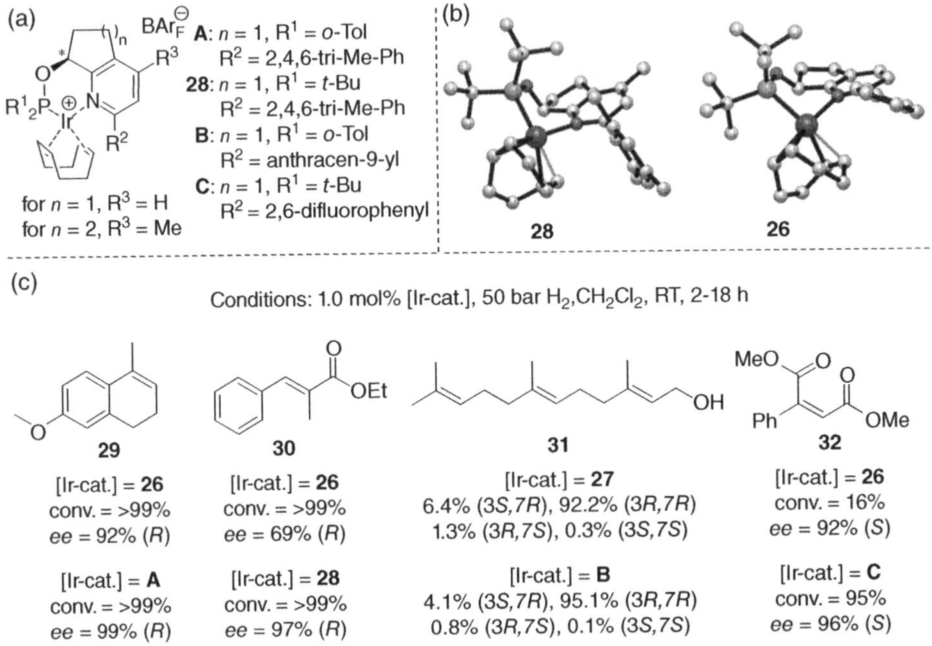

Figure 7 *Comparison of the different generations of pyridine-based Ir complexes. (See insert for color/color representation of this figure)*

complex. In this model the aryl substituent at C(2) of the pyridine ring shields the lower right quadrant of the coordination sphere, creating a chiral pocket similar to that in the Ir-PHOX complex shown in Scheme 13. The model implies that this aryl substituent should be mainly responsible for enantiocontrol. Therefore, several more sterically demanding aryl groups were introduced that generate a more congested environment near the coordinated substrate (see Figure 7a and the comparison of structures **26** and **28** in Figure 7b).

Indeed, variation of the aryl substituent had a significant effect on the enantioselectivity. Catalysts derived from ligands with substituted phenyl groups such as mesityl, anthracen-9-yl, or 2,6-difluorophenyl showed superior results for several model substrates (Figure 7c). Hydrogenation of the cyclic substrate **29**, the α,β-unsaturated carboxylic ester **30**, and (2*E*,6*E*)-farnesol (**31**) exhibited significantly increased enantioselectivity [45]. Furthermore, higher conversion as well as higher *ee* were observed in the hydrogenation of dimethyl 2-phenylmaleate **32** [46].

An analogous model was successfully used for the development and optimization of O,P ligands derived from proline (Scheme 14). After initial encouraging results obtained with Tomioka's ligand (**33**) [47], which induced 70% *ee* in the hydrogenation of (*E*)-1,2-diphenylprop-1-ene (**10**), the ligand structure was systematically modified, guided by the model shown in Scheme 14, which was derived from crystal structures [48]. As predicted by the model, higher *ee* values were obtained by increasing the size of the amide or urea group. For example, in the hydrogenation of ethyl (*E*)-3-phenylbut-2-enoate (**34**) the *ee* was raised from 33 up to 98% by increasing

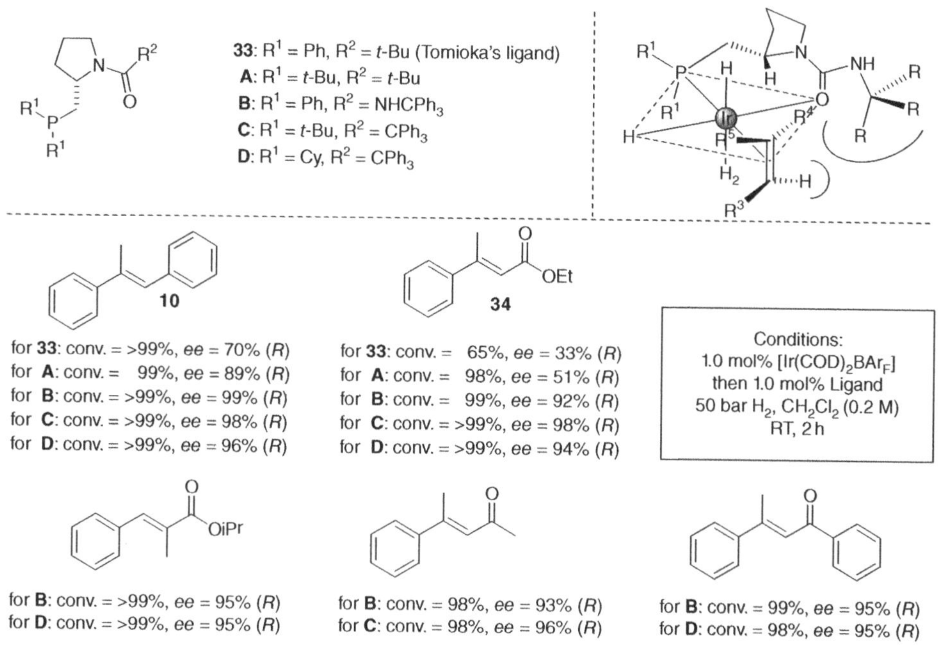

Scheme 14 *Proline-based Ir complexes as catalysts in the hydrogenation of different substrates*

the steric bulk of the ligand. Overall, this easily accessible ligand class proved to be particularly effective for the asymmetric hydrogenation of α,β-unsaturated carboxylic esters and ketones.

3.4 Conclusion

Since the first successful application of an Ir complex in the enantioselective hydrogenation of alkenes reported in 1998, cationic Ir complexes derived from heterobidentate ligands have emerged as highly efficient versatile catalysts for asymmetric hydrogenation of a wide range of functionalized and unfunctionalized olefins. In contrast to Rh and Ru catalysts, they do not require a coordinating group near the C=C bond and, therefore, exhibit significantly broader substrate scope. A key feature of these catalysts is an electronically unsymmetrical ligand core with two different coordinating heteroatoms. The application range has been continuously increased by the development of modular ligand classes that allow broad variation of individual structural parts, in order to optimize the catalyst structure for a given application. Phosphines connected to an oxazoline or pyridine ring, or a related *N*-heterocycle have emerged as the most versatile ligands. Although purely rational design based on computational methods is not possible yet, a qualitative model has been devised that may serve as a useful guide for the design and optimization of new ligands.

References

[1] (a) J. G. de Vries, C. J. Elsevier (eds), *Handbook of Homogeneous Hydrogenation, Vols 1–3*, Wiley-VCH, Weinheim, **2007**; (b) D. J. Ager, A. H. M. de Vries, J. G. de Vries, *Chem. Soc. Rev.* **2012**, *41*, 3340–3380; (c) N. B. Johnson, I. C. Lennon, P. H. Moran, J. A. Ramsden, *Acc. Chem. Res.* **2007**, *40*, 1291–1299; (d) H. U. Blaser, F. Spindler, M. Studer, *Appl. Catal., A* **2001**, *221*, 119–143.

[2] (a) R. Noyori, *Angew. Chem., Int. Ed.* **2002**, *41*, 2008–2022; (b) W. S. Knowles, *Angew. Chem. Int. Ed.* **2002**, *41*, 1998–2007.

[3] (a) J. K. Whitesell, *Chem. Rev.* **1989**, *89*, 1581–1590; (b) H. B. Kagan, P. Dang Tuan, *J. Am. Chem. Soc.* **1972**, *94*, 6429–6433.

[4] K. Inoguchi, S. Sakuraba, K. Achiwa, *Synlett* **1992**, 169–178.

[5] (a) H.-U. Blaser, *Adv. Synth. Catal.* **2002**, *344*, 17–31; (b) H.-U. Blaser, H.-P. Buser, K. Coers, R. Hanreich, H.-P. Jalett, E. Jelsch, B. Pugin, H.-D. Schneider, F. Spindler, A. Wegmann, *Chimia* **1999**, *53*, 275–280.

[6] (a) J. Sprinz, G. Helmchen, *Tetrahedron Lett.* **1993**, *34*, 1769–1772; (b) P. von Matt, A. Pfaltz, *Angew. Chem., Int. Ed. Engl.* **1993**, *32*, 566–568; (c) G. J. Dawson, C. G. Frost, J. M. J. Williams, S. J. Coote, *Tetrahedron Lett.* **1993**, *34*, 3149–3150.

[7] (a) G. Helmchen, A. Pfaltz, *Acc. Chem. Res.* **2000**, *33*, 336–345; (b) C. Bausch, A. Pfaltz, in *Privileged Chiral Ligands and Catalysts*, Wiley-VCH, Weinheim, **2011**, pp. 221–256.

[8] R. Crabtree, *Acc. Chem. Res.* **1979**, *12*, 331–337.

[9] (a) A. Lightfoot, P. Schnider, A. Pfaltz, *Angew. Chem. Int. Ed.* **1998**, *37*, 2897–2899; (b) S. J. Roseblade, A. Pfaltz, *Acc. Chem. Res.* **2007**, *40*, 1402–1411.

[10] C. Mazet, S. P. Smidt, M. Meuwly, A. Pfaltz *J. Am. Chem. Soc.* **2004**, *126*, 14176–14181.

[11] S. Gruber, A. Pfaltz, *Angew. Chem. Int. Ed.* **2014**, *53*, 1896–1900.

[12] (a) R. H. Crabtree, H. Felkin, T. Fillebeen-Khan, G. E. Morris, *J. Organomet. Chem.* **1979**, *168*, 183–195; (b) R. H. Crabtree, R. J. Uriarte, *Inorg. Chem.* **1983**, *22*, 4152–4154; (c) R. H. Crabtree, P. C. Demou, D. Eden, J. M. Mihelcic, C. A. Parnell, J. M. Quirk, G. E. Morris, *J. Am. Chem. Soc.* **1982**, *104*, 6994–7001.

[13] J. Mazuela, P.-O. Norrby, P. G. Andersson, O. Pàmies, M. Diéguez, *J. Am. Chem. Soc.* **2011**, *133*, 13634–13645.

[14] D. G. Blackmond, A. Lightfoot, A. Pfaltz, T. Rosner, P. Schnider, N. Zimmermann, *Chirality* **2000**, *12*, 442–449.

[15] S. P. Smidt, A. Pfaltz, E. Martínez-Viviente, P. S. Pregosin, A. Albinati, *Organometallics* **2003**, *22*, 1000–1009.

[16] D. F. Chodosh, R. H. Crabtree, H. Felkin, G. E. Morris, *J. Organomet. Chem.* **1978**, *161*, C67–C70.

[17] S. P. Smidt, N. Zimmermann, M. Studer, A. Pfaltz, *Chem.– Eur. J.* **2004**, *10*, 4685–4693.

[18] (a) J. Blankenstein, A. Pfaltz, *Angew. Chem. Int. Ed.* **2001**, *40*, 4445–4447; (b) F. Menges, A. Pfaltz, *Adv. Synth. Catal.* **2002**, *344*, 40–44.

[19] (a) S. McIntyre, E. Hörmann, F. Menges, S. P. Smidt, A. Pfaltz, *Adv. Synth. Catal.* **2005**, *347*, 282–288; (b) A. Baeza, A. Pfaltz, *Chem.– Eur. J.* **2010**, *16*, 2036–2039.

[20] O. Pàmies, P. G. Andersson, M. Diéguez, *Chem.– Eur. J.* **2010**, *16*, 14232–14240.

[21] (a) J. Mazuela, J. J. Verendel, M. Coll, B. Schäffner, A. Börner, P. G. Andersson, O. Pàmies, M. Diéguez, *J. Am. Chem. Soc.* **2009**, *131*, 12344–12353; (b) M. Dieguez, J. Mazuela, O. Pamies, J. J. Verendel, P. G. Andersson, *Chem. Commun.* **2008**, 3888–3890.

[22] R. D. Broene, S. L. Buchwald, *J. Am. Chem. Soc.* **1993**, *115*, 12569–12570.

[23] M. G. Schrems, E. Neumann, A. Pfaltz, *Angew. Chem. Int. Ed.* **2007**, *46*, 8274–8276.

[24] (a) S.-F. Zhu, J.-B. Xie, Y.-Z. Zhang, S. Li, Q.-L. Zhou, *J. Am. Chem. Soc.* **2006**, *128*, 12886–12891; (b) S. Li, S.-F. Zhu, C.-M. Zhang, S. Song, Q.-L. Zhou, *J. Am. Chem. Soc.* **2008**, *130*, 8584–8585.

[25] A. Cadu, P. G. Andersson, *Dalton Trans.* **2013**, *42*, 14345–14356.

[26] M.-A. Müller, A. Pfaltz, *Angew. Chem. Int. Ed.* **2014**, *53*, 8668–8671.

[27] S. Song, S.-F. Zhu, L.-Y. Pu, Q.-L. Zhou, *Angew. Chem., Int. Ed.* **2013**, *52*, 6072–6075.

[28] S. Li, S.-F. Zhu, J.-H. Xie, S. Song, C.-M. Zhang, Q.-L. Zhou, *J. Am. Chem. Soc.* **2010**, *132*, 1172–1179.

[29] S. Song, S.-F. Zhu, S. Yang, S. Li, Q.-L. Zhou, *Angew. Chem. Int. Ed.* **2012**, *51*, 2708–2711.

[30] S. Song, S.-F. Zhu, Y.-B. Yu, Q.-L. Zhou, *Angew. Chem., Int. Ed.* **2013**, *52*, 1556–1559.

[31] M. T. Powell, D.-R. Hou, M. C. Perry, X. Cui, K. Burgess, *J. Am. Chem. Soc.* **2001**, *123*, 8878–8879.

[32] Y. Zhu, Y. Fan, K. Burgess, *J. Am. Chem. Soc.* **2010**, *132*, 6249–6253.

[33] (a) Y. Zhu, K. Burgess, *Adv. Synth. Catal.* **2008**, *350*, 979–983; (b) P. Cheruku, S. Gohil, P. G. Andersson, *Org. Lett.* **2007**, *9*, 1659–1661.

[34] A. Schumacher, M. Bernasconi, A. Pfaltz, *Angew. Chem. Int. Ed.* **2013**, *52*, 7422–7425.

[35] (a) T. L. Church, T. Rasmussen, P. G. Andersson, *Organometallics* **2010**, *29*, 6769–6781; (b) Y. Fan, X. Cui, K. Burgess, M. B. Hall, *J. Am. Chem. Soc.* **2004**, *126*, 16688–16689; (c) K. H. Hopmann, A. Bayer, *Organometallics* **2011**, *30*, 2483–2497.

[36] R. Dietiker, P. Chen, *Angew. Chem. Int. Ed.* **2004**, *43*, 5513–5516.

[37] J. Halpern, *Science* **1982**, *217*, 401–407.

[38] K. Källström, C. Hedberg, P. Brandt, A. Bayer, P. G. Andersson, *J. Am. Chem. Soc.* **2004**, *126*, 14308–14309.

[39] C. Hedberg, K. Källström, P. Brandt, L. K. Hansen, P. G. Andersson, *J. Am. Chem. Soc.* **2006**, *128*, 2995–3001.

[40] W. J. Drury, N. Zimmermann, M. Keenan, M. Hayashi, S. Kaiser, R. Goddard, A. Pfaltz, *Angew. Chem. Int. Ed.* **2004**, *43*, 70–74.

[41] S. Kaiser, S. P. Smidt, A. Pfaltz, *Angew. Chem., Int. Ed.* **2006**, *45*, 5194–5197.

[42] (a) A. Baeza, A. Pfaltz, *Chem.– Eur. J.* **2009**, *15*, 2266–2269; (b) A. Wang, B. Wüstenberg, A. Pfaltz, *Angew. Chem. Int. Ed.* **2008**, *47*, 2298–2300; (c) A. Wang, R. P. A. Fraga, E. Hörmann, A. Pfaltz, *Chem. – Asian J.* **2011**, *6*, 599–606; (d) A. Ganić, A. Pfaltz, *Chem.– Eur. J.* **2012**, *18*, 6724–6728.

[43] S. Bell, B. Wüstenberg, S. Kaiser, F. Menges, T. Netscher, A. Pfaltz, *Science* **2006**, *311*, 642–644.

[44] (a) M. C. Pischl, C. F. Weise, M.-A. Müller, A. Pfaltz, C. Schneider, *Chem.– Eur. J.* **2014**, 20, 17360–17374; (b) M. C. Pischl, C. F. Weise, M.-A. Müller, A. Pfaltz, C. Schneider, *Angew. Chem. Int. Ed.* **2013**, *52*, 8968–8972; (c) T. Yoshinari, K. Ohmori, M. G. Schrems, A. Pfaltz, K. Suzuki, *Angew. Chem. Int. Ed.* **2010**, *49*, 881–885; (d) G. G. Bianco, H. M. C. Ferraz, A. M. Costa, L. V. Costa-Lotufo, C. Pessoa, M. O. de Moraes, M. G. Schrems, A. Pfaltz, L. F. Silva, *J. Org. Chem.* **2009**, *74*, 2561–2566; (e) H. J. Jessen, A. Schumacher, F. Schmid, A. Pfaltz, K. Gademann, *Org. Lett.* **2011**, *13*, 4368–4370.

[45] D. H. Woodmansee, M.-A. Müller, M. Neuburger, A. Pfaltz, *Chem. Sci.* **2010**, *1*, 72–78.

[46] M. Bernasconi, M.-A. Müller, A. Pfaltz, *Angew. Chem. Int. Ed.* **2014**, *53*, 5385–5388.

[47] M. Kuriyama, K. Nagai, K. Yamada, Y. Miwa, T. Taga, K. Tomioka, *J. Am. Chem. Soc.* **2002**, *124*, 8932–8939.

[48] D. Rageot, D. H. Woodmansee, B. Pugin, A. Pfaltz, *Angew. Chem. Int. Ed.* **2011**, *50*, 9598–9601.

4

Spiro Ligands for Asymmetric Catalysis

Shou-Fei Zhu and Qi-Lin Zhou

State Key Laboratory and Institute of Elemento-organic Chemistry, Nankai University, Collaborative Innovation Center of Chemical Science and Engineering (Tianjin), Tianjin 300071, China

4.1 Development of chiral spiro ligands

Catalytic asymmetric synthesis is one of the most active research areas in modern chemistry [1]. Asymmetric catalysis with enzymes, chiral metal complexes, and chiral organic molecules have emerged as successful and powerful tools for the synthesis of optically active compounds. Among these three catalytic asymmetric processes, those catalyzed by chiral metal complexes developed tremendously fast and the Nobel Prize in Chemistry in 2001 was awarded to Knowles and Noyori, and to Sharpless for their work. Metal catalyzed asymmetric synthesis is stimulated mainly by the development of chiral ligands, which tune the steric and electric properties of the metal catalysts and can induce chirality in the products. A large number of metal catalysts modified with chiral ligands have been prepared and used in both academic research as well as industrial production. Although there is no universal chiral catalyst for diverse reactions with different mechanisms, a few core structures can be regarded as truly successful as they demonstrate proficiency in a variety of mechanistically unrelated reactions, which were named "privileged chiral catalysts" by Jacobsen [2].

Ligand Design in Metal Chemistry: Reactivity and Catalysis, First Edition. Edited by Mark Stradiotto and Rylan J. Lundgren.

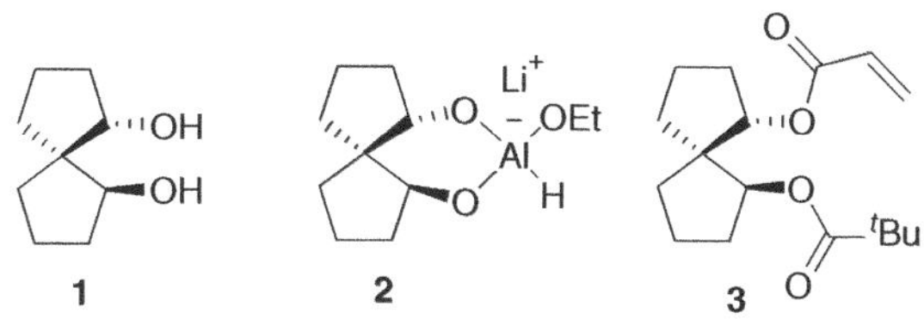

Figure 1 *Spiro[4,4]nonane-1,6-diol and related chiral auxiliaries*

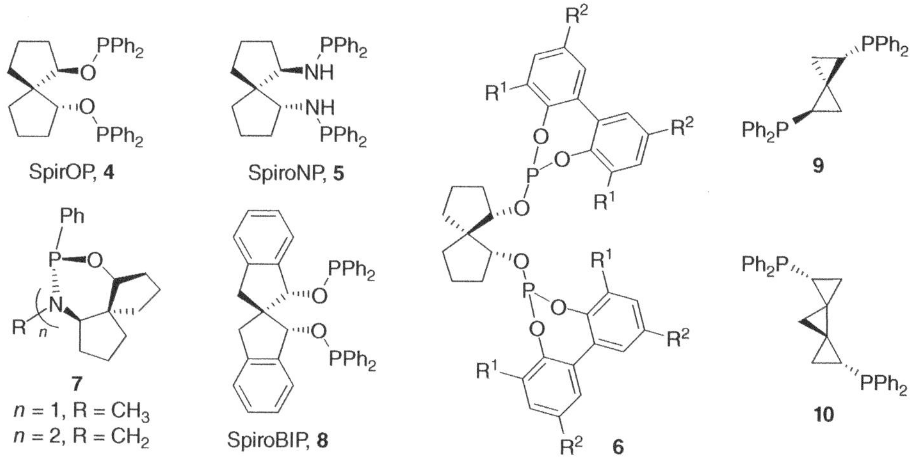

Figure 2 *Chiral spiro phosphorous ligands with spiro[4.4]nonane, spiro[2.2]pentane and dispiro[2.0.2.1]heptane scaffolds*

Molecules containing a spiro framework are ubiquitous in nature and there is a long history of the synthesis and applications of spiro compounds [3]. Beginning in the 1990s, chiral spiro[4,4]nonane-1,6-diol (**1**), a chiral spiro compound with two hydroxy groups at 1,6-positions (also called the "bay position") of spiro[4.4]nonane (Figure 1) was used as a chiral modifier of lithium aluminum hydride (**2**) for the reduction of ketones [4]. Later, compound **3**, the ester form of **1**, was successfully applied as a chiral auxiliary in the asymmetric Diels–Alder reaction [5]. These spiro compounds have inherent molecular rigidity. The quaternary structure of the spiro carbon atom makes the racemization of chiral spiro compounds virtually impossible. Given these special properties of chiral spiro compounds, they are ideal scaffold candidates for the design of chiral ligands.

Since 1997, a number of chiral spiro ligands have been developed, which can be classified from their backbones [6]. 1,6-Substituted spiro[4.4]nonane is the first spiro backbone used in the design of chiral ligands. By using spiro diol **1**, Chan *et al.* [7] developed the chiral spiro ligand SpirOP (**4**) (Figure 2), which was successfully applied in the Rh-catalyzed asymmetric hydrogenation of α-dehydroamino acid derivatives. Several chiral spiro phosphorous ligands with 1,6-substituted spiro[4.4]nonane structure were also developed, including the bisphosphinamine ligand SpiroNP (**5**) [8], the diphosphite ligands **6** [9], and the monodentate or bidentate P-stereogenic oxazaphosphorine ligands **7** [10]. Some of these chiral spiro ligands were demonstrated to be efficient for the Rh-catalyzed hydrogenation of α-dehydroamino acid derivatives and hydroformylation of styrenes. A bisphospinite ligand SpiroBIP (**8**) [11] having a 2,2′-spirobiindane-1,1′-diol structure, which can be regarded as a benzo spiro diol **1**, was also developed and afforded a high level of enantioselectivity in the Rh-catalyzed asymmetric hydrogenation of α-dehydroamino acids. It is noteworthy that all the chiral ligands with spiro[4.4]nonane backbone need a O or N as a tether to introduce the coordinating P atom. The direct introduction of the P atom is difficult because the high steric hindrance that exists at the bay positions. Recently, Khlebnikov *et al.* [12] developed chiral spiro diphosphane ligands **9** and **10** having spiro[2.2]pentane and dispiro[2.0.2.1]heptane backbones, which allow the direct introduction of PPh_2 groups to the spirocycles. These ligands exhibited moderate enantioselectivity in the Pd-catalyzed asymmetric allylic alkylation reaction.

Sasai's group [13] developed a series of bisnitrogen ligands with fused hetero spiro cyclic backbones, including spiro bis(isoxazoline) **11** and **12**, spiro isoxazole-isoxazoline **13**, spiro bis(oxazoline) **14**, spiro bis(pyrazole) **15**, and spiro bis(isoxazole) **16** (Figure 3). Ligands **11** [14] and **13** [15] exhibited excellent reactivity and enantioselectivity in the Pd-catalyzed oxidative cyclization reactions.

Although the above-mentioned two types of chiral spiro ligands afforded high enantioselectivites in several asymmetric reactions, they generally have a number of stereoisomers, which makes the preparation of the optically pure ligands challenging [13, 16]. Moreover, the modification of the coordination groups in these ligands is difficult, which markedly limited their applications in metal-catalyzed asymmetric reactions.

Zhou's group designed a class of chiral spiro ligands with a 1,1′-spirobiindane scaffold (Figure 4), which brings a number of advantages [6]. First, the 1,1′-spirobiindane has only one point of axial chirality, which significantly facilitates the preparation of optically pure ligands. For instance, 1,1′-spirobiindane-7,7′-diol (SPINOL), the

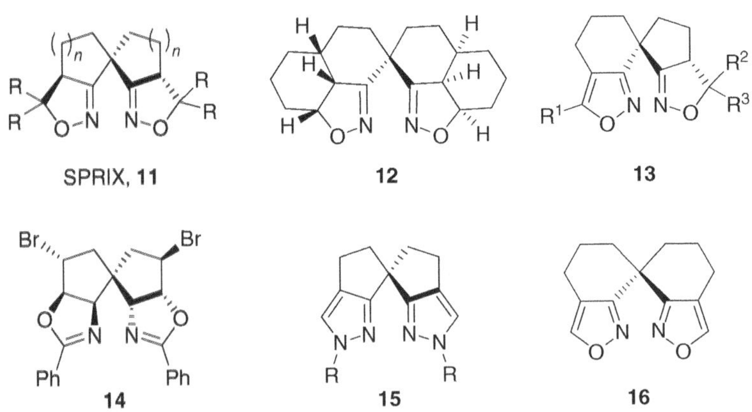

Figure 3 Chiral dinitrogen ligands with fused hetero spiro backbone

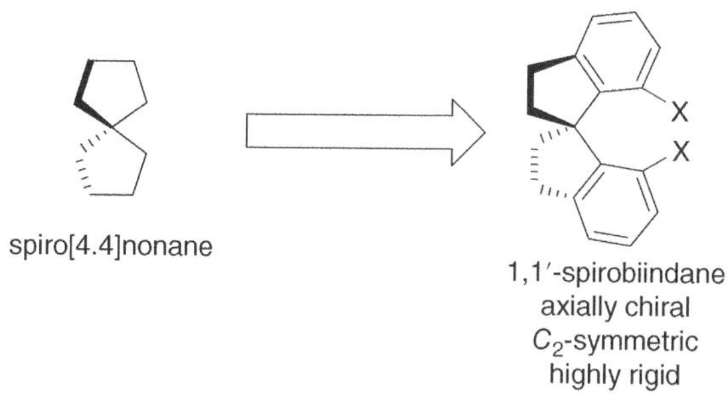

Figure 4 *The design of 1,1′-spirobiindane scaffold*

common starting material for spiro ligand synthesis can be prepared [17] and resolved on a kilogram scale [18]. Secondly, the benzo groups of 1,1′-spirobiindane increase the chemical robustness and conformational rigidity of the backbone and benefit the chiral inducement of the ligands. The functional groups (X) and benzo rings of the 1,1′-spirobiindane backbone leave positions for further structure modifications of ligands to meet different requirements of various reactions. Starting with optically pure SPINOL, more than 100 chiral spiro ligands with 1,1′-spirobiindane scaffold have been prepared through a single or multiple steps (Figure 5).

The ligands can be classified by their coordinating atoms to: monodentate phosphorous ligands (P ligands), such as phosphoramidite SIPHOS (**17**) [19], phosphite ShiP (**18**) [20], phosphinite FuP (**19**) [21], phosphine SITCP (**20**) [22]; bidentate phosphine ligands (P–P ligands), including diphosphine ligand SDP (**21**) [23] and diphosphite ligand SDPO (**22**) [24]; bidentate nitrogen ligands (N–N ligands), including bis(oxazoline) ligand SpiroBOX (**23**) [25] and diimine ligand SIDIM (**24**) [26]; bidentate phosphorous–nitrogen ligands (P–N ligands), including phosphine-oxazoline ligand SIPHOX (**25**) [27], phosphine–amine ligand SpiroBAP (**26**) [28], SpiroAP (**27**) [29], SpiroPAP (**28**) [30]; and other phosphine-containing ligands (P–X ligands), including phosphine–olefin ligand **29** [31] and phosphine–oxygen ligand **30** [23]. Some of these ligands are now commercially available from Aldrich, Strem, or Jiuzhou Pharmaceutical Co. Ltd. Procedures for introducing substituents at the 4,4′-position or 6,6′-position of the spirobiindane framework have been developed in order to tune the electronic or steric properties of the ligands [32].

Compared with biaryl ligands, the chiral spirobiindane ligands tolerated relatively hash conditions during their preparation and application in catalysis because the chiral spirobiindane scaffold does not undergo racemization. It is worth mentioning that all the chiral spirobiindane ligands with trivalent phosphorous atoms are stable and capable of being purified through chromatography on a silica gel.

The high rigidity is the key feature of the chiral spirobiindane ligands, which exhibited special advantages in many asymmetric reactions. One example is the Cu complex of chiral spiro bisoxazoline ligand (S_a,S,S)-**23a** (R = Ph). Zhu *et al.* [33] analyzed the X-ray structures of Cu(I)-(S_a,S,S)-**23a** with various anions (PF_6^-, ClO_4^-, and BAr_F^-). All the complexes have an unexpected binuclear Cu structure, as shown

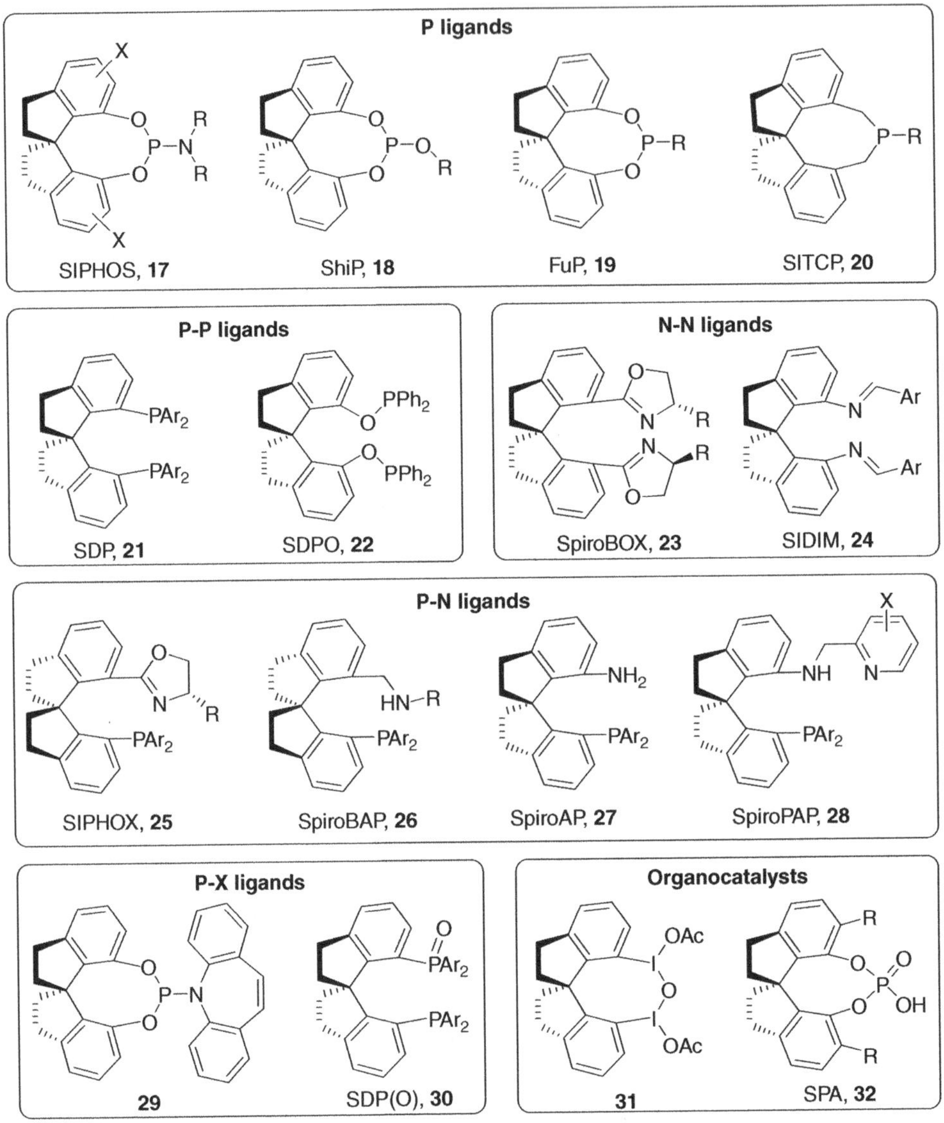

Figure 5 *Chiral ligands with 1,1′-spirobiindane backbone*

in Figure 6. In the complex Cu-(S_a,S,S)-**23a**, each Cu(I) atom is coordinated by two N atoms from the two spiro bisoxazoline ligands. The N1–Cu1–N3 and N2–Cu2–N4 angles are 169.0(2) and 169.5(2)°, respectively, which means that the two N atoms coordinate to a Cu(I) atom in a *trans* orientation. The phenyl groups of the oxazoline ligands form a perfect C_2-symmetric chiral environment (so-called "chiral pocket") around the Cu center. The Cu atoms in this complex have a 14-electron structure. The distance between the two Cu centers is 2.7828(10) Å, which implies an interaction

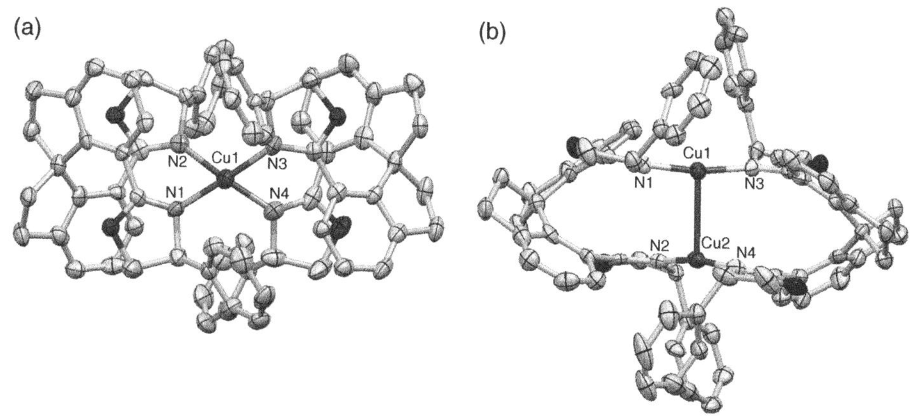

Figure 6 *Single crystal structure of complex Cu-(S_a,S,S)-**23a**. (The hydrogen atoms and anion PF_6^- are omitted for clarity)*

between the two Cu atoms. The unique structure of the complex Cu-(S_a,*S*,*S*)-**23a** is apparently caused by the rigid backbone of the ligand (S_a,*S*,*S*)-**23a**. The special structures lead to excellent enantioselectivity in the challenging carbene insertion reactions with heteroatom–hydrogen bonds (X–H) [34].

The chiral ligands based on the 1,1′-spirobiindane skeleton feature high rigidity, facile modification, and perfect C_2 symmetry, exhibiting high efficiency and enantioselectivity in a number of mechanistically unrelated transformations. Due to these exceptional characteristics, the chiral spirobiindane ligands are regarded as one of the "privileged chiral ligands" and have found widespread application.

Compounds with a 1,1′-spirobiindane backbone have also been used as chiral organocatalysts in enantioselective transformations. For instance, Fu's group [35] used the spiro monophosphine (*R*)-**20** to promote various highly enantioselective γ-addition reactions of allenoates or alkynoates through vinyl phosphonium intermediates. Wang *et al.* [36] used (*S*)-**20** as catalysts for the asymmetric intramolecular ylide annulation to prepare benzobicyclo[4.3.0] compounds with three continuous stereogenic centers with high enantioselectivity and diastereoselectivity. Wang *et al.* [37] also realized a highly enantioselective [3+2] annulation of allenoates with alkylidene azlactones by using (*R*)-**20** as organocatalyst. Dohi *et al.* [38] developed a hypervalent iodine(III) reagent (*R*)-**31** with a 1,1′-spirobiindane backbone, which afforded high enantioselectivity in iodine(III)-mediated dearomatization of phenols. Recently, chiral spiro phosphoric acids **32** have been developed and used to promote diverse asymmetric reactions with excellent enantioselectivities [39].

Following spirobiindane ligands, various chiral spiro ligands having benzo spiro rings have been developed (Figure 7). Huo *et al.* [40] developed a monophosphoramidite ligand (**33**) having a spirobitetraline scaffold and the diphosphine ligand SFDP (**34**) with a spirobifluorene scaffold, which were proven to be efficient for Ru-catalyzed asymmetric hydrogenation of α,β-unsaturated carboxylic acids [41]. Zhang and co-workers

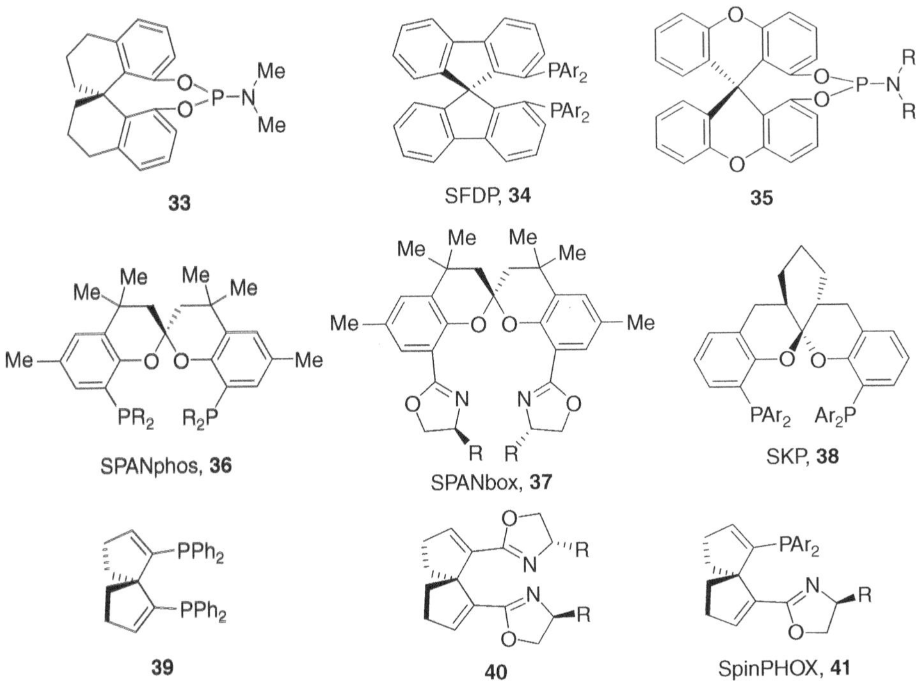

Figure 7 *Ligands with benzo spiro ring and spiro[4.4]nona-1,6-diene backbones*

prepared the spiro monophosphoramidite ligand **35** with a spirobixanthene backbone and applied it in Rh-catalyzed asymmetric hydrogenation of α-dehydroamino acid derivatives and Cu-catalyzed conjugate addition of diethyl zinc to cycloenones [42].

van Leeuwen and co-workers [43] developed chiral ligands with spiro-2,2′-bischroman scaffold (**36**). Although the coordinating atoms of ligands **36** are distanced from each other compared with those ligands derived from spirobiindane scaffold, the skewed configuration of spiro-2,2′-bis(chroman) structure makes the chelating coordination with transition metals possible, which facilitated the Pd-catalyzed asymmetric fluorination of 2-cyano-2-arylacetates [43f]. Ligand **37** exhibited high enantioselecitvity in Zn- or Cu-catalyzed asymmetric hydroxylation or chlorination of 1,3-dicarbonyl compounds [44]. Ding and co-workers [45] prepared the spiro-2,2′-bis(chroman) ligands SKP (**38**) with a fused ring through an Ir-catalyzed asymmetric hydrogenation of conjugated ketones and a consequent spiroketalization. The ligands **38** were successfully applied in Pd-catalyzed asymmetric allylic amination reactions [46] and Au-catalyzed asymmetric cyclopropanation reactions [47], respectively.

The chiral diphosphine ligand **39** [48], bisoxazoline ligand **40** [49], and phosphine-oxazoline ligand SpinPHOX (**41**) [50] having a spiro[4.4]nona-1,6-diene framework were also developed by Ding and co-workers. The existence of carbon–carbon double bonds in the spiro[4.4]nona-1,6-diene reduces the number of possible stereoisomers and increases the rigidity of the ligands. The phosphine-oxazoline ligand **41** exhibited high enantioselectivity in Ir-catalyzed hydrogenations of imines [50] and polar olefins [51].

4.2 Asymmetric hydrogenation

4.2.1 Rh-catalyzed hydrogenation of enamides

Rh-catalyzed asymmetric hydrogenation of enamides provides one of the most efficient methods for the preparation of optically active amines, which are important building blocks for the synthesis of biologically active compounds. Chiral bidentate phosphorous ligands have been predominantly used in asymmetric hydrogenation of enamides. Hu *et al.* [52] first introduced monodentated chiral phosphorous ligands into the Rh-catalyzed asymmetric hydrogenation of α-arylethenyl acetamides and found that the spiro phosphoramidite ligand (*S*)-**17a** was highly enantioselective (91–99.7% ee) (Scheme 1). A highly enantioselective Rh-catalyzed hydrogenation of both (*Z*)- and (*E*)-β-arylenamides was also developed by using monodentate chiral spiro phosphite ligand (*S*)-**18d** and phosphine ligand (*R*)-**20b**, respectively (Scheme 2 and Scheme 3) [53].

4.2.2 Rh- or Ir-catalyzed hydrogenation of enamines

The *N*-acyl group of enamides, which can form a chelate complex with the metal of the catalyst in the transition state, is important for obtaining high enantioselectivity in asymmetric hydrogenation of enamides. The lack of chelating *N*-acyl group results in the asymmetric hydrogenation of unprotected enamines challenging. The Rh complexes of ligand (*S*)-**19a** showed exceptionally high enantioselectivities [up to 99.9% enantiomeric excess (ee)] (Scheme 4) for the hydrogenation of unprotected enamines, (*E*)-1-(1-pyrrolidinyl)-1,2-diarylethenes [21b].

Ir/(R_a,*S*,*S*)-**17b** was found to be a highly efficient catalyst for the asymmetric hydrogenation of cyclic *N*,*N*-dialkyl enamines to optically active cyclic tertiary amines, essential structural units in natural products and drugs, with good to excellent enantioselectivities

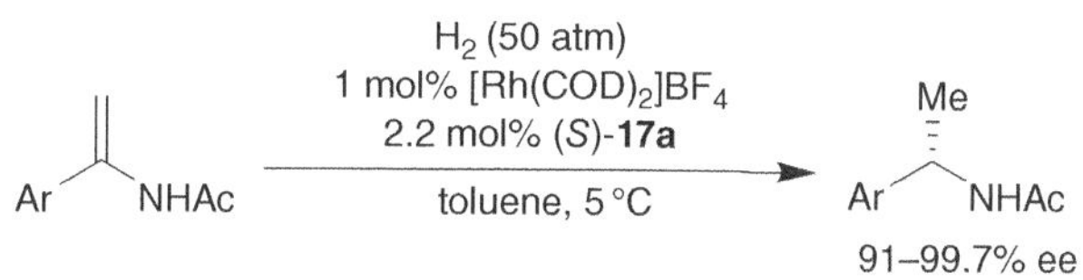

Scheme 1

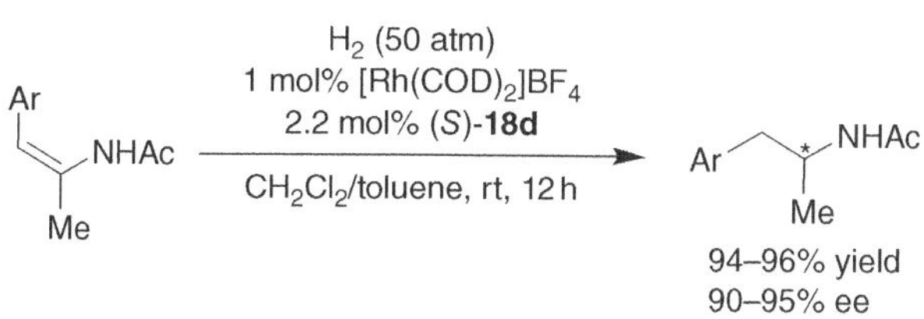

Scheme 2

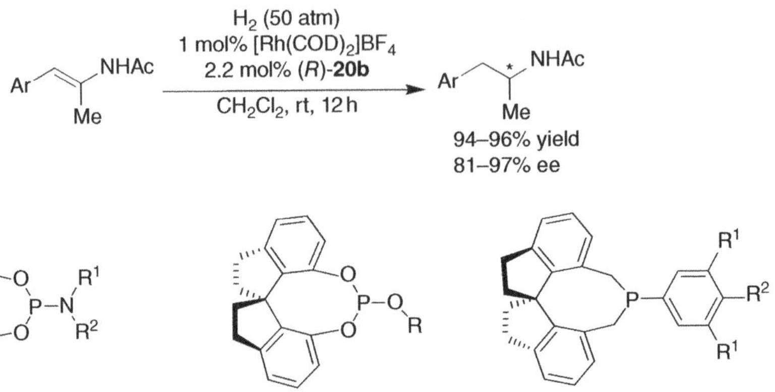

(*S*)-**17a** $R^1 = R^2$ = Me
(S_a,*S*,*S*)-**17b** $R^1 = R^2$ = (*S*)-CH(Me)Ph
(*S*)-**17c** R^1 = Me, R^2 = Ph

(*S*)-**18a** R = Ph
(*S*)-**18b** R = 4-$MeOC_6H_4$
(*S*)-**18c** R = 2-Np
(*S*)-**18d** R = Et
(*S*)-**18e** R = 2,6-(^{t}Bu)$_2$-4-MeC_6H_2

(*R*)-**20a** $R^1 = R^2$ = H
(*R*)-**20b** R^1 = ^{t}Bu, R^2 = OMe

Scheme 3

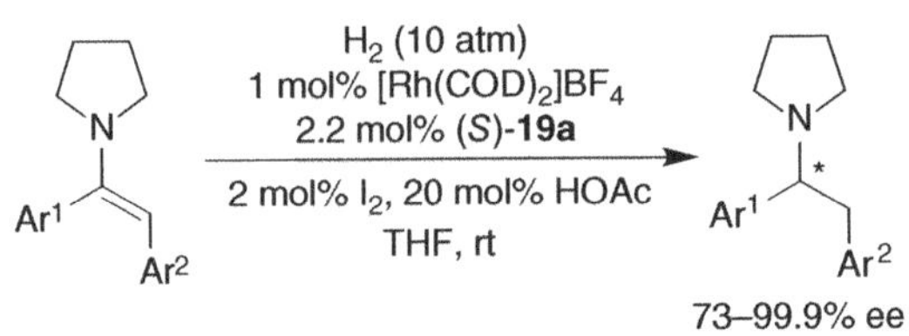

Scheme 4

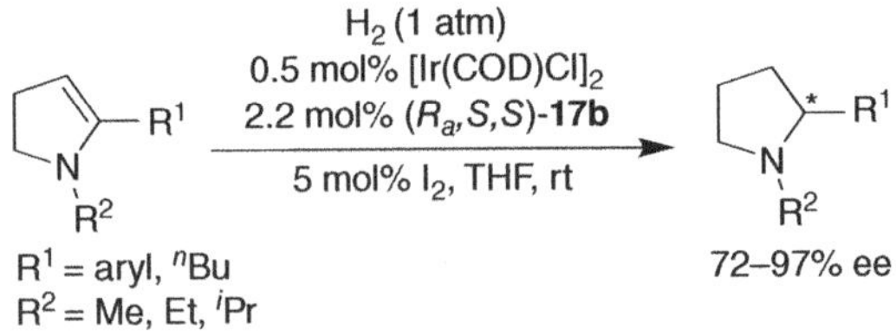

Scheme 5

(72–97% ee) (Scheme 5) [54]. This reaction provided an efficient methodology for the synthesis of natural product crispine A. The same catalyst accomplished the asymmetric hydrogenation of *N*-alkyl-1-alkylidenetetrahydroisoquinolines with an exocyclic double bond to produce *N*-alkyl-tetrahydroisoquinolines, including natural product carnegine, in high yields with up to 98% ee (Scheme 6) [55].

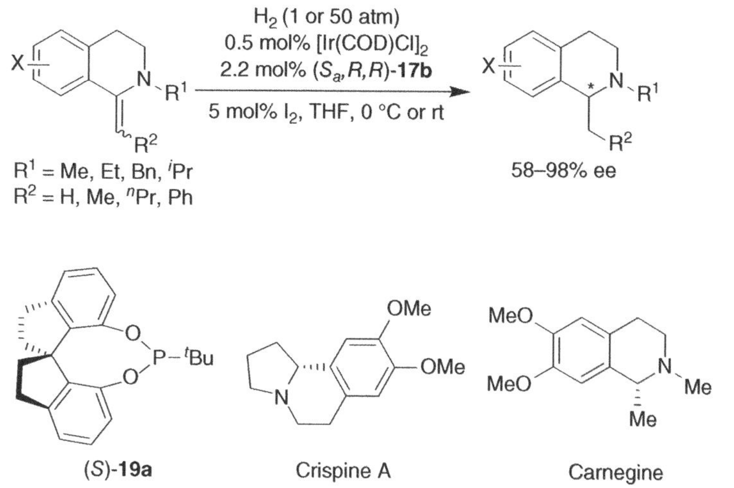

Scheme 6

4.2.3 Ir-catalyzed hydrogenation of α,β-unsaturated carboxylic acids

The transition metal catalyzed enantioselective hydrogenation of α,β-unsaturated carboxylic acids is a straightforward method for the synthesis of enantiopure carboxylic acids, which are related to biologically active compounds, such as ibuprofen and naproxen. The Ru and Rh complexes with chiral mono- or diphosphine ligands are predominant catalysts for this transformation, however, the efficiency of these catalysts does not reach the requirement of practical use [56]. Zhou and co-workers recently introduced the Ir complexes of spiro phosphine-oxazoline ligands SIPHOX (**25**) for the asymmetric hydrogenation of α,β-unsaturated carboxylic acids. In view of the extremely high activity and enantioselectivity, broad substrate scope, and mild reaction conditions, the Ir/SIPHOX complex represents one of the most efficient catalysts for asymmetric hydrogenation of α,β-unsaturated carboxylic acids [57].

Chiral cationic iridium complex Ir-**25** showed excellent reactivity [turnover frequency (TOF) up to 800 h^{-1}] and enantioselectivity (90–99.4% ee) in the hydrogenation of α-substituted cinnamic acids and aliphatic α,β-unsaturated acids (Scheme 7) [27b]. Using this asymmetric hydrogenation as a key step the catalytic enantioselective total synthesis of (*S*)-equol was accomplished starting from commercially available materials in six steps with 48.4% overall yield [58]. The same catalyst showed a high efficiency in the asymmetric hydrogenation of α-aryloxy and α-alkoxy substituted α,β-unsaturated acids (Scheme 8) [59]. Under mild reaction conditions, a broad range of α-aryloxy and α-alkoxy substituted α,β-unsaturated acids were hydrogenated with exceptional enantioselectivities (up to 99.8% ee) at high turnover numbers (TONs; TON up to 10 000). Using this asymmetric hydrogenation, the key intermediate for the syntheses of rupintrivir, the rhinovirus protease inhibitor, was prepared. A highly enantioselective hydrogenation of α,β-unsaturated carboxylic acids with tetrasubstituted olefins, a challenging substrate with high steric hindrance, was also achieved by using the catalyst Ir-(S_a,*S*)-**25b** (Scheme 9) [60].

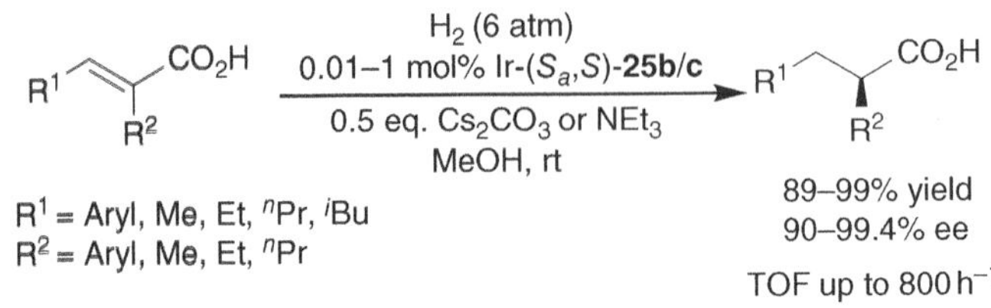

Scheme 7

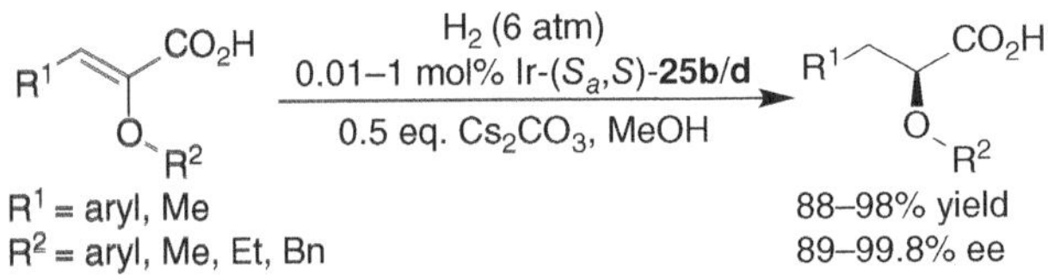

Scheme 8

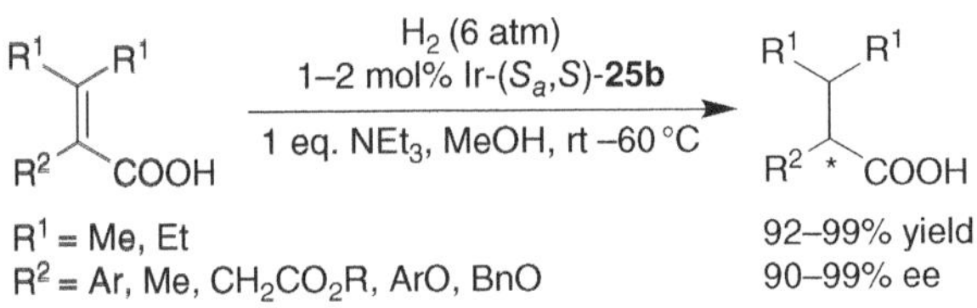

Scheme 9

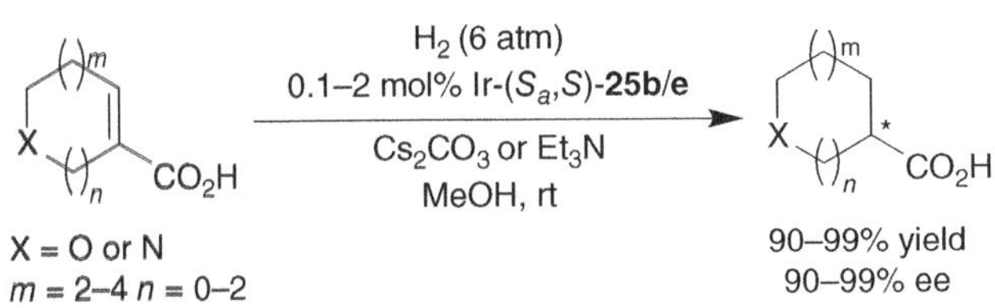

Scheme 10

The transition metal catalyzed enantioselective hydrogenation of unsaturated heterocyclic carboxylic acids provides a direct approach to the chiral heterocyclic acids, which are present in various pharmaceuticals. However, satisfactory methods for the asymmetric hydrogenation of cyclic unsaturated carboxylic acids are rare. Song *et al.* [61] realized the highly enantioselective hydrogenation of unsaturated heterocyclic acids by means of Ir-**25** (Scheme 10). The concise synthesis of (*R*)-tiagabine, a γ-aminobutyric acid reuptake inhibitor marketed for the treatment of epilepsy, was accomplished through this hydrogenation.

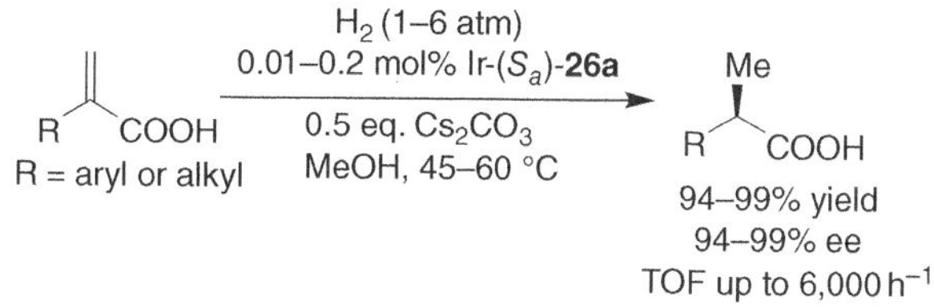

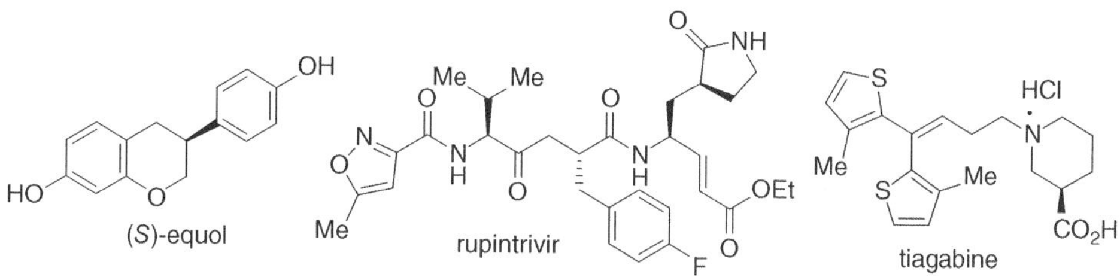

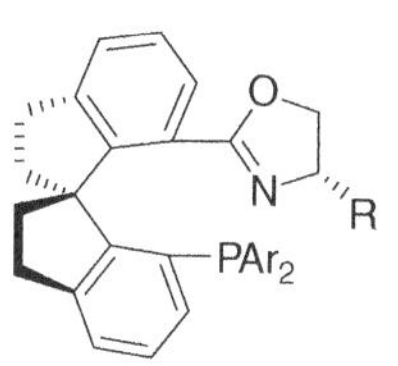

25	Ar, R
25a	3,5-$(CH_3)_2C_6H_3$, Bn
25b	3,5-$^tBu_2C_6H_3$, Bn
25c	3,5-$^tBu_2C_6H_3$, ^{i}Pr
25d	3,5-$^tBu_2C_6H_3$, Me
25e	3,5-$^tBu_2C_6H_3$, H
25f	3,5-$^tBu_2C_6H_3$, ^{i}Bu
25g	3,5-$Me_2C_6H_3$, CH_2-1-Np

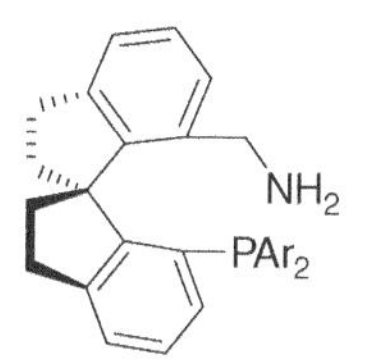

(S_a)-**26a** Ar = 3,5-$^tBu_2C_6H_3$

Scheme 11

A novel class of complexes, Ir-(S_a)-**26a**, with chiral spiro aminophosphine ligands was found to be effective catalyst for the asymmetric hydrogenation of α-substituted acrylic acids (Scheme 11) [62]. Under mild reaction conditions and at ambient pressure, various α-aryl and alkyl propionic acids were produced with extremely high efficiency (TONs up to 10 000; TOFs up to 6000 h^{-1}) and excellent enantioselectivity (up to 99% ee). This reaction provides a practically useful method for the preparation of α-aryl propionic acids, a popular class of non-steroid anti-inflammtory reagents.

A crystallographic study on the structure of catalyst Ir-(S_a,S)-**25b** disclosed that the spiro phosphine-oxazoline ligand (S_a,S)-**25b** creates a rigid and sterically hindered chiral pocket around the Ir center, which results in the high stability of the catalyst by preventing auto-aggregation. Moreover, the crowded chiral pocket of the catalyst minimizes the number of possible transition states in the reaction and assists chiral induction. In a stereo-recognition model, part of the spirobiindane backbone of ligand (S_a,S)-**25b** blocks one of the quadrants in front of the central Ir. One of the aryl groups on the P atom and the *tert*-butyl group on another P-phenyl block another two quadrants. This structure of the catalyst directs the olefin double bond of the α,β-unsaturated acid substrate coordinated to Ir by its *Re* face, leading to the hydrogenation product with *S* configuration, which is consistent with the experimental result (Figure 8) [59].

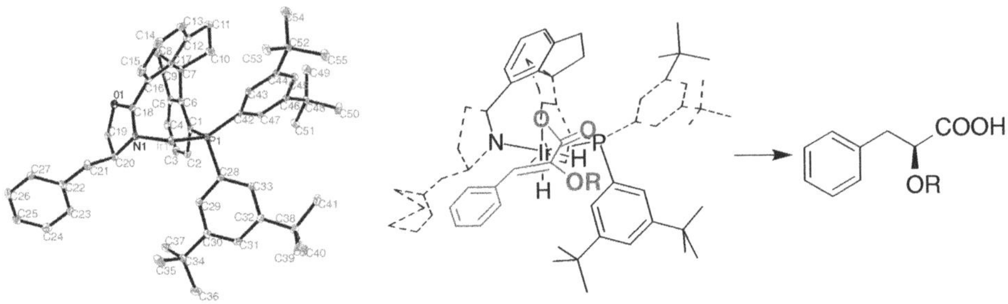

Figure 8 The single crystal structure of Ir-(S_a,S)-**25b** and stereo-recognition model. (See insert for color/color representation of this figure)

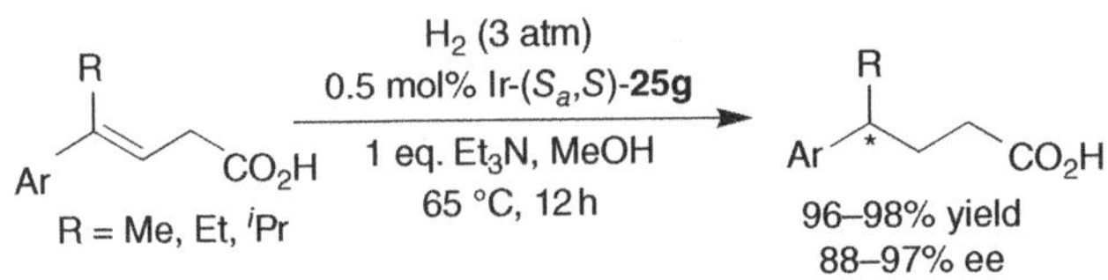

Scheme 12

4.2.4 Ir-catalyzed hydrogenation of olefins directed by the carboxy group

The catalytic asymmetric hydrogenation of β,γ-unsaturated acids is challenging because of the difficulties of remote control of enantioselectivity. Song *et al.* [63] developed a carboxy-directed strategy for the asymmetric hydrogenation of remote olefins. The iridium catalyst Ir-**25g** showed high enantioselectivity (up to 97% ee) for the hydrogenation of 4-alkyl-4-aryl-3-butenoic acids (Scheme 12). Mechanism studies imply that the carboxylic group of substrate acts as an anchor coordinating to Ir and makes the hydrogenation possible.

The differentiation of *Re* and *Si* faces of olefin in enantioselective hydrogenation is difficult if the substituents at the double bond are similar in size, as is in the case of 1,1-dialkylethenes or 1,1-diarylethenes. Zhou and co-workers [64] applied the carboxy-directed strategy in the Ir-catalyzed asymmetric hydrogenation of these two types of olefins. The carboxylic groups in the substrates functionalized as a directing group and increased the enantioselectivity of the hydrogenation reaction, and could be easily removed or transformed to other useful functional groups. By using catalysts Ir-(S_a,S)-**25f** and Ir-(S_a)-**25e**, a variety of chiral γ-methyl fatty acids and chiral diarylethanes, which are core structures for many biologically active compounds, were prepared in high yields and excellent enantioselectivities (Scheme 13 and Scheme 14). The carboxy-directing strategy was further extended to the asymmetric hydrogenation of α-alkyl-α-aryl terminal olefins catalyzed by Ir-(S_a)-**25e**, providing a highly efficient approach to the compounds with a chiral benzylmethyl center [65].

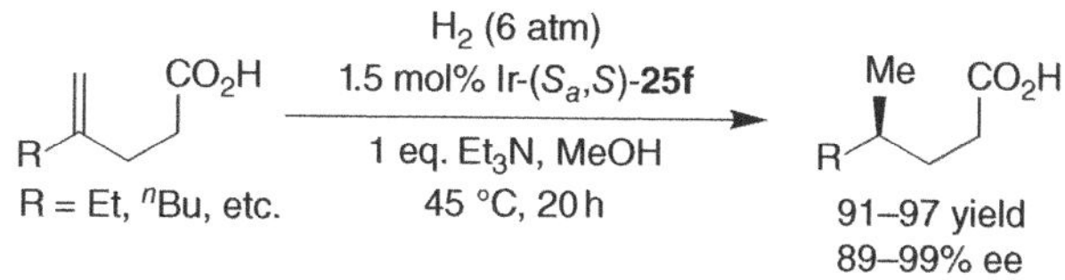

Scheme 13

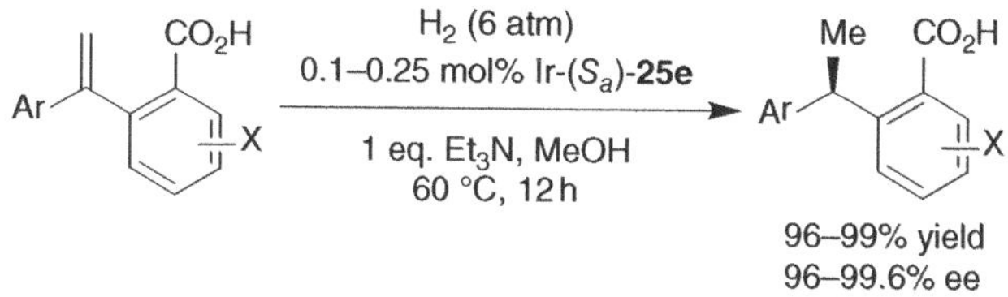

Scheme 14

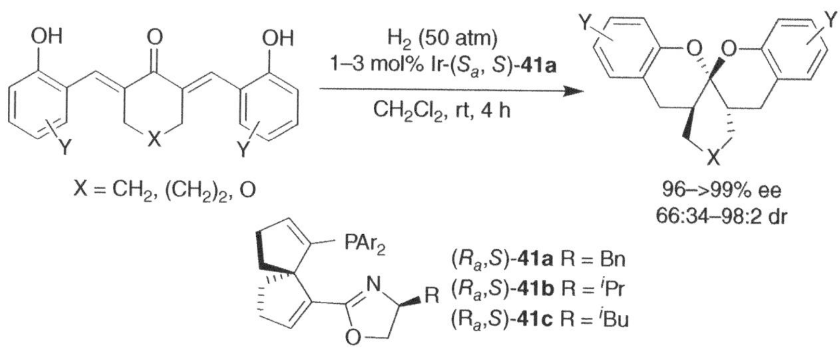

Scheme 15

4.2.5 Ir-catalyzed hydrogenation of conjugate ketones

Wang *et al.* [45a] developed a catalytic enantioselective synthesis of aromatic spiroketals through the asymmetric hydrogenation of α,α′-bis(2-hydroxyarylidene)ketones catalyzed by Ir-(S_a,S)-**41a** and the spiroketalization of the resulting hydrogenation product, bisphenolic ketones (Scheme 15). The Ir complexes were found to play a dual catalytic role in the reaction, acting as catalysts for both the hydrogenation of C=C bonds and the spiroketalization of the hydrogenated ketone-bearing bis(phenol) moieties without racemization of the chiral α-carbon centers. This methodology was applied for the synthesis of chiral spiro ligands (**38**).

4.2.6 Ir-catalyzed hydrogenation of ketones

Although the transition metal catalyzed enantioselective hydrogenation of carbonyl compounds is a highly efficient protocol for producing optically active alcohols, only few catalysts could reach TON values over a million [66]. An example is the Ru catalyst $RuCl_2$[(diphosphine)(1,2-diamine)] reported by Noyori and Ohkuma [67], which gave a TON of 2 400 000 with 80% ee. Xie *et al.* [30] developed an Ir catalyst Ir-(*R*)-**28a** with a tridentate spiro ligand, which exhibited extremely high reactivity in the hydrogenation of ketones, affording chiral alcohols in up to 99.9% ee with TONs as high as 4 550 000 (Scheme 16). This is the highest TON for hydrogenation with molecular catalysts reported to date [66]. The additional coordinating group (pyridine moiety) in the ligand is considered to play an important role in stabilizing the catalysts.

The catalyst Ir-(*R*)-**28a** also showed good performance in the asymmetric hydrogenation of β-aryl β-ketoesters with excellent enantioselectivity (up to 99.8% ee) and extremely high TONs (as high as 1 230 000) under mild reaction conditions (Scheme 17) [68].

When the catalyst Ir-(*R*)-**28a** was employed for the hydrogenation of the δ-aryl δ-ketoesters, both the keto and ester groups were hydrogenated, yielding chiral 1-arylpentane-1,5-diols with high activity (TONs up to 100 000) and excellent enantioselectivity (up to 99.9% ee) (Scheme 18) [69]. This was the first example of highly efficient Ir-catalyzed hydrogenation of esters.

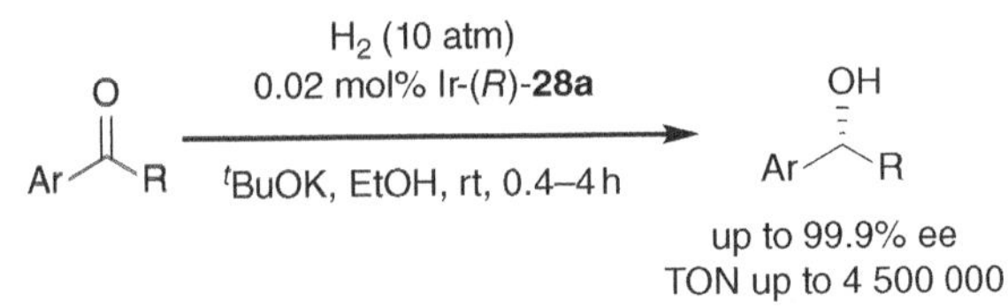

Scheme 16

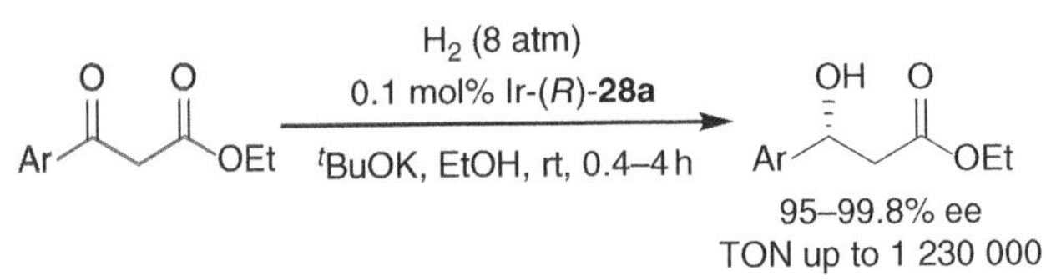

Scheme 17

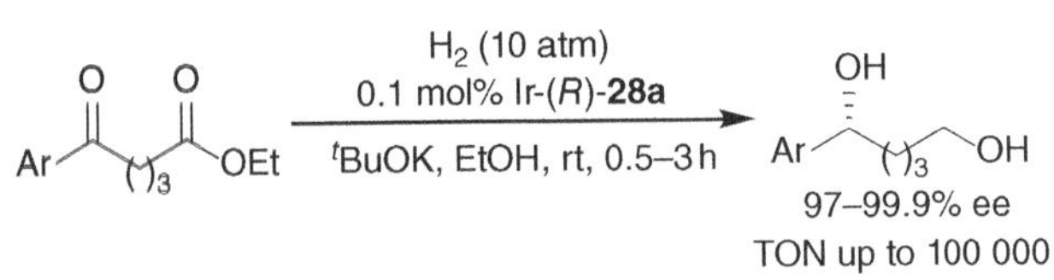

Scheme 18

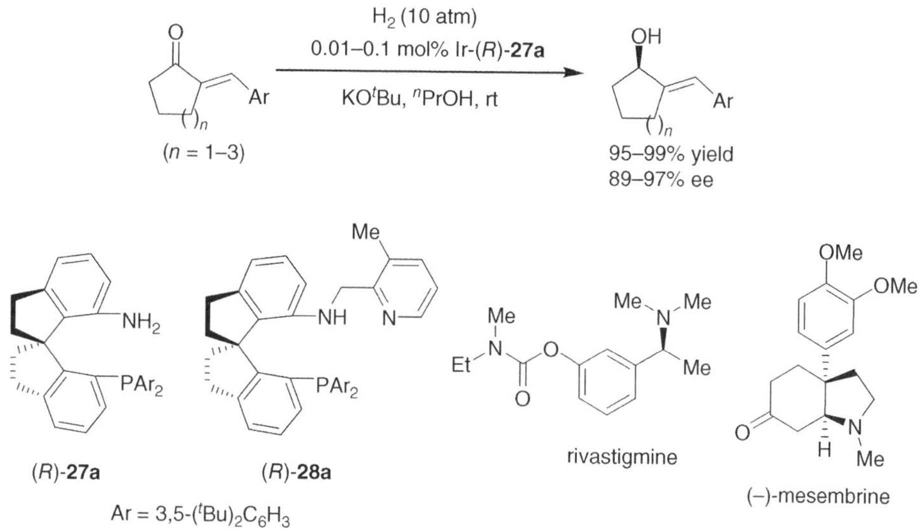

Scheme 19

Two efficient processes for the synthesis of rivastigmine, one of the potent drugs for the treatment of Alzheimer's disease, have been developed for industrial applications with Ir-(*S*)-**28a**-catalyzed asymmetric hydrogenation as the key step [70]. Additionally, the Ir-**28a**-catalyzed asymmetric hydrogenation of ketones was also used as a key step in the synthesis of natural products of (−)-mesembrine [71] and (−)-centrolobine [72].

The catalytic asymmetric hydrogenation of the carbonyl group of α,β-unsaturated ketones is a direct method to produce chiral allylic alcohols, versatile intermediates in organic synthesis. While the asymmetric hydrogenation of exocyclic α,β-unsaturated ketones is a challenge, an asymmetric hydrogenation of (*E*)-α-arylmethene cycloalkanones was accomplished with high activity as well as enantioselectivity by using chiral Ir complexes of spiro aminophosphine ligands (*R*)-**27a** (Scheme 19) [29a]. The hydrogenation reaction was used for the synthesis of the key intermediate of the active form of the anti-inflammatory loxoprofen.

4.2.7 Ru-catalyzed hydrogenation of racemic 2-substituted aldehydes via dynamic kinetic resolution

The Ru complexes of chiral spiro diphosphine ligands, $RuCl_2$[(SDPs)(1,2-diamine)] (**42**), were proven to be efficient catalysts for the asymmetric hydrogenation of racemic α-branched aldehydes via dynamic kinetic resolution (DKR) [73] to provide chiral primary alcohols in high ee (up to 96% ee) (Scheme 20) [74]. In this hydrogenation, the catalyst can selectively hydrogenate one of two enantiomers of aldehyde and the remaining enantiomer can be rapidly racemized under the reaction conditions, the two enantiomers of aldehyde ultimately converting to chiral alcohol enantioselectively.

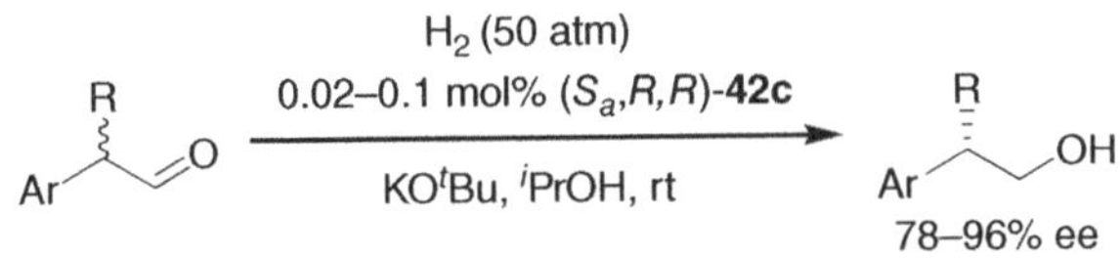

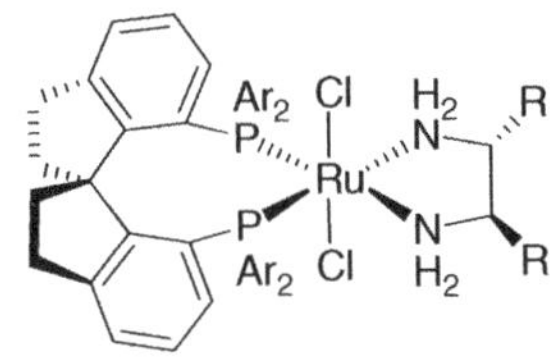

(S_a,R,R)-**42a** Ar = Ph, R = Ph

(S_a,R,R)-**42b** Ar = 3,5-Me_2C6H_3, R = Ph

(S_a,R,R)-**42c** Ar = 3,5-Me_2-4-$MeOC_6H_2$, R-R = $(CH_2)_4$

Scheme 20

This protocol was applied in the preparation of a leukotriene receptor antagonist (*S*,*S*)-fenvalerate and a lipoxygenase inhibitor BAY X 1005.

4.2.8 Ru-catalyzed hydrogenation of racemic 2-substituted ketones via DKR

By using catalysts **42** a number of aryl-, aryloxy- and amino-substituted ketones were hydrogenated via DKR in extremely high reactivity and enantioselectivity. The hydrogenation of α-amino ketones via DKR provides a direct access to chiral amino alcohols. However, very limited progress has been achieved for the asymmetric hydrogenation of α-amino ketones via DKR. The difficulty for this reaction presumably arose from the amino group, which coordinated to the metal center of the catalyst, lowering activity of the catalyst. By using catalyst (S_a,*R*,*R*)-**42a**, the asymmetric hydrogenations of racemic α-aminocycloalkanones [75] and acyclic α-amino ketones [76] via DKR were realized (Scheme 21) with excellent enantioselectivities (up to 99.9% ee) and diastereoselectivities (*cis*/*trans* >99:1), as well as high TON (up to 30 000). In the asymmetric hydrogenation of racemic acyclic α-*N*-alkyl/arylamino aliphatic ketones with an unprotected α-amino group, the catalyst (S_a,*R*,*R*)-**42b** showed high enantioselectivity and diastereoselectivity [77]. This protocol provided a practical approach for the synthesis of all four isomers of piperidine alkaloid conhydrine. Various *racemic* α-aryloxydialkyl ketones underwent the (S_a,*R*,*R*)-**42a**-catalyzed hydrogenation via DKR to produce β-aryloxy alcohols with perfect *anti*/*syn* selectivity (>99:1) and enantioselectivity (up to 99.1% ee for the *anti* isomer) (Scheme 22) [78].

Liu *et al.* [79] further developed a Ru-catalyzed hydrogenation of racemic α,α′-disubstituted cycloketones through DKR for the one-step synthesis of chiral diols with three contiguous stereocenters with high diastereoselectivity and enantioselectivity (Scheme 23). This new strategy facilitates the enantioselective total synthesis of alkaloid (+)-γ-lycorane.

The Ru-catalyzed asymmetric hydrogenation of racemic α-substituted cyclic ketones was applied to the synthesis of various chiral natural products and analogs, such as (−)-galanthamine [80], (−)-Δ^9-THC [81], CP 55,940 [82], and (−)-α-lycorane [83].

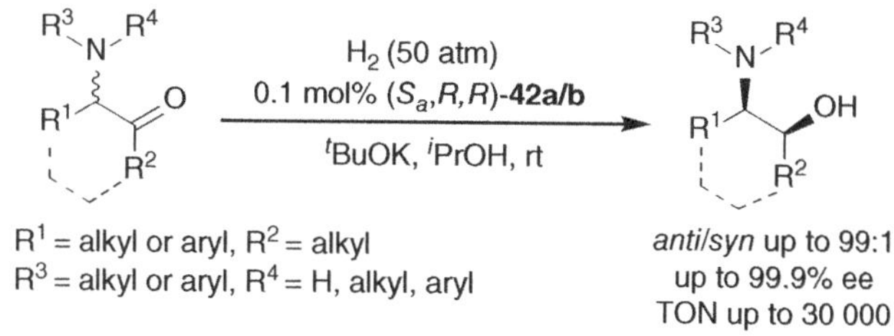

Scheme 21

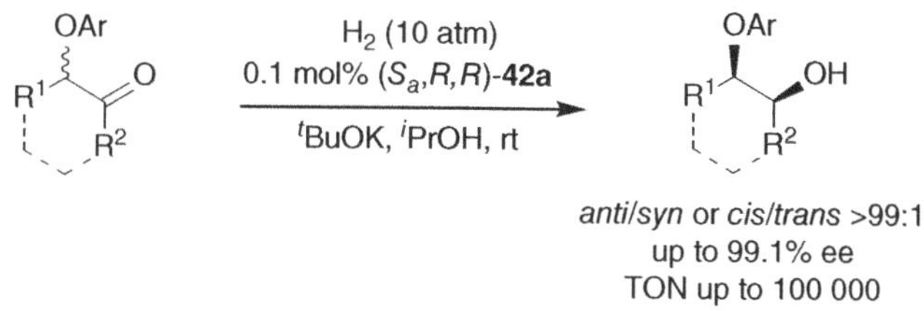

Scheme 22

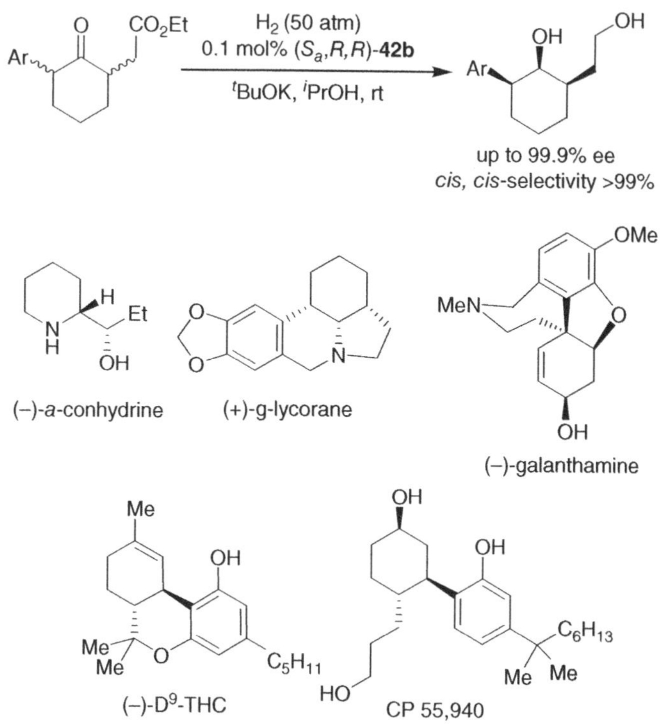

Scheme 23

4.2.9 Ir-catalyzed hydrogenation of imines

The asymmetric hydrogenation of imines provides an efficient and direct route for the synthesis of chiral amines. Zhu *et al.* [27a] reported a highly efficient asymmetric hydrogenation of acyclic *N*-aryl ketimines catalyzed by Ir-(S_a,*S*)-**25a** under ambient pressure with excellent enantioselectivities (up to 97% ee) (Scheme 24). The high rigidity and bulkiness of the ligands **25** efficiently prevents the trimerization of the catalysts Ir-**25** and maintains the stability of the catalysts under hydrogen atmosphere.

By using catalyst Ir-**41**, Han *et al.* [50] hydrogenated *N*-aryl and *N*-alkyl ketimines with good to excellent enantioselectivities (up to 98% ee) under mild reaction conditions (Scheme 25).

The hydrogenation reaction was applied in the synthesis of the chiral antidepressant drug sertraline.

The asymmetric hydrogenation of 1-alkyl 3,4-dihydroisoquinolines catalyzed by Ir/(S_a,*R*,*R*)-**17b** was developed, providing chiral 1-alkyl tetrahydroisoquinolines with high yields (88–96%) and good to excellent enantioselectivities (85–99% ee) (Scheme 26) [84].

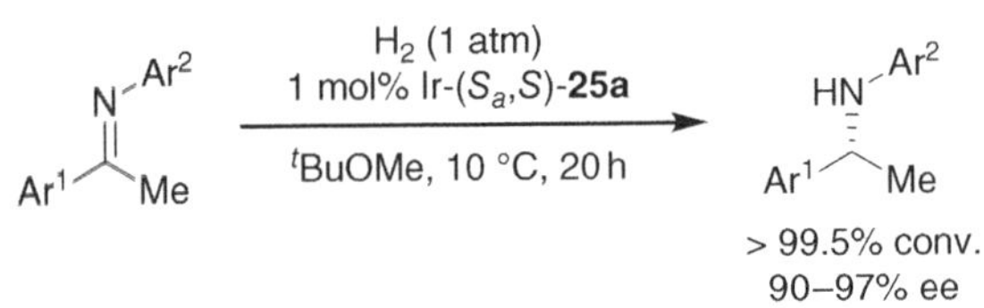

Scheme 24

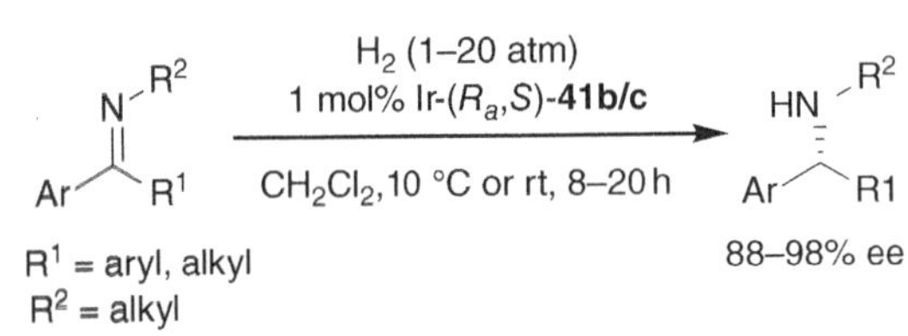

Scheme 25

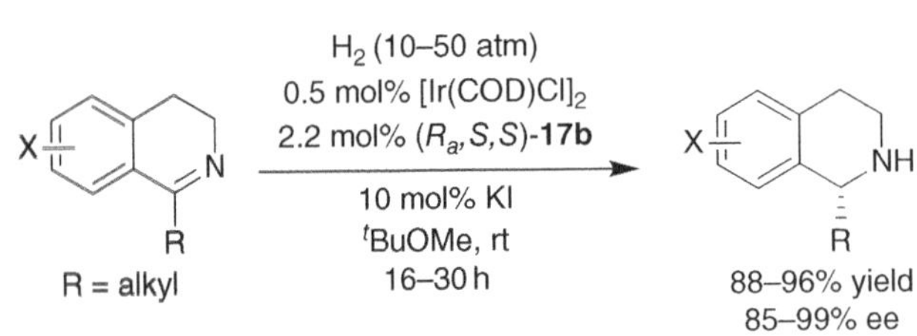

Scheme 26

4.3 Carbon–carbon bond-forming reactions

4.3.1 Ni-catalyzed hydrovinylation of olefins

Ni-catalyzed asymmetric hydrovinylation of vinylarenes is an important carbon–carbon bond-forming reaction [85]. However, most investigations have focused on the hydrovinylation of vinylarenes, yielding chiral 3-arylbut-1-enes with a chiral tertiary carbon center. By using Ni complexes of spiro phosphoramidite ligand (S_a,*R*,*R*)-**17b**, Shi *et al.* [86] developed an asymmetric hydrovinylation of α-alkyl vinylarenes, giving a hydrovinylation product bearing a chiral quaternary center in high enantioselectivity (Scheme 27). The reaction provides a new method for the construction of an all-carbon quaternary center from very simple feedstock materials. The Ni/(S_a,*R*,*R*)-**17b** can also catalyze the asymmetric hydrovinylation reaction of trimethyl(2-arylallyloxy)silanes, a type of functionalized olefin, to produce homoallylic alcohols with a chiral quaternary center in high yields and high enantioselectivities [87].

4.3.2 Rh-catalyzed hydroacylation

Coulter *et al.* [88] developed an Rh/(R_a,*R*,*R*)-**17b**-catalyzed, regio- and enantioselective intermolecular hydroacylation of homoallylic sulfides with salicylaldehydes to give α-branched ketones in high regioselectivity and up to 97% ee (Scheme 28). The spiro ligand (R_a,*R*,*R*)-**17b** exhibited significant advantages on enantiocontrol compared with other chiral ligands in this reaction.

Hoffman and Carreira [31] reported a Rh-catalyzed asymmetric intramolecular hydroacylation of pent-4-enal substrates, providing β-substituted cyclopentanones in good yield and excellent selectivity by using chiral spiro phosphoramidite–alkene ligand (*R*)-**29** (Scheme 29).

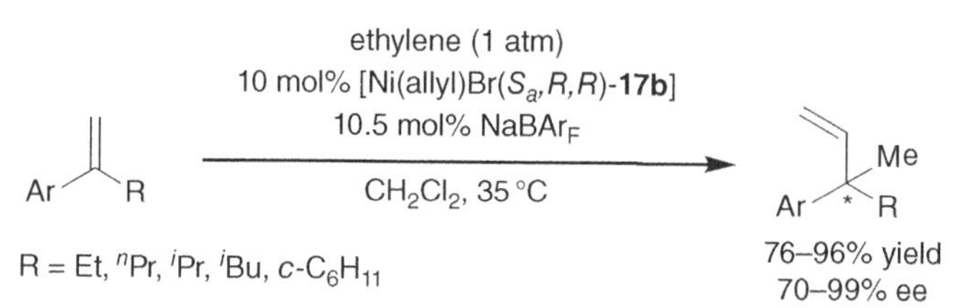

Scheme 27

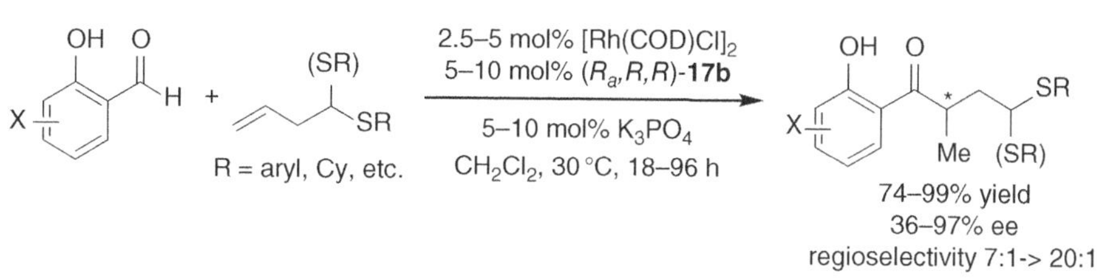

Scheme 28

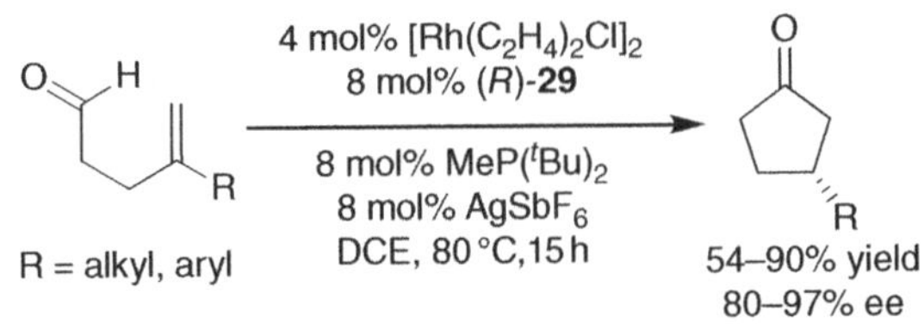

Scheme 29

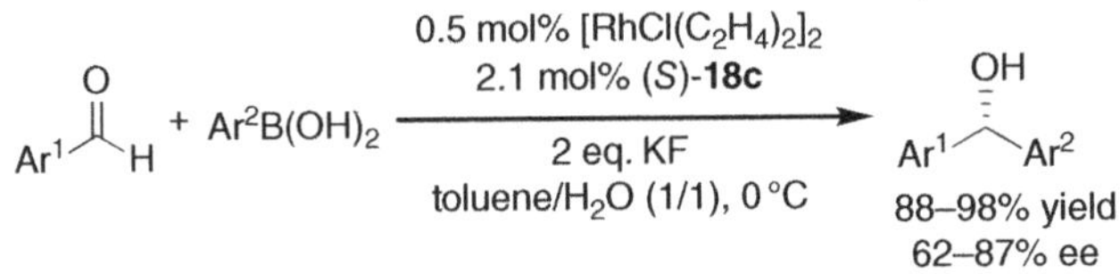

Scheme 30

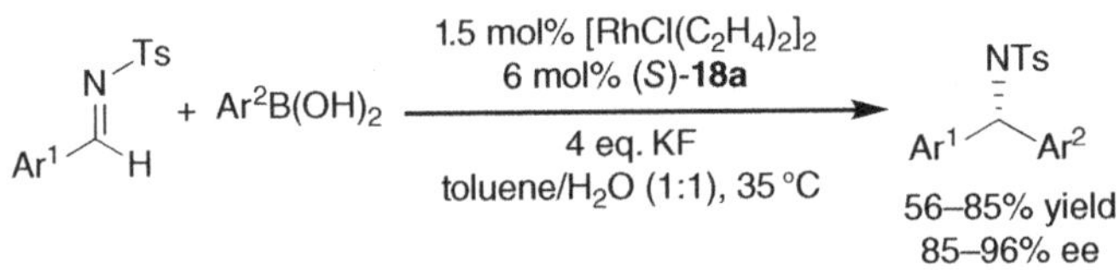

Scheme 31

4.3.3 Rh-catalyzed arylation of carbonyl compounds and imines

The transition metal catalyzed asymmetric addition of aryl organometallic reagents to aldehydes, ketones, and imines has provided efficient access to chiral aryl alcohols or aryl amines [89]. Arylboronic acids are less toxic, stable toward air and moisture, and tolerant towards a variety of functional groups, and are ideal reagents for the addition to aldehydes. However, when Sakai *et al.* [90] attempted the enantioselective Rh-catalyzed addition of phenylboronic acid to naphthaldehyde, only 41% ee was obtained. Chiral spiro phosphite complex (*S*)-**18c** was found to be an efficient catalyst for asymmetric addition reactions of arylboronic acids to aldehydes, providing diarylmethanols in excellent yields (88–98%) with up to 87% ee (Scheme 30) [20c].

The Rh/(*S*)-**18a** exhibited high enantioselectivity for the addition of arylboronic acids to *N*-tosylarylimines (Scheme 31) [91]. The reaction proceeded in aqueous toluene to give diarylmethylamines in good yields (56–85%) with up to 96% ee. By using catalyst Rh/(*S*)-**18b**, the enantioselective addition of arylboronic acids to the α-ketoesters was realized, yielding chiral tertiary α-hydroxyesters in good yields and high ee (Scheme 32) [92].

The X-ray analysis of the crystal structure of Rh-(*S*)-**18a** showed that two coordinating spiro phosphite ligands (*S*)-**18a** create an effective asymmetric environment around the Rh (Figure 9). The rigid chiral pocket may be the reason for its highly efficient chiral induction in the addition reactions. It can be seen from the model that the transfer of the

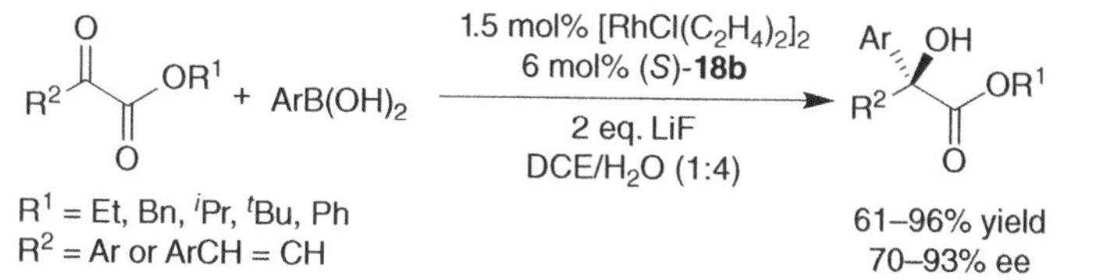

Scheme 32

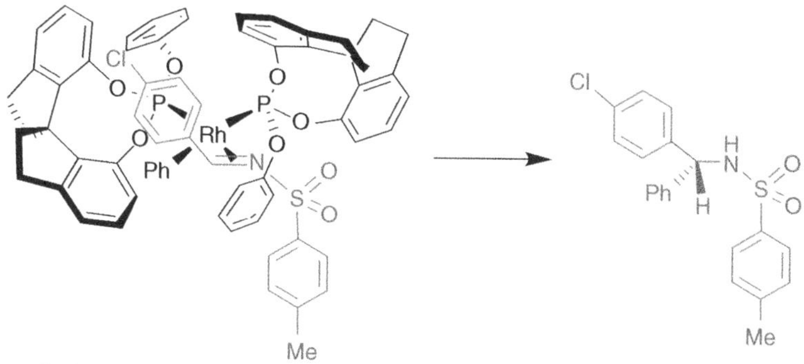

Figure 9 *The single crystal structure of Rh-(S)-**18a** and stereo-recognition model. (See insert for color/color representation of this figure)*

phenyl group to the *Re* face of *N*-tosylarylimine to generate the (*R*)-product is much more favorable, which is consistent with the experimental result.

4.3.4 Pd-catalyzed umpolung allylation reactions of aldehydes, ketones, and imines

Compared with well-established electrophilic π-allylpalladium chemisty, the catalytic asymmetric reaction via umpolung of π-allylpalladium has received very limited exploration [93]. Zhou and co-workers investigated the Pd-catalyzed asymmetric umpolung allylation reactions of aldehydes [22a, 94], activated ketones [95], and imines [96] by using chiral spiro ligands (*S*)-**18e**, (*S*)-**17c**, and (*S*)-**17a**, respectively. One representative example is that of the Pd/(*S*)-**18e**-catalyzed umpolung allylation of aldehydes with allylic alcohols and their derivatives, which provided synthetically useful homoallylic alcohols from readily available allylic alcohols, with high yields and excellent enantioselectivities (Scheme 33).

4.3.5 Ni-catalyzed three-component coupling reaction

Ni-catalyzed multi-component coupling reactions are an efficient protocol for carbon–carbon bond formation [97]. The Ni-catalyzed asymmetric intermolecular reductive coupling of 1,3-dienes and aldehydes with Et_2Zn as a reducing reagent was realized

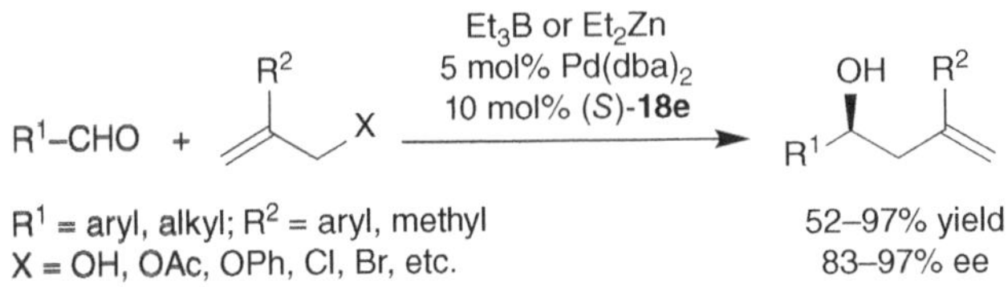

Scheme 33

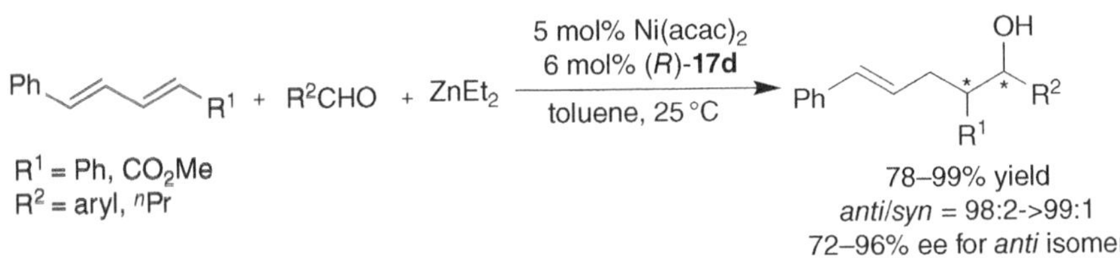

Scheme 34

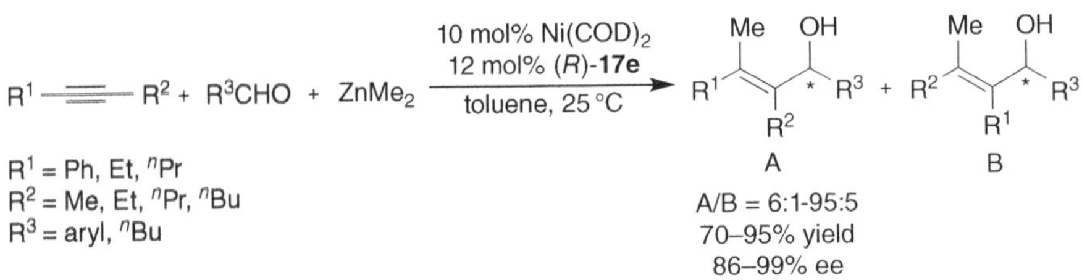

Scheme 35

by using 6,6′-substituted phosphoramidite ligand (*R*)-**17d** (Scheme 34) [32b]. The coupling products, chiral bishomoallylic alcohols, were produced in high yields (78–99%) with excellent diastereoselectivity (*anti*/*syn* = 98:2–99:1) and enantioselectivity (up to 96% ee).

Ni complexes of (*R*)-**17e** were demonstrated to be efficient catalysts for the asymmetric alkylative coupling reaction of alkyne, aldehyde and Me_2Zn (Scheme 35) [98]. A variety of chiral allylic alcohols with tetrasubstituted olefin moiety were produced in high yields (70–95%), high regioselectivity (6:1–95:5), and excellent enantioselectivity (up to 99% ee). The groups on the 6,6′-position of spiro phosphoramidite ligand (*R*)-**17e** are crucial for obtaining high enantioselectivity. It can be seen from the proposed transition state model that the phenyl groups on the 6,6′-position of the ligand direct the Ni-activated alkyne to approach the aldehyde from its *Si* face (model TS-1), leading to the formation of allylic alcohols with *R* configuration (Figure 10).

Zhou *et al.* [99] developed a Ni-catalyzed reductive coupling of alkynes and imines with Et_2Zn as a reductant by using electron-rich spiro phosphine ligands, which afforded various allylic amines with high yields and excellent chemoselectivities. The asymmetric version of this reaction was also accomplished by using

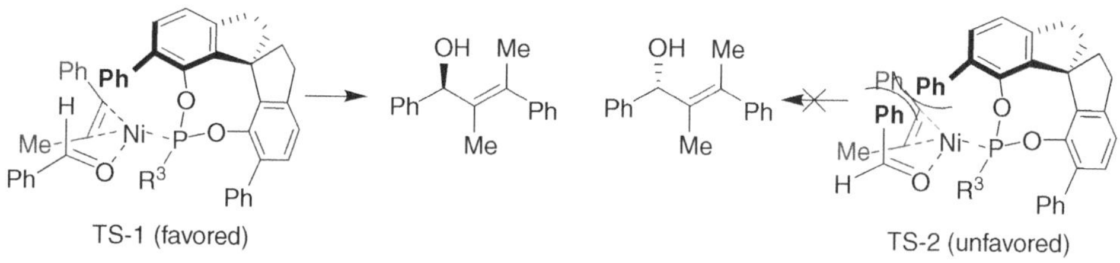

Figure 10 Chiral inducing models for Ni-catalyzed three-component coupling reactions

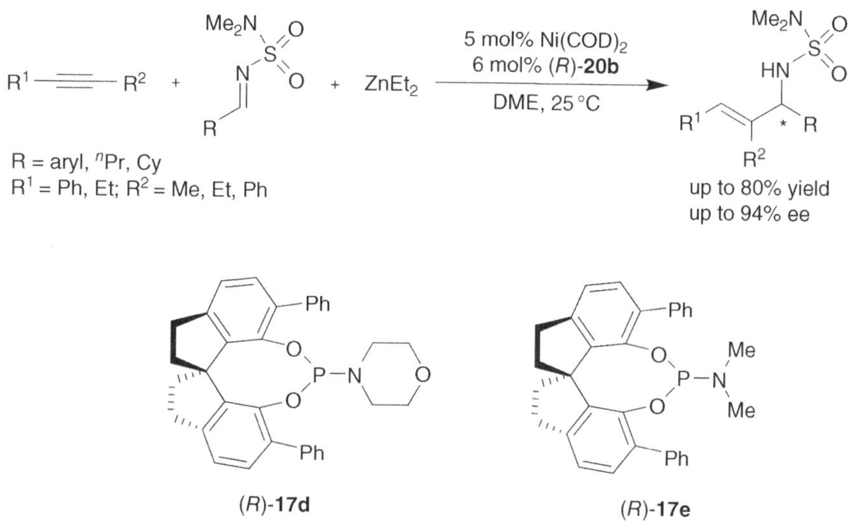

Scheme 36

chiral spiro phosphine ligand (*R*)-**20b** with high enantioselectivities (Scheme 36). An isotope-labeling experiment showed that the transferred hydrogen was most likely from the ethyl group of Et_2Zn.

4.3.6 Au-catalyzed Mannich reactions of azlactones

Melhado *et al.* [100] reported an enantioselective Au(I)-catalyzed Mannich reaction employing spiro bisphosphines ligands (*R*)-**21b**, which provides direct access to aliphatic and aromatic α,β-diamino acid derivatives in high diastereo- and enantioselectivities (Scheme 37).

4.3.7 Rh-catalyzed hydrosilylation/cyclization reaction

The Rh complex of spiro diphosphine (*R*)-**21a** was proven to be an efficient catalyst for the hydrosilylation/cyclization of 1,6-enynes and afforded unprecedented high enantioselectivity (89–99% ee) for a wide range of substrates (Scheme 38) [101].

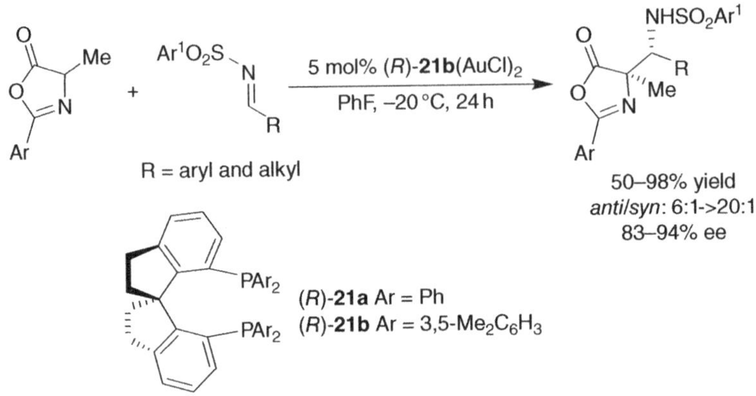

Scheme 37

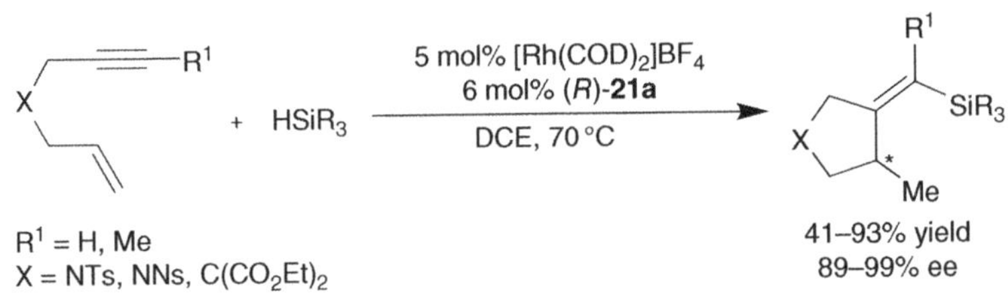

Scheme 38

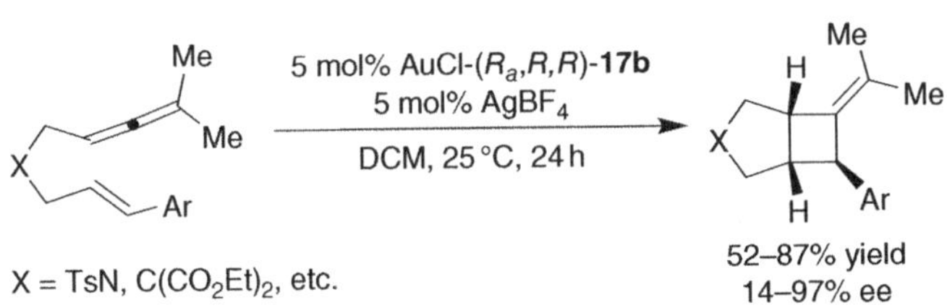

Scheme 39

4.3.8 Au-catalyzed [2+2] cycloaddition

González *et al.* [102] disclosed that Au(I)-(R_a,*R*,*R*)-**17b** is a highly selective catalyst for the cycloadditions of allene-enes (Scheme 39). The chiral spiro ligand (R_a,*R*,*R*)-**17b** exhibited much higher enantioselectivity than other chiral ligands in this reaction. An asymmetric Au(I)-catalyzed intermolecular [2+2] cycloaddition of *N*-allenylsulfonamides and styrene derivatives was also realized by using chiral spiro phosphoramidite ligand (S_a,*R*,*R*)-**17b** to furnish chiral cyclobutanes in both high yield and enantioselectivity at –70 °C (Scheme 40) [103].

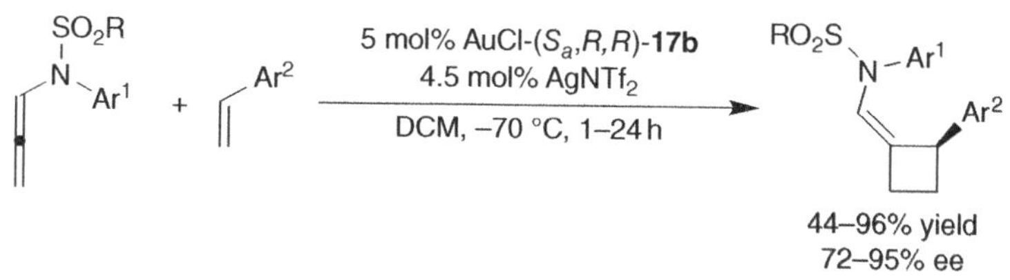

Scheme 40

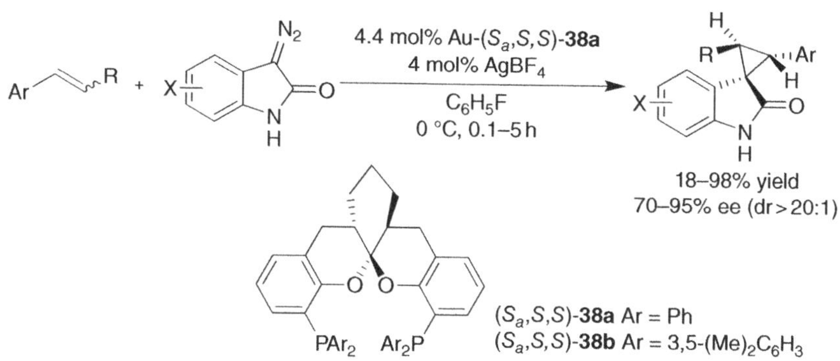

Scheme 41

4.3.9 Au-catalyzed cyclopropanation

A spiroketal bisphosphine (R_a,*R*,*R*)-**38a** derived chiral Au complex was found to be an efficient catalyst for asymmetric cyclopropanation of diazooxindoles with a broad range of alkenes, providing a highly diastereo- and enantioselective approach for spiro cyclopropyloxindoles (Scheme 41) [47]. These results further demonstrate the special advantage of rigid spiro ligands in Au-catalyzed reactions.

4.3.10 Pd-catalyzed Heck reactions

Hu *et al.* [104] disclosed that the spiro phosphine oxide ligand (*R*)-**30a** is highly efficient for the Pd-catalyzed Heck reaction of cyclic and heterocyclic olefins with aryl triflates. An asymmetric domino Heck cyclization was established using catalyst Pd/(*R*)-**30a** to form fused carbo- and heterocycles with excellent stereoselectivity (Scheme 42) [105]. This method was used in a short synthesis of (–)-martinellic acid, a primary ingredient in traditional eye medicine in South America.

4.4 Carbon–heteroatom bond-forming reactions

4.4.1 Cu-catalyzed N–H bond insertion reactions

The catalytic insertion of α-diazocarbonyl compounds into X–H (X=C, N, O, S) bonds is a powerful organic transformation; however, only limited success has been

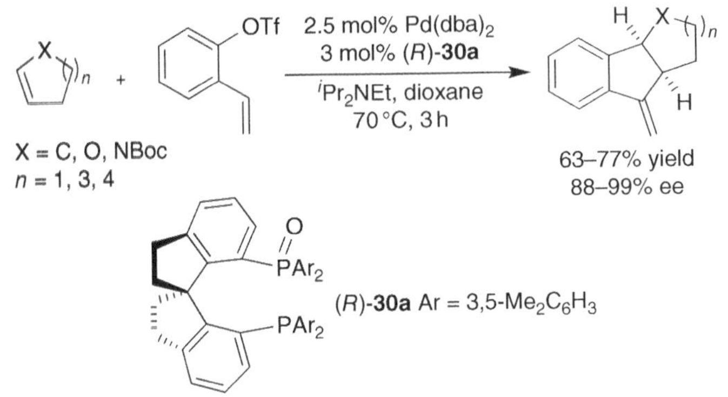

Scheme 42

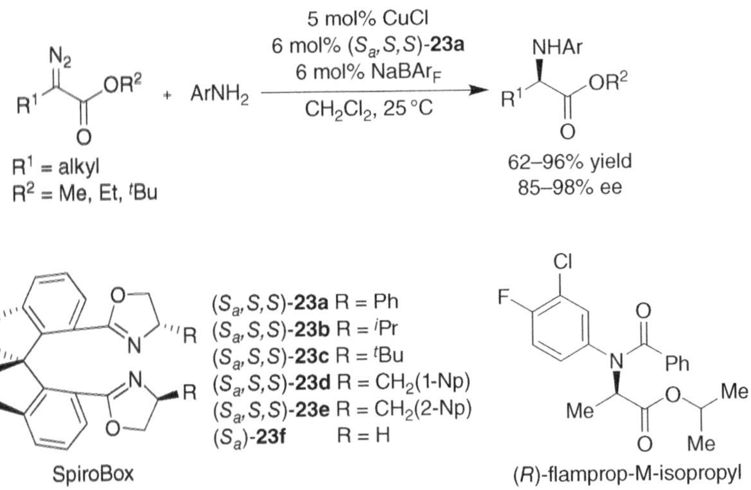

Scheme 43

achieved for asymmetric versions of this reaction [106]. For instance, the concept of metal–carbenoid insertion into N—H bonds has been known for more than five decades, but only two catalytic asymmetric versions of this topic have been documented, and both of them gave low enantioselectivity (<50% ee) [107]. Zhu and Zhou [34] disclosed that the chiral spiro bisoxazoline ligands exhibit excellent enantioselectivity in Cu-catalyzed N—H bond insertion reactions.

By using Cu complexes of chiral spiro bis(oxazoline) ligand (S_a,*S*,*S*)-**23a** as catalysts, the enantioselective catalytic insertion of α-diazoesters into N—H bond of aromatic amines was realized in high yields and high enantioselectivities (Scheme 43) [25b,108]. The chiral spiro Cu catalysts have unique binuclear structures, as showing in Figure 6, which may address the excellent performance of the Cu-(S_a,*S*,*S*)-**23a** catalyst for this challenging reaction. With the Cu-catalyzed asymmetric N—H insertion as a key step, a chiral arylalanine herbicide (*R*)-flamprop-M-isopropyl was concisely prepared from easily available material.

4.4.2 Cu-, Fe-, or Pd-catalzyed O—H insertion reactions

The catalyst Cu/(S_a,*S*,*S*)-**23a** was also efficient for the insertion of O—H bonds of phenols (Scheme 44) [109]. Under similar conditions as that for N—H insertion reactions, a wide range of phenol derivatives underwent O—H bond insertion with α-diazopropionates, providing α-aryloxypropionates with excellent enantioselectivities (95–99.6% ee). The Pd/(S_a,*S*,*S*)-**23a**-catalyzed asymmetric O—H bond insertion reaction between α-aryl-α-diazoacetates and phenols provided the first enantioselective method for the preparation of chiral α-aryloxy-α-arylacetates, which are ubiquitous in biologically active molecules (Scheme 45) [110].

It was found that the iron catalysts prepared in situ from $FeCl_2 \cdot 4H_2O$ and chiral spiro bisoxazoline ligands **23** exhibited excellent enantioselectivity as well as reactivity for the insertion of O—H bonds of various saturated alcohols and allylic alcohols (Scheme 46) [111]. The catalyst Fe-(S_a,*S*,*S*)-**23c** showed higher yields and higher enantioselectivities than any other metal catalysts.

The asymmetric insertion of α-diazoesters into the O—H bond of water provides an extremely simple approach for the synthesis of chiral α-hydroxyesters in an efficient and atom-economical way. The challenges of asymmetric O—H insertion of water are mainly attributed to two considerations: first, the active metal carbene intermediates are generally sensitive to water; and secondly, the small molecular structure of water makes chiral discrimination quite difficult. Zhou and co-workers discovered a highly enantioselective O—H insertion of water catalyzed by chiral spiro Cu [112] and Fe catalysts [111]. Under mild conditions, both Cu and Fe complexes of ligand (S_a,*S*,*S*)-**23a**

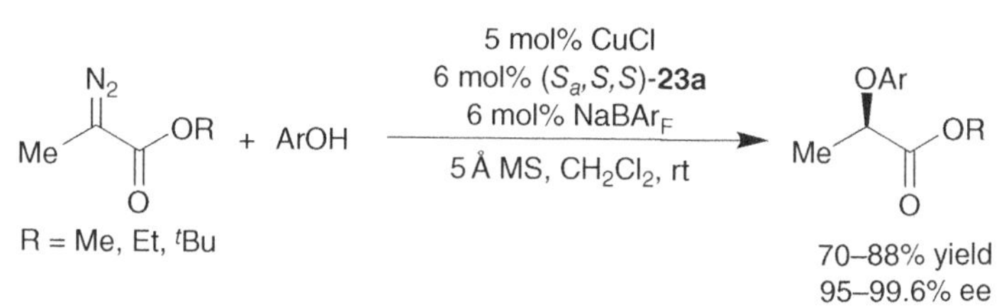

Scheme 44

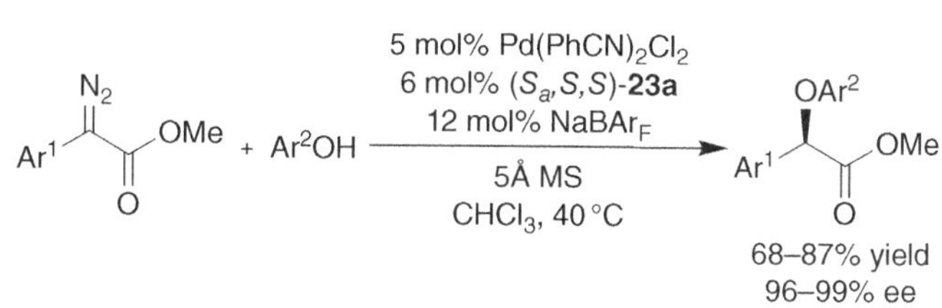

Scheme 45

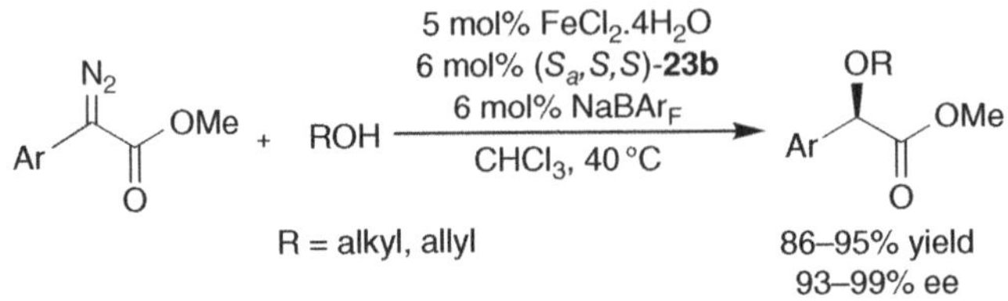

Scheme 46

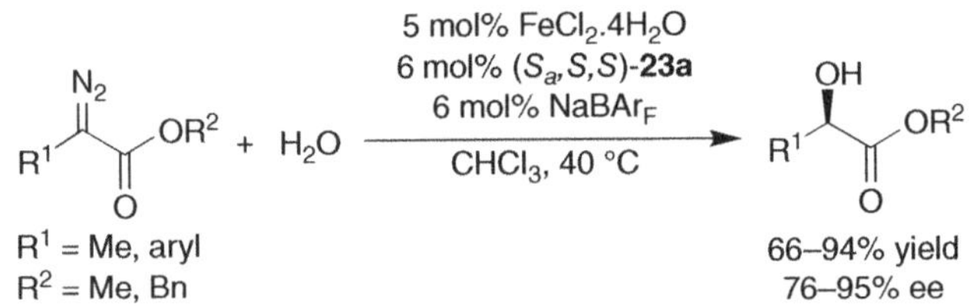

Scheme 47

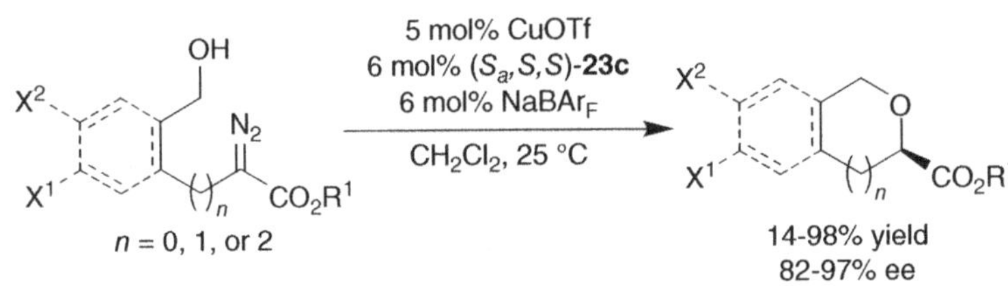

Scheme 48

exhibited high enantioselectivity for the O—H insertion of water (Scheme 47). Notably, the insertions of α-aryl-α-diazoacetates having a coordinating group at the *ortho* position were less enantioselective in Cu-catalyzed reactions, but gave high enantioselection in Fe-catalyzed reactions. For instance, in the reaction of methyl α-diazo-2-chlorophenylacetate with water, the catalyst Fe-(S_a,*S*,*S*)-**23a** produced methyl (*R*)-*o*-chloromandelate with 95% ee, while the analogous Cu-(S_a,*S*,*S*)-**23a** catalyst gave the same product with only 36% ee. The facile conversion of methyl (*R*)-*o*-chloromandelate to clopidogrel, a platelet aggregation inhibitor, demonstrated that the Fe-catalyzed asymmetric insertion with water has potential for applications in the preparation of chiral non-racemic drugs.

Catalytic intramolecular O—H insertion is a useful reaction for the construction of cyclic ethers and esters. The Cu-catalyzed highly enantioselective intramolecular O—H insertion of δ- or ε-hydroxy-α-diazoesters (Scheme 48) [113] and phenolic O—H bonds (Scheme 49) [114] were achieved by using ligand (S_a,*S*,*S*)-**23**. The substrates of intramolecular O—H insertions can be easily modified at the side chain, and thus provides a useful method for preparing chiral 2-carboxy cyclic ethers with different ring sizes and substitution patterns.

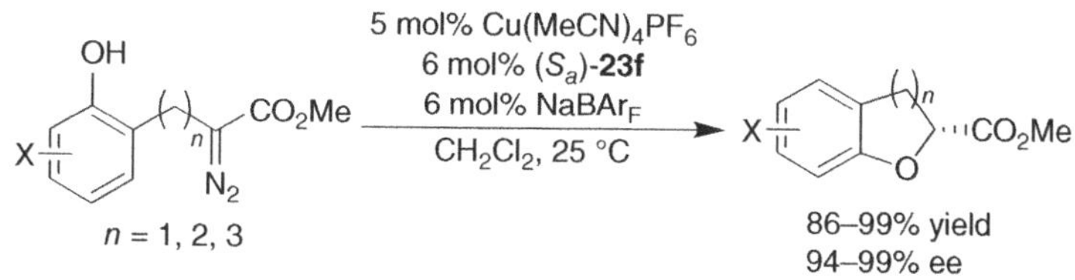

Scheme 49

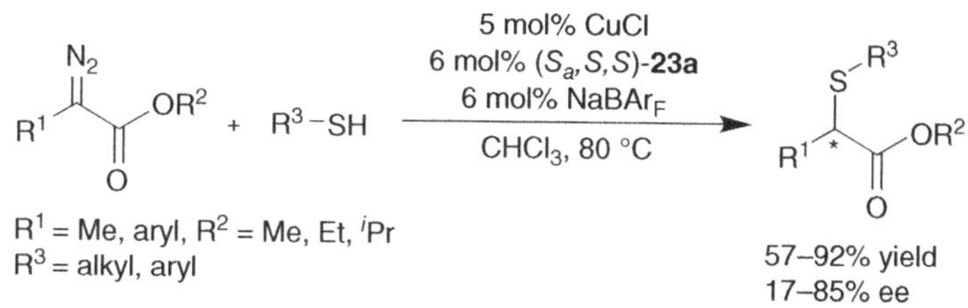

Scheme 50

4.4.3 Cu-catalyzed S—H, Si—H and B—H insertion reactions

Transition metal catalyzed S—H bond insertion has been extensively studied for forming C—S bonds, however progress on the asymmetric version of this reaction had been limited until catalyst Cu-(S_a,*S*,*S*)-**23a** was found to promote the reaction with up to 85% ee (Scheme 50) [115].

Transition metal catalyzed carbene insertion into the Si—H bond provides a direct and efficient method for the synthesis of organosilicon compounds. When chiral spiro diimine ligand (*R*)-**24a** was applied in Cu-catalyzed asymmetric insertion of α-diazo-α-arylacetates with silanes, the Si—H insertion products were obtained in high yields (85–97%) and excellent enantioselectivities (90–99% ee) (Scheme 51) [26a].

The development of methodologies for preparation of versatile organoboron compounds has been an actively pursued topic in organic chemistry. The B—H bond of borane is electron-deficient; thus it is inert to the metal carbene insertion reaction. However, upon forming adducts with amine or phosphine, the borane can undergo B—H bond insertion reactions. Cheng *et al.* [116] developed Cu-catalyzed B—H bond insertion reactions of the amine- or phosphine-borane adducts with α-diazo-α-arylacetates. The asymmetric version of this reaction was also established by using ligand (R_a,*S*,*S*)-**23a**, which provides an efficient approach to chiral organoboranes (Scheme 52).

4.4.4 Pd-catalyzed allylic amination

The Pd complexes of (R_a,*R*,*R*)-**38b** were found to be highly efficient in the enantioselective allylic amination of the esters of racemic Morita–Baylis–Hillman (MBH) adducts with aromatic amines, affording optically active β-arylamino acid esters in high activity (TON up to 4750) with high region- and enantioselectivities (Scheme 53)

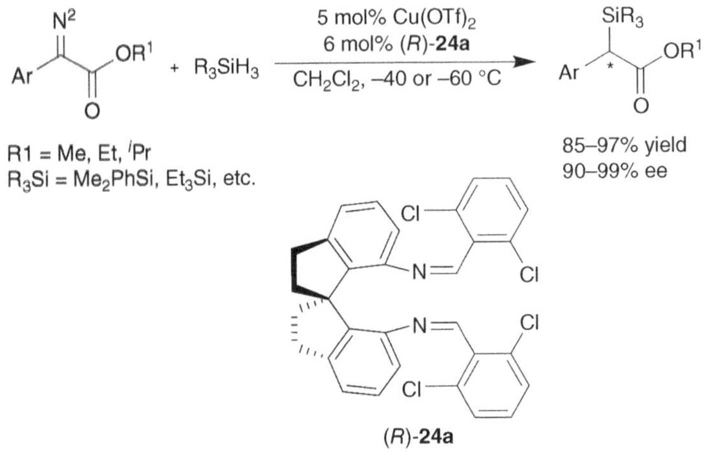

Scheme 51

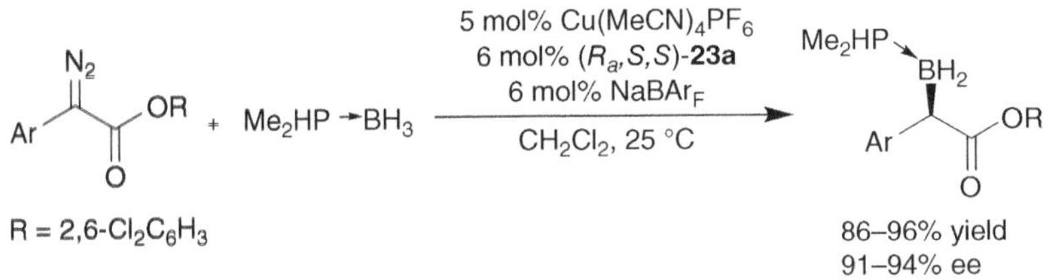

Scheme 52

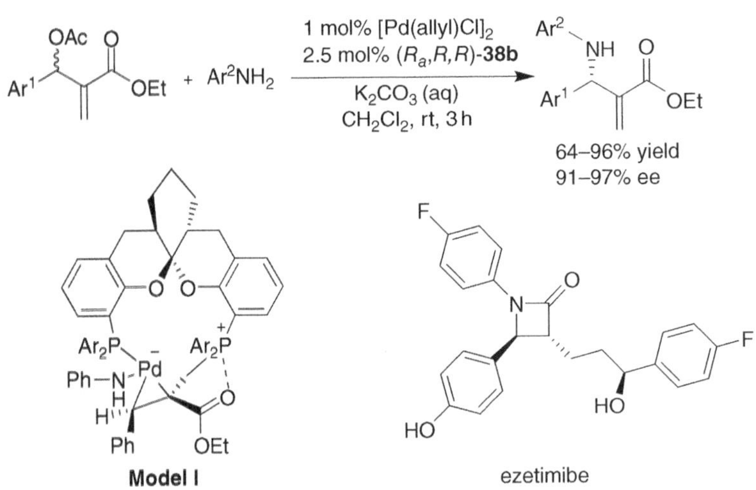

Scheme 53

[46]. The β-arylamino acid esters can be readily transformed into β-lactam derivatives retaining enantioselectivity, thus providing an efficient access to various biologically important molecules, such as ezetimibe.

Mechanistic studies indicated that the unique structure of the ligand **38**, with a long P—P distance, was favorable for allowing two P atoms to play a bifunctional role in the catalysis. Herein, one of the P atoms of the ligand forms a C—P σ bond with the terminal C atom of the allyl moiety as a Lewis base, and another P atom of the ligand coordinates to the Pd atom (Model I, Scheme 53). The cooperative action of organo- and organometallic catalysis discovered in this reaction is most likely responsible for its high activity, as well as excellent regio- and enantioselectivities.

4.4.5 Pd-catalyzed allylic cyclization reactions with allenes

Shu and Ma [25c] found that the spiro bisoxazoline (R_a,*S*,*S*)-**23d** containing α-naphthylmethyl groups is a suitable ligand for the Pd-catalyzed enantioselective cyclization of 3,4-allenyl hydrazines with organic halides, affording optically active 3-substituted pyrazolidines in good yields with 92–95% ee (Scheme 54). The β-naphthylmethyl-substituted spiro bisoxazoline ligand (R_a,*S*,*S*)-**23e** was also successfully applied to the Pd-catalyzed enantioselective cyclization reaction of allenes with *o*-aminoiodobenzenes, yielding 3-alkylideneindolines in good yields with excellent ee values (Scheme 55) [25d]. The spiro ligand **23** exhibited significantly higher enantioselectivity than other chiral ligands in these reactions.

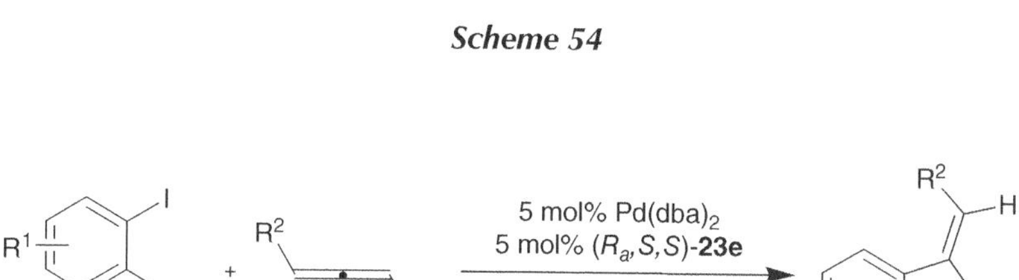

Scheme 54

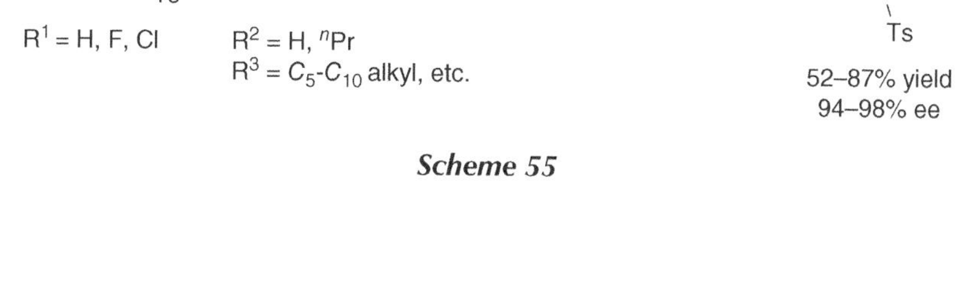

Scheme 55

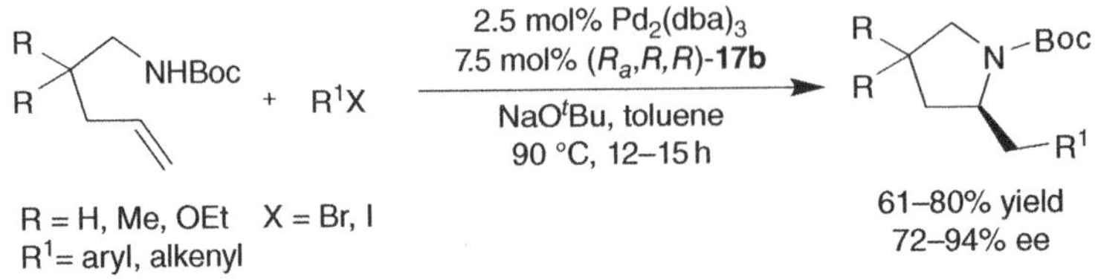

Scheme 56

4.4.6 Pd-catalyzed alkene carboamination reactions

Wolfe and co-workers [117] developed a series of Pd-catalyzed enantioselective carboaminations of alkenes with up to 95% ee by using chiral spiro phosphoramidite ligand ((R_a,R,R)-**17b** (Scheme 56). These transformations provide new methods for the asymmetric construction of nitrogen heterocycles including phenanthroindolizidine alkaloids such as (−)-tylophorine and (+)-aphanorphine. Moreover, Babij and Wolfe [118] applied the Pd-catalyzed desymmetrization carboamination reaction of mesodiallylpyrrolidinyl ureas to prepare enantiomerically enriched bicyclic ureas bearing three stereocenters with good levels of diastereoselectivity (up to 20:1 diastereomeric ratio) and enantioselectivity (up to 90% ee).

4.5 Conclusion

Although the history for the application of chiral spiro ligands in asymmetric catalysis is not long, many chiral spiro ligands have been synthesized and applied in a variety of seemingly unrelated reactions including hydrogenation of unsaturated bonds (carbon–carbon double bonds, carbonyl, imines, etc.), carbon–carbon bond-forming and carbon–heteroatom bond-forming reactions accompanied by the metals Ru, Rh, Ir, Pd, Cu, Fe, Ni, and so on. In most of these reactions, spiro ligands exhibit a remarkable capability for enantiocontrol. As more and more groups are involved in developing or applying chiral spiro ligands and catalysts, it can be expected that the chiral spiro ligands will have a broad application in asymmetric catalysis in the future.

The features that make spiro ligands "privileged" include their extreme chemical robustness, high conformational rigidity, perfect C_2 symmetry, and ease of modification. The exciting results obtained with chiral spiro ligands will stimulate future efforts to understand the features that account for the wide applicability of these ligands, which are of significance for designing new chiral ligands and catalysts.

References

[1] (a) I. Ojima (ed.), *Catalytic Asymmetric Synthesis, 3rd edn*, John Wiley & Sons, Inc., Hoboken, **2010**; (b) G.-Q. Lin, Y.-M. Li, A. S. C. Chan, *Priciples and Applications of Asymmetric Synthesis*, John Wiley & Sons, Inc., New York, **2001**; (c) E. N. Jacobsen, A. Pfaltz, H. Yamamoto (eds), *Comprehensive Asymmetric Catalysis*, Springer, Berlin, **1999**.

[2] T. P. Yoon, E. N. Jacobsen, *Science* **2003**, *299*, 1691–1693.
[3] A. P. Krapcho, *Synthesis* **1974**, 383–419
[4] N. Srivastava, A. Mital, A. Kumar, *J. Chem. Soc., Chem. Commun.* **1992**, 493–494.
[5] J. A. Nieman, B. A. Keay, *Tetrahedron: Asymmetry* **1996**, *7*, 3521–3526.
[6] (a) S.-F. Zhu, Q.-L. Zhou, in *Privileged Chiral Ligands and Catalysts* (ed. Q.-L. Zhou), Wiley-VCH, Weinheim, **2011**, pp. 137–170; (b) J.-H. Xie, Q.-L. Zhou, *Acta Chim. Sinica* **2014**, *72*, 778–797; (c) J.-H. Xie, Q.-L. Zhou, *Acc. Chem. Res.* **2008**, *41*, 581–593; (d) K.-L. Ding, Z.-B. Han, Z. Wang, *Chem. Asian J.* **2009**, *4*, 32–41.
[7] A. S. C. Chan, W.-H. Hu, C.-C. Pai, C.-P. Lau, Y.-Z. Jiang, A.-Q. Mi, M. Yan, J. Sun, R.-L. Lou, J.-G. Deng, *J. Am. Chem. Soc.* **1997**, *119*, 9570–9571.
[8] C. W. Lin, C.-C. Lin, L. F.-L. Lam, T. T.-L. Au-Yeung, A. S. C. Chan, *Tetrahedron Lett.* **2004**, *45*, 7379–7381.
[9] Y.-Z. Jiang, S. Xue, Z. Li, J.-G. Deng, A.-Q. Mi, A.S.C. Chan, *Tetrahedron: Asymmetry* **1998**, *9*, 3185–3189.
[10] W. L. Benoit, M. Parvez, B. A. Keay, *Tetrahedron: Asymmetry* **2009**, *20*, 69–77.
[11] Z.-Q. Guo, X.-Y. Guan, Z.-Y. Chen, *Tetrahedron: Asymmetry* **2006**, *17*, 468–473.
[12] A. F. Khlebnikov, S. I. Kozhushkov, D. S. Yufit, H. Schill, M. Reggelin, V. Spohr, A. de Meijere, *Eur. J. Org. Chem.* **2012**, 1530–1545.
[13] (a) M. A. Arai, T. Arai, H. Sasai, *Org. Lett.* **1999**, *1*, 1795–1797; (b) G. B. Bajracharya, M. A. Arai, P. S. Koranne, T. Suzuki, S. Takizawa, H. Sasai, *Bull. Chem. Soc. Jpn.* **2009**, *82*, 285–302.
[14] (a) M. A. Arai, M. Kuraishi, T. Arai, H. Sasai, *J. Am. Chem. Soc.* **2001**, *123*, 2907–2908; (b) T. Tsujihara, K. Takenaka, K. Onitsuka, M. Hatanaka, H. Sasai, *J. Am. Chem. Soc.* **2009**, *131*, 3452–3453.
[15] P. S. Koranne, T. Tsujihara, M. A. Arai, G. B. Bajracharya, T. Suzuki, K. Onitsuka, H. Sasai, *Tetrahedron: Asymmetry* **2007**, *18*, 919–923.
[16] A. S. C. Chan, C.-C. Lin, J. Sun, W.-H. Hu, Z. Li, W.-D. Pan, A.-Q. Mi, Y.-Z. Jiang, T.-M. Huang, T.-K. Yang, J.-H. Chen, Y. Wang, G.-H. Lee, *Tetrahedron: Asymmetry* **1995**, *6*, 2953–2959.
[17] V. B. Birman, A. L. Rheingold, K.-C. Lam, *Tetrahedron: Asymmetry* **1999**, *10*, 125–131.
[18] J.-H. Zhang, J. Liao, X. Cui, K.-B. Yu, J. Zhu, J.-G. Deng, S.-F. Zhu, L.-X. Wang, Q.-L. Zhou, L.W. Chung, T. Ye, *Tetrahedron: Asymmetry* **2002**, *13*, 1363–1366.
[19] (a) Y. Fu, J.-H. Xie, A.-G. Hu, H. Zhou, L.-X. Wang, Q.-L. Zhou, *Chem. Commun.* **2002**, 480–481; (b) H. Zhou, W.-H. Wang, Y. Fu, J.-H. Xie, W.-J. Shi, L.-X. Wang, Q.-L. Zhou, *J. Org. Chem.* **2003**, *68*, 1582–1584.
[20] (a) W.-J. Shi, L.-X. Wang, Y. Fu, S.-F. Zhu, Q.-L. Zhou, *Tetrahedron: Asymmetry* **2003**, *14*, 3867–3872; (b) W.-J. Shi, J.-H. Xie, Q.-L. Zhou, *Tetrahedron: Asymmetry* **2005**, *16*, 705–710; (c) H.-F. Duan, J.-H. Xie, W.-J. Shi, Q. Zhang, Q.-L. Zhou, *Org. Lett.* **2006**, *8*, 1479–1481.
[21] (a) Y. Fu, G.-H. Hou, J.-H. Xie, L. Xing, L.-X. Wang, Q.-L. Zhou, *J. Org. Chem.* **2004**, *69*, 8157–8160; (b) G.-H. Hou, J.-H. Xie, L.-X. Wang, Q.-L. Zhou, *J. Am. Chem. Soc.* **2006**, *128*, 11774–11775.
[22] (a) S.-F. Zhu, Y. Yang, L.-X. Wang, B. Liu, Q.-L. Zhou, *Org. Lett.* **2005**, *7*, 2333–2335; (b) W. Zhang, S.-F. Zhu, X.-C. Qiao, Q.-L. Zhou, *Chem. Asian J.* **2008**, *3*, 2105–2111.
[23] (a) J.-H. Xie, L.-X. Wang, Y. Fu, S.-F. Zhu, B.-M. Fan, H.-F. Duan, Q.-L. Zhou, *J. Am. Chem. Soc.* **2003**, *125*, 4404–4405; (b) J.-H. Xie, H.-F. Duan, B.-M. Fan, X. Cheng, L.-X. Wang, Q.-L. Zhou, *Adv. Synth. Catal.* **2004**, *346*, 625–632.

[24] W.-J. Tang, S.-F. Zhu, L.-J. Xu, Q.-L. Zhou, Q.-H. Fan, H.-F. Zhou, K. Lam, A. S. C. Chan, *Chem. Commun.* **2007**, 613–615.

[25] (a) B. Liu, S.-F. Zhu, L.-X. Wang, Q.-L. Zhou, *Tetrahedron: Asymmetry* **2006**, *17*, 634–641; (b) B. Liu, S.-F. Zhu, W. Zhang, C. Chen, Q.-L. Zhou, *J. Am. Chem. Soc.* **2007**, *129*, 5834–5835; (c) W. Shu, S.-M. Ma, *Chem. Commun.* **2009**, 6198–6200; (d) W. Shu, Q. Yu, S.-M. Ma, *Adv. Synth. Catal.* **2009**, *351*, 2807–2810.

[26] (a) Y.-Z. Zhang, S.-F. Zhu, L.-X. Wang, Q.-L. Zhou, *Angew. Chem. Int. Ed.* **2008**, *47*, 8496–8498; (b) C. Chen, S.-F. Zhu, X.-Y. Wu, Q.-L. Zhou, *Tetrahedron: Asymmetry* **2006**, *17*, 2761–2767.

[27] (a) S.-F. Zhu, J.-B. Xie, Y.-Z. Zhang, S. Li, Q.-L. Zhou, *J. Am. Chem. Soc.* **2006**, *128*, 12886–12891; (b) S. Li, S.-F. Zhu, C.-M. Zhang, S. Song, Q.-L. Zhou, *J. Am. Chem. Soc.* **2008**, *130*, 8584–8585.

[28] S.-F. Zhu, Y.-B. Yu, S. Li, L.-X. Wang, Q.-L. Zhou, *Angew. Chem. Int. Ed.* **2012**, *51*, 8872–8875.

[29] (a) J.-B. Xie, J.-H. Xie, X.-Y. Liu, W.-L. Kong, S. Li, Q.-L. Zhou, *J. Am. Chem. Soc.* **2010**, *132*, 4538–4539; (b) J.-B. Xie, J.-H. Xie, X.-Y. Liu, Q.-Q. Zhang, Q.-L. Zhou, *Chem. – Asian J.* **2011**, *6*, 899–908.

[30] J.-H. Xie, X.-Y. Liu, J.-B. Xie, L.-X. Wang, Q.-L. Zhou, *Angew. Chem. Int. Ed.* **2011**, *50*, 7329–7332.

[31] T. J. Hoffman, E. M. Carreira, *Angew. Chem. Int. Ed.* **2011**, *50*, 10670–10674.

[32] (a) S.-F. Zhu, Y. Fu, J.-H. Xie, B. Liu, L. Xing, Q.-L. Zhou, *Tetrahedron: Asymmetry* **2003**, *14*, 3219–3224; (b) Y. Yang, S.-F. Zhu, H.-F. Duan, C.-Y. Zhou, L.-X. Wang, Q.-L. Zhou, *J. Am. Chem. Soc.* **2007**, *129*, 2248–2249.

[33] S.-F. Zhu, B. Xu, G.-P. Wang, Q.-L. Zhou, *J. Am. Chem. Soc.* **2012**, *134*, 436–442.

[34] S.-F. Zhu, Q.-L. Zhou, *Acc. Chem. Res.* **2012**, *45*, 1355–1377.

[35] (a) Y. K. Chung, G. C. Fu, *Angew. Chem. Int. Ed.* **2009**, *48*, 2225–2227; (b) R. J. Lundgren, A. Wilsily, N. Marion, C. Ma, Y. K. Chung, G. C. Fu, *Angew. Chem. Int. Ed.* **2013**, *52*, 2525–2528.

[36] Q.-G. Wang, S.-F. Zhu, L.-W. Ye, C.-Y. Zhou, X.-L. Sun, Y. Tang, Q.-L. Zhou, *Adv. Synth. Catal.* **2010**, *352*, 1914–1919.

[37] D. Wang, Y. Wei, M. Shi, *Chem. Commun.* **2012**, *48*, 2764–2766.

[38] T. Dohi, A. Maruyama, N. Takenaga, K. Senami, Y. Minamitsuji, H. Fujioka, S. B. Caemmerer, Y. Kita, *Angew. Chem. Int. Ed.* **2008**, *47*, 3787–3790.

[39] (a) F. Xu, D. Huang, C. Han, W. Shen, X.-F. Lin, Y.-G. Wang, *J. Org. Chem.* **2010**, *75*, 8677–8680; (b) I. Čorić, S. Müller, B. List, *J. Am. Chem. Soc.* **2010**, *132*, 17370–17373; (c) C.-H. Xing, Y.-X. Liao, J. Ng, Q.-S. Hu, *J. Org. Chem.* **2011**, *76*, 4125–4131; (d) B. Xu, S.-F. Zhu, X.-L. Xie, J.-J. Shen, Q.-L. Zhou, *Angew. Chem. Int. Ed.* **2011**, *50*, 11483–11486; (e) S. Müller, M. J. Webber, B. List, *J. Am. Chem. Soc.* **2011**, *133*, 18534–18537; (f) D. M. Rubush, M. A. Morges, B. J. Rose, D. H. Thamm, T. Rovis, *J. Am. Chem. Soc.* **2012**, *134*, 13554–13557; (g) Z.-L. Chen, B.-L. Wang, Z.-B. Wang, G.-Y. Zhu, J.-W. Sun, *Angew. Chem. Int. Ed.* **2013**, *52*, 2027–2031; (h) X.-J. Li, Y.-Y. Zhao, H.-J. Qu, Z.-J. Mao, X.-F. Lin, *Chem. Commun.* **2013**, *49*, 1401–1403; (i) Q. Cai, X.-W. Liang, S.-G. Wang, S.-L. You, *Org. Biomol. Chem.* **2013**, *11*, 1602–1605; (j) B. Xu, S.-F. Zhu, Z.-C. Zhang, Q.-L. Zhou, *Chem. Sci.* **2014**, *5*, 1442–1448; (k) B. Xu, S.-F. Zhu, X.-D. Zuo, Z.-C. Zhang, Q.-L. Zhou, *Angew. Chem. Int. Ed.* **2014**, *53*, 3913–3916.

[40] X.-H. Huo, J.-H. Xie, Q.-S. Wang, Q.-L. Zhou, *Adv. Synth. Catal.* **2007**, *349*, 2477–2484.

[41] X. Cheng, Q. Zhang, J.-H. Xie, L.-X. Wang, Q.-L. Zhou, *Angew. Chem. Int. Ed.* **2005**, *44*, 1118–1121.

[42] (a) S.-L. Wu, W.-C. Zhang, Z.-G. Zhang, X.-M. Zhang, *Org. Lett.* **2004**, *6*, 3565–3567; (b) W.-C. Zhang, C.-J. Wang, W.-Z. Gao, X.-M. Zhang, *Tetrahedron Lett.* **2005**, *46*, 6087–6090.

[43] (a) Z. Freixa, M. S. Beentjes, G. D. Batema, C. B. Dieleman, G. P. F. van Strijdonck, J. N. H. Reek, P. C. J. Kamer, J. Fraanje, K. Goubitz, P. W. N. M. van Leeuwen, *Angew. Chem. Int. Ed.* **2003**, *42*, 1284–1287; (b) Z. Freixa, P. C. J. Kamer, M. Lutz, A. L. Spek, P. W. N. M. van Leeuwen, *Angew. Chem. Int. Ed.* **2005**, *44*, 4385–4388; (c) C. Jiménez-Rodríguez, F. X. Roca, C. Bo, J. Benet-Buchholz, E. C. Escudero-Adán, Z. Freixa, P. W. N. M. van Leeuwen, *Dalton Trans.* **2006**, 268–278; (d) X. Sala, E. J. G. Suárez, Z. Freixa, J. Benet-Buchholz, P. W. N. M. van Leeuwen, *Eur. J. Org. Chem.* **2008**, 6197–6205; (e) O. Jacquet, N. D. Clément, Z. Freixa, A. Ruiz, C. Claver, P. W. N. M. van Leeuwen, *Tetrahedron: Asymmetry* **2011**, *22*, 1490–1498; (f) O. Jacquet, N. D. Clément, C. Blanco, M. M. Belmonte, J. Benet-Buchholz, P. W. N. M. van Leeuwen, *Eur. J. Org. Chem.* **2012**, 4844–4852.

[44] (a) J. Li, G. Chen, Z. Wang, R.-Z. Zhang, X.-M. Zhang, K.-L. Ding, *Chem. Sci.* **2011**, *2*, 1141–1144; (b) J. Li, W. Pan, Z. Wang, X.-M. Zhang, K.-L. Ding, *Adv. Synth. Catal.* **2012**, *354*, 1980–1986.

[45] (a) X.-M. Wang, Z.-B. Han, Z. Wang, K.-L. Ding, *Angew. Chem. Int. Ed.* **2012**, *51*, 936–940; (b) X.-B. Wang, P.-H. Guo, X.-M. Wang, Z. Wang, K.-L. Ding, *Adv. Synth. Catal.* **2013**, *355*, 2900–2907.

[46] (a) X.-M. Wang, F.-Y. Meng, Y. Wang, Z.-B. Han, Y.-J. Chen, L. Liu, Z. Wang, K.-L. Ding, *Angew. Chem. Int. Ed.* **2012**, *51*, 9276–9282; (b) X.-M. Wang, P.-H. Guo, Z.-B. Han, X.-B. Wang, Z. Wang, K.-L. Ding, *J. Am. Chem. Soc.* **2013**, *136*, 405–411.

[47] Z.-Y. Cao, X.-M. Wang, C. Tan, X.-L. Zhao, J. Zhou, K.-L. Ding, *J. Am. Chem. Soc.* **2013**, *135*, 8197–8200.

[48] Z.-B. Han, Z. Wang, X.-M. Zhang, K.-L. Ding, *Sci. Sinica Chim.* **2010**, *40*, 950–955.

[49] Z.-B. Han, Z. Wang, X.-M. Zhang, K.-L. Ding, *Chinese Sci. Bull.* **2010**, *55*, 2840–2846.

[50] Z.-B. Han, Z. Wang, X.-M. Zhang, K.-L. Ding, *Angew. Chem. Int. Ed.* **2009**, *48*, 5345–5349.

[51] (a) Y. Zhang, Z.-B. Han, F.-Y. Li, K.-L. Ding, A. Zhang, *Chem. Commun.* **2010**, *46*, 156–158; (b) J. Shang, Z.-B. Han, Y. Li, Z. Wang, K.-L. Ding, *Chem. Commun.* **2012**, *48*, 5172–5174.

[52] A.-G. Hu, Y. Fu, J.-H. Xie, H. Zhou, L.-X. Wang, Q.-L. Zhou, *Angew. Chem. Int. Ed.* **2002**, *41*, 2348–2350.

[53] S.-F. Zhu, T. Liu, S. Yang, S. Song, Q.-L. Zhou, *Tetrahedron* **2012**, *68*, 7685–7690.

[54] G.-H. Hou, J.-H. Xie, P.-C. Yan, Q.-L. Zhou, *J. Am. Chem. Soc.* **2009**, *131*, 1366–1367.

[55] P.-C. Yan, J.-H. Xie, G.-H. Hou, L.-X. Wang, Q.-L. Zhou, *Adv. Synth. Catal.* **2009**, *351*, 3243–3250.

[56] W.-J. Tang, X.-M. Zhang, *Chem. Rev.* **2003**, *103*, 3029–3069.

[57] S. Khumsubdee, K. Burgess, *ACS Catal.* **2013**, *3*, 237–249.

[58] S. Yang, S.-F. Zhu, C.-M. Zhang, S. Song, Y.-B. Yu, S. Li, Q.-L. Zhou, *Tetrahedron* **2012**, *68*, 5172–5178.

[59] S. Li, S.-F. Zhu, J.-H. Xie, S. Song, C.-M. Zhang, Q.-L. Zhou, *J. Am. Chem. Soc.* **2010**, *132*, 1172–1179.

[60] S. Song, S.-F. Zhu, Z.-Y. Li, Q.-L. Zhou, *Org. Lett.* **2013**, *15*, 3722–3725.

[61] S. Song, S.-F. Zhu, L.-Y. Pu, Q.-L. Zhou, *Angew. Chem. Int. Ed.* **2013**, *52*, 6072–6075.

[62] S.-F. Zhu, Y.-B. Yu, S. Li, L.-X. Wang, Q.-L. Zhou, *Angew. Chem. Int. Ed.* **2012**, *51*, 8872–8875.

[63] S. Song, S.-F. Zhu, S. Yang, S. Li, Q.-L. Zhou, *Angew. Chem. Int. Ed.* **2012**, *51*, 2708–2711.

[64] (a) S. Song, S.-F. Zhu, Y.-B. Yu, Q.-L. Zhou, *Angew. Chem. Int. Ed.* **2013**, *52*, 1556–1559; (b) T. Besset, R. Gramage-Doria, J. N. H. Reek, *Angew. Chem. Int. Ed.* **2013**, *52*, 8795–8797.
[65] S. Yang, S.-F. Zhu, N. Guo, S. Song, Q.-L. Zhou, *Org. Biomol. Chem.* **2014**, *12*, 2049–2052.
[66] N. Arai, T. Ohkuma, *Chem. Rec.* **2012**, *12*, 284–289.
[67] R. Noyori, T. Ohkuma, *Angew. Chem. Int. Ed.* **2001**, *40*, 40–73.
[68] J.-H. Xie, X.-Y. Liu, X.-H. Yang, J.-B. Xie, L.-X. Wang, Q.-L. Zhou *Angew. Chem. Int. Ed.* **2012**, *51*, 201–203.
[69] X.-H. Yang, J.-H. Xie, W.-P. Liu, Q.-L. Zhou, *Angew. Chem. Int. Ed.* **2013**, *52*, 7833–7836.
[70] P.-C. Yan, G.-L. Zhu, J.-H. Xie, X.-D. Zhang, Q.-L. Zhou, Y.-Q. Li, W.-H. Shen, D.-Q. Che, *Org. Process. Res. Dev.* **2013**, *17*, 307–312.
[71] Q.-Q. Zhang, J.-H. Xie, X.-H. Yang, J.-B. Xie, Q.-L. Zhou, *Org. Lett.* **2012**, *14*, 6158–6161.
[72] J.-H. Xie, L.-C. Guo, X.-H. Yang, L.-X. Wang, Q.-L. Zhou, *Org. Lett.* **2012**, *14*, 4758–4761.
[73] (a) R. Noyori, M. Tokunaga, M. Kitamura, *Bull. Chem. Soc. Jpn.* **1995**, *68*, 36–56; (b) E. Vedejs, M. Jure, *Angew. Chem. Int. Ed.* **2005**, *44*, 3974–4001; (c) H. Pellissier, *Tetrahedron* **2008**, *64*, 1563–1601.
[74] J.-H. Xie, Z.-T. Zhou, W.-L. Kong, Q.-L. Zhou, *J. Am. Chem. Soc.* **2007**, *129*, 1868–1869.
[75] S. Liu, J.-H. Xie, L.-X. Wang, Q.-L. Zhou, *Angew. Chem. Int. Ed.* **2007**, *46*, 7506–7508.
[76] J.-H. Xie, S. Liu, W.-L. Kong, W.-J. Bai, X.-C. Wang, L.-X. Wang, Q.-L. Zhou, *J. Am. Chem. Soc.* **2009**, *131*, 4222–4223.
[77] S. Liu, J.-H. Xie, W. Li, W.-L. Kong, L.-X. Wang, Q.-L. Zhou, *Org. Lett.* **2009**, *11*, 4994–4997.
[78] W.-J. Bai, J.-H. Xie, Y.-L. Li, S. Liu, Q.-L. Zhou, *Adv. Synth. Catal.* **2010**, *352*, 81–84.
[79] C. Liu, J.-H. Xie, Y.-L. Li, J.-Q. Chen, Q.-L. Zhou, *Angew. Chem. Int. Ed.* **2013**, *52*, 593–596.
[80] J.-Q. Chen, J.-H. Xie, D.-H. Bao, S. Liu, Q.-L. Zhou, *Org. Lett.* **2012**, *14*, 2714–2717.
[81] L.-J. Cheng, J.-H. Xie, Y. Chen, L.-X. Wang, Q.-L. Zhou, *Org. Lett.* **2013**, *15*, 764–767.
[82] L.-J. Cheng, J.-H. Xie, L.-X. Wang, Q.-L. Zhou, *Adv. Synth. Catal.* **2012**, *354*, 1105–1113.
[83] G. Li, J.-H. Xie, J. Hou, S.-F. Zhu, Q.-L. Zhou, *Adv. Synth. Catal.* **2013**, *355*, 1597–1604.
[84] J.-H. Xie, P.-C. Yan, Q.-Q. Zhang, K.-X. Yuan, Q.-L. Zhou, *ACS Catal.* **2012**, *2*, 561–564.
[85] T.V. RajanBabu, *Chem. Rev.* **2003**, *103*, 2845–2860.
[86] W.-J. Shi, Q. Zhang, J.-H. Xie, S.-F. Zhu, G.-H. Hou, Q.-L. Zhou, *J. Am. Chem. Soc.* **2006**, *128*, 2780–2781.
[87] Q. Zhang, S.-F. Zhu, Y. Cai, L.-X. Wang, Q.-L. Zhou, *Sci. China Chem.* **2010**, *53*, 1899–1906.
[88] M. M. Coulter, K. G. M. Kou, B. Galligan, V. M. Dong, *J. Am. Chem. Soc.* **2010**, *132*, 16330–16333.
[89] K. Fagnou, M. Lautens, *Chem. Rev.* **2003**, *103*, 169–196.
[90] M. Sakai, M. Ueda, N. Miyaura, *Angew. Chem. Int. Ed.* **1998**, *37*, 3279–3281.
[91] H.-F. Duan, Y.-X. Jia, L.-X. Wang, Q.-L. Zhou, *Org. Lett.* **2006**, *8*, 2567–2569.
[92] H.-F. Duan, J.-H. Xie, X.-C. Qiao, L.-X. Wang, Q.-L. Zhou, *Angew. Chem. Int. Ed.* **2008**, *47*, 4351–4353.
[93] G. Zanoni, A. Pontiroli, A. Marchetti, G. Vidari, *Eur. J. Org. Chem.* **2007**, 3599–3611.
[94] S.-F. Zhu, X.-C. Qiao, Y.-Z. Zhang, L.-X. Wang, Q.-L. Zhou, *Chem. Sci.* **2011**, *2*, 1135–1140.
[95] X.-C. Qiao, S.-F. Zhu, Q.-L. Zhou, *Tetrahedron: Asymmetry* **2009**, *20*, 1254–1261.
[96] X.-C. Qiao, S.-F. Zhu, W.-Q. Chen, Q.-L. Zhou, *Tetrahedron: Asymmetry* **2010**, *21*, 1216–1220.
[97] R. M. Moslin, K. Miller-Moslin, T. F. Jamison, *Chem. Commun.* **2007**, 4441–4449.
[98] Y. Yang, S.-F. Zhu, C.-Y. Zhou, Q.-L. Zhou, *J. Am. Chem. Soc.* **2008**, *130*, 14052–14053.
[99] C.-Y. Zhou, S.-F. Zhu, L.-X. Wang, Q.-L. Zhou, *J. Am. Chem. Soc.* **2010**, *132*, 10955–10957.

[100] A. D. Melhado, G. W. Amarante, Z. J. Wang, M. Luparia, F. D. Toste, *J. Am. Chem. Soc.* **2011**, *133*, 3517–3527.

[101] B.-M. Fan, J.-H. Xie, S. Li, L.-X. Wang, Q.-L. Zhou, *Angew. Chem. Int. Ed.* **2007**, *46*, 1275–1277.

[102] A. Z. González, D. Benitez, E. Tkatchouk, W. A. Goddard III, F. D. Toste, *J. Am. Chem. Soc.* **2011**, *133*, 5500–5507.

[103] S. Suárez-Pantiga, C. Hernández-Díaz, E. Rubio, J. M. González, *Angew. Chem. Int. Ed.* **2012**, *51*, 11552–11555.

[104] J. Hu, Y. Lu, Y. Li, J. Zhou, *Chem. Commun.* **2013**, *49*, 9425–9427.

[105] J. Hu, H. Hirao, Y. Li, J. Zhou, *Angew. Chem. Int. Ed.* **2013**, *52*, 8676–8680.

[106] M. P. Doyle, M. A. McKervey, T. Ye, *Modern Catalytic Methods for Organic Synthesis with Diazocompounds*, John Wiley & Sons, Inc., New York, **1998**.

[107] (a) C. F. García, M. A. McKervey, T. Ye, *Chem. Commun.* **1996**, 1465–1466; (b) S. Bachmann, D. Fielenbach, K. A. Jørgensen, *Org. Biomol. Chem.* **2004**, *2*, 3044–3049.

[108] S.-F. Zhu, B. Xu, G.-P. Wang, Q.-L. Zhou, *J. Am. Chem. Soc.* **2012**, *134*, 436–442.

[109] C. Chen, S.-F. Zhu, B. Liu, L.-X. Wang, Q.-L. Zhou, *J. Am. Chem. Soc.* **2007**, *129*, 12616–12617.

[110] X.-L. Xie, S.-F. Zhu, J.-X. Guo, Y. Cai, Q.-L. Zhou, *Angew. Chem. Int. Ed.* **2014**, *53*, 2978–2981.

[111] S.-F. Zhu, Y. Cai, H.-X. Mao, J.-H. Xie, Q.-L. Zhou, *Nature Chem.* **2010**, *2*, 546–551.

[112] S.-F. Zhu, C. Chen, Y. Cai, Q.-L. Zhou, *Angew. Chem. Int. Ed.* **2008**, *47*, 932–934.

[113] S.-F. Zhu, X.-G. Song, Y. Li, Y. Cai, Q.-L. Zhou, *J. Am. Chem. Soc.* **2010**, *132*, 16374–16376.

[114] X.-G. Song, S.-F. Zhu, X.-L. Xie, Q.-L. Zhou, *Angew. Chem. Int. Ed.* **2013**, *52*, 2555–2558.

[115] Y.-Z. Zhang, S.-F. Zhu, Y. Cai, H.-X. Mao, Q.-L. Zhou, *Chem. Commun.* **2009**, 5362–5364.

[116] Q.-Q. Cheng, S.-F. Zhu, Y.-Z. Zhang, X.-L. Xie, Q.-L. Zhou, *J. Am. Chem. Soc.* **2013**, *135*, 14094–14097.

[117] (a) D. N. Mai, J. P. Wolfe, *J. Am. Chem. Soc.* **2010**, *132*, 12157–12159; (b) D. N. Mai, B. R. Rosen, J. P. Wolfe, *Org. Lett.* **2011**, *13*, 2932–2935; (c) B. A. Hopkins, J. P. Wolfe, *Angew. Chem. Int. Ed.* **2012**, *51*, 9886–9890;

[118] N. R. Babij, J. P. Wolfe, *Angew. Chem. Int. Ed.* **2013**, *52*, 9247–9250.

5

Application of Sterically Demanding Phosphine Ligands in Palladium-Catalyzed Cross-Coupling leading to C(sp^2)—E Bond Formation (E = NH_2, OH, and F)

Mark Stradiotto[1] and Rylan J. Lundgren[2]

[1]*Department of Chemistry, Dalhousie University, 6274 Coburg Road, PO Box 15000, Halifax, Nova Scotia, Canada B3H 4R2*
[2]*Department of Chemistry, Univeristy of Alberta, Edmonton, Alberta, Canada T6G 2G2*

5.1 Introduction

The development of palladium-catalyzed cross-coupling reactions involving (hetero) aryl (pseudo)halide and various nucleophilic reaction partners has revolutionized modern chemical synthesis, as acknowledged by the awarding of the Nobel Prize for Chemistry in 2010 to Richard F. Heck,[1] Ei-ichi Negishi,[2] and Akira Suzuki[3] for their pioneering work in establishing effective methodologies for the formation of carbon–carbon bonds.[4, 5] In the ensuing years a diversity of complementary palladium-catalyzed cross-coupling reactions leading to carbon–carbon and carbon–heteroatom bond formation have also been developed, with perhaps the most notable example of the latter being the arylation of NH-containing substrates (i.e., Buchwald–Hartwig

Ligand Design in Metal Chemistry: Reactivity and Catalysis, First Edition. Edited by Mark Stradiotto and Rylan J. Lundgren.

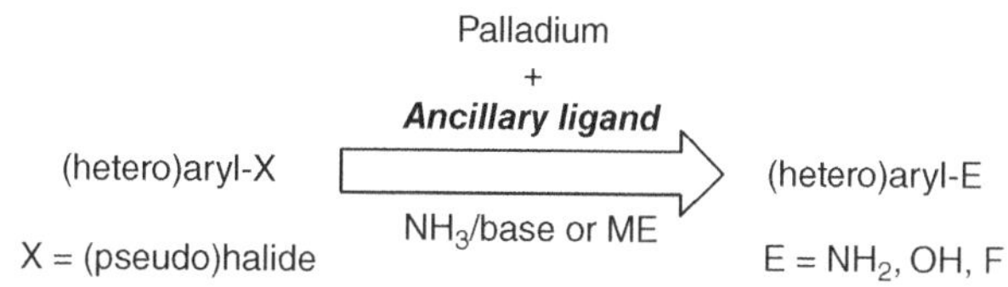

Figure 1 *Palladium-catalyzed C(sp^2)—E bond formation (E=NH_2, OH, and F)*

amination, BHA).[6–9] Collectively, such palladium-catalyzed carbon-element bond-forming methodologies enjoy widespread use in both academic and industrial settings in the synthesis of a diverse array of pharmaceuticals, natural products, and conjugated organic materials.[10]

Early breakthroughs within the field of palladium-catalyzed cross-coupling reactions typically employed catalysts featuring relatively simple ligands, such as triphenylphosphine. However, in the ensuing years an increased understanding of the mechanisms of such transformations has guided the development of much more structurally complex ancillary ligands that enable new and/or challenging substrate transformations to be achieved with greater selectivity and/or under increasingly mild conditions. Ancillary ligand design continues to play a central role in enabling advances in such palladium-catalyzed transformations,[11] and represents an ever-expanding area of research that spans the fields of organometallic chemistry, organic chemistry, and catalysis. In this context, any reasonable attempt to comprehensively document advances in ancillary ligand design within the domain of palladium-catalyzed cross-coupling chemistry would require a multi-volume monograph – *well beyond the scope herein.* In an alternative approach, we provide in this chapter a focused examination of ancillary ligand design within specifically in the context of palladium-catalyzed C(sp^2)—E bond formation (E=NH_2, OH, and F; Figure 1). Such sought-after cross-couplings involving small, rather challenging nucleophilic reaction partners have emerged only within the past decade, enabled by the judicious design and/or selection of supporting phosphorus-based ancillary ligands. We envisioned that a survey of ancillary ligand advances in this area would serve as an instructive case study regarding the development of effective palladium-catalyst systems where monoarylation selectivity and/or product-forming C—E reductive elimination may be limiting factors. Moreover, the featured examples also serve to highlight the utility of "repurposing" phosphine ligands developed in the context of mechanistically unrelated catalytic applications, and to illustrate the potential complications that can arise in "rational" ancillary ligand design when the ancillary ligand itself is transformed by the substrate in the course of palladium-catalyzed cross-coupling chemistry.

5.1.1 General mechanistic overview and ancillary ligand design considerations

The development of palladium-catalyzed BHA chemistry has been described in a number of comprehensive reviews[8, 9, 12, 13] and the mechanisms[14–17] of these transformations have been examined, including in the case of ammonia cross-couplings;[18] the

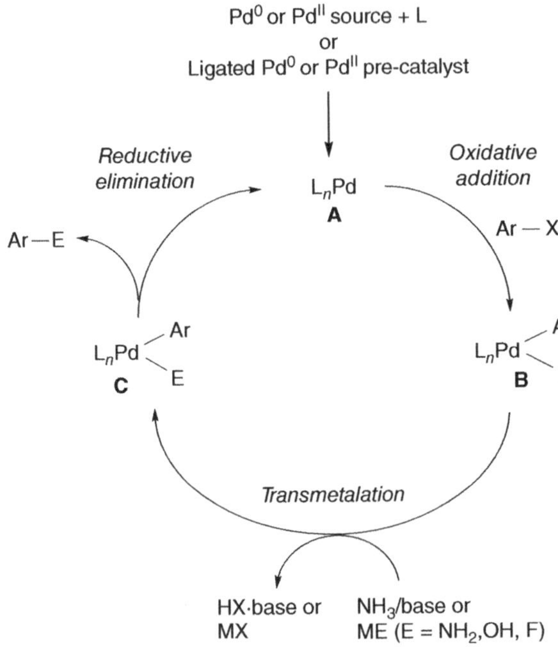

Scheme 1 *Presumed catalytic cycle for palladium-catalyzed C(sp^2)—E bond formation (E = NH_2, OH, and F). L_n = ancillary ligand(s); Ar—X = (hetero)aryl (pseudo)halide*

mechanisms of related nucleophilic C(sp^2)—OH and C(sp^2)—F bond-forming reactions are presumed to be conceptually related. Notwithstanding the fact that the precise nature of the elementary steps and catalytic intermediates can vary on the basis of the reactants and other experimental conditions employed, the mechanistic pathways of such reactions commonly proceed as outlined in Scheme 1. Oxidative addition of the (hetero)aryl (pseudo)halide (i.e., Ar-X) to a L_nPd(0) species (**A**) affords the Pd(II) intermediate **B**. In the case of BHA involving ammonia, transmetalation involves a stepwise process in which an ammine adduct of **B** is formed, followed by HX extrusion by an external base to afford a Pd—NH_2 intermediate of type **C**; alternatively, the pre-formed metal amide (i.e., MNH_2) can be employed directly. Subsequent C—N bond reductive elimination affords the aniline derivative with concomitant regeneration of **A**. Reports to date of the palladium-catalyzed nucleophilic hydroxylation or fluorination of (hetero)aryl (pseudo)halides have involved the use of MOH and MF reagents, which similarly intercept the oxidative addition intermediate **B**, followed by product-forming C—O/C—F bond reductive elimination.

As with all such palladium-catalyzed carbon–heteroatom bond-forming chemistry, the ancillary ligand(s) (i.e., L_n; often featuring phosphine or *N*-heterocyclic carbene donors[11, 19, 20]) employed have a direct influence over the course of the elementary transformations. Electron-rich and sterically demanding ligands promote the formation of low-coordinate compounds of type **A** that are predisposed to undergo Ar—X oxidative addition to give **B**. The steric and electronic properties of the ligands can serve to circumvent unwanted dimerization within low-coordinate intermediates of type **B**, as well as providing monoarylation selectivity by discouraging unwanted subsequent arylation reactions involving the primary aniline or phenol products. Product-forming C(sp^2)—E (E=NH_2, OH, and F) reductive elimination is faciliated by sterically demanding, yet less electron-donating ancillary ligands. From an electronic perspective, this requirement is orthogonal to the ligand demands associated with the Ar—X oxidative addition step. As depicted in Scheme 1, entrance into the catalytic cycle to afford **A** can be achieved by combining a Pd(II) source and the ancillary ligand, which requires that both L—Pd bond formation and reduction to Pd(0) occur smoothly. In some cases phosphines can be employed in excess as both the ancillary ligand and the sacrificial reductant, although the costly nature of some phosphine ligands makes such methods unattractive. While the use of a Pd(0) source circumvents the need for reduction, such complexes are often cumbersome to prepare, costly, and/or exhibit air-sensitivity, thereby preventing routine handling under benchtop conditions. Regardless of whether one uses a Pd(0) or Pd(II) precursor, in situ catalyst formation, while operationally simple, requires the efficient binding of the added ancillary ligand to palladium. When such reactions do not proceed cleanly, significant decomposition can occur such that the actual quantity of **A** successfully formed and available for catalysis is very low. In response, the use of pre-formed, well-characterized L_nPd(II) pre-catalyst complexes that can be reduced cleanly to **A** without consumption of the ancillary ligand has attracted considerable attention as a means of circumventing the aforementioned problems during catalyst activation.[21] Alternatively, the direct use of a pre-formed catalyst intermediate (e.g., **B** or structural analog), while in some cases more labor-intensive than the aforementioned protocols, serves to deliver all of the added palladium directly into the catalytic cycle. In all cases, ligands and/or pre-catalysts that can be handled and used under non-inert benchtop conditions are particularly valued by end-users.

5.1.2 Reactivity challenges

With the aforementioned mechanistic picture in mind, it is understandable how nucleophilic reaction partners that lack steric demand or that are highly basic, such as parent amide, hydroxide and fluoride, can present difficulties under standard palladium-catalyzed cross-coupling conditions. Such species can readily deactivate the metal center by binding irreversibly to palladium, by forming catalytically inactive bridging structures, or by displacing ancillary ligands that are required in order to achieve desired catalytic performance. As alluded to above, an additional and significant challenge associated with the pursuit of monoarylation selectivity is that the product primary aniline or phenol are potentially contending, and for most palladium catalysts are superior, cross-coupling reaction partners. The result is that reactions can proceed in an uncontrolled manner to form polyarylated products even in the presence of limiting (hetero)aryl electrophile. Given that reductive eliminations from Pd(II) intermediates of type **C** to liberate product molecules is driven in part by the release of steric strain, smaller nucleophilic fragments (i.e., E=NH_2, OH, F in **C**) will have a reduced propensity to undergo such processes efficiently. Despite the abovementioned challenges, effective protocols for the challenging palladium-catalyzed cross-couplings depicted in Scheme 1 have been developed. In all cases the judicious choice of ancillary ligand that prevents catalyst deactivation and promotes selectivity proved critical in enabling such reactivity advances.

5.2 Palladium-catalyzed selective monoarylation of ammonia

Palladium-catalyzed amine arylation by use of (hetero)aryl (pseudo)halides (i.e., BHA) has evolved into a powerful tool for the synthesis of aniline derivatives.[8, 9] Prior to the development of BHA chemistry, the construction of C(sp^2)—N bonds was limited primarily to arene nitration–reduction reaction sequences, as well as nucleophilic aromatic substitutions employing selected amine nucleophiles in combination with electron-poor, and thus highly activated, (hetero)aryl halides. While these conventional synthetic methods can be employed with success in the synthesis of C(sp^2)—N linkages, they suffer from a number of important limitations, including low substrate scope and poor functional group tolerance. The emergence of BHA protocols circumvented these limitations, enabling reactions to be conducted efficiently and under mild, user-friendly conditions employing amines directly in combination with structurally diverse (hetero)aryl (pseudo)halides.

Primary (hetero)anilines represent common molecular fragments in biologically active compounds, and can also serve as synthons in the rational construction of unsymmetrical secondary and tertiary aryl amines with varied application;[10, 22, 23] two selected examples are presented in Figure 2. In this regard it is not surprising that the pursuit of BHA protocols for the construction of primary (hetero)anilines has attracted considerable attention. Whereas the use of "ammonia equivalents"[24] has been applied successfully in BHA chemistry, including by Alabanza *et al.*[25] at Hoffmann-La Roche in the assembly of the synthetic intermediate depicted in Figure 2, the direct and more atom-economical use of ammonia – the most abundant N—H reagent from which all synthetic nitrogen compounds are derived[26] – in the synthesis of primary (hetero)anilines has proven to be particularly challenging. Indeed, many metal-catalyzed chemical transformations, including BHA, for which ample precedent exists involving other

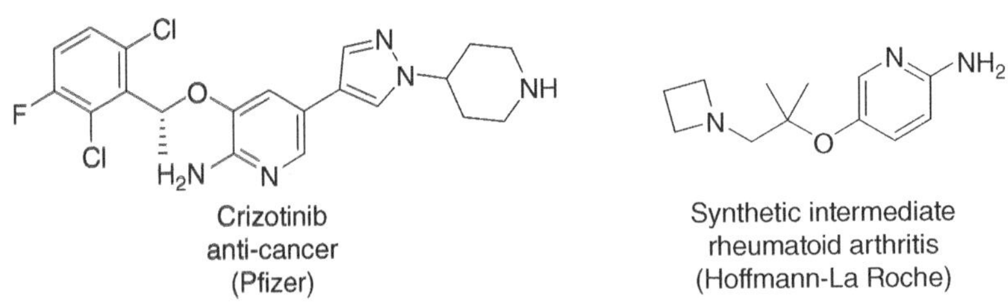

Figure 2 Selected examples of pharmaceutically relevant primary (hetero)anilines

amine classes do not proceed with useful efficiency and selectivity when employing ammonia under commonly employed reaction conditions.[27–29] Building on the themes outlined in Section 5.1.2, potential challenges associated with the use of ammonia in BHA chemistry include but are not restricted to: deactivation of the catalyst resulting from ancillary ligand dissociation and the formation of Werner-type ammine adducts; aggregation of amido intermediates of type **C** (Scheme 1, where E=NH_2) leading to inactive polynuclear complexes; the slow rate of reductive elimination from sterically unencumbered intermediates of type **C**;[18] and, uncontrolled polyarylation arising due to the competitive nature of the product (hetero)anilines relative to ammonia when using most commonly employed BHA catalysts.[27–29] While the use of copper-based catalysts does allow for the direct cross-coupling of (hetero)aryl bromides and iodides with ammonia,[29, 30] such catalysts have thus far proven incapable of accommodating analogous chloride or sulfonate reagents in a useful manner, thereby limiting the applicability of such methods. Notwithstanding these notable challenges, the development of useful palladium (and nickel, see below) catalysts that enable the selective monoarylation of ammonia with a broad range of (hetero)aryl (pseudo)halides has been realized in recent years, enabled by the design and/or application of appropriately configured ancillary ligands.

5.2.1 Initial development

The palladium-catalyzed selective monoarylation of ammonia was first described in 2006 (Scheme 2).[31] In this report, Shen and Hartwig successfully employed the Pd(II) pre-catalyst (CyPF-*t*Bu)$PdCl_2$ featuring the commercially available, air-stable JosiPhos ligand, CyPF-*t*Bu (**L1**). The JosiPhos ligand family was developed initially by Solvias for use in the asymmetric hydrogenation of alkenes on an industrial scale; as a result, several members of the JosiPhos ligand family are now commercially available.[32] In this regard, the use of **L1** in mechanistically unrelated BHA chemistry represents an

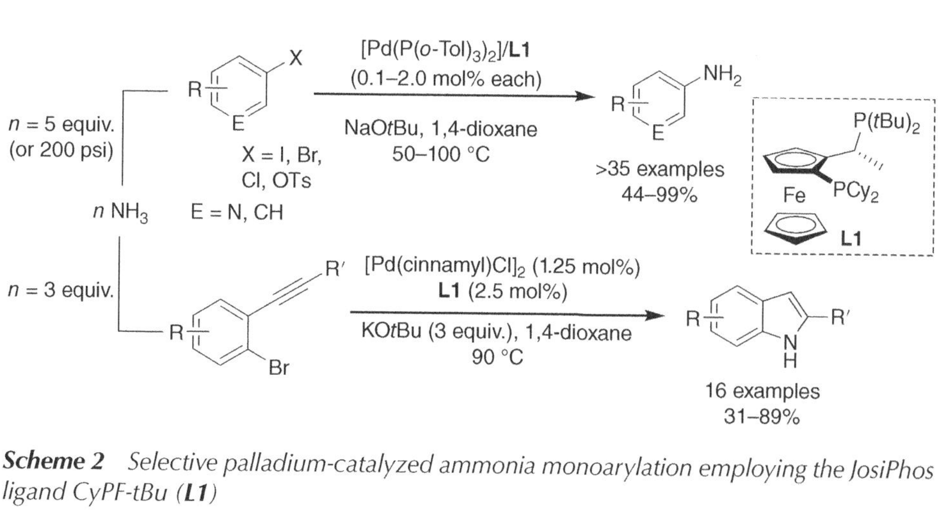

Scheme 2 *Selective palladium-catalyzed ammonia monoarylation employing the JosiPhos ligand CyPF-tBu (**L1**)*

effective "repurposing" of ancillary ligands, and underscores the scientific benefit of commercialization, whereby ligands can be purchased and screened for catalytic utility in a manner that is divergent from their initial intended use, leading to fundamental new breakthroughs in reactivity. While the application of the chiral (non-racemic) ligand **L1** represents a curious choice in BHA given the lack of stereocenters in the cross-coupling products, it is worthy of note that the use of alternative monodentate phosphine ($P(tBu)_3$, XPhos, QPhos) or *N*-heterocyclic carbene (IPr) ligands, or bisphosphines (DPPF, BINAP) resulted in negligible product formation under the catalyst screening conditions employed (80 psi ammonia, 90 °C).[31] These observations suggest that the rigid, sterically demanding, and electron-rich nature of **L1** discourages unwanted polyarylation, as well as possibly increasing catalyst lifetime. Improvements to this chemistry were achieved subsequently through the use of $Pd(P(o\text{-}Tol)_3)_2$/**L1** as the catalyst mixture,[33] thereby allowing for the efficient monoarylation of ammonia using aryl bromides, chlorides, iodides, and tosylates, including substrates featuring base-sensitive groups, without the routine need for high ammonia pressures (Scheme 2); the ability to utilize ammonium salts in place of ammonia was also subsequently demonstrated.[34] Nonethelsss, the need for relatively high reaction temperatures and the small demonstrated scope in the heteroaryl (pseudo)halide reaction partner left room for improvement in terms of ancillary ligand design. Stoichiometric reactivity and kinetics investigations of ammonia cross-coupling reactions employing Pd/**L1** established that the catalyst resting state is an (**L1**)Pd(aryl)(NH_2) complex of type **C** ($E=NH_2$, Scheme 1).[18] Notably, the first reports of nickel-catalyzed ammonia monoarylation appeared in 2015, featuring catalyst systems supported by JosiPhos ancillary ligands.[35, 36]

5.2.2 Applications in heterocycle synthesis

Indoles are among the most scrutinized core structures in all of medicinal chemistry;[37, 38] in this regard there is significant interest in developing efficient synthetic routes to such complexes beyond the classical Fischer indole synthesis.[39] Inspired by the successful application of **L1** in palladium-catalyzed ammonia monoarylation (see Section 5.2.1), Stradiotto and co-workers demonstrated that $[Pd(cinnamyl)Cl]_2$/**L1** catalyst mixtures could be applied successfully in analogous reactions employing functionalized 2-bromoarylacetylenes,[40] with K*Ot*Bu-catalyzed hydroamination of the putative 2-aminoarylacetylene intermediates affording NH-indoles in a one-pot process (Scheme 2). Notably, this represented the first reported synthesis of the indole framework directly from ammonia employing metal-catalyzed cross-coupling. However, the lack of success in this chemistry when using 2-chloroarylacetylenes, heteroaryl halides, or 2-bromoarylacetylenes featuring sp^3-substituents at the alkynyl terminus highlighted important limitations of this $[Pd(cinnamyl)Cl]_2$/**L1** catalyzed protocol.

The air-stable BippyPhos ligand (**L2**) was developed initially by Singer and co-workers[41,42] at Pfizer as a non-proprietary ligand for use in palladium-catalyzed cross-coupling applications. Stradiotto and co-workers subsequently reported on the successful application of **L2** in a broad spectrum of BHA applications.[43] Included in the substrate scope were challenging NH-containing reagents featuring divergent steric and electronic properties, including ammonia and NH indoles. The unique ability of the $[Pd(cinnamyl)Cl]_2$/**L2** catalyst system to accommodate such divergent nucleophilic partners was exploited in the development of a novel one-pot synthesis of *N*-aryl indoles and related heterocyclic derivatives involving three C—N bond-forming steps, the first of which being the selective monoarylation of ammonia (Scheme 3). Two complementary routes for accessing *N*-arylated indoles from ammonia in this manner were developed: (a) ammonia monoarylation with a 2-halo(hetero)arylacetylene in the presence of excess base to form an NH indole that was subsequently cross-coupled with an aryl halide to form the corresponding *N*-arylated indole (Method A, Scheme 3); and (b) monoarylation of ammonia with an aryl halide to form an aniline that in turn was cross-coupled with a 2-halo(hetero)arylacetylene in the presence of excess base to form the corresponding substituted *N*-arylated indole (Method B, Scheme 3). These protocols provided access to a range of functionalized *N*-arylated indoles and related heterocyclic compounds in synthetically useful isolated yields. It is worthy of mention that this ammonia monoarylation chemistry involving 2-halo(hetero)arylacetylenes does not exhibit the substrate scope limitations encountered when using Pd(cinnamyl)

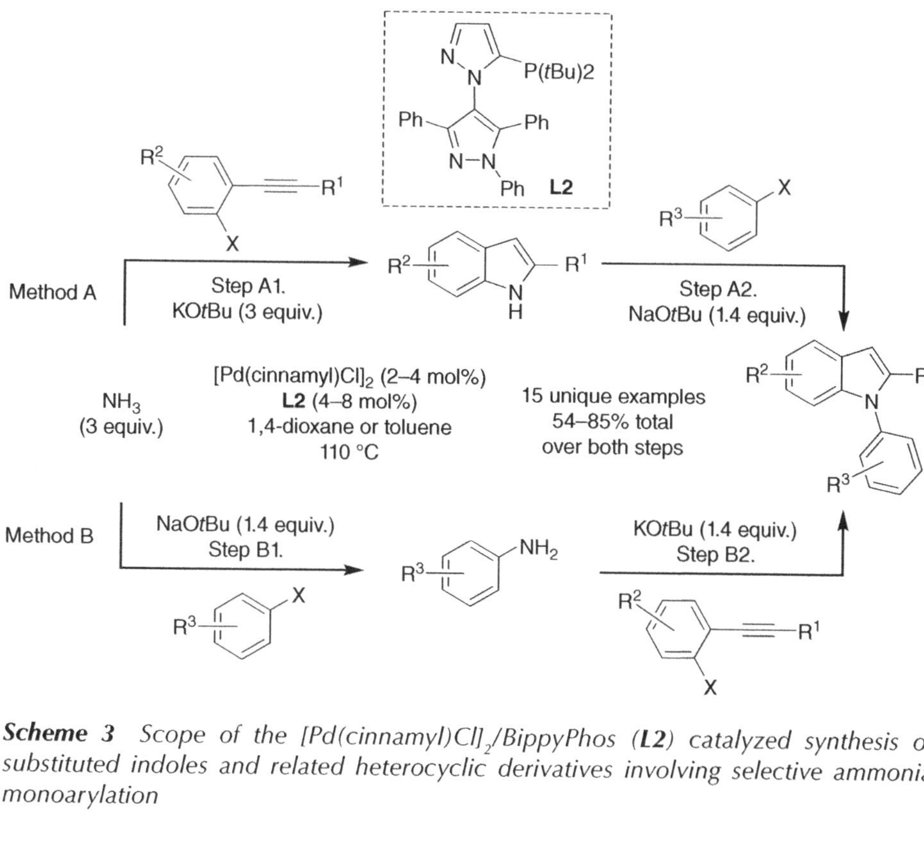

Scheme 3 *Scope of the [Pd(cinnamyl)Cl]$_2$/BippyPhos (**L2**) catalyzed synthesis of substituted indoles and related heterocyclic derivatives involving selective ammonia monoarylation*

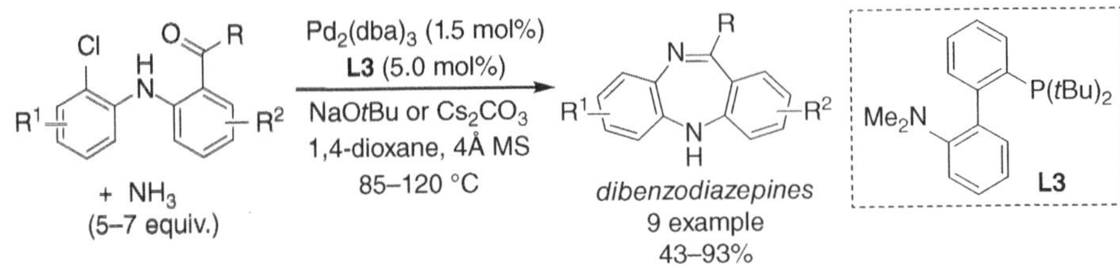

Scheme 4 *Application of ammonia monoarylation in the synthesis of dibenzodiazepines*

$Cl]_2$/**L1** (Scheme 2),[40] in that 2-chloro(hetero)arylacetylenes and substrates featuring sp^3-substituents at the alkynyl terminus were successfully transformed in the presence of [Pd(cinnamyl)$Cl]_2$/**L2** catalyst mixtures.[43]

Air-stable biaryl monophosphine ligands developed by Buchwald and co-workers have played a central role in the advancement of palladium-catalyzed cross-coupling reactions, including BHA chemistry;[9, 12, 13] the utility of appropriately configured ancillary ligands of this type in the selective monoarylation of ammonia has been examined.[44–46] Following on preliminary experimentation that established the capability of the $Pd_2(dba)_3$/**L3** catalyst system (dba = dibenzylideneacetone) to promote the monoarylation of ammonia at elevated temperature (80 °C),[44] Tsvelikhovsky and Buchwald[45] applied such reactivity in the synthesis of dibenzodiazepines and related biologically active structural analogs (Scheme 4). The propensity of biaryl monophosphine ligands including **L3**[47] to bind Pd(0) or Pd(II) via phosphorus and one or more carbon atoms of the lower flanking arene ring has been established;[9] from an ancillary design perspective, the specific involvement of the presumably uncoordinated dimethylamino group in **L3** in promoting ammonia monoarylation selectivity remains unclear.

5.2.3 Application of Buchwald palladacycles and imidazole-derived monophosphines

As mentioned in Section 5.1.1, the use of well-characterized pre-catalyst complexes featuring pre-formed ancillary ligand–palladium connectivity is advantageous, relative to alternative methods employing admixtures of an ancillary ligand and a palladium source, in terms of optimizing the amount of active catalyst present at the onset of cross-coupling.[21] Despite the added effort associated with the preparation of such pre-catalysts, the derived reactivity benefits have resulted in considerable interest in this area of catalyst development. Aminobiphenyl palladacyclic pre-catalysts featuring bulky biaryl monophosphine ligands, such as **P1–P3** in Scheme 5, are among the most widely used in the field of cross-coupling.[48, 49] Such air-stable pre-catalyst complexes are conveniently reduced to a mixture of putative "LPd(0)" (i.e., **A** in Scheme 1) and carbazole, simply via HX extrusion by the base used under the cross-coupling conditions. In this way, no external reductant is required – although the NH carbazole by-product can be problematic.[50] The application of such palladacyclic pre-catalysts toward the

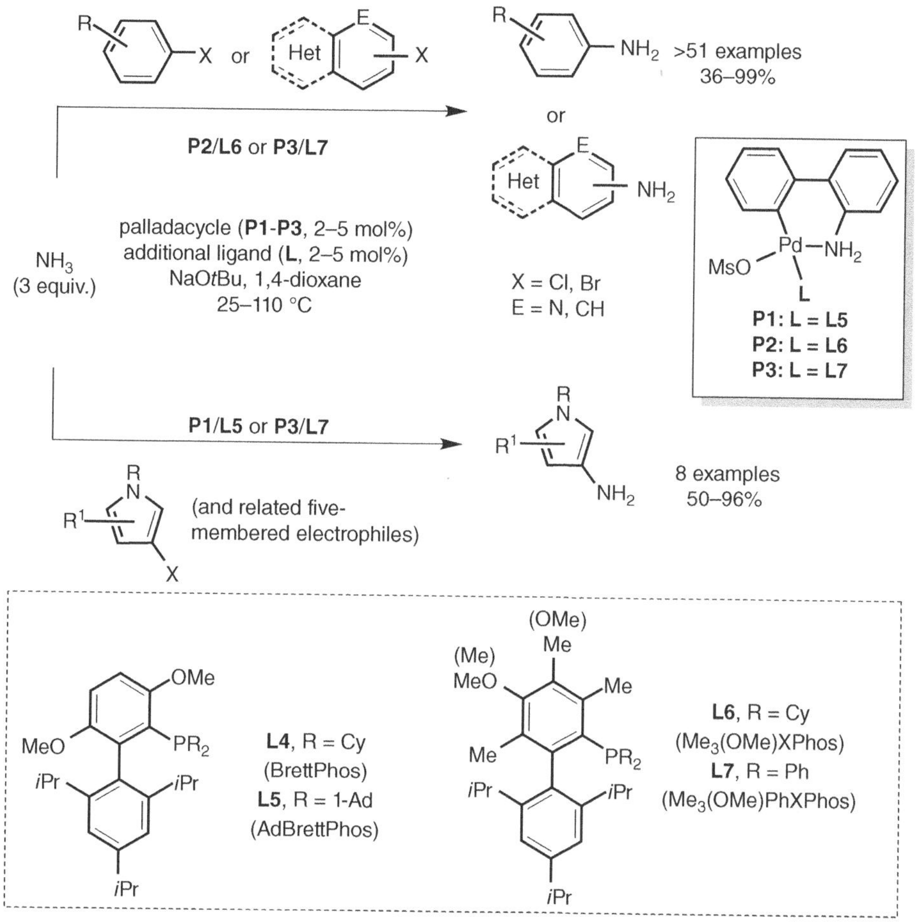

Scheme 5 Expanding the scope of ammonia monoarylation by use of biaryl monophosphine-ligated palladacyclic pre-catalysts. 1-Ad = 1-adamantyl

selective monoarylation of ammonia was examined by Buchwald and co-workers in 2013.[46] While BrettPhos (**L4**) had been described in a previous publication by Fors and Buchwald[51] as being ineffective for the monoarylation of ammonia, catalyst screening studies confirmed the superiority of AdBrettPhos (**L5**), Me_3(OMe)XPhos (**L6**), Me_3(OMe)PhXPhos (**L7**) and other structurally related biaryl monophosphine ligands relative to **L3** in the selective monoarylation of ammonia with chlorobenzene using $Pd_2(dba)_3$ as the palladium source (3 equiv. NH_3; 2 mol% Pd; 5 mol% ligand).[46] Further improvements in catalytic performance were achieved through the use of the pre-catalysts **P1**, **P2**, and **P3** (featuring **L5**, **L6**, and **L7**, respectively, Scheme 5). The use of **P2/L6** (2 mol% each; i.e., pre-catalyst with added ancillary ligand) in place of $Pd_2(dba)_3$/**L6** (1 and 4 mol%, respectively) afforded faster rates of reaction and higher monoarylation product yields in the cross-coupling of ammonia with (hetero)aryl halides. The **P2/L6** catalyst system proved useful in the selective monoarylation of ammonia employing electron-rich, -neutral, and -deficient aryl chlorides and bromides (Scheme 5). For more sterically hindered (hetero)aryl chlorides where the performance of **P2/L6** was shown to be relatively poor, the **P3/L7** catalyst mixture was employed as a means of obtaining higher yields of the target (hetero)aniline derivative. Reaction protocols employing **P2/L6** or **P3/L7** catalyst mixtures also proved effective with regard to the cross-coupling of six-membered heteroaryl bromides and chlorides with ammonia, although in some particularly challenging cases diminished monoarylation selectivity was observed. A diverse array of aminopyridines, aminoquinolines, and related NH_2-functionalized heterocycles including benzothiophene, indole, benzothiazole, benzoxazole, pyrazine, quinoxaline, pyrimidine, pyridazine, and carbazole rings were prepared (Scheme 5). While base-sensitive functionalities such as cyano and carbonyl groups, as well as heterocyclic addenda, were reasonably well-tolerated in this chemistry, room temperature reactions were limited to a relatively small number of examples featuring primarily aryl bromides and electronically activated (hetero)aryl chlorides at higher catalyst loading (typically 5 mol%).[46]

The development of new and effective cross-coupling protocols involving five-membered heteroaryl halide reaction partners arises from the utility of such transformations in the synthesis of biologically active functionalized heterocycles. However, such substrates have proven to be particularly challenging in the context of BHA chemistry. In the pursuit of a catalyst system capable of enabling the hitherto unknown BHA of five-membered heteroaryl halides with ammonia, Buchwald and co-workers conducted further ligand screenings.[46] Whereas catalysts based on Me_3(OMe)XPhos (**L6**) performed rather poorly in the cross-coupling of ammonia with 4-bromo-1-(4-fluorophenyl)-pyrazole (9% monoarylation), the di(1-adamantyl)phosphino-functionalized BrettPhos ligand variant **L5** afforded high yield (78%) of the desired ammonia monoarylation product. The use of **P1/L5** catalyst mixtures (2 mol% each) was exploited in ammonia monoarylation employing a range of five-membered heteroaryl bromides and chlorides including benzothiazoles, indazoles, imidazoles, and pyrazoles (Scheme 5). The cross-coupling of the rather hindered 4-bromo-1,3,5-trimethylpyrazole with ammonia proved challenging when using the **P1/L5** catalyst system (40%); for this and another tri-substituted pyrazole substrate, the use of **P3/L7** afforded the desired monoarylation product in good yield (78 and 82%).[46] Collectively, the aforementioned work by the Buchwald group in the area of palladium-catalyzed ammonia monoarylation highlights the benefits employing a highly tunable ancillary ligand motif, whereby structural changes to the ancillary ligand can be introduced as a means of modifying catalytic behavior so as to circumvent substrate scope challenges that may arise.

The Beller group has contributed to the development of effective ancillary ligands for use in the palladium-catalyzed monoarylation of ammonia. In a pair of publications[52, 53] it was demonstrated that appropriately constructed imidazole-derived monophosphine ligands (including **L8**) are capable of supporting active complexes for the monoarylation of ammonia, albeit under somewhat forcing conditions (≥120 °C; 10 bar N_2).

5.2.4 Heterobidentate κ²-P,N ligands: chemoselectivity and room temperature reactions

In light of the successful application of sterically demanding monophosphine and bisphosphine ligands in BHA chemistry, including with applications in ammonia monoarylation (see above), it is surprising that what can be viewed as electronically intermediate heterobidentate ligands featuring pairings of soft and hard donor atoms have received little attention in such applications. This is especially true in the context of the proven utility of κ²-P,N ligands in alternative late metal-catalyzed chemical transformations (e.g., PHOX[54] in alkene hydrogenation and allylic substitution). Sterically demanding κ²-P,N ligands represent attractive targets of inquiry in palladium-catalyzed ammonia monoarylation chemistry, given their potential to discourage unwanted dimerization of catalytic intermediates (e.g., **B** and **C** in Scheme 1), which can be problematic when using some monophosphine ancillary ligands. Moreover, while challenging (hetero)aryl (pseudo)halide oxidative additions can be enabled via incorporation of an electron-rich dialkylphosphino ligand donor fragment, a κ²-P,N ligand of this type featuring an adjacent nitrogen donor should also render the palladium centers in intermediates of type **C** (Scheme 1) less electron-rich than their bisphosphine-ligated counterparts, thereby providing an electronic means of promoting challenging reductive eliminations in addition to sterically promoted processes (see Section 5.1.1). Sterically demanding κ²-P,N ligands may also offer an effective means of acheiving selectivity in ammonia monoarylation reactions, by favoring the binding of ammonia over the aniline product.

The successful application of sterically demanding κ²-P,N ligands in the palladium-catalyzed selective monoarylation of ammonia was first described by Lundgren *et al.*[55, 56] In an initial report,[55] air-stable Me-DalPhos (**L9**) was shown to be effective in BHA chemistry involving a wide range of (hetero)aryl chlorides and NH-containing substrates, including ammonia. However, while **L9** afforded high conversions and good monoarylation selectivities in the cross-coupling of ammonia with *ortho*-substituted aryl chlorides, the use of aryl chlorides lacking steric bias resulted in diminished monoarylation selectivity.[55] Subsequent ancillary ligand structural optimization, enabled by the modular synthesis of DalPhos ligands, gave rise to air-stable Mor-DalPhos (**L10**), which is highly effective for palladium-catalyzed ammonia monoarylation involving aryl chlorides and tosylates, including those lacking *ortho* substitution; included in this report are the first examples of room temperature BHA chemistry involving ammonia (Scheme 6).[56] Electron-rich aryl chlorides that had proven to be challenging in previous reports of ammonia monoarylation chemistry were effectively cross-coupled, as were substrates containing N-, O-, F- or S- heteroatoms. Sterically biased *ortho*-substituted aryl chlorides were also found to be suitable reaction partners, as were some heteroaryl chlorides. The propensity of the $[Pd(cinnamyl)Cl]_2$/**L10** catalyst system for ammonia monoarylation was exploited in unprecedented chemoselective aminations involving aryl chloride substrates featuring potentially competitive NH- functionalities, including those featuring contending primary aryl- or alkylamino groups. While the use of $[Pd(cinnamyl)Cl]_2$/**L10** catalyst mixtures in the monoarylation

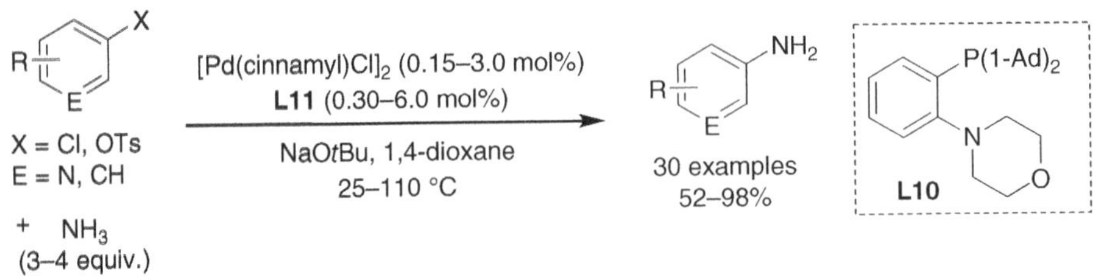

Scheme 6 *Ammonia monoarylation employing the* κ^2*-P,N ligand Mor-DalPhos (**L10**; 1-Ad = 1-adamantyl)*

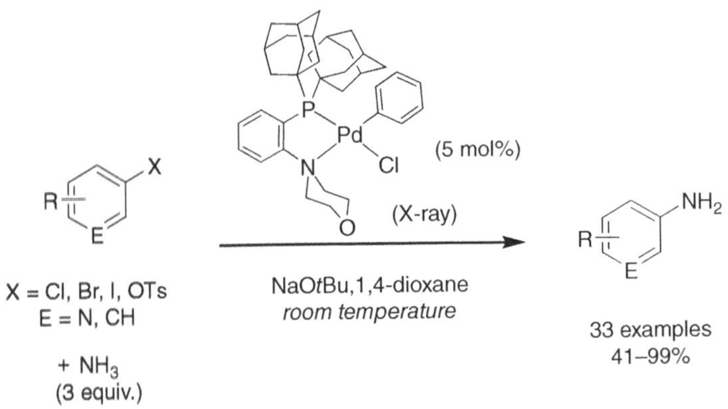

Scheme 7 *Room temperature ammonia monoarylation employing (*κ^2*-P,N-**L10**)Pd(Ph)Cl as a pre-catalyst*

of ammonia with (hetero)aryl chlorides required heating (≥50 °C), analogous reactions involving aryl tosylates were found to proceed at room temperature with good yields (69–83%). The practical nature of ammonia monoarylation employing the [Pd(cinnamyl) Cl]$_2$/**L10** catalyst system was further demonstrated by Frontier and co-workers in the first synthesis of (+/-)-tetrapetalone A-Me aglycon.[57]

Stradiotto and co-workers also explored the coordination chemistry of **L10**, including the synthesis of putative catalytic intermediates.[56] In using the rationally prepared and air-stable oxidative addition complex (κ^2-P,N-**L10**)Pd(Ph)Cl (cf. **B** in Scheme 1) as a pre-catalyst, the first examples of room temperature ammonia monoarylation employing (hetero)aryl chlorides was achieved. Stradiotto and co-workers subsequently established expanded scope for room temperature ammonia monoarylation reactivity when using (κ^2-P,N-**L10**)Pd(Ph)Cl as a pre-catalyst (5 mol%), including a range of (hetero)aryl (pseudo)halides (X = Cl, Br, I, OTs) with diverse substituents (alkyl, aryl, ether, thioether, ketone, amine, fluoro, trifluoromethyl, and nitrile), as well as chemoselective monoarylations (Scheme 7).[58] While a complete understanding of the properties of (κ^2-P,N-**L10**)Pd(Ph)Cl that promote efficient room temperature ammonia monoarylation with (hetero)aryl (pseudo)halides is currently lacking, including kinetic data, it is conceivable that the direct use of this putative catalytic intermediate serves to by-pass deleterious side reactions that may otherwise occur during catalyst activation steps. Moreover, it is feasible that the κ^2-P,N ancillary ligand motif provides the correct balance in terms of enabling both Ar-X oxidative addition and Ar-NH_2 reductive elimination steps relative to alterative monophosphine and bisphosphine ancillary ligand classes.

5.2.5 Summary

Over the past ten years, several highly effective palladium catalyst systems for the otherwise challenging monoarylation of ammonia using (hetero)aryl (pseudo)halide reaction partners have been identified. Such catalysts have enabled a broad spectrum of aryl electrophiles to be accommodated, including transformations that are highly chemoselective, proceed at room temperature, and/or that can be exploited in the assembly of synthetically important heterocyclic frameworks including natural products. The appropriate selection of ancillary ligand proved critical in achieving such reactivity breakthroughs, with sterically demanding and electron-rich monophosphines, heterobidentate κ^2-P,N ligands, and bisphosphines each proving successful in this regard. The reactivity advantages that can be derived from employing structurally well-defined pre-catalyst complexes, in which the ancillary ligand is coordinated to palladium, are also demonstrated in this chemistry.

5.3 Palladium-catalyzed selective hydroxylation of (hetero)aryl halides

Phenols, much like anilines, represent important synthons for the construction of biologically active compounds and functional materials.[59–61] Moreover, the phenol moiety itself is present in a number of top selling pharmaceuticals with diverse function; selected examples are depicted in Figure 3. The use of water, or related MOH salts, as nucleophiles in palladium-catalyzed cross-couplings involving (hetero)aryl halides conceptually represents a mild and selective route to phenols that is complementary to more established phenol syntheses,[59–61] including oxidative protocols. However, the difficulties associated with realizing such transformations mirror those

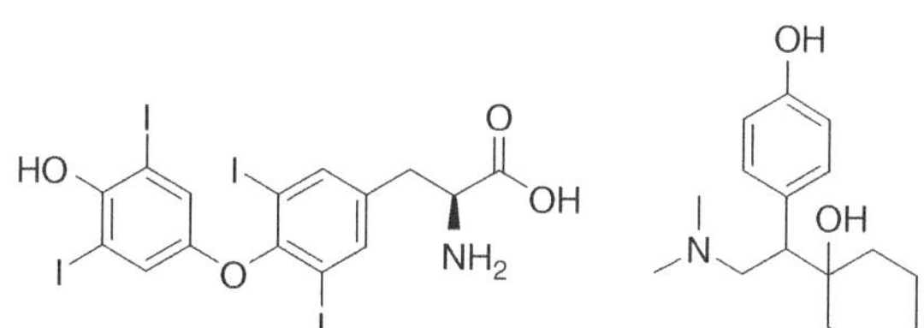

Levothyroxine
thyroid preparation
(Merck)

Desvenlafaxine
anti-depressant
(Pfizer)

Figure 3 Selected examples of pharmaceutically relevant phenol derivatives

of the ammonia-based transformations (see Section 5.2).[62] As such, the catalyst employed must be judiciously chosen so as to overcome potential challenges, including: catalyst inhibition by free hydroxide anions; difficult C—OH bond reductive elimination owing to the small size of the hydroxide group; and uncontrolled arylation of the target phenol to afford the undesired diaryl ether. While copper catalysts have been identified for the hydroxylation of (hetero)aryl halides,[61,63] the need for high metal/ligand loadings and harsh reaction conditions, as well as their typically poor performance with synthetically useful aryl chlorides, represent important practical drawbacks. Despite these considerable challenges, the palladium-catalyzed hydroxylation of (hetero)aryl halides has been established, owing in part to the availability of suitably constructed ancillary ligands.

5.3.1 Initial development

The first examples of the selective palladium-catalyzed hydroxylation of (hetero)aryl halides were reported by Anderson *et al.*[64] in 2006, by using XPhos-type biaryl monophosphines *t*BuXPhos (**L11**) or Me_4*t*BuXPhos (**L12**) in combination with Pd_2dba_3 in the presence of KOH in H_2O/1,4-dioxane solvent mixtures at 100 °C (Scheme 8). Some interesting reactivity trends with regard to ancillary ligand design were observed in comparing **L11** and **L12**. The less hindered **L11** proved to be more effective than **L12** with bulky substrates such as 2-bromomesitylene (88% isolated yield for **L11**; <10% conversion of the aryl bromide when using **L12**). In such transformations, the smaller

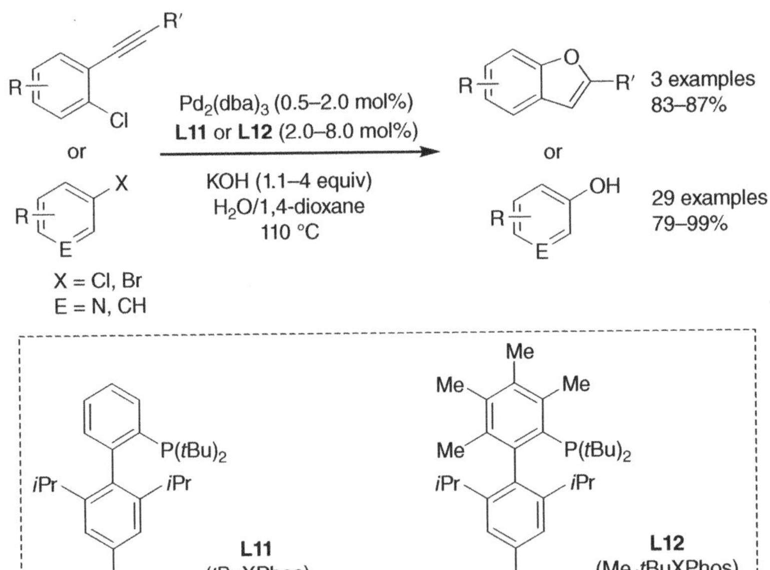

Scheme 8 *Palladium-catalyzed cross-coupling of (hetero)aryl bromides and chlorides with KOH employing XPhos-type biaryl monophosphines*

ligand **L11** apparently is better suited to accommodate the oxidative addition of the hindered aryl bromide, with C—OH reductive elimination being facilitated by *ortho* substitution within the mesityl group. However, catalysts incorporating the bulkier **L12** in general proved to be more robust and efficient than those of **L11**, as evidenced by the lower catalyst loadings that could be employed and the reduced observed formation of Pd-black. The authors postulate that the increased efficacy of **L12**-derived catalysts can be attributed to a faster rate of C—OH reductive elimination within (**L**)Pd(aryl)OH intermediates (cf. **C** in Scheme 1). Moreover, preliminary mechanistic experimentation suggests that the desired phenol product predominates when using KOH (2.0 equiv.) by virtue of the fact that attack on the (**L**)Pd(aryl)X intermediate by hydroxide is faster than attack by the phenoxide that is generated in situ under the reaction conditions employed. Tandem transformations of the newly formed phenol to generate benzofurans or alkyl aryl ethers were also established. While nitrile, methoxy, trifluoromethyl, carboxylic acid, methyl ketone, and aldehyde functionalities were each well-tolerated in this chemistry, only two transformations involving heteroaryl halides were presented.[64]

A subsequent report by Cheung and Buchwald[65] focused on the application of a *t*BuBrettPhos (**L13**) ligated palladacyclic pre-catalyst for the hydroxylation of (hetero) aryl halides (Scheme 9). The use of this **L13**-based catalyst system, which exploits in part developments by Sergeev *et al.*[66] including the use of CsOH in enabling room temperature reactivity (see below), allows for the use of lower palladium loadings and a more broad substrate scope of heteroaryl halide coupling partners in comparison catalyst systems based on mixtures of Pd_2dba_3 and either **L11** or **L12** (Scheme 8). Buitrago *et al.*[67] at Merck subsequently demonstrated that the use of a phosphazene

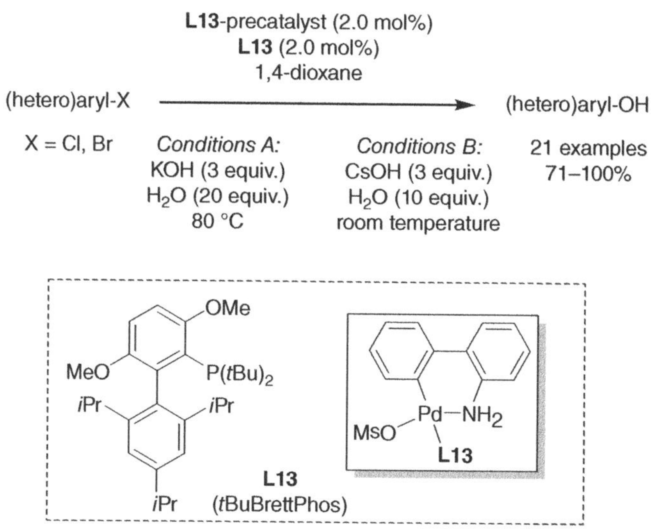

Scheme 9 *Palladium-catalyzed hydroxylation of (hetero)aryl bromides and chlorides employing an **L13**-ligated palladacyclic pre-catalyst*

base in combination with this **L13**-ligated palladacyclic pre-catalyst allowed for the successful hydroxylation of 3-bromo-5-phenylpyridine in the presence of alkyl and aryl esters, in the absence of ester reactivity.

The application of biaryl monophosphines in the palladium-catalyzed hydroxylation of (hetero)aryl halides has attracted the attention of other research groups. Yu *et al.*[68] reported on related palladium-catalyzed selective hydroxylation chemistry using **L11** under microwave irradiation that enables the application of K_2CO_3 or Cs_2CO_3 using a DMF/H_2O (9:1) co-solvent, while Dong *et al.*[69] have applied the biaryl monophosphine JohnPhos in the palladium-catalyzed hydroxylation of chiral 6-bromo- and 6,6′-dibromo- 1,1′-binaphthols. Chen *et al.*[70] have shown that Pd_2dba_3/P(*t*Bu)$_3$ catalyst mixtures can effect the selective hydroxylation of aryl halides. However, the use of P(*t*Bu)$_3$ is disadvantageous given the pyrophoric nature of this phosphine, and only nitro-substituted electrophilic coupling partners proved useful in this chemistry, thereby providing additional support for the idea that more complex ancillary ligation is required in order to achieve optimal catalyst performance under user-friendly conditions. Nonetheless, the ability of palladium nanoparticles supported on polyaniline nanofibers to catalyze the hydroxylation of simple aryl halides (4 examples, 80–90%) under conditions analogous to those described by Anderson *et al.*[64] suggests that alternative catalyst classes for such transformations are viable.[71]

5.3.2 Application of alternative ligand classes

In 2009, Schulz *et al.*[72] reported on the development of a new imidazole-based P,N ancillary ligand set (**L14**; Scheme 10) that enabled the first examples of the hydroxylation of aryl halides at room temperature. Preliminary experimentation established that the use of Pd_2dba_3/**L14** catalyst mixtures, under conditions similar to those employed by Anderson *et al.*[64] (Scheme 8), enabled the selective hydroxylation of substituted aryl halides (15 examples, 50–99%). Building on this success, Sergeev *et al.*[66] observed that bromomesitylene could be converted to the corresponding phenol in near quantitative yield at room temperature by employing Pd(COD)($CH_2SiMe_3)_2$]/**L14** catalyst mixtures (Scheme 10). Stoichiometric reactivity studies confirmed that the crystallographically characterized (κ^2-P,N-**L14**)Pd(Mes)Br (cf. **B** in Scheme 1) is formed at room temperature upon addition of bromomesitylene to Pd(COD)($CH_2SiMe_3)_2$]/**L14** mixtures; subsequent treatment at room temperature with a hydroxide source afforded 2,4,6-trimethylphenol, presumably via reductive elimination involving the unobserved intermediate (κ^2-P,N-**L14**)Pd(Mes)OH (cf. **C** in Scheme 1). Collectively, these results suggested that the room temperature hydroxylation of bromomesitylene using Pd(COD)($CH_2SiMe_3)_2$]/**L14** or (κ^2-P,N-**L14**) Pd(Mes)Br should be feasible. This indeed proved to be true (Scheme 10) with both catalysts affording 2,4,6-trimethylphenol in near quantitative yield when using CsOH•H_2O as the hydroxide source. The use of other hydroxide sources, including KOH, proved to be significantly less effective. The inability of related ligands such as P(*t*Bu)$_3$, QPhos, *t*BuXPhos (**L11**) and Me$_4$*t*BuXPhos (**L12**) to afford appreciable

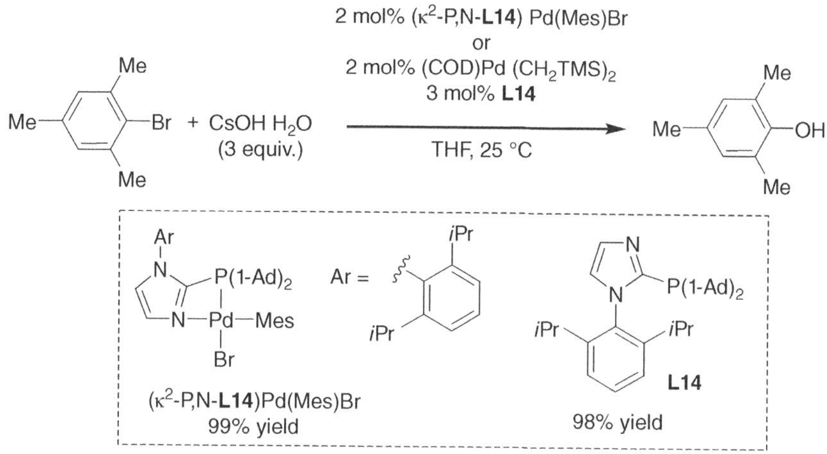

Scheme 10 *Room temperature palladium-catalyzed hydroxylation of bromomesitylene employing* ***L14****. COD = 1,5-cyclooctadiene*

amounts of 2,4,6-trimethylphenol under analogous room temperature conditions confirmed the importance of the ancillary ligand structure in this room temperature chemistry; when binding in a κ^2-P,N fashion, **L14** can be viewed as giving rise to high steric congestion at the metal center that may promote C—OH reductive elimination. Employing the Pd(COD)(CH_2SiMe_3)$_2$]/**L14** catalyst system allowed for the selective room temperature hydroxylation of various substituted aryl halides (1.0–4.0 mol% Pd, 1.5–6.0 mol% **L14**; 17 examples, 67–99%), including those featuring nitrile and trifluoromethyl substituents, which can be prone to hydrolysis at elevated temperatures under typical catalytic conditions. Notwithstanding the exceptionally mild conditions under which such reactions proceed, reactivity with electron-rich aryl halides, and synthetically useful (hetero)aryl halides, was not well-demonstrated in this system.

Lavery *et al.*[73] reported on the use of BippyPhos (**L2**) for the palladium-catalyzed hydroxylation of (hetero)aryl chlorides and bromides. Preliminary ligand screening focused on the use of Pd_2dba_3/**L** catalyst mixtures (2.0 and 8.0 mol%, respectively) for the room temperature hydroxylation of bromomesitylene in THF using CsOH•H_2O as the hydroxide source. Despite the efficacy of both JosiPhos CyPF-*t*Bu (**L1**; Section 5.2.1) and Mor-DalPhos (**L10**; Section 5.2.4) in analogous ammonia monoarylation chemistry, these ligands failed to produce appreciable quantities of 2,4,6-trimethylphenol under the screening conditions employed; the failure of **L10** in this regard is particularly perplexing from a ligand design perspective, given the structural relationship to Beller's ligand (**L14**; Scheme 10). However, BippyPhos (**L2**; Section 5.2.2) emerged from this ligand screening campaign as being highly effective for such transformations (Scheme 11). While in some cases elevated reaction temperatures were employed, a significant number of the reported reactions were found to proceed with success at room temperature, including cyclization reactions leading to benzofurans. The feasibility of conducting

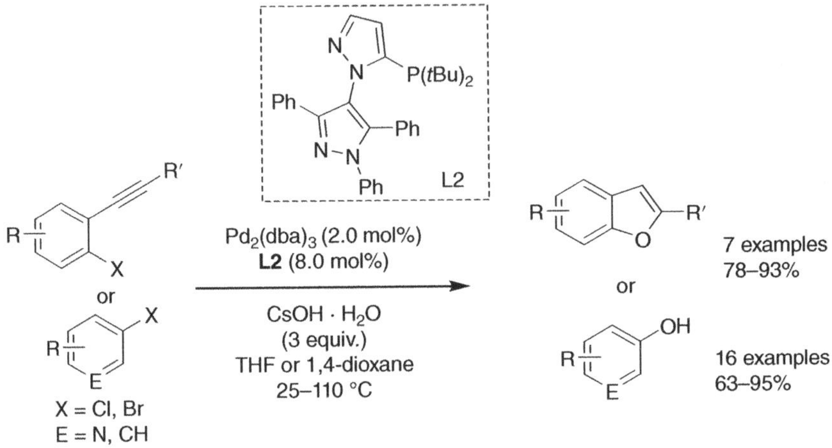

Scheme 11 *Palladium-catalyzed hydroxylation of (hetero)aryl bromides and chlorides employing **L2***

such reactions on the benchtop under air using unpurified solvents with negligible loss in reactivity was also demonstrated. From a ligand design perspective, the crystallographic characterization of (κ^2-P,C-**L2**)$PdCl_2$ provided in this report established for the first time the ability of BippyPhos to adopt a bidentate binding motif reminiscent of Buchwald's biarylphosphine ligand class.

5.3.3 Summary

The palladium-catalyzed hydroxylation of (hetero)aryl halides has emerged as an effective strategy for the construction of phenols that is complementary to other existing synthetic methodologies. The successful catalyst systems that have been identified to date encompass a broad spectrum of aryl electrophiles; such chemistry can also be exploited in the one-pot synthesis of substituted ethers including benzofuran derivatives. Ancillary ligand design has played a key role in the advancement of this chemistry, in some cases enabling room temperature transformations under non-inert reaction conditions. Given the isoelectronic relationship between NH_2- and OH-, it is perhaps not surprising that some of the ligands that perform well in ammonia monoarylation chemistry also excel in related selective hydroxylation reactions, including BippyPhos, BrettPhos variants, and imidazole-derived phosphines. However, the failure of JosiPhos CyPF-tBu and Mor-DalPhos in the palladium-catalyzed hydroxylation of (hetero)aryl halides, when placed in the context of the efficacy of these ligands in ammonia monoarylation chemistry, serves as a reminder that ancillary ligand design is a complex and inexact science, in which a priori predictivity of structure–reactivity relationships remains a challenge.

5.4 Palladium-catalyzed nucleophilic fluorination of (hetero)aryl (pseudo)halides

Fluorinated (hetero)aryl compounds are highly sought-after in a number of chemical applications, including in medicinal chemistry.[74, 75] While the substitution of an $C(sp^2)$—H for an $C(sp^2)$—F group does not alter the steric properties of the molecule of interest, such a modification can lead to significant changes in solubility and other physical/reactivity properties. In some cases, substitution of an $C(sp^2)$—H for an $C(sp^2)$—F group can influence metabolic stability, thus altering the biological properties of the molecule upon fluorination. This phenomenon has been used to advantage in the development of highly effective active pharmaceutical ingredients, with about one-third of top-performing drugs currently on the market featuring a C—F bond. In addition to the example of Crizotinib presented in Figure 2, two other prominent pharmaceuticals featuring an $C(sp^2)$—F linkage are presented in Figure 4; notably Atorvastatin (Lipitor; now off-patent) was the world's top-selling drug from 1996 to 2012.[75] Notably, radioactive ^{18}F-labeled organic compounds are also employed as contrast agents in positron emission tomography.[74]

A number of methodologies for preparing (hetero)aryl fluoride compounds have been developed, some of which involve palladium catalysis;[76, 77] nonetheless, synthetically useful protocols for the synthesis of such compounds under mild conditions and with high selectivity are still lacking. In this regard, the palladium-catalyzed nucleophilic fluorination of (hetero)aryl (pseudo)halides conceptually represents a promising route to $C(sp^2)$—F bond formation that is complementary to methods that employ electrophilic fluorine reagents.[77] Unfortuately, beyond the challenges associated with using fluoride that are shared with amide and hydroxide nucleophiles, such as difficult $C(sp^2)$—F bond reductive elimination (Section 5.1.2), the particularly strong propensity of the fluoride ion to form strong hydrogen bonds can lead to poor reactivity.[78] Despite these daunting challenges, the palladium-catalyzed nucleophilic fluorination of (hetero)aryl (pseudo)halides has been achieved by the Buchwald group.[79] Key to the successful development of this chemistry was the application of appropriately constructed biaryl monophosphines. However, in the course of such

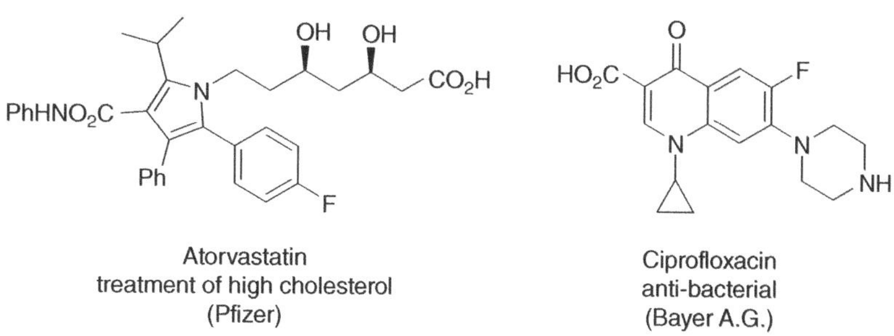

Figure 4 *Selected examples of pharmaceuticals featuring fluorinated aryl groups*

investigations, careful experimentation revealed that such ancillary ligands are not simply inert spectators during the C(sp^2)—F cross-coupling process.

5.4.1 Development of palladium-catalyzed C(sp^2)—F coupling employing (hetero)aryl triflates

In 2009, Watson *et al.*[79] disclosed the palladium-catalyzed cross-coupling of (hetero) aryl triflates with fluoride sources, thereby addressing a long-standing challenging in cross-coupling chemistry. The use of the sterically demanding *t*BuBrettPhos (**L13**; Section 5.3.1) ancillary ligand proved key to establishing a high-yielding protocol. In the test reaction between 1-naphthol triflate and rigorously dried CsF, using $[Pd(cinnamyl)Cl]_2$ as the palladium source, the use of BrettPhos (**L4**; Section 5.2.3) as the ancillary ligand resulted in a 30% yield of the desired aryl fluoride at 90% conversion (10 mol% Pd with Pd:L 1:1; 110 °C, 18 h), revealing low target product selectivity. Relatively poor performance was also observed when using BippyPhos (**L2**), with 52% yield of the target aryl fluoride obtained along with 11% naphthalene (10 mol% Pd with Pd:L 1:1; 150 °C, 12 h). Conversely, under similar conditions to those employed with **L4**, the use of **L13** afforded a 71% yield of the desired aryl fluoride at 100% conversion. In subsequently applying $[Pd(cinnamyl)Cl]_2$/**L13** catalyst mixtures, a range of (hetero)aryl triflates were successfully converted into the corresponding (hetero)aryl fluoride species, including sterically hindered substrates, electron poor arenes, as well as certain classes of heterocycles (Scheme 12). The accommodation of nucleophilic functional groups within the observed substrate scope demonstrates the complementary nature of such protocols to those employing electrophilic fluorine sources, which do not routinely tolerate such functionalities.[77] Notably, in some cases regioisomeric aryl fluoride products were obtained in the observed cross-couplings, the proportion of which was found to increase with increasing electron-richness of the aryl triflate reactant. Evidence for C(sp^2)—F reductive elimination from

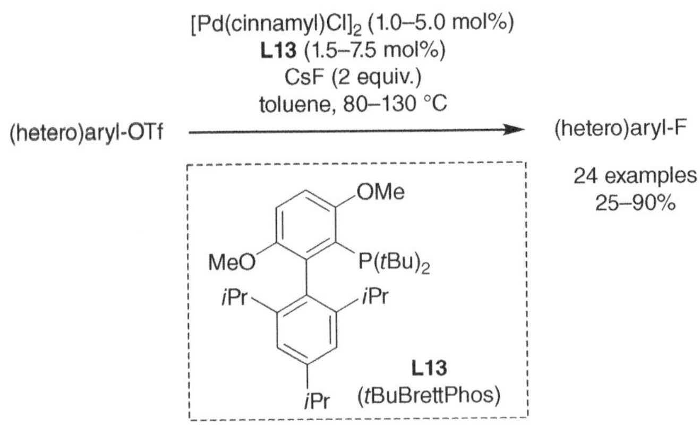

Scheme 12 *Palladium-catalyzed cross-coupling of (hetero)aryl triflates with fluoride employing $[Pd(cinnamyl)Cl]_2$/**L13** catalyst mixtures*

a crystallographically characterized three-coordinate T-shaped Pd(II) intermediate was provided in the case of **L4**-based model compounds.[79] The adaptation of such cross-couplings to flow conditions has been reported,[80] as has the one-pot nonaflation/fluorination of phenols under microwave conditions employing a $Pd_2(dba)_3$/**L13** catalyst system.[81]

5.4.2 Discovery of biaryl monophosphine ancillary ligand modification

Remarkably, in the course of mechanistic investigations pertaining to the newly developed palladium-catalyzed cross-coupling of (hetero)aryl triflates with fluoride, Maimone *et al.*[82] acertained that the active palladium catalyst involved in such transformations featured a *t*BuBrettPhos (**L13**) ancillary ligand that had undergone lower-ring arylation in situ under the catalytic conditions employed. Initial studies focused on evaluating further the viability of aryl—F reductive elimination processes involving putative (**L13**)Pd(aryl)F intermediates, where the aryl group is electron rich; as part of this study, they established that the closely related ligand, RockPhos (**L15**, Scheme 13), exhibits a similar reactivity profile to **L13** in the palladium-catalyzed nucleophilic

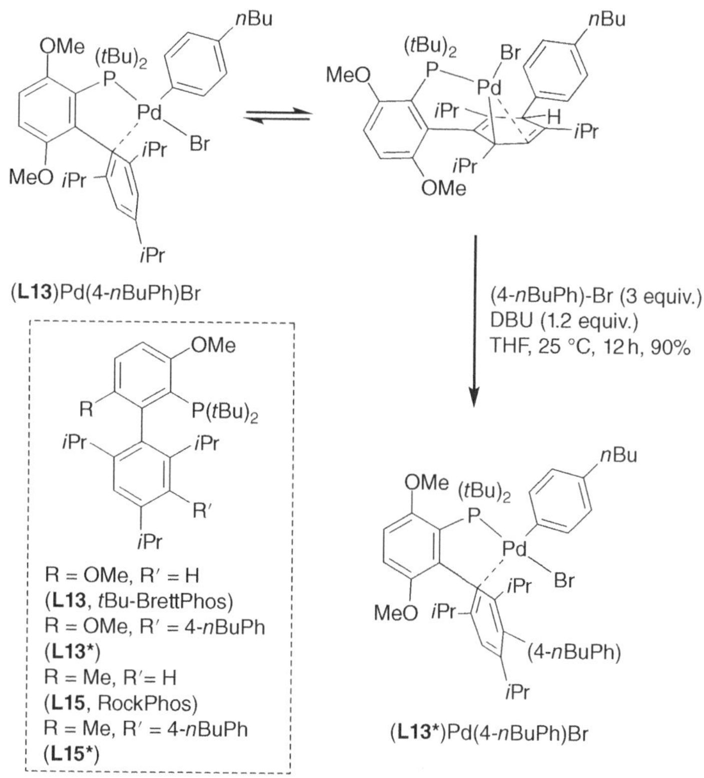

Scheme 13 *Previously unobserved aryl transfer from palladium to a coordinated biaryl monophosphine ancillary ligand resulting in lower ring dearomatization*

fluorination of aryl triflates.[82] Surprisingly, heating of (**L15**)Pd(4-*n*BuPh)F failed to generate the anticipated product of C(sp^2)—F reductive elimination, despite the apparent efficacy of the $[Pd(cinnamyl)Cl]_2$/**L15** catalyst system. Moreover, (**L13**) Pd(4-*n*BuPh)Br was observed to undergo a previously undocumented rearrangement process involving transfer of the palladium-bound aryl group to the lower flanking ring of the biarylphosphine ligand in a net-dearomatization process; exposure of (**L13**) Pd(4-*n*BuPh)Br to base in the presence of (4-*n*BuPh)Br afforded the ring-arylated complex (**L13***)Pd(4-*n*BuPh)Br (Scheme 13). An analogous aryl-transfer rearrangement was detected under catalytic conditions when using **L15**, thereby affording **L15***.[82]

Use of the ring-arylated RockPhos (**L15***) in the palladium-catalyzed fluorination of (4-*n*BuPh)-OTf led to a modest (13%) improvement in yield compared with that obtained by use of the parent RockPhos ancillary ligand (**L15**). Furthermore, heating of (**L13***)Pd(4-*n*BuPh)F in the presence of 10 equiv. of (4-*n*BuPh)OTf produced a mixture of (3-*n*BuPh)F and (4-*n*BuPh)F that closely mirrored the ratio of products obtained under catalytic conditions when using **L13** as part of the pre-catalyst mixture. Collectively, these observations establish that **L13**, **L15** and likely other biaryl monophosphine ancillary ligands are susceptible to structural modification by the aryl electrophile coupling partner under catalytic cross-coupling conditions.[82] A subsequent mechanistic investigation into such dearomative rearrangements within palladium biaryl monophosphine complexes established that such processes proceed in a concerted fashion involving the previously unknown palladium-mediated insertion of an aryl group into an unactivated arene, whereby the steric parameters of both the substituents on the phosphorus-functionalized "upper" ring as well as the alkyl groups on phosphorus influence the rate and extent of dearomatization.[83] Notably, the structural features that render bulky biaryl monophosphine ancillary ligands such as *t*BuXPhos (**L11**), *t*BuBrettPhos (**L13**), and RockPhos (**L15**) effective in promoting challenging reductive eliminations from Pd(II) also appear to enable the rearrangement of their oxidative addition complexes to the corresponding dearomatized isomers.

The implications of these observations by Buchwald and co-workers[82, 83] with respect to "rational" ligand design are significant: the success or failure of a given biaryl monophosphine ligand in this context may in part be attributable to the way in which the substrate modifies the ancillary ligand in situ, with each substrate generating a uniquely modified ancillary ligand under catalytic conditions. A complementary report by Allgeier *et al.*,[84] in which two stable palladium degradation coordination complexes derived from *t*BuXPhos (**L11**) are characterized, provides additional support for the view that such in situ biaryl monophosphine ancillary ligand modification may be somewhat common. It is worthy of mention that the serendipitous formation of highly effective ancillary ligands by way of in situ modification is not a new phenomenon, as evidenced by the transmutation of $(\eta^5\text{-}C_5H_5)Fe(\eta^5\text{-}C_5H_4P(t\text{Bu})_2)$ into $(\eta^5\text{-}C_5Ph_5)Fe(\eta^5\text{-}C_5H_4P(t\text{Bu})_2)$ (i.e., QPhos) under palladium-catalyzed C—O cross-coupling conditions.[85]

In 2013, Lee *et al.*[86] developed an improved protocol for the palladium-catalyzed nucleophilic fluorination of (hetero)aryl triflates that makes use of the $(\textbf{L5})_2Pd_2(COD)$

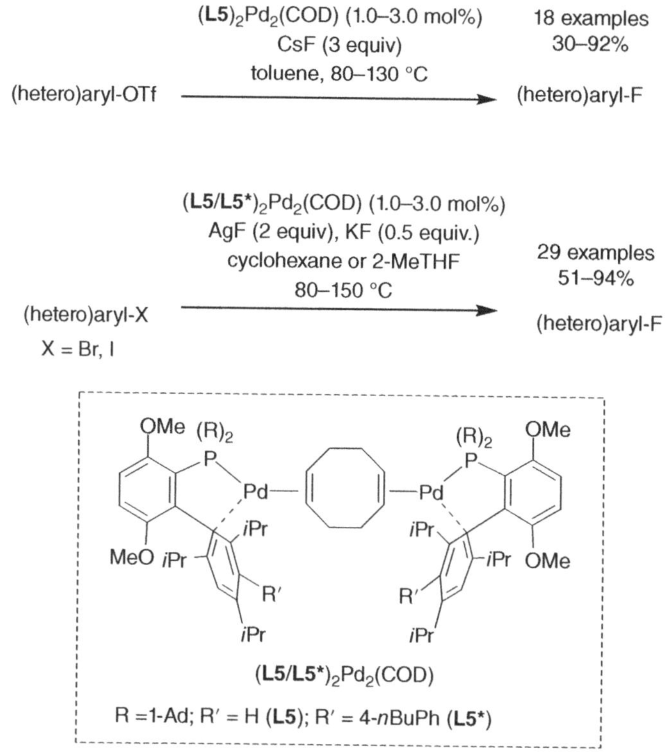

Scheme 14 *Palladium-catalyzed cross-coupling of (hetero)aryl triflates with fluoride employing the ($\mathbf{L5}$)$_2$Pd$_2$(COD) pre-catalyst. Ad = 1-adamantyl; COD = 1,5-cyclooctadiene*

pre-catalyst (**L5** = AdBrettPhos, COD = 1,5-cyclooctadiene, Scheme 14). While somewhat cumbersome to prepare and having a half-life on the order of days in air, this pre-catalyst has the advantage of giving rise to "(**L5**)Pd(0)" (cf. **A** in Scheme 1) without the formation of reactive by-products that can either inhibit catalysis or complicate product mixtures, including chloride, HF, dba, or carbazole; as mentioned in Section 5.2.3, the last of these is generated as a by-product when using aminobiphenyl palladacyclic pre-catalysts. This new catalyst system proved to be particularly useful for the fluorination of (hetero)aryl triflates derived from biologically active and heteroaryl phenols, which proved to be challenging substrates when using [Pd(cinnamyl)Cl]$_2$/**L13** (see above).

5.4.3 Extending reactivity to (hetero)aryl bromides and iodides

Lee *et al.*[87] subsequently reported the first examples of the palladium-catalyzed nucleophilic fluorination of (hetero)aryl bromides and iodides (Scheme 14). In seeking to achieve such transformations, they envisioned two required modifications to their previous catalytic protocols employed with (hetero)aryl triflates: the use of a more reactive metal fluoride source to facilitate transmetalation (i.e., the conversion of **B** to **C** in Scheme 1), and the use of substoichiometric amounts of base to facilitate in situ ancillary ligand arylation (Scheme 13). After a brief optimization campaign, AgF and KF (respectively) were found to give rise to a suitably effective catalyst system in combination with $(\mathbf{L5})_2Pd_2(COD)$, allowing for the fluorination of a variety of substituted aryl bromides and iodides, including those featuring base-sensitive functional groups as well as nitrile, ester, and amide addenda. Substrates that are not amenable to electrophilic fluorination, such as alkyl sulfides and electron-rich amines, were also successfully accomodated. As was observed in the palladium-catalyzed nucleophilic fluorination of aryl triflates, the electron-rich and unhindered 4-bromoanisole afforded a mixture of regioisomeric products. Efforts to expand the substrate scope to include (hetero)aryl halides failed when using the $(\mathbf{L5})_2Pd_2(COD)$ pre-catalyst; the authors suggest that the origin of such poor reactivity may be associated with inefficient in situ ancillary ligand arylation (to give heteroaryl variants **L5***) or poor performance of the palladium catalyst arising from the thus-formed modified ancillary ligand.[87] In response, synthetic routes to $(\mathbf{L5^*})_2Pd_2(COD)$, featuring a pre-arylated AdBrettPhos ancillary ligand were established as a means of circumventing in situ ancillary ligand modification altogether. This pre-catalyst exhibited significantly improved reactivity with (hetero)aryl bromide substrates relative to $(\mathbf{L5})_2Pd_2(COD)$, where pyridine, indole, quinoline, isoquinoline, pyrimidine, indazole, and quinoxaline core structures were each well-tolerated. In principle, the use of $(\mathbf{L5^*})_2Pd_2(COD)$ should obviate the need for KF, given that the ancillary ligand has been "pre-arylated". Nonetheless, the use of AgF alone in this chemistry proved inferior to the use of AgF in combination with KF, suggesting that the latter plays a more complex role in the observed catalysis.

5.4.4 Summary

The palladium-catalyzed nucleophilic fluorination of (hetero)aryl (pseudo)halides has been achieved, thus providing a complementary protocol for the synthesis of sought-after (hetero)aryl fluorides. The evolution of this chemistry represents one of the most interesting and informative stories in the domain of ancillary ligand design for use in palladium catalysis. Whereas the initial development of this novel $C(sp^2)$—F bond-forming chemistry was enabled by the availability of appropriately constructed biaryl monophosphine ancillary ligands that were prepared for use in other cross-coupling applications, careful mechanistic analysis and reactivity studies revealed processes whereby these ancillary ligands underwent structural modification in situ under catalytic conditions. These insights provided the basis for the rational design and preparation of new modified ancillary ligand variants and pre-catalysts that offered improved catalytic performance and expanded substrate scope. All reports to date regarding the palladium-catalyzed nucleophilic fluorination of (hetero)aryl (pseudo)halides have featured catalysts supported by biaryl monophosphine ancillary ligation; however, it is unlikely that this ancillary ligand class is uniquely capable of promoting such transformations. In this regard, continued catalyst development with particular emphasis on the application of alternative ancillary ligand frameworks that are less-susceptible to in situ transmutation will likely lead to further practical advances [e.g., lower reaction temperatures and catalyst loading, transformations of (hetero)aryl chlorides, etc.].

5.5 Conclusions and outlook

Ancillary ligand design has played a key role with respect to the rapid advancement of palladium-catalyzed cross-coupling chemistry in recent years. It is our hope that this has been illustrated through the case studies presented in this chapter regarding C(sp^2)—E bond formation (E = NH_2, OH, and F), where issues of catalyst inhibition and product selectivity represent formidable challenges for most commonly employed palladium catalyst systems. In each of these otherwise difficult cross-couplings, more mild experimental conditions, expanded substrate scope, and increased functional group tolerance have been achieved through the judicious, and in some cases iterative, design and application of supporting ancillary ligands. The ancillary ligands discussed in this chapter are collected in Figure 5, along with their common names where available, and include: a bisphosphine (**L1**); monophosphines based on a heterocyclic ligand backbone (**L2**, **L8**, and **L14**); phenylene P,N ligands (**L9** and **L10**); and biaryl monophosphines (**L3–L7**, **L11–L13**, **L15**, and the arylated variants **L13*** and **L15***). From a practical perspective, all of the ancillary ligands presented in this chapter are air-stable, and in many cases are commercially available, thereby facilitating uptake by end-users in both industrial and academic settings. While these various high-performing ancillary ligands differ in structure, the most successful variants feature sterically demanding, electron-rich phosphorus donor fragments, and exhibit the capacity for bidentate connectivity with Pd(II); the latter may serve to enhance catalyst selectivity and lifetime. The utility of employing a modular ligand design as a means of addressing task-specific reactivity challenges through tailoring of the ligand architecture has been demonstrated in the examples provided, as have the benefits of employing pre-catalysts in which palladium is coordinated to the ancillary ligand of choice. While no single ancillary ligand or class of ancillary ligands has demonstrated superiority across all of the cross-coupling applications discussed herein, the biaryl monophosphine motif appears to be privileged. Nonetheless, careful scrutiny of the behavior of this dominant ancillary ligand class has revealed unexpected in situ modification of the ancillary ligand structure under catalytic conditions. These observations serve to highlight the ease with which unexpected ancillary ligand transmutation can occur under catalytic conditions, as well as the need for new and robust ancillary ligand frameworks that circumvent such processes. Such reports also underscore the importance of obtaining comprehensive supporting experimental data when making definitive mechanistic claims regarding ancillary ligand involvement throughout the course of a particular palladium-catalyzed process. Indeed, "rational" ancillary ligand design remains a complex and inexact science; a more profound understanding of the specific design criteria that enable ligands to confer desirable reactivity properties to the metal center is needed. Such insights will play a central role in the quest to address outstanding reactivity goals in palladium-catalyzed cross-coupling chemistry.

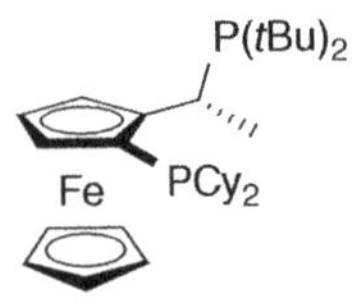

L1
(JosiPhos; CyPF-*t*Bu)

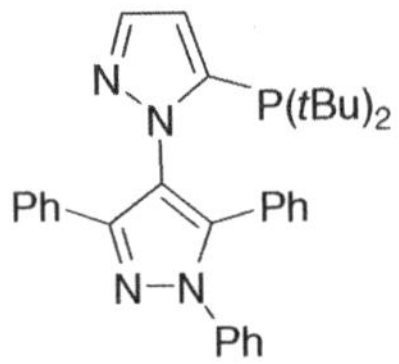

L2
(BippyPhos)

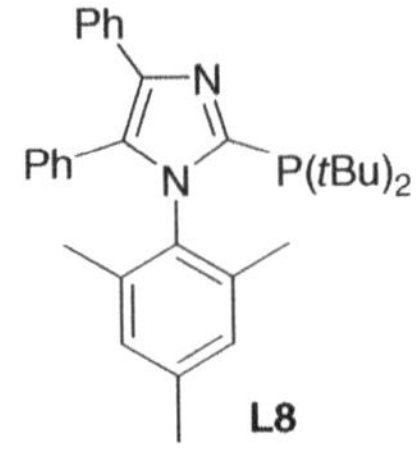

L8

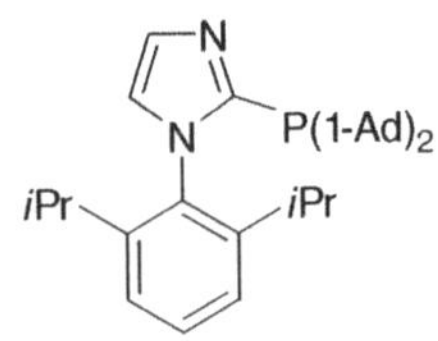

L14

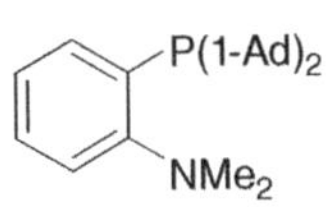

L9
(Me-DalPhos)

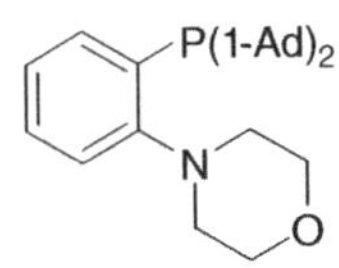

L10
(Mor-DalPhos)

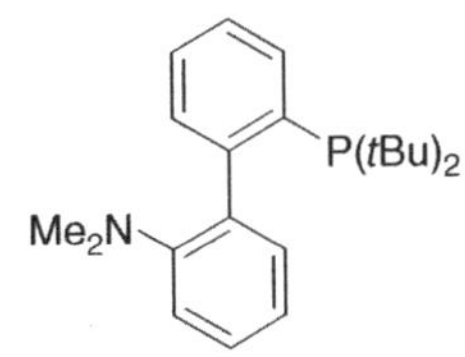

L3
(*t*BuDavePhos)

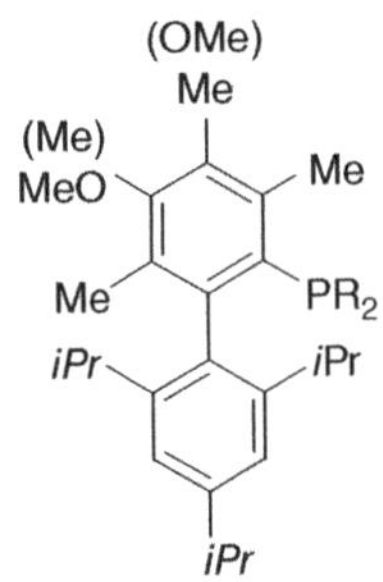

L6, R = Cy
(Me_3(OMe)XPhos)
L7, R = Ph
(Me_3(OMe)PhXPhos)

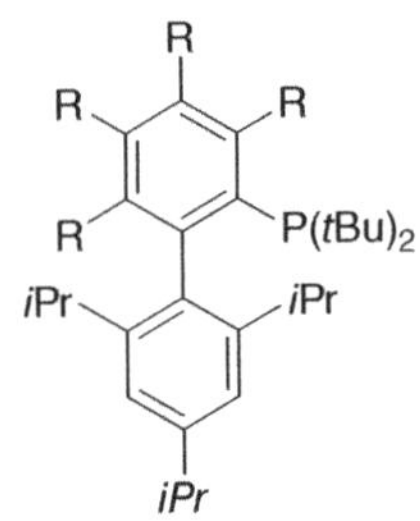

L11, R = H
(*t*BuXPhos)
L12, R = Me
(Me_4*t*BuXPhos)

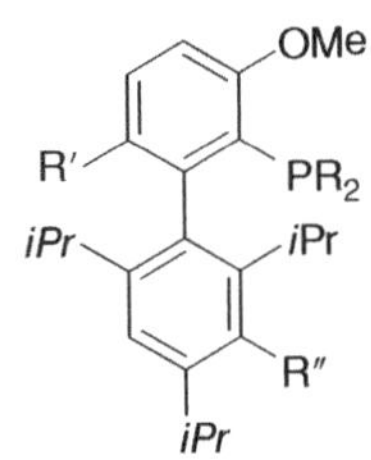

L4, R = Cy, R′ = OMe, R″ = H
(BrettPhos)
L5, R = 1-Ad, R′ = OMe, R″= H
(AdBrettPhos)
L13, R = *t*Bu, R′ = OMe, R″ = H
(*t*BuBrettPhos)
L13*, R = *t*Bu, R′ = OMe, R″ = 4-*n*BuPh
(arylated *t*BuBrettPhos)
L15, R = *t*Bu, R′ = Me, R″= H
(RockPhos)
L15*, R = *t*Bu, R′ = Me, R″ = 4-*n*BuPh
(arylated RockPhos)

Figure 5 *The ancillary ligands discussed in this chapter*

Acknowledgments

M.S. acknowledges support from the Natural Sciences and Engineering Research Council of Canada, Canadian Institutes of Health Research, the Killam Trusts, and Dalhousie University (Alexander McLeod Professorship). R.J.L acknowledges support from the Natural Sciences and Engineering Research Council of Canada, the Canadian Foundation for Innovation, and the University of Alberta.

References

[1] R. F. Heck, *Organic Reactions* **1982**, *27*, 345–390.
[2] E. I. Negishi, *Angew. Chem. Int. Ed.* **2011**, *50*, 6738–6764.
[3] A. Suzuki, *Angew. Chem. Int. Ed.* **2011**, *50*, 6722–6737.
[4] X. F. Wu, P. Anbarasan, H. Neumann, M. Beller, *Angew. Chem. Int. Ed.* **2010**, *49*, 9047–9050.
[5] C. C. C. J. Seechurn, M. O. Kitching, T. J. Colacot, V. Snieckus, *Angew. Chem. Int. Ed.* **2012**, *51*, 5062–5085.
[6] A. S. Guram, R. A. Rennels, S. L. Buchwald, *Angew. Chem. Int. Ed.* **1995**, *34*, 1348–1350.
[7] J. Louie, J. F. Hartwig, *Tetrahedron Lett.* **1995**, *36*, 3609–3612.
[8] J. F. Hartwig, *Acc. Chem. Res.* **2008**, *41*, 1534–1544.
[9] D. S. Surry, S. L. Buchwald, *Angew. Chem., Int. Ed.* **2008**, *47*, 6338–6361.
[10] S. A. Lawrence, *Amines: Synthesis, Properties and Applications*, Cambridge University Press, Cambridge, **2006**.
[11] R. J. Lundgren, M. Stradiotto, *Chem. Eur. J.* **2012**, *18*, 9758–9769.
[12] D. Maiti, B. P. Fors, J. L. Henderson, Y. Nakamura, S. L. Buchwald, *Chem. Sci.* **2011**, *2*, 57–68.
[13] D. S. Surry, S. L. Buchwald, *Chem. Sci.* **2011**, *2*, 27–50.
[14] M. S. Driver, J. F. Hartwig, *J. Am. Chem. Soc.* **1995**, *117*, 4708–4709.
[15] R. A. Widenhoefer, S. L. Buchwald, *Organometallics* **1996**, *15*, 2755–2763.
[16] M. S. Driver, J. F. Hartwig, *J. Am. Chem. Soc.* **1997**, *119*, 8232–8245.
[17] S. Shekhar, P. Ryberg, J. F. Hartwig, J. S. Mathew, D. G. Blackmond, E. R. Strieter, S. L. Buchwald, *J. Am. Chem. Soc.* **2006**, *128*, 3584–3591.
[18] J. L. Klinkenberg, J. F. Hartwig, *J. Am. Chem. Soc.* **2010**, *132*, 11830–11833.
[19] C. Valente, S. Calimsiz, K. H. Hoi, D. Mallik, M. Sayah, M. G. Organ, *Angew. Chem. Int. Ed.* **2012**, *51*, 3314–3332.
[20] G. C. Fortman, S. P. Nolan, *Chem. Soc. Rev.* **2011**, *40*, 5151–5169.
[21] H. B. Li, C. C. C. J. Seechurn, T. J. Colacot, *ACS Catal.* **2012**, *2*, 1147–1164.
[22] A. G. MacDiarmid, *Angew. Chem. Int. Ed.* **2001**, *40*, 2581–2590.
[23] M. Liang, J. Chen, *Chem. Soc. Rev.* **2013**, *42*, 3453–3488.
[24] X. H. Huang, S. L. Buchwald, *Org. Lett.* **2001**, *3*, 3417–3419.
[25] L. M. Alabanza, Y. Dong, P. Wang, J. A. Wright, Y. C. Zhang, A. J. Briggs, *Org. Process Res. Dev.* **2013**, *17*, 876–880.
[26] M. Appl, *Ammonia: Principles and Industrial Practice*, Wiley-VCH, Weinheim, **1999**.
[27] Y. Aubin, C. Fischmeister, C. M. Thomas, J. L. Renaud, *Chem. Soc. Rev.* **2010**, *39*, 4130–4145.
[28] J. I. van der Vlugt, *Chem. Soc. Rev.* **2010**, *39*, 2302–2322.
[29] J. L. Klinkenberg, J. F. Hartwig, *Angew. Chem. Int. Ed.* **2011**, *50*, 86–95.
[30] N. Xia, M. Taillefer, *Angew. Chem. Int. Ed.* **2009**, *48*, 337–339.
[31] Q. L. Shen, J. F. Hartwig, *J. Am. Chem. Soc.* **2006**, *128*, 10028–10029.
[32] H. U. Blaser, W. Brieden, B. Pugin, F. Spindler, M. Studer, A. Togni, *Top. Catal.* **2002**, *19*, 3–16.
[33] G. D. Vo, J. F. Hartwig, *J. Am. Chem. Soc.* **2009**, *131*, 11049–11061.
[34] R. A. Green, J. F. Hartwig, *Org. Lett.* **2014**, *16*, 4388–4391.
[35] A. Borzenko, N. L. Rotta-Loria, P. M. MacQueen, C. M. Lavoie, R. McDonald, M. Stradiotto, *Angew. Chem. Int. Ed.* **2015**, *54*, 3773–3777.
[36] R. A. Green, J. F. Hartwig, *Angew. Chem. Int. Ed.* **2015**, *54*, 3768–3772.
[37] D. A. Horton, G. T. Bourne, M. L. Smythe, *Chem. Rev.* **2003**, *103*, 893–930.

[38] G. R. Humphrey, J. T. Kuethe, *Chem. Rev.* **2006**, *106*, 2875–2911.
[39] B. Robinson, *Chem. Rev.* **1963**, *63*, 373–401.
[40] P. G. Alsabeh, R. J. Lundgren, L. E. Longobardi, M. Stradiotto, *Chem. Commun.* **2011**, *47*, 6936–6938.
[41] R. A. Singer, M. L. Dore, J. E. Sieser, M. A. Berliner, *Tetrahedron Lett.* **2006**, *47*, 3727–3731.
[42] G. J. Withbroe, R. A. Singer, J. E. Sieser, *Org. Process Res. Dev.* **2008**, *12*, 480–489.
[43] S. M. Crawford, C. B. Lavery, M. Stradiotto, *Chem. Eur. J.* **2013**, *19*, 16760–16771.
[44] D. S. Surry, S. L. Buchwald, *J. Am. Chem. Soc.* **2007**, *129*, 10354–10355.
[45] D. Tsvelikhovsky, S. L. Buchwald, *J. Am. Chem. Soc.* **2011**, *133*, 14228–14231.
[46] C. W. Cheung, D. S. Surry, S. L. Buchwald, *Org. Lett.* **2013**, *15*, 3734–3737.
[47] U. Christmann, D. A. Pantazis, J. Benet-Buchholz, J. E. McGrady, F. Maseras, R. Vilar, *J. Am. Chem. Soc.* **2006**, *128*, 6376–6390.
[48] N. C. Bruno, M. T. Tudge, S. L. Buchwald, *Chem. Sci.* **2013**, *4*, 916–920.
[49] A. Bruneau, M. Roche, M. Alami, S. Messaoudi, *ACS Catal.* **2015**, *5*, 1386–1396.
[50] N. C. Bruno, N. Niljianskul, S. L. Buchwald, *J. Org. Chem.* **2014**, *79*, 4161–4166.
[51] B. P. Fors, S. L. Buchwald, *J. Am. Chem. Soc.* **2010**, *132*, 15914–15917.
[52] T. Schulz, C. Torborg, S. Enthaler, B. Schaffner, A. Dumrath, A. Spannenberg, H. Neumann, A. Borner, M. Beller, *Chem. Eur. J.* **2009**, *15*, 4528–4533.
[53] A. Dumrath, C. Lubbe, H. Neumann, R. Jackstell, M. Beller, *Chem. Eur. J.* **2011**, *17*, 9599–9604.
[54] G. Helmchen, A. Pfaltz, *Acc. Chem. Res.* **2000**, *33*, 336–345.
[55] R. J. Lundgren, A. Sappong-Kumankumah, M. Stradiotto, *Chem. Eur. J.* **2010**, *16*, 1983–1991.
[56] R. J. Lundgren, B. D. Peters, P. G. Alsabeh, M. Stradiotto, *Angew. Chem. Int. Ed.* **2010**, *49*, 4071–4074.
[57] P. N. Carlsen, T. J. Mann, A. H. Hoveyda, A. J. Frontier, *Angew. Chem. Int. Ed.* **2014**, *53*, 9334–9338.
[58] P. G. Alsabeh, R. J. Lundgren, R. McDonald, C. C. C. J. Seechurn, T. J. Colacot, M. Stradiotto, *Chem. Eur. J.* **2013**, *19*, 2131–2141.
[59] Z. Rappoport, *The Chemistry of Phenols*, John Wiley & Sons, Inc., Hoboken, NJ, **2003**.
[60] D. A. Alonso, C. Najera, I. M. Pastor, M. Yus, *Chem. Eur. J.* **2010**, *16*, 5274–5284.
[61] S. Enthaler, A. Company, *Chem. Soc. Rev.* **2011**, *40*, 4912–4924.
[62] M. C. Willis, *Angew. Chem. Int. Ed.* **2007**, *46*, 3402–3404.
[63] A. Tlili, N. Xia, F. Monnier, M. Taillefer, *Angew. Chem. Int. Ed.* **2009**, *48*, 8725–8728.
[64] K. W. Anderson, T. Ikawa, R. E. Tundel, S. L. Buchwald, *J. Am. Chem. Soc.* **2006**, *128*, 10694–10695.
[65] C. W. Cheung, S. L. Buchwald, *J. Org. Chem.* **2014**, *79*, 5351–5358.
[66] A. G. Sergeev, T. Schulz, C. Torborg, A. Spannenberg, H. Neumann, M. Beller, *Angew. Chem. Int. Ed.* **2009**, *48*, 7595–7599.
[67] A. Buitrago Santanilla, M. Christensen, L. C. Campeau, I. W. Davies, S. D. Dreher, *Org. Lett.* **2015**, *17*, 3370–3373.
[68] C. W. Yu, G. S. Chen, C. W. Huang, J. W. Chern, *Org. Lett.* **2012**, *14*, 3688–3691.
[69] J. Dong, T. Li, C. Dai, W. Weng, X. Xue, Y. Zhang, Q. Zeng, *Appl. Organometal. Chem.* **2013**, *27*, 337–340.
[70] G. Chen, A. S. C. Chan, F. Y. Kwong, *Tetrahedron Lett.* **2007**, *48*, 473–476.
[71] B. J. Gallon, R. W. Kojima, R. B. Kaner, P. L. Diaconescu, *Angew. Chem. Int. Ed.* **2007**, *46*, 7251–7254.
[72] T. Schulz, C. Torborg, B. Schaffner, J. Huang, A. Zapf, R. Kadyrov, A. Borner, M. Beller, *Angew. Chem. Int. Ed.* **2009**, *48*, 918–921.

[73] C. B. Lavery, N. L. Rotta-Loria, R. McDonald, M. Stradiotto, *Adv. Synth. Catal.* **2013**, *355*, 981–987.
[74] T. Liang, C. N. Neumann, T. Ritter, *Angew. Chem. Int. Ed.* **2013**, *52*, 8214–8264.
[75] J. Wang, M. Sanchez-Rosello, J. L. Acena, C. del Pozo, A. E. Sorochinsky, S. Fustero, V. A. Soloshonok, H. Liu, *Chem. Rev.* **2014**, *114*, 2432–2506.
[76] A. C. Albéniz, J. A. Casares, *Adv. Organomet. Chem.* **2014**, *62*, 1–110.
[77] M. G. Campbell, T. Ritter, *Chem. Rev.* **2015**, *115*, 612–633.
[78] D. J. Adams, J. H. Clark, *Chem. Soc. Rev.* **1999**, *28*, 225–231.
[79] D. A. Watson, M. Su, G. Teverovskiy, Y. Zhang, J. Garcia-Fortanet, T. Kinzel, S. L. Buchwald, *Science* **2009**, *325*, 1661–1664.
[80] T. Noel, T. J. Maimone, S. L. Buchwald, *Angew. Chem. Int. Ed.* **2011**, *50*, 8900–8903.
[81] J. Wannberg, C. Wallinder, M. Unlusoy, C. Skold, M. Larhed, *J. Org. Chem.* **2013**, *78*, 4184–4189.
[82] T. J. Maimone, P. J. Milner, T. Kinzel, Y. Zhang, M. K. Takase, S. L. Buchwald, *J. Am. Chem. Soc.* **2011**, *133*, 18106–18109.
[83] P. J. Milner, T. J. Maimone, M. Su, J. Chen, P. Muller, S. L. Buchwald, *J. Am. Chem. Soc.* **2012**, *134*, 19922–19934.
[84] A. M. Allgeier, B. J. Shaw, T. L. Hwang, J. E. Milne, J. S. Tedrow, C. N. Wilde, *Organometallics* **2012**, *31*, 519–522.
[85] Q. Shelby, N. Kataoka, G. Mann, J. Hartwig, *J. Am. Chem. Soc.* **2000**, *122*, 10718–10719.
[86] H. G. Lee, P. J. Milner, S. L. Buchwald, *Org. Lett.* **2013**, *15*, 5602–5605.
[87] H. G. Lee, P. J. Milner, S. L. Buchwald, *J. Am. Chem. Soc.* **2014**, *136*, 3792–3795.

6

Pd-*N*-Heterocyclic Carbene Complexes in Cross-Coupling Applications

Jennifer Lyn Farmer, Matthew Pompeo, and Michael G. Organ

Department of Chemistry, York University, 4700 Keele Street, Toronto, Ontario, Canada M3J 1P3

6.1 Introduction

The recognition by Kharasch and Fields[1] in 1941 that small quantities of transition metal salts could catalyze carbon–carbon bond formation between Grignard reagents and organic halides in high yields laid the groundwork for the explosion of cross-coupling methodology that has persisted to the present day. The ability to make bonds between sp^2-hybridized carbon centers represented a monumental advance in the preparation of substituted aromatic compounds and culminated in the 2010 Nobel Prize in Chemistry, awarded to Ei-ichi Negishi, Richard Heck, and Akira Suzuki for 'Palladium-catalyzed cross-couplings in organic synthesis'.[2] The widespread use of cross-coupling reactions, which has grown far beyond its origins using Grignard nucleophiles, has driven the widespread investigation of multiple aspects of the general cross-coupling mechanism in order to improve the ease and generality of this highly useful area of catalysis. Reaction parameters that have been systematically probed

Ligand Design in Metal Chemistry: Reactivity and Catalysis, First Edition. Edited by Mark Stradiotto and Rylan J. Lundgren.

include the structure and suitability of various nucleophiles and electrophiles as substrates, solvents, additives (e.g., bases), reaction conditions (e.g., reaction time and temperature), suitable transition metals, and the area that has seen the most rapid development over the last decade: ancillary ligands.

Ligands have evolved considerably from simple amine and phosphane derivatives used primarily to maintain the metal as a homogeneous complex. Indeed, with a deeper mechanistic insight into the catalytic cycle, advances in ligand design have produced, in some cases, intricate, highly tailored ligands that greatly enhance catalytic performance. In the case of cross-coupling chemistry, phosphanes have been the primary workhorse and ligand modifications have historically focused on optimizing the three identical substituents that adorn the phosphorus atom (e.g., triphenylphosphine or tricyclohexylphosphine). More recently however, considerable effort has been spent in the preparation of more architecturally complex phosphane ligands, including mixed alkyl/aryl phosphanes such as Buchwald's dialkyl biaryl phosphines[3] and Stradiotto's P,N ligands.[4]

N-heterocyclic carbenes (NHCs) have also been shown to possess a number of important characteristics that make them attractive as ligands, not only for cross-coupling, but also catalysis in general. These desirable attributes have led to high interest in the field that has dramatically accelerated the development of a plethora of new NHC ligands whose Pd complexes are at least equal to, or surpass the reactivity of the most reactive Pd-phosphine complexes. This chapter focuses on the use of NHC-Pd complexes in Kumada–Tamao–Corriu (KTC), Suzuki–Miyaura, and Negishi cross-coupling and rather than being exhaustively comprehensive, will focus primarily on contributions that have most significantly advanced the mechanistic understanding of C—C bond formation leading to breakthroughs in reactivity and thus application of this methodology.

6.2 *N*-heterocyclic carbenes as ligands for catalysis

N-heterocyclic carbenes were initially reported independently by Wanzlick[5a], Wanzlick and Schönherr[5b], and Öfele[6] as free salts and as mercury and chromium complexes. Over two decades later, Arduengo *et al.*[7] reported the isolation of the first stable free carbene after which Herrmann and co-workers demonstrated that Pd-NHC complexes were suitable for use in catalysis.[8] Arduengo *et al.*[9] subsequently reported the isolation of a number of free carbenes, and with an increasing number of available NHCs, wider interest for their use in catalysis began to develop. Some Pd-NHC complexes were demonstrated to possess excellent thermal stability and it was recognized that their reactivity could rival that of phosphines for some Pd-catalyzed cross-coupling applications.[10] As with any ligand system, considerable effort has been extended to understand the key structural features of the NHC core that leads to improved catalytic performance and broader usage. A selection of commonly used NHC ligands and Pd-NHC complexes mentioned throughout this chapter is shown in Figure 1.

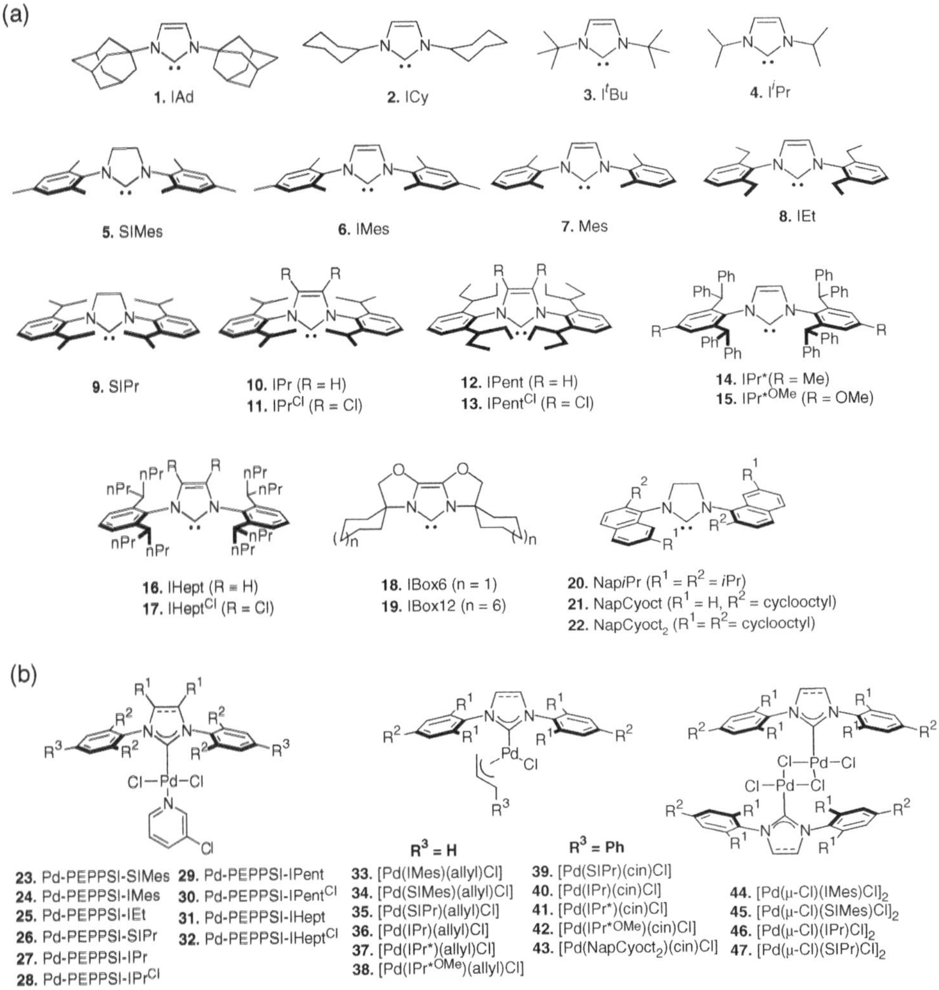

Figure 1 *Selected examples of (a) five-membered imidazole and imidazoline-based NHC ligands and (b) Pd-NHC pre-catalysts used in cross-coupling*

6.3 The relationship between *N*-heterocyclic carbene structure and reactivity

6.3.1 Steric parameters of NHC ligands

Unlike phosphines, the steric bulk of NHC ligands is projected downward towards the metal center, thus measurement of NHC steric bulk using Tolman's cone angle[11] was deemed insufficient (Figure 2). The Tolman cone angle is a method used to quantify the steric properties of phosphine ligands and is obtained by measuring the angle of the cone that contains all of the ligand atoms, when the apex of the cone is at a metal–phosphorus distance of 2.28 Å.

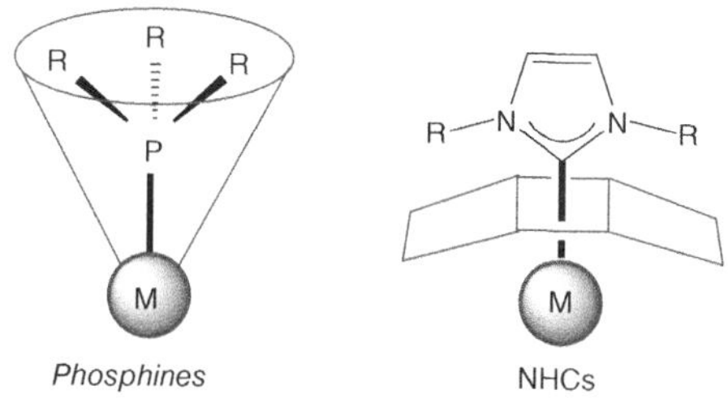

Figure 2 *General steric topology of phosphine ligands and NHC ligands*

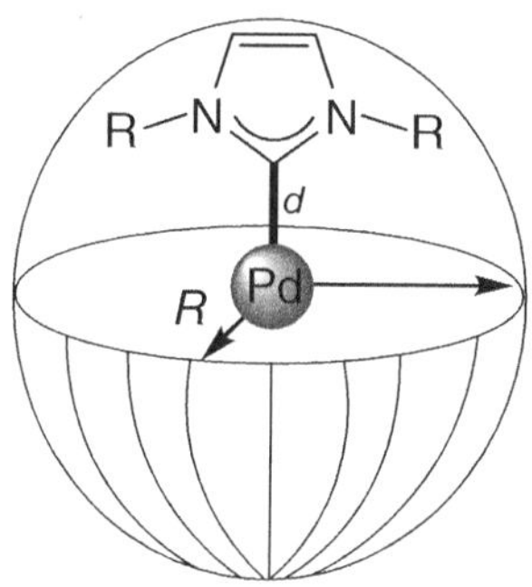

NHC	$\%V_{Bur}$
IAd (**1**)	36.1
I^tBu (**3**)	35.5
IMes (**6**)	31.6
IPr (**10**)	33.6
SIMes (**5**)	32.7
SIPr (**9**)	35.7
PPh_3	30.5
PCy_3	35.3

Figure 3 *Percent buried volume ($\%V_{Bur}$), steric parameter of some common NHC ligands computed from the DFT-optimized geometries of [(NHC)Ir(CO)$_2$Cl] complexes in which d = 2.10 Å and R = 3.5 Å.[12b] For comparison purposes, the $\%V_{Bur}$ values of two popular phosphines are also listed*

To address this difficulty, Nolan, Cavallo, and co-workers introduced the percent buried volume ($\%V_{Bur}$), a steric parameter specifically tailored to the steric topology of NHCs. The $\%V_{Bur}$ is defined as the fraction of the volume of the first coordination sphere around a metal occupied by a given ligand (Figure 3).[12] The value of the parameter can be computed from density functional theory (DFT)-optimized structures or from X-ray crystal structures of the desired NHC-metal complex. The standard NHC-metal complex from which the parameter could be computed was initially the same [(L)Ni(CO)$_3$]-type complexes from which the Tolman cone angle of phosphines was calculated, however difficulties were encountered in the preparation of some bulkier NHC-ligated congeners.[13] As a result, it was determined that [(NHC)Ir(CO)$_2$Cl] complexes were optimal for the determination of NHC $\%V_{Bur}$ since they are readily accessible square planar complexes that simulate a reasonably bulky environment that might be encountered during a cross-coupling catalytic cycle, for example. The $\%V_{Bur}$ parameter was found to be sufficiently generic that the steric properties of phosphines could also be quantified using this technique.[14] The $\%V_{Bur}$ of some common NHCs is show in Figure 3.

6.3.2 Electronic parameters of NHC ligands

Since Tolman's seminal 1977 review, the use of IR spectroscopy in concert with a coordinated trans-ligated carbonyl (CO) as a "reporting group" has become a standard approach for probing the electronic properties of transition metals in complexes of widely varying structure.[15] It is not surprising then that a number of workers have sought to extend Tolman's phosphine methodology to NHC ligands. In 2003, Chianese *et al.*[16] reported on the use of $[(L)Ir(CO)_2Cl]$ complexes in order to compare a series of phosphine and NHC ligands in terms of their electron donating ability. It was noted that by correlating the average IR stretching frequency of the Ir system and the A_1 stretching frequency from Tolman's $[(L)Ni(CO_3)]$ complexes, a linear relationship was obtained for a series of ligands where data were available for both systems. Using this correlation, it was then possible to evaluate the Tolman electronic parameter (TEP) for these new NHC ligands by simple extrapolation. The discovery by Chianese *et al.*[16] was instrumental in that it allowed evaluation of the TEP without the need for the corresponding $[(NHC)Ni(CO)_3]$ complexes whose synthesis requires the extremely toxic $[Ni(CO)_4]$ precursor. Chianese *et al.*,[17] Altenhoff *et al.*,[18] and Frey *et al.*[19] expanded on this work and used the same Ir system to explore the donating properties of other NHC ligands including the bisoxazoline-derived NHCs (**18** and **19**) found to be highly active in challenging Suzuki–Miyaura couplings. The most significant contribution to date in this area was made by Kelly *et al.*[20], who in 2008 published a detailed account of the correlation between the TEP and mean CO stretching frequencies of $[(NHC)Ir(CO)_2Cl]$ complexes bearing a variety of structurally diverse NHC ligands. The larger sample size enabled the calculation of a more precise linear regression equation between the two parameters (Figure 4).

From these studies, Crabtree and Nolan confirmed that typical imidazole-2-ylidine NHCs are significantly more donating than even the most electron-rich phosphines, such as PCy_3, thus rendering the metal to which they are coordinated much more electron rich relative to their phosphine counterparts. In addition, it became evident that the nature of the *ortho*-alkyl substituents on the aryl ring did not significantly affect the donating ability of the carbene, demonstrating independent tunability of NHC steric and electronic properties. For example, a difference of only 0.8 cm^{-1} was observed between the TEP of IMes (**6**) and the profoundly bulkier IPr (**10**). This electronic invariability of NHC ligands starkly contrasts with trialkylphosphines, whose donor abilities vary greatly as the substituent on the P atom is varied. For example, comparing PEt_3 (**48**) and $P(^iPr)_3$ (**49**), a simple change from ethyl to isopropyl substituents results in a much larger change in the TEP of 2.8 cm^{-1}. This can be rationalized by considering that the substituents that confer steric bulk to an NHC are far removed from the carbene carbon, whereas in phosphines, they are bonded directly to the phosphorus atom. In this way, the steric and electronic parameters of this class of phosphine are intimately linked in a way that limits the tunability of the ligand class.

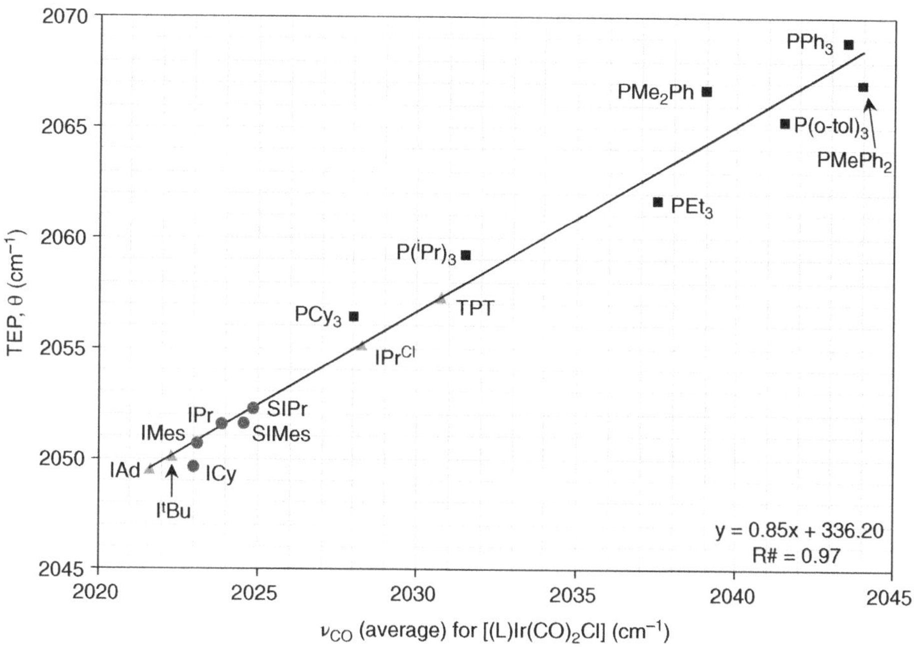

Figure 4 *Nolan and co-workers' correlation of average* ν_{CO} *values for* $[(L)Ir(CO)_2Cl]$ *complexes with the Tolman electronic parameter (TEP).*[20] *(■) Experimental values for phosphines; (•) experimental values for NHCs; and (▲) values obtained by linear regression. (See insert for color/color representation of this figure)*

6.3.3 Tuning the electronic properties of NHC ligands

Unlike steric tuning of NHCs, which has largely been focused on increasing the size of the N-substituents (compare NHCs **6**, **10**, and **12**), electronic modification has been relatively less explored.[21] The critical structural parameters that can be modified to tune the donating ability of NHCs are: (a) the NHC skeleton; (b) the nature of the substituents on the ligand backbone; and (c) the N-substituents. From the TEP studies by Nolan's group, it is evident that alterations to the ligand backbone result in the most significant changes to ligand donicity. For example, the IPr^{Cl} ligand **11** with electron-withdrawing Cl atoms on the ligand backbone and the triazole-based NHC **50** were found to be significantly less donating than their unmodified counterparts (Figure 5). Khramov *et al.*[22] also observed a similar trend when systematically varying the backbone substituent of related NHC ligands. Interestingly, Leuthäußer *et al.*[23] discovered that the donating ability of these ligands could be similarly modulated by varying the *para*-substituents on the aryl ring of the N-substituents, a somewhat unexpected finding considering that the imidazole ring of the NHC is not in conjugation with the N-aryl substituent. It is important to note that in all cases, the electronic effects observed during these modifications are significant and result in changes in the TEP of up to 9 cm^{-1} in the case of backbone modification.

Backbone modification:

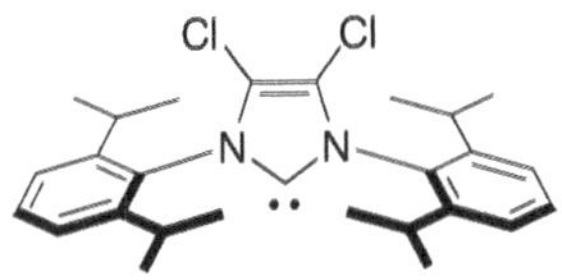

11. IPr^{Cl}
TEP = 2055.1 cm^{-1}

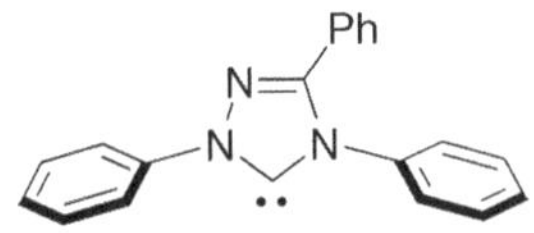

50. ITPT
TEP = 2057.3 cm^{-1}

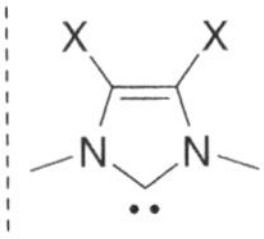

51–54

	X	TEP (cm^{-1})
51	H	2056.4
52	Cl	2060.1
53	NO_2	2061.5
54	CN	2065.2

Modification of N-Substituents:

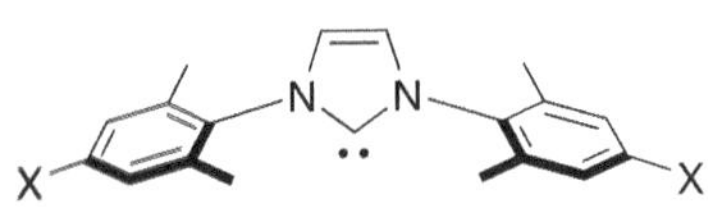

55–60

	X	TEP (cm^{-1})
55	NEt_3	2052.2
56	CH_3	2053.6
57	H	2054.3
58	Br	2055.4
59	SOAr	2057.6
60	SO_2Ar	2058.6

Figure 5 *Electronic modification of NHC ligands*

6.4 Cross-coupling reactions leading to C—C bonds that proceed through transmetalation

Cross-coupling reactions are formally double displacement reactions. In the presence of Pd catalysts, a nucleophilic organometallic reagent reacts with an electrophilic organohalide (i.e., Cl, Br, I) or pseudohalide (i.e., OTf and OTs) reagent to form a new carbon–carbon bond and an inorganic salt by-product.[24] The general mechanism of a cross-coupling reaction typically features at least three elementary steps: oxidative addition; transmetalation; and reductive elimination (Scheme 1). When Pd(II) complexes are used, the first step in the catalytic cycle is catalyst activation. Here the Pd(II) pre-catalyst is reduced to the active Pd(0) species presumably by a double transmetalation step of the organometallic reagent followed by reductive elimination. The next step is oxidative addition of Pd(0) to the organohalide or pseudohalide. The mechanism of oxidative addition has been well studied and is proposed to proceed through σ-bond coordination, followed by nucleophilic insertion into the carbon–halide bond. This process transfers the first organic fragment onto the Pd center and by doing so oxidizes the Pd(0) catalyst to Pd(II). The order of reactivity of the halide leaving group decreases in the following order I > OTf > Br > Cl and can be correlated with R—X bond dissociation enthalpies.[25] Although organochlorides are much less reactive than their bromide or iodide analogs, their low cost, stability, and ready accessibility make them more attractive reagents to work with and thus, much research has gone into improving their reactivity.[26] The next step in the catalytic cycle is transmetalation. It is here that the second organic fragment is transferred to the metal. Since the Pd(0)/(II) catalyzed cross-coupling reaction shares the common reaction steps of oxidative addition and reductive elimination, the transmetalation step offers the most opportunity for reaction

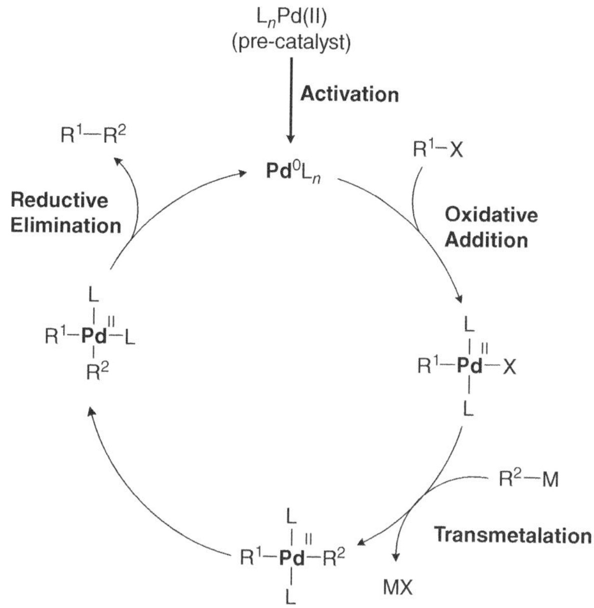

X = I, Br, Cl, OTf, OMs, OTs; **R** = aryl, alkenyl, alkyl
M = Li (Murahashi), Mg (Kumada–Tamao–Corriu), Zn (Negishi), B (Suzuki–Miyaura), Si (Hiyama), Sn (Stille–Migata)

Scheme 1 *General mechanism for Pd-catalyzed cross-coupling reaction forming C—C bonds involving organometallic transmetalators*

diversity. Depending on the nature of the transmetalating agent used, several cross-coupling protocols have developed and the names of these reactions are directly linked to the chemists that either discovered them or developed their use: the Kumada–Tamao–Corriu reaction (organomagnesium), the Negishi reaction (organozinc), the Suzuki–Miyaura reaction (organoboron), the Hiyama reaction (organosilicon), and the Stille–Migita reaction (organostannanes).[27] The mechanism of transmetalation can be composed of a single elementary step, as is the case when pre-formed organometallic reagents are used (e.g., organozincs) or may be composed of two or more elementary reactions and require the presence of a base additive, as is the case when organoboron reagents are used in Suzuki–Miyaura cross-coupling.[24] The last step in the catalytic cycle is reductive elimination, where the desired cross-coupled product is released from the cycle and the active Pd(0) catalyst is regenerated.

6.5 Kumada–Tamao–Corriu

The cross-coupling of an organomagnesium reagent (Grignard reagent) with an organohalide or pseudohalide serves as a direct method for assembling C—C bonds. The first transition metal catalyzed cross-coupling of Grignard reagents was reported

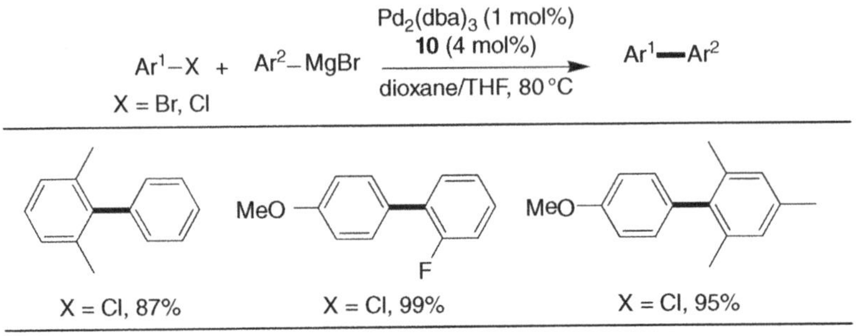

Scheme 2 *Reactivity of in situ-generated catalyst system in KTC coupling*

in 1972 independently by the groups of Kumada and Tamao[28] and Corriu.[29] In these seminal reports, the authors successfully employed Ni-based catalyst systems to couple a variety of aryl and vinyl halides with alkyl and aryl Grignard reagents. A few years later, Yamamura *et al.*[30] reported the first KTC reaction catalyzed by Pd in 1975. Despite the potential for straightforward C—C bond construction using readily available Grignard reagents, the KTC reaction is much less common relative to those employing other organometallic reagents (e.g., boron, zinc, tin). This is largely due to the enhanced functional group tolerance observed with other organometallics, such as organoboron (Suzuki–Miyaura reaction), organozinc (Negishi reaction), and organostannane (Stille–Migita reaction) reagents, relative to Grignard reagents, which are both strongly basic and nucleophilic.[31] As a result, only modest development of this reaction has taken place in recent years, with very few reports involving the use of Pd-NHC systems.

In 1999, Huang and Nolan[10d] were the first to investigate NHCs in the KTC reaction. Using $Pd_2(dba)_3$ (1 mol%) and IPr•HCl (4 mol%) (**10**) as the catalyst system, unactivated aryl iodides, bromides and chlorides were coupled with a variety of aryl Grignard reagents. Compared with the less bulky IMes ligand (**6**), IPr (**10**) was found to dramatically enhance catalyst performance. These results are consistent with earlier studies by Arduengo *et al.*[32] and Huang *et al.*[33] in which bulky *ortho*-substituents on the imidazole N-aryl moieties helps facilitate fast reductive elimination. Under the optimized conditions, sterically hindered biaryls were generated in excellent yield, however tetra-*ortho*-substituted biaryls could not be obtained (Scheme 2).

In 2003, Beller and co-workers investigated different NHC-Pd naphthoquinone complexes in the KTC coupling between aryl Grignard reagents and alkyl chlorides.[34] In particular, in situ generated NHC-Pd species were compared against well-defined catalyst systems, with the latter exhibiting higher catalytic activity in this reaction. IMes (**6**) was found to be the optimal ligand for this reaction, outperforming the bulkier IPr (**10**) ligand. Furthermore, when higher concentrations of the NHC ligand was used for the in situ generated systems, catalyst deactivation was observed, thus demonstrating the importance of a free and accessible coordination site on Pd in these reactions. Well-defined mono-carbene Pd(0) naphthoquinone complex (**61**) proved to be the optimal catalyst, coupling primary alkyl chlorides and aryl magnesium bromides

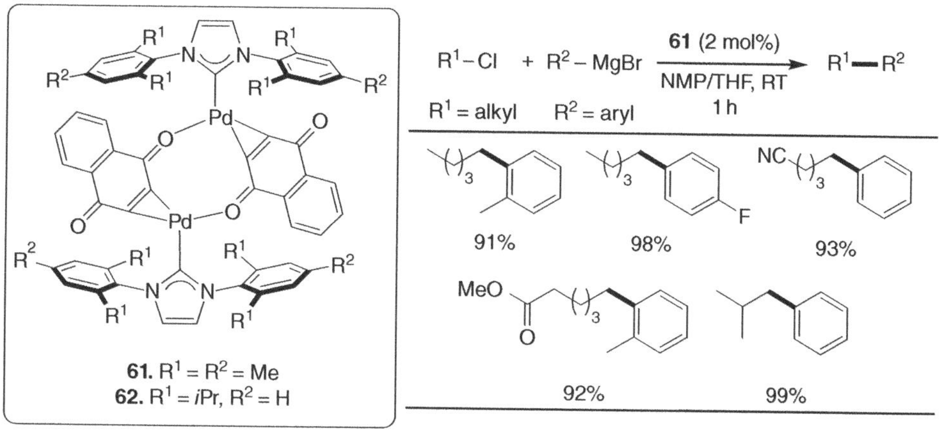

Scheme 3 Kumada–Tamao–Corriu coupling using well-defined Pd-NHC complex **61**

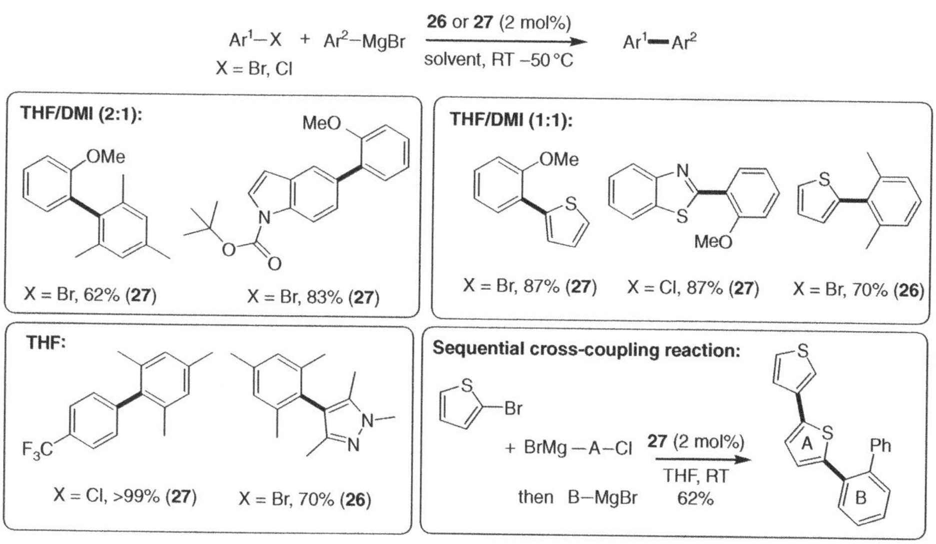

Scheme 4 Reactivity of Pd-PEPPSI-SIPr (**26**) and Pd-PEPPSI-IPr (**27**) in KTC coupling

possessing various functional groups under mild reaction conditions. Unfortunately, this reaction could not be extended to include secondary alkyl chlorides (Scheme 3).

In 2007, Organ and co-workers evaluated Pd-PEPPSI complexes **24–27** in the KTC reaction between (hetero)aryl halides and aryl Grignard reagents.[35] Both Pd-PEPPSI pre-catalysts SIPr (**26**) and IPr (**27**) were equally highly effective, producing a range of hindered biaryls and drug-like heterocycles in high yields under mild conditions (temperature ranging from room temperature to 50 °C) (Scheme 4). In particular, tetra-*ortho*-substituted biaryls could, for the first time, be synthesized at room temperature using catalyst **27**. Further, PEPPSI pre-catalysts **26** and **27** also tolerated amine

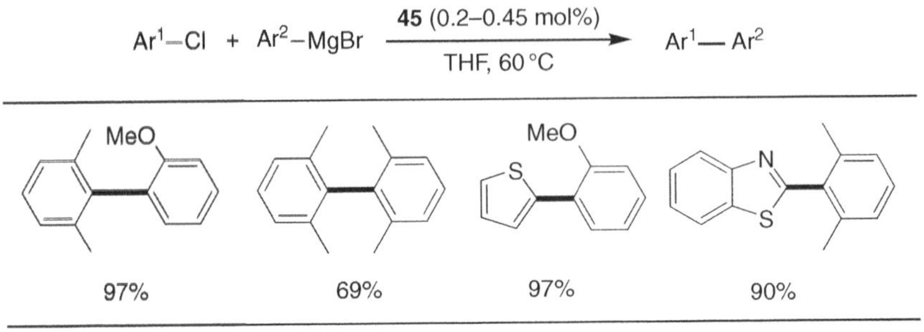

Scheme 5** Synthesis of tri- and tetra-ortho-substituted biaryl products in KTC coupling catalyzed by Pd-NHC dimer **45

protecting groups, such as Boc and sulfones as well as unprotected phenols using modified conditions. Organ and co-workers also demonstrated that catalyst **27** could be used in a multi-component reaction, allowing for the generation of polyaryl and polyheteroaryl architectures in a single operation (Scheme 4).

Later in 2009, Nolan and co-workers demonstrated that tri- and tetra-*ortho* biaryl products could also be generated using well-defined dimer complexes of the formula $[Pd(\mu{-}Cl)(NHC)Cl]_2$ (Figure 1, **44–47**).[36] Four NHC ligands were screened (IPr, SIPr, IMes, SIMes) and the SIPr ligand was found to give optimal results. Under the optimized conditions, $[Pd(\mu{-}Cl)(SIPr)Cl]_2$ (**45**) proved to be highly efficient, coupling Grignard reagents with a variety of functionalized aryl, heteroaryl and sterically hindered aryl chlorides using low catalyst loadings (between 0.2 mol% and 0.45 mol%). In particular, a number of heterobiaryl products were produced in moderate to excellent yields (Scheme 5).

In 2010, Organ and co-workers further investigated the KTC reaction using a series of Pd-PEPPSI complexes. In this structure–activity relationship study, the NHC, halide, and pyridine (also known as "throw-away") ligands were varied to study their effect on catalyst activation and overall catalyst performance.[37] Consistent with previous findings,[36] the IPr ligand was found to give optimal results, as complexes containing less bulky ligands, such as IMes (e.g., **24**, Figure 1), failed to give the cross-coupled product. However, there appears to be a limit to the amount of bulk that can surround the Pd center as the bulkier adamantyl ligand (**65** and **66**, Figure 6) also failed to give the cross-coupled product. Introducing *ortho*-substituents on the pyridine ligand was also found to negatively impact the activity of the catalyst. Kinetic experiments showed that pre-catalyst **64** is activated more slowly than **63** and **27**, leading to lower conversion to product **67** (Figure 6). The anionic ligands (Br, Cl) on the Pd were found to have no significant impact on catalyst activity.

Later in 2010, Wu and co-workers developed a series of carbene adducts of cyclopalladated ferrocenylimine (Figure 7) and evaluated their activity in the KTC reaction between aryl halides and aryl Grignard reagents.[38] Palladacycle **68** was identified as the most active catalyst, generating a number of di- and tri-*ortho*-substituted biaryls in good to excellent yield using 0.5 mol% catalyst loading and 2 equiv. of LiCl (Scheme 6).

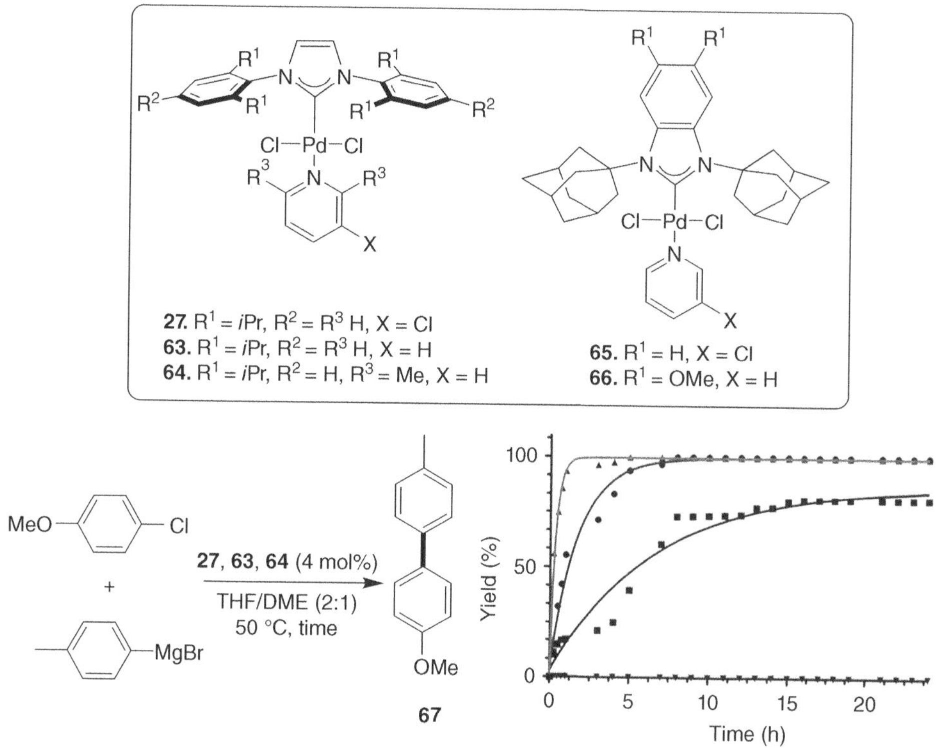

Figure 6 *Structures of some Pd-PEPPSI complexes used in this study. Rate of formation of **67** by using catalysts **29** (•), **63** (■), **64** (▲) and $PdCl_2$ (▼; control). Conversion was determined by using GC/MS against a calibrated internal standard (undecane)*

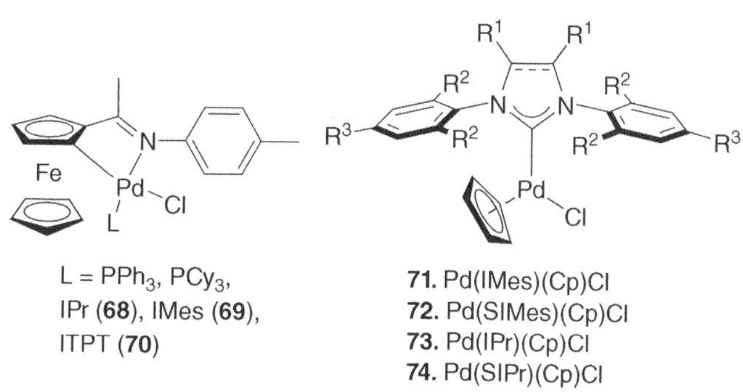

Figure 7 *Structures of well-defined Pd-NHC complexes used in KTC coupling*

Complexes containing the less bulky IMes ligand or bulky phosphine ligands (PPh_3, PCy_3) were found to be less active, especially at lower catalyst loadings.

Then in 2011, Jin *et al.*[39] reported the use of well-defined NHC-Pd complexes with the formula [Pd(NHC)(Cp)Cl] (Figure 7) in the KTC coupling of aryl and heteroaryl

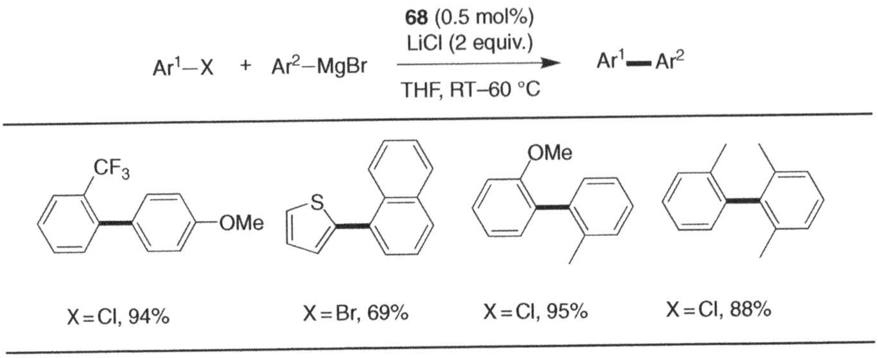

Scheme 6 *Reactivity of palladacycle **68** in KTC coupling generating di- and tri-ortho-substituted biaryls*

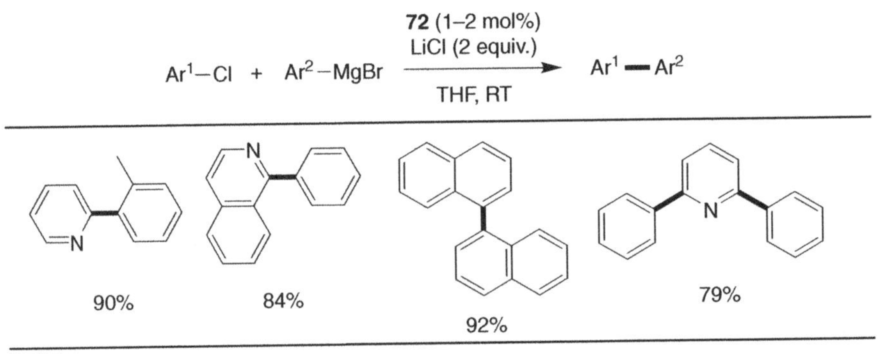

Scheme 7 *Reactivity of [Pd(SIMes)(Cp)Cl] **72** in KTC coupling*

chlorides with aryl Grignard reagents at room temperature. In this study, Jin and co-workers found that NHC ligands with less steric bulk (**71** and **72**) showed higher catalytic activity than NHC ligands with increased steric bulk (**73** and **74**). This is in stark contrast to previous reports by Nolan's group,[33] Organ's group,[35] and Cazin's group,[36] where catalyst systems containing the more sterically bulky IPr and SIPr ligands were shown to increase catalyst efficiency. The authors attribute the difference in trend to the presence of the cyclopentadienyl (Cp) ligand, stating that the less bulky ligand helps facilitate the formation of Pd(0) faster. Complex **72** was identified as the most active catalyst, coupling a variety of sterically encumbered aryl and heteroaryl chlorides using 1 mol% catalyst loading (Scheme 7).

Later that same year, Lui and co-workers introduced a dinuclear Pd complex (**75**), which contains a 2,7-bis(mesitylimidazolylidenyl)naphthyridine (IMes-NP) ligand and evaluated its activity in the coupling of aryl halides with alkyl Grignard reagents.[40] Complex **75** was found to couple cyclohexylmagnesium bromide with a variety of aryl bromides in good to moderate yields; however, only trace amounts of product were obtained with aryl chlorides (Scheme 8). Nevertheless, this was the first example of an alkyl Grignard reagent being coupled with aryl bromides using NHC-Pd complexes.

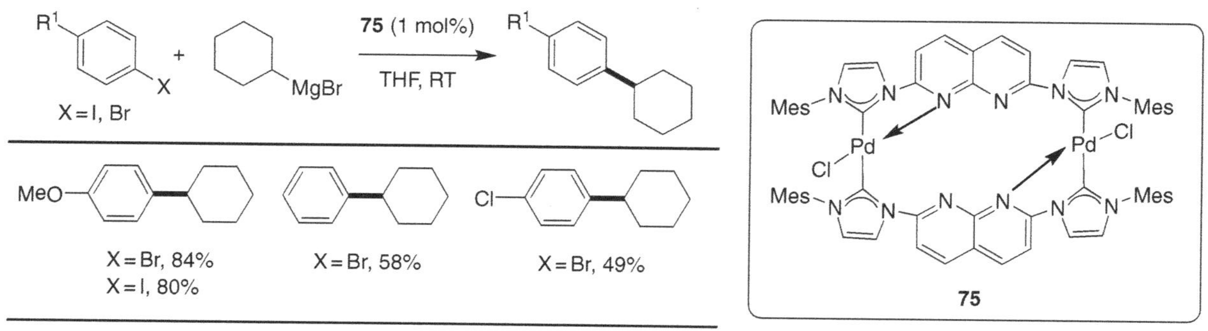

Scheme 8 *Kumada–Tamao–Corriu coupling of cyclohexylmagnesium bromide with aryl halides using **75***

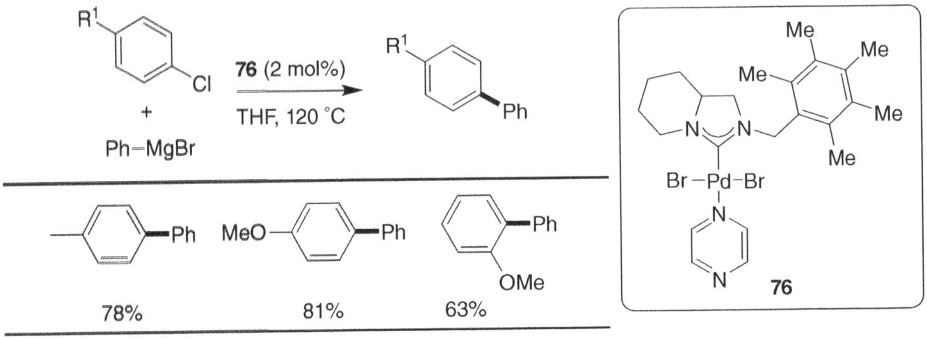

Scheme 9 *Reactivity of **76** in KTC coupling*

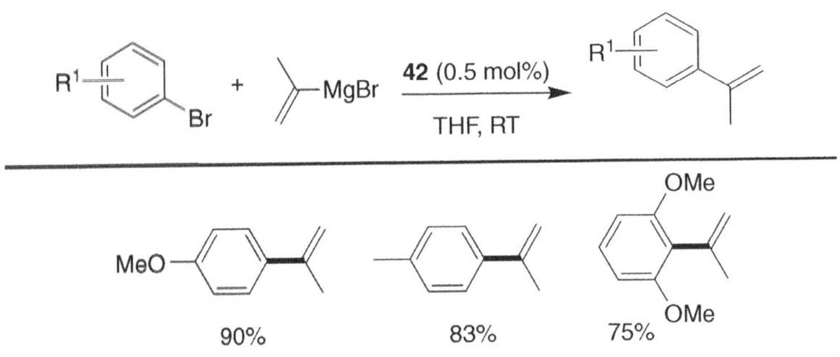

Scheme 10 *Kumada–Tamao–Corriu coupling of isopropenylmagnesium bromide with aryl bromides catalyzed by [Pd(IPr*OMe)(cin)Cl] (**42**)*

Then in 2013, Türkmen and Kani developed mixed Pd(II) complexes containing NHC ligand piperidoimidazolin-2-ylidene and "throw away ligands" pyrazine and pyridine.[41] Complex **76** was tested in the KTC reaction and was found to couple aryl chlorides with phenylmagnesium bromide in good to excellent yield in THF at 50 °C using 1 mol% catalyst loading (Scheme 9).

More recently, Nolan and co-workers developed [Pd(IPr^{*OMe})(cin)Cl] (**42**) and evaluated its activity in KTC coupling.[42] Complex **42** was found to be highly active, requiring only 0.5 mol% catalyst loading to couple aryl bromides with isopropenylmagnesium bromide in high yield at room temperature. Unfortunately, when substituted vinylmagnesium bromides (e.g., substituted in the β and/or β′ positions) were employed, no cross-coupling was observed (Scheme 10).

6.6 Suzuki–Miyaura

The Suzuki–Miyaura reaction is perhaps one of the most well studied and commonly employed cross-coupling protocols for the construction of C—C bonds. First reported in 1979 by Miyaura *et al.*,[43] this reaction involves the coupling of an organoboron reagent (e.g., boronic acid) with an organo halide or pseudohalide in the presence of a base. Organoboron reagents offer many advantages over other organometallic reagents, including high reactivity as transmetallating agents, air and moisture stability, commercial availability, thermal stability, and the low toxicity of the reagents themselves and their by-products.[44] Furthermore, both excess organoboron reagent and the inorganic by-products formed can be easily separated from the desired product at the end of the reaction, making purification techniques easy and straightforward. The Suzuki–Miyaura reaction can also tolerate a number of reaction conditions (i.e., solvents, bases, temperatures, etc.) and functional groups, thus enhancing the scope and utility of this transformation. Throughout its development over the past 40 years, many excellent reports and reviews highlighting the major advances in NHC-Pd-catalyzed Suzuki–Miyaura coupling have been published.[45] Therefore, this section will primarily focus on recent developments in Suzuki–Miyaura coupling involving the use of new Pd-NHC complexes in the formation of challenging C—C bonds under mild reaction conditions.

6.6.1 The formation of tetra-*ortho*-substituted (hetero)biaryl compounds

While the synthesis of biaryls under mild reaction conditions has been widely explored and seen much progress over the years, the formation of multi-*ortho*-substituted biaryls, in particular tetra-*ortho*-substituted biaryls, via Suzuki–Miyaura coupling still remains a challenge, especially under mild reaction conditions.[26] This is unfortunate since tetra-*ortho*-substituted biaryls are common motifs found in the structure of many important biologically active compounds and organic materials.[46] Therefore, the development of catalyst systems capable of facilitating the coupling of sterically hindered di-*ortho*-substituted coupling partners is highly desirable and has been the subject of intense study.

In 1997, Johnson and Foglesong[47] reported the synthesis of an unsymmetrical biaryl with tetra-*ortho*-substitution in 12% overall yield using $Pd(PPh_3)_4$ and Na_2CO_3 as the base. In 2002, Buchwald and co-workers reported the first catalyst system capable of preparing a library of tetra-*ortho*-substituted biaryls.[48] Using $Pd_2(dba)_3$ in combination with a tertiary phosphine ligand, a variety of aryl bromides were coupled with aryl boronic acids in high yield in refluxing toluene. A couple of years later, Altenhoff *et al.*[18a] reported the first example of a NHC ligand used to prepare sterically congested biaryls. Using a sterically bulky, yet flexible derivative of their bioxazoline-derived NHC ligand (IBox12•HOTf) (**19**) with $Pd(OAc)_2$, sterically hindered aryl chlorides and boronic acids were coupled to give a variety of tetra-*ortho*-substituted biaryls in high yield. Although this catalyst system was capable of coupling aryl chlorides, which are less expensive and more readily available than their bromide and iodide analogs, strictly anhydrous conditions were required to avoid proto-deboronation of the boronic acid, a common side reaction observed in Suzuki–Miyaura coupling. Nevertheless, the report highlighted for the first time the importance of "flexible steric bulk" around the metal center necessary to promote this challenging reaction. Unfortunately, the need for high reaction temperatures (>100 °C), the use of excess

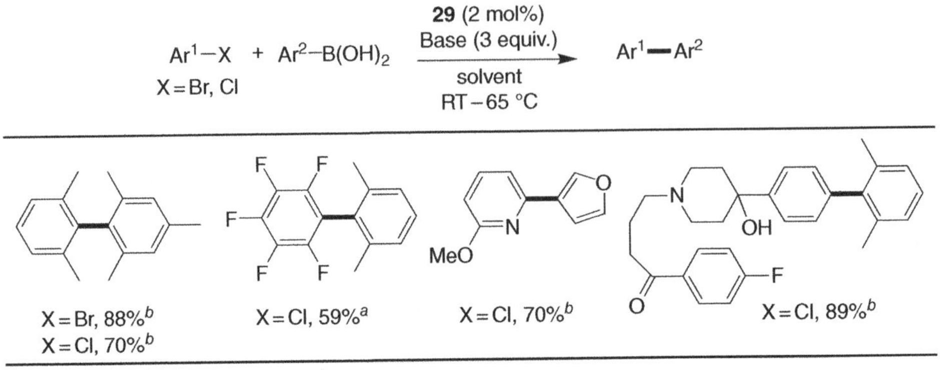

aKOH, dioxane; bKOtBu, tBuOH, 4 Å M.S.

Scheme 11 *Preparation of tetra-ortho-substituted and functionalized biaryls using* ***29***

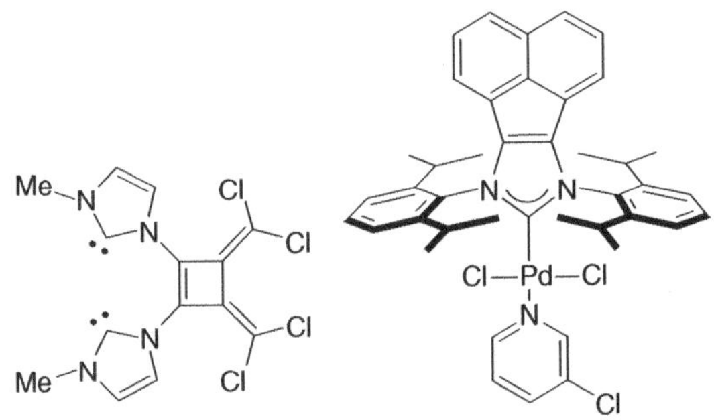

77 **78**

Figure 8 *Structures of NHC ligand* ***77*** *and Pd-NHC pre-catalyst* ***78*** *used in preparation of tetra-ortho-substituted biaryls*

ligand, and high catalyst loadings with the early protocols was a major drawback and placed significant limitations on this useful methodology.

More recently, the use of well-defined Pd-NHC pre-catalysts has been demonstrated to be extremely effective in the construction of sterically encumbered and functionalized biaryls. In 2009, Organ and co-workers were able to improve on the reaction temperature, coupling a variety of sterically hindered aryl halides with boronic acids at 65 °C using their sterically bulky Pd-PEPPSI-IPent pre-catalyst (**29**) (Scheme 11).[49] The Organ group also noted the importance of "flexible steric bulk" since when the cyclopentyl analog of **12** was employed, only traces of product were observed compared with 91% conversion when using the IPent ligand. Further, the mild reaction conditions were tolerant of sensitive functional groups, such as alcohols and ketones, thus significantly widening the substrate scope of the reaction compared with previous reports.

A year later, Schmidt and Rahimi[50] reported ligand **77** (Figure 8) to be highly efficient in Suzuki–Miyaura coupling. Combining **77** with $Pd(OAc)_2$, this in situ generated

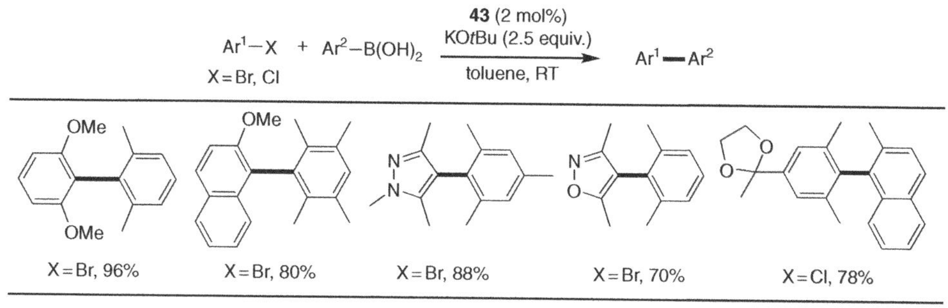

Scheme 12 *Reactivity of **43** in the Suzuki–Miyaura coupling of tetra-ortho-substituted biaryls*

system was able to generate various hindered biaryls at room temperature and tetra-*ortho*-substituted biaryls at elevated temperatures (60–110 °C).

A significant advancement in the construction of sterically encumbered tetra-*ortho*-substituted biaryls came a year later in 2011, when Wu *et al.*[51] reported the first example of an NHC ligand capable of promoting this challenging reaction at room temperature via Suzuki–Miyaura coupling. Employing 2 mol% of complex **43** (Figure 1) along with KO*t*Bu as the base, a variety of bulky tetra-*ortho*-substituted biaryls were formed in high yield (Scheme 12). When directly compared against other Pd-NHC-based pre-catalysts reported to be highly efficient in this transformation, complex **43** was found to significantly out-perform all other complexes. The authors attribute the enhanced catalytic activity of this catalyst system to the unique steric properties of the bulky NHC ligand. Advanced DFT calculations showed that the naphthyl side chains on the NHC are twisted with respect to the metal center, leaving two out of the four faces of the ligand open. The less hindered quadrants of the metal–ligand coordination sphere would promote the oxidative addition and transmetalation steps, whereas the bulkier quadrants would facilitate fast reductive elimination.

Shortly thereafter, in 2012, Chartoire *et al.*[52] reported improvements in catalyst efficiency using [Pd(IPr*)(cin)Cl] (**41**). With only 1 mol% of pre-catalyst **41** and KOH as the base, a variety of tetra-*ortho*-substituted biaryls were produced either at room temperature or at 65 °C (Scheme 13). The % V_{Bur} was calculated for the IPr* ligand and was found to be higher than all other reported [Pd(NHC)(cin)Cl] complexes (Table 1).

Also in 2012, Tu and co-workers investigated the use of acenaphthoimidazolylidene-based PEPPSI pre-catalysts in Suzuki–Miyaura couplings with sterically encumbered substrates.[53] Complex **78** (Figure 8) was found to be the most efficient, generating tetra-*ortho*-substituted biaryls in excellent yield with low catalyst loadings (0.05 mol%); however, elevated temperatures (80 °C) and a harsh base (KO*t*Bu) were still required to achieve full conversation. Interestingly, the protocol could also be extended to polychloro aromatics to generate a variety of polyarylbenzenes in good to excellent yields (Scheme 14).

More recently, Bastug and Nolan[42] reported the synthesis of tetra-*ortho*-substituted biaryls using [Pd(IPr*^{OMe})(cin)Cl] (**42**). While the catalytic activity of **42** was found to be

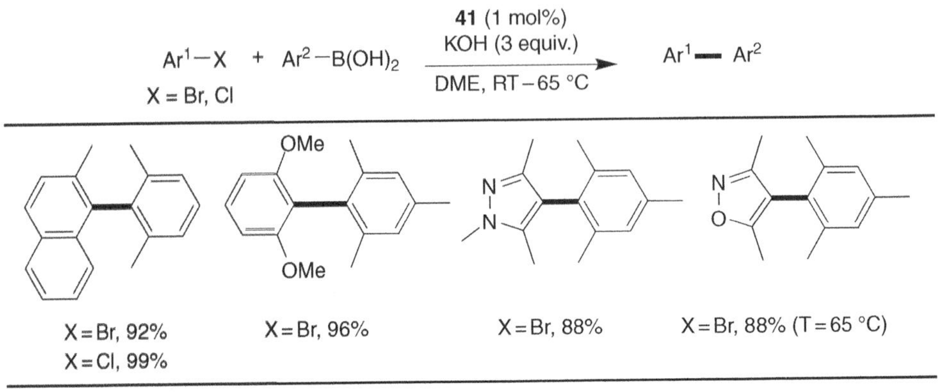

Scheme 13 *Preparation of tetra-ortho-substituted biaryls using **41***

Table 1 *Comparison of the percent buried volume (%V_{Bur}) values in [Pd(NHC)(cin)Cl] complexes*

NHC	IPr (**10**)	SIPr (**9**)	NapCyoct (**21**)	IPr* (**14**)
%V_{Bur} [a]	36.7	37.0	42.0	44.6

[a] %V_{Bur} calculated for Pd–C1 = 2.00 Å. Values taken from ref. [52].

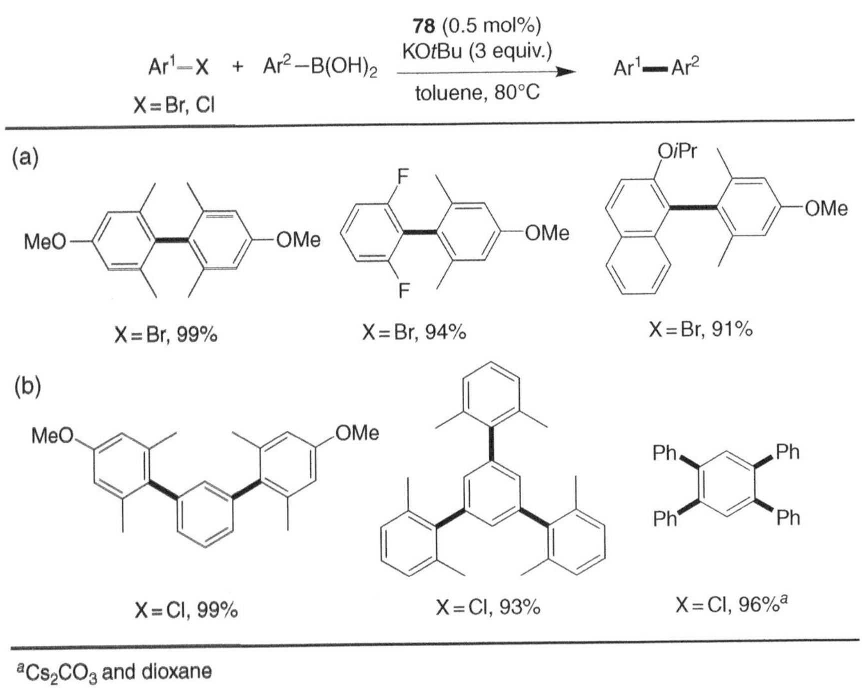

Scheme 14 *Preparation of (a) tetra-ortho-substituted biaryls and (b) polyarylbenzenes using **78***

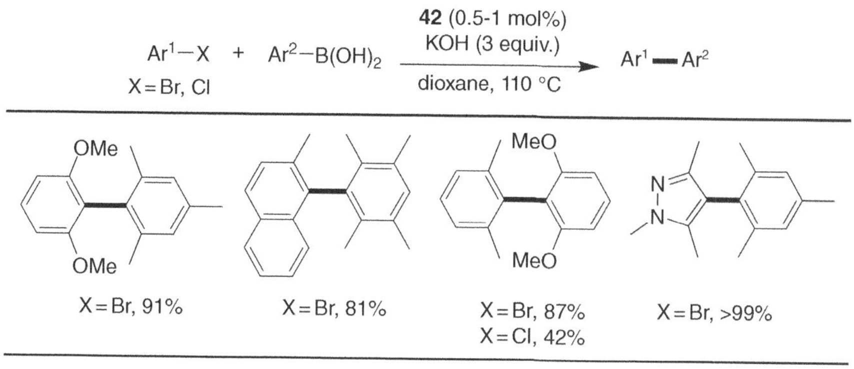

Scheme 15 *Reactivity of* ***42*** *in the Suzuki–Miyaura coupling of tetra-ortho-substituted biaryls*

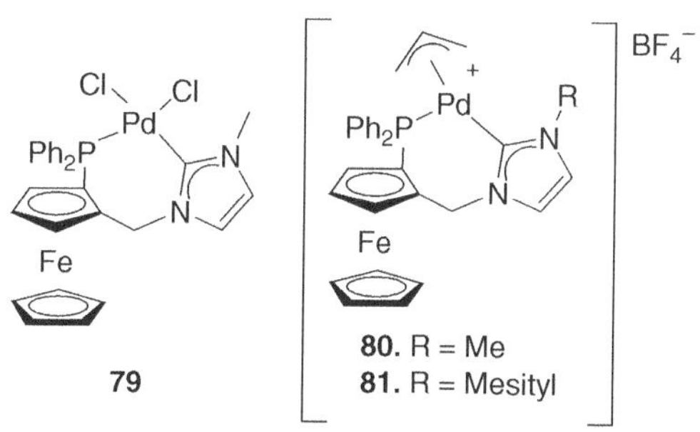

Figure 9 *Structures of well-defined Pd-NHC complexes used in asymmetric Suzuki–Miyaura coupling*

good, requiring low catalyst loadings (0.5–1 mol%), elevated reaction temperatures (110 °C) were still required for the reaction to go to completion. Nevertheless, a variety of hindered aryl halides and boronic acids were coupled in high yield (Scheme 15).

6.6.2 Enantioselective Suzuki–Miyaura coupling

The development of catalyst systems promoting efficient asymmetric couplings would be valuable for accessing configurationally stable derivatives in a highly direct manner, however the development of such protocols has been rare. In 2010, Debono *et al.*[54] were the first to report the use of a Pd-NHC pre-catalyst in asymmetric Suzuki–Miyaura coupling. A series of neutral and cationic palladacycles bearing chiral phosphine-NHC ligands with planar chirality were prepared (Figure 9, **79–81**) and evaluated in the asymmetric coupling of aryl bromides with aryl boronic acids. Complex **79** was found to be the most active and was used to prepare a variety of chiral binaphthyl compounds in good yield but with relatively low enantioselectivities (enantiomeric excess, *ee*, up to 42%, Table 2).

Table 2 *Preparation of chiral binaphthyls in asymmetric Suzuki–Miyaura coupling catalyzed by* ***79***

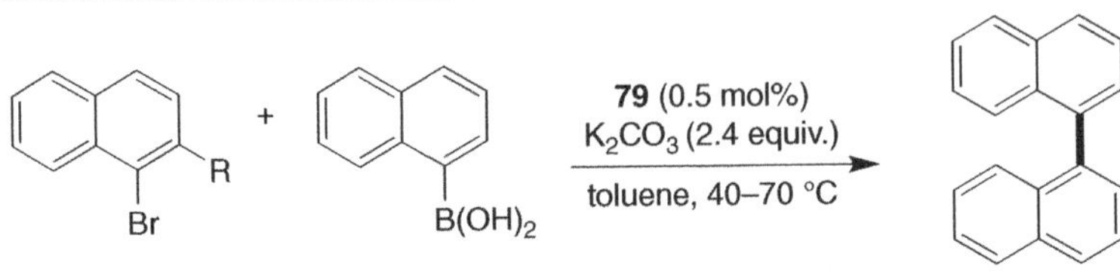

Entry	R	Temperature (°C)	Yield (%)[a]	ee (%)[b]
1	Me	70	89	38
2	OMe	70	86	33
3	OEt	70	89	30
4	Me	40	88	42
5	OMe	40	93	33
6	OEt	40	92	24

[a] Isolated yield.
[b] Determined by HPLC with a Chiracel-OJ column.

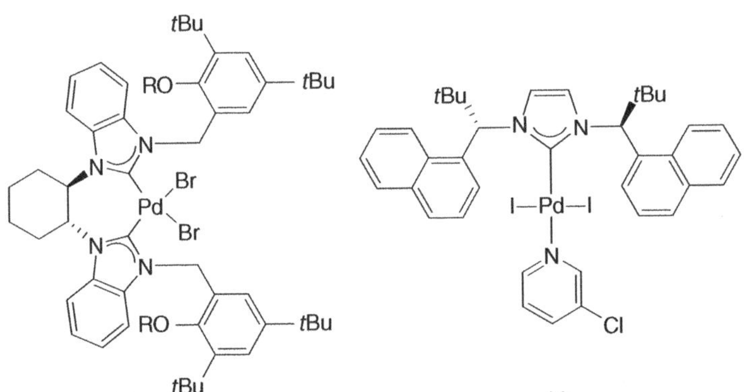

82 **83**

Figure 10 *Structures of well-defined chiral Pd-NHC complexes used in asymmetric Suzuki–Miyaura coupling*

In 2014, Zhang and co-workers examined a series of chiral bis-NHC Pd-pre-catalysts in the enantioselective Suzuki–Miyaura coupling of naphthyl halides with naphthyl boronic acids.[55] Chiral binaphthyl compounds were produced in good yields (up to 95%) and moderate enantioselectivities (up to 64% *ee*) using 3 mol% of complex **82** (Figure 10 and Table 3). While changing the halide ligands was found to have no influence on the reaction outcome, steric effects associated with the coupling partners were observed to impact the enantioselectivity of the reaction.

Later in 2014, Benhamou *et al.*[56] also reported the synthesis of chiral biaryl products using the new bulky chiral PEPPSI pre-catalyst **83** (Figure 10). The combination

Table 3 *Preparation of chiral binaphthyls in asymmetric Suzuki–Miyaura coupling catalyzed by* ***82***

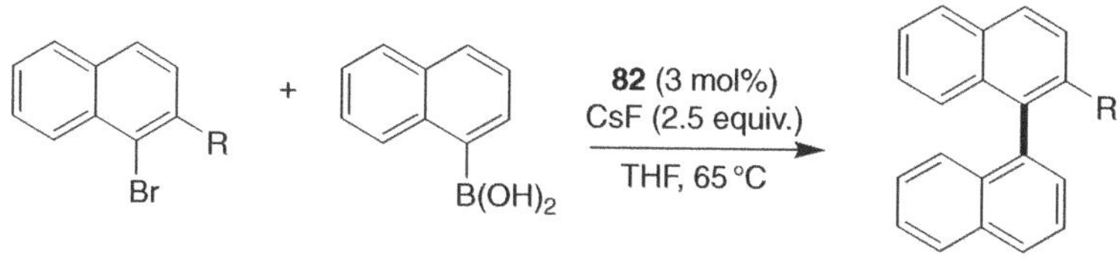

Entry	X	R	Yield (%)	ee (%)
1	Br	OMe	83	45
2	Cl	OMe	52	64
3	Br	OCH_2Ph	45	57

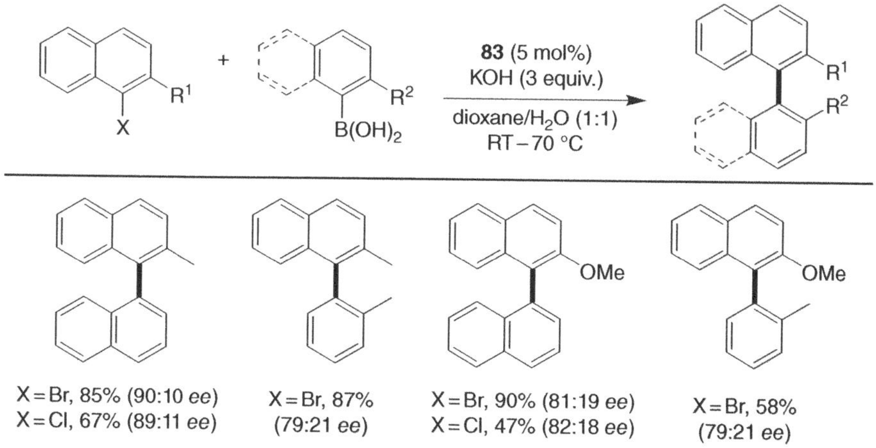

Scheme 16 *Reactivity of* ***83*** *in asymmetric Suzuki–Miyaura coupling*

of KOH and a 1:1 mixture of dioxane:H_2O in the presence of 5 mol% of complex **83** was found to be effective, allowing the preparation of a variety of biaryls in good yields and with moderate to good enantioselectivities (up to 88% *ee*). Although the results were highly substrate-dependent, bulkier substituents at the 2-position of the aryl halide generally led to better selectivities (Scheme 16).

More recently, Takeda *et al.*[57] reported [Pd(SIPr)(cin)Cl)] (**39**) to be highly efficient in the regioselective and stereospecific cross-coupling of enantiopure 2-arylaziridines with arylboronic acids. Using 4 mol% of the pre-catalyst, a variety of chiral 2-aryl-phenethylamine derivatives were produced under mild reaction conditions in high yield and with excellent enantioselectivity (up to 99% *ee*, Scheme 17). Electron neutral and electron deficient aziridines and sterically encumbered and/or functionalized aryl boronic acids were all well tolerated under the developed conditions.

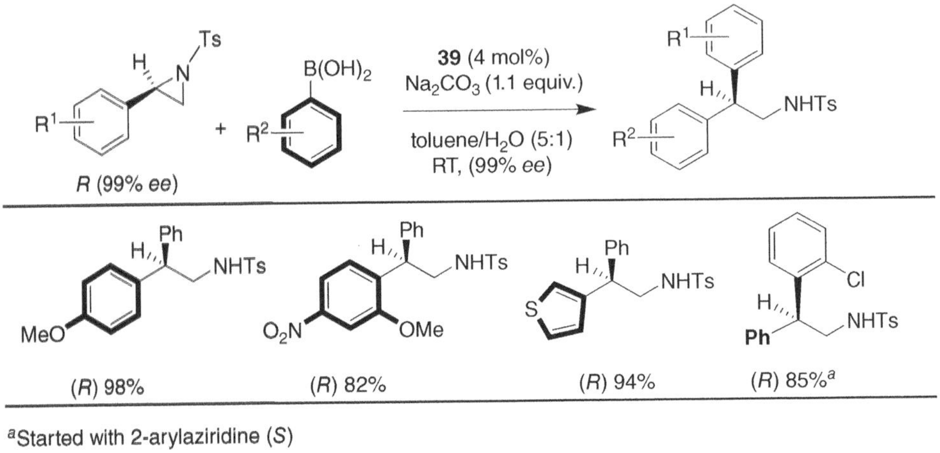

Scheme 17 *Asymmetric Suzuki–Miyaura arylation of 2-arylaziridines using* **39**

6.6.3 Formation of sp^3—sp^3 or sp^2—sp^3 bonds

While the construction of $C(sp^2)$—$C(sp^2)$ bonds has been widely explored in Suzuki–Miyaura coupling, the formation of $C(sp^3)$—$C(sp^3)$ or $C(sp^2)$—$C(sp^3)$ bonds has attracted much less attention, with very few examples involving the use of NHC ligands. The main difficulties with these reactions are slow oxidative addition of alkyl halides, compared with aryl or vinyl halides, and β-hydride elimination (BHE), which leads to formation of undesired by-products.[58] In a series of publications, Fu and co-workers demonstrated that these problems can be overcome by employing electron-rich, sterically bulky phosphine ligands; however, when NHC ligand **6** was employed, no cross-coupled product was observed.[59] The first example of an NHC ligand used in alkyl–alkyl Suzuki–Miyaura coupling was reported by Arentsen *et al.*[60] in 2004. The active NHC-based catalyst was generated in situ by combining imidazolium salt IPr•HCl (8 mol%) with $Pd_2(dba)_3$ (4 mol%), and was shown to couple various alkyl bromides with alkyl-9-BBN reagents at 40 °C with modest yields.

In 2008, Valente *et al.*[61] reported that Pd-PEPPSI-IPr (**27**) was the first well-defined pre-catalyst used to efficiently construct both $C(sp^3)$—$C(sp^3)$ and $C(sp^2)$—$C(sp^3)$ bonds. After extensive optimization, the optimal base and catalyst loading were shown to be K_3PO_4•H_2O and 4 mol%, respectively, alkylating a variety of alkyl bromides and aryl bromides and chlorides at room temperature in high yield. A variety of functional groups, including phenols, ester, anilines, carbamates, and aldehydes were well tolerated under the mild conditions. Furthermore, the reaction conditions could also be extended to include heteroaryl halides, generating interesting drug-like compounds (Scheme 18).

In 2010, Kantchev and co-workers developed a series of palladacycle complexes of the formula [Pd(NHC)(palladacycle)] and evaluated their activity in Suzuki–Miyaura coupling.[62] Complex **84** was found to be the most active pre-catalyst, efficiently forming a number of C—C bonds, including sp^2—sp^2, sp^2—sp^3, sp^3—sp^2, and sp^3—sp^3, in good-to-excellent yield (Scheme 19). The high reactivity of complex **84** in

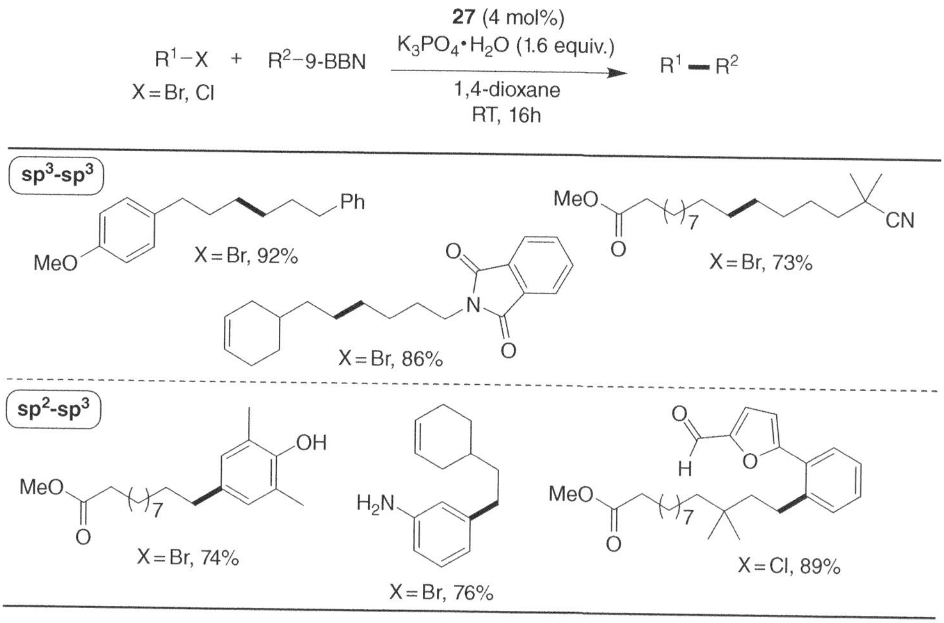

Scheme 18 *Reactivity of* ***27*** *in alkyl Suzuki–Miyaura coupling*

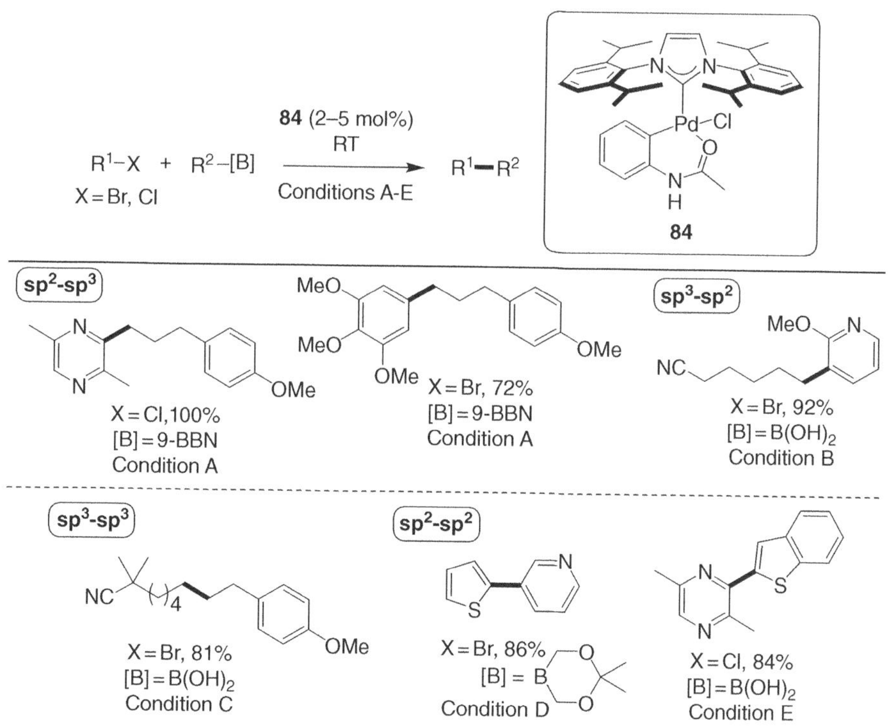

Conditions: (A) K_3PO_4, THF/H_2O, **84** (4 mol%); (B) KO*t*Bu, dioxane/MeOH, **84** (4 mol%); (C) $K_3PO_4 \cdot H_2O$, dioxane, **84** (4 mol%); (D) K_3PO_4, dioxane/H_2O,100 °C, **84** (5 mol%); (E) NaOH, THF/MeOH, **84** (2 mol%).

Scheme 19 *Reactivity of* ***84*** *in Suzuki–Miyaura coupling*

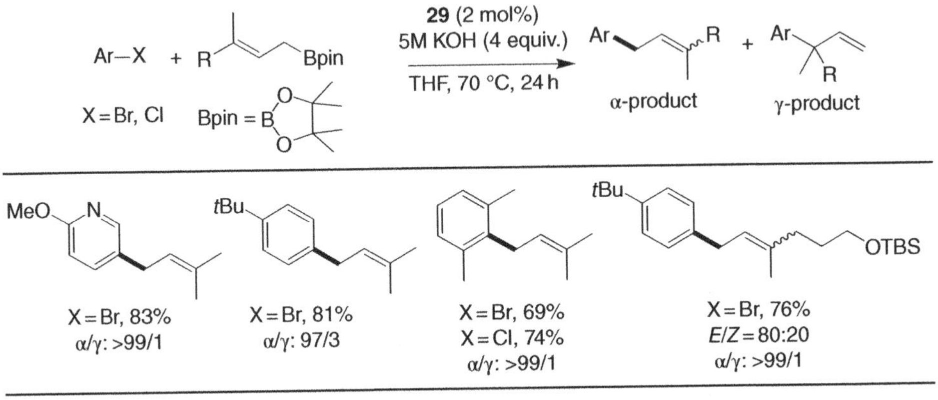

Scheme 20 *Reactivity of **29** in Suzuki–Miyaura coupling of allylboronic acid pinacol ester derivatives*

a variety of Suzuki–Miyaura couplings was attributed to faster activation of the Pd(II) pre-catalyst to form the active Pd(0) species.

Another example of a challenging Suzuki–Miyaura reaction is the regioselective allylation of aryl halides. The difficulty in this transformation lies in controlling the regiochemical outcome associated with the allylboronate reagent. In 2012, Farmer *et al.*[63] disclosed that Pd-PEPPSI-IPent (**29**) could effectively catalyze the cross-coupling of various allylboronate derivatives with a variety of aryl and heteroaryl halides to generate the desired linear isomer in a highly regioselective manner (i.e., >97% α–selectivity) under mild reaction conditions (Scheme 20). Furthermore, in the case of trisubstituted allylboronates with different substituents on the olefin, only minor olefin geometry isomerization was observed (*E/Z* ≈ 80/20).

More recently, Lu and co-workers examined the use of [Pd(IPr)(Im)Cl_2] (**85**) in aqueous Suzuki–Miyaura coupling of benzyl chlorides with (hetero)arylboronic acids or potassium trifluroborate salts.[64] Reactions were carried out using 1.5 mol% of complex **85** in neat water at 60 °C. Using this methodology, a variety of diarylmethane products were generated in high yield. This protocol provided an alternative route for accessing diarylmethane derivatives, which are important motifs found in many active pharmaceutical ingredients and supramolecules (Scheme 21).

6.6.4 The formation of (poly)heteroaryl compounds

Heterobiaryls are common motifs found in the structure of many important compounds, including pharmaceutical agents, ligands in metal catalysis, and functional polymers. The Suzuki–Miyaura reaction has proven to be one of the most powerful and reliable techniques for the formation of (poly)heterobiaryls. The major challenge in such reactions is that heterocyclic motifs often cause catalyst poisoning or deactivation, which is why high catalyst loadings are often required. While several Pd-phosphine catalysts have been reported to effectively promote this challenging reaction,[65] the use

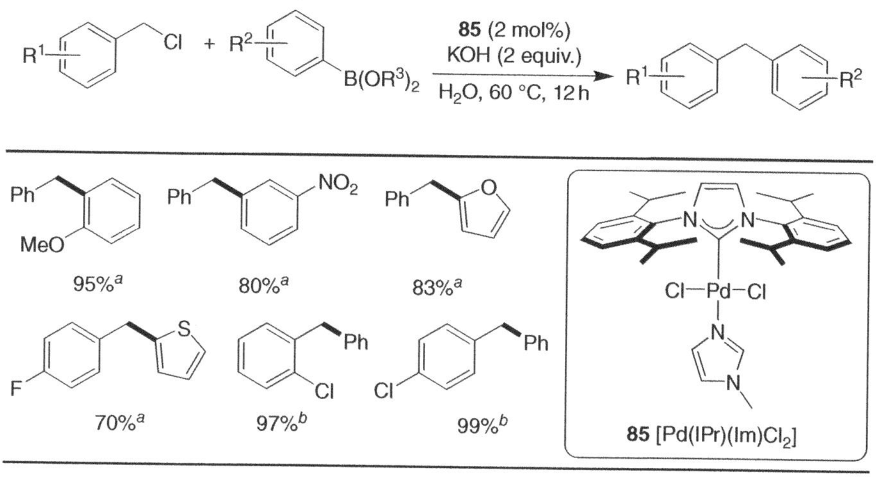

${}^{a}B(OH)_2$, ${}^{b}BF_3K$

Scheme 21 *Reactivity of **85** in the preparation of diarylmethane products using Suzuki–Miyaura coupling*

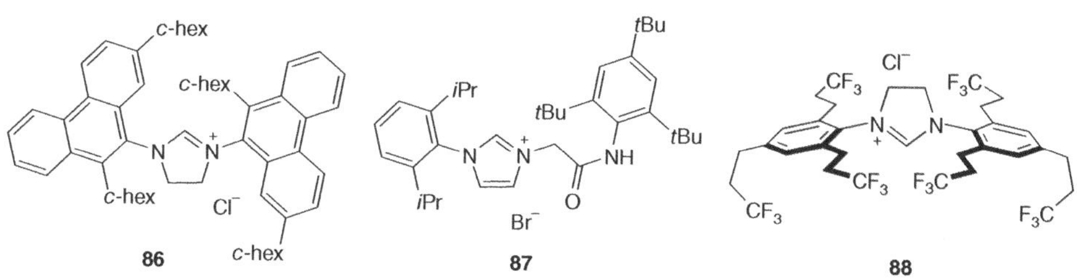

Figure 11 *Structures of NHC ligands used in the preparation of (poly)heterobiaryls by Suzuki–Miyaura coupling*

of Pd-NHC catalysts for the synthesis of heterobiaryls from heteroaryl halides, in particular chlorides, and/or heteroaryl boron reagents has been less developed. In 2005, Andrus and co-workers demonstrated that bis-phenanthryl NHC salt **86** (H_2-ICP•HCl, Figure 11) in combination with $Pd(OAc)_2$ could effectively couple 2-chloropyridine with a variety of sterically encumbered arylboronic acids, generating heterobiaryls in high yield at room temperature.[66] A year later, Organ and co-workers used the well-defined Pd-PEPPSI-IPr (**27** in Figure 1) to form (poly)heterobiaryls in good-to-excellent yield under a variety of reaction conditions (Scheme 22).[67]

Later in 2010, Kantchev and co-workers examined palladacycle **84** in Suzuki–Miyaura coupling and found it to be a highly active catalyst, forming a variety of C-C bonds, including the formation of (poly)heteroaromatic compounds at elevated temperatures (Scheme 19, see above).[62] Also in 2010, Lee and co-workers developed amido-*N*-imidazolium salt **87** and evaluated it in the formation of heteroaromatic

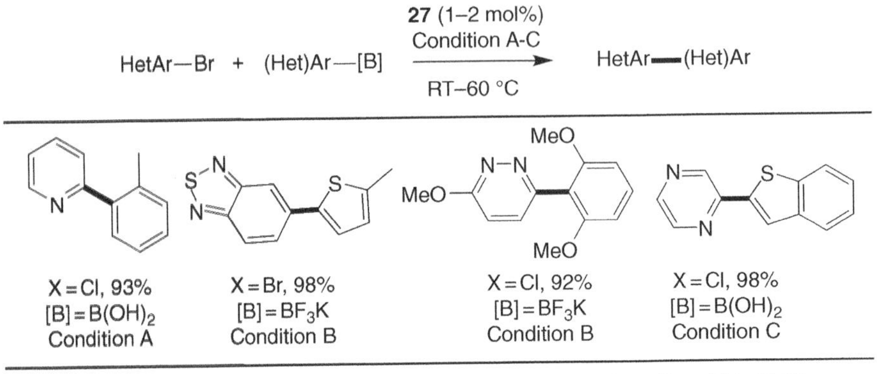

Conditions: (A) KO*t*Bu, *i*PrOH, **27** (1 mol%), RT; (B) K_2CO_3, MeOH, **27** (2 mol%), 60 °C; (C) dioxane, **27** (2 mol%), 60 °C.

Scheme 22 *Preparation of (poly)heterobiaryls using* ***27***

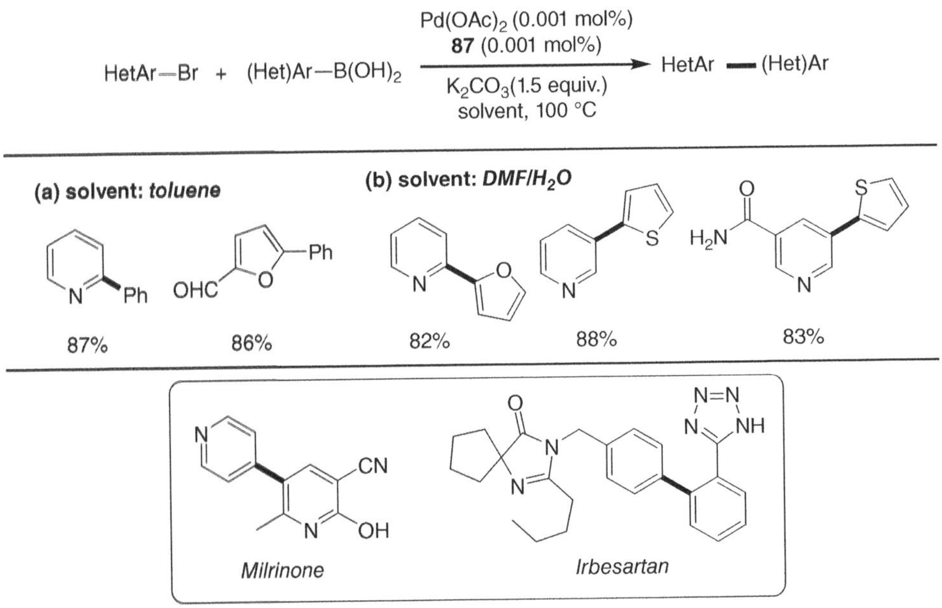

Scheme 23 *Preparation of poly(hetero)biaryls and natural products milrinone and irbesartan at low catalyst loadings using* ***87***

compounds via Suzuki–Miyaura coupling.[68] The catalyst system (**87**/Pd(OAc)$_2$) was found to be extremely active, generating a variety of heterobiaryls using very low catalyst loadings (as low as 0.0001 mol%). Most notably, pharmaceutical compounds milrinone and irbesartan were prepared using this methodology (Scheme 23).

In 2012, Tu *et al.*[53] demonstrated that robust acenaphthoimidazolydiene PEPPSI-based Pd-complex **78** (0.5 mol%) could be used to effectively couple (hetero)arylboronic

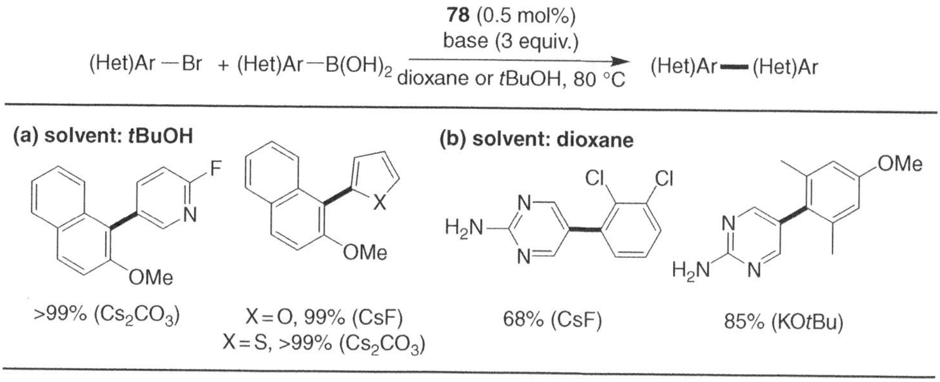

Scheme 24 *Reactivity of **78** in Suzuki–Miyaura coupling*

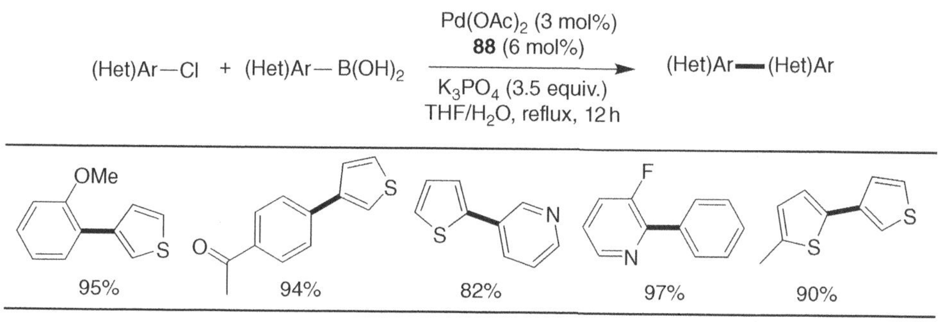

Scheme 25 *Preparation of (poly)heterobiaryls using an in situ generated Pd-NHC complex Pd(OAc)$_2$/**88***

acids with (hetero)aryl bromides at 80 °C using a various bases and solvents. Furthermore, this protocol could also be extended to construct other (poly)heterobiaryl compounds that display important biological activity (Scheme 24).

Also in 2012, Lu and co-workers developed an in situ generated system using ligand **88** and evaluated it in the formation of sterically hindered biaryls and (poly) heterobiaryls via Suzuki–Miyaura coupling.[69] The catalyst system was found to be highly efficient in coupling (hetero)aryl chlorides with a variety of (hetero)arylboronic acids generating a variety of (poly)heterobiaryl products in high yield using K_3PO_4 as the base (Scheme 25).

More recently, Küçükbay and co-workers investigated the microwave-assisted Suzuki–Miyaura coupling of 2- and 3-halopyridines using a Pd(OAc)$_2$/benzimidazolium salt catalyst system and K_2CO_3 as the base.[70] All complexes were reported to have similar activity with the exception of complex **93,** which was found to be the least active in Suzuki–Miyaura coupling. In general, heteroaryl chlorides were found to be less reactive than the bromide analogs. In addition, 3-halopyridines were found to couple more efficiently than 2-halopyridines (Table 4).

Table 4 Microwave-assisted Suzuki–Miyaura coupling of 2- and 3-halopyridines

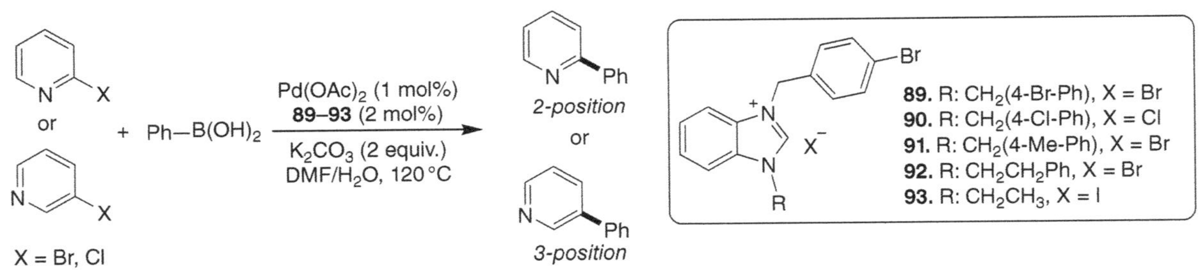

Entry	NHC salt	Yield (%) X = Br	Yield (%) X = Cl
		2-Halopyridine	
1	**91**	70	62
2	**89**	65	64
3	**90**	67	65
4	**92**	72.5	66
5	**93**	67	52
		3-Halopyridine	
6	**89**	90	67
7	**90**	89	77
8	**91**	91	70
9	**92**	93	73.5
10	**93**	90	58

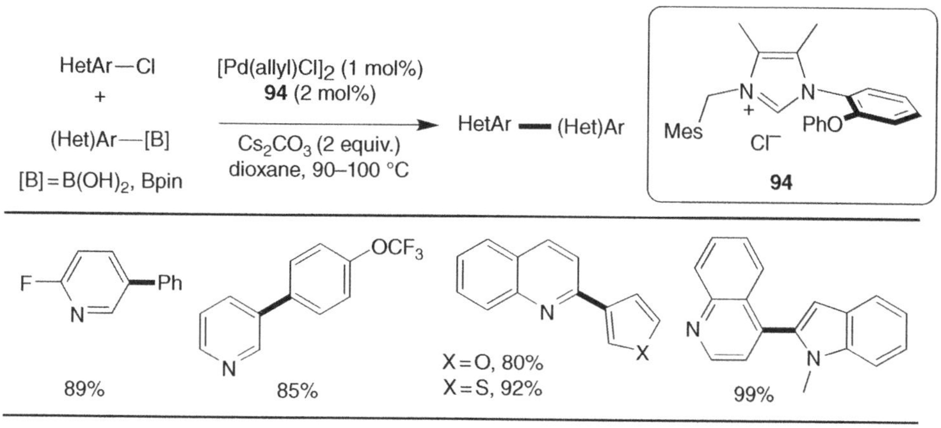

Scheme 26 *Preparation of (poly)heterobiaryls using NHC ligand* ***94***

Also in 2013, Kuriyama and co-workers developed a series of ether-imidazolium salts **94** for the construction of (poly)heterobiaryls.[71] $Pd(OAc)_2$/**94** proved to be the most effective catalyst system, coupling various heteroaryl chlorides and (hetero)arylboronic acids in good yields (Scheme 26).

6.7 Negishi coupling

Since its discovery in 1977, the Negishi reaction has become a highly practical and reliable method for the construction of C—C bonds.[72] The reaction involves the coupling of an organohalide or pseudohalide with an organozinc compound. Although organozincs are not as air and moisture stable as their organoboron counterparts, their high reactivity and excellent functional group compatibility make the Negishi reaction an attractive alternative to other cross-coupling transformations.[73]

The first successful NHC-Pd catalyzed Negishi cross-coupling reaction was reported by Hadei *et al.*[74] in 2005, in which an in situ generated system composed of $Pd_2(dba)_3$ (2 mol %) and **10** (8 mol%) was used for room temperature alkyl-alkyl cross-coupling. $Pd_2(dba)_3$/**10** proved to be an effective catalyst system, coupling various unactivated alkyl bromides and alkyl organozinc reagents with a variety of functionality in good to excellent yields (Scheme 27).

A year later, Organ *et al.*[75] reported the synthesis and characterization of well-defined NHC-Pd pre-catalysts **23**, **24**, and **27**. Pd-PEPPSI-IPr (**27**) was found to be the most active, successfully coupling all combinations of alkyl and aryl centers (i.e., sp^3—sp^3, sp^3—sp^2, sp^2—sp^3, sp^2—sp^2) (Scheme 28). Since the σ–donor abilities of IMes and IPr carbenes are similar, improvements in catalyst performance were attributed to the increased steric bulk around the metal centre.[68] Organohalides (i.e., Cl, Br, I) and pseudohalides (i.e., triflates, tosylates, mesylates) were used in this study and proved to be excellent oxidative addition partners, all resulting in high yield of the

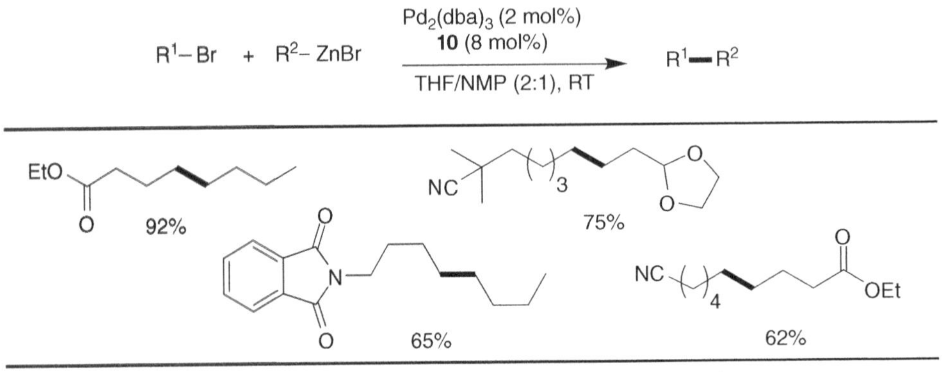

Scheme 27 Reactivity of ligand **10** in Negishi coupling

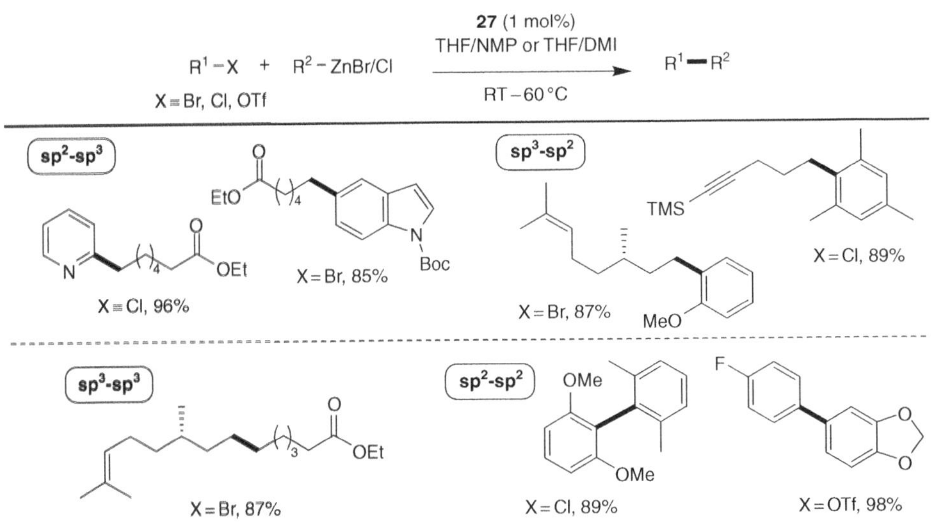

Scheme 28 Reactivity of **27** in Negishi coupling

cross-coupled product. Additionally, sterically encumbered biaryls and drug-like heteroaromatic systems were easily synthesized using this catalyst system. Furthermore, this well-defined, air stable pre-catalyst (**27**) was found to significantly increase the scope, reliability, and ease-of-use of the Negishi reaction relative to in situ generated systems reported previously.[68]

In 2010, Organ and co-workers investigated the activity of Pd-PEPPSI-IPr (**27**) and Pd-PEPPSI-IPent (**29**) in the synthesis of di-, tri-, and tetra-*ortho*-substituted biaryl and heterobiaryl products.[76] With a few exceptions, **29** was found to be the optimal pre-catalyst system, generating various biaryl and heterobiaryl products possessing a variety of functional groups and/or *ortho*-substituents (Scheme 29). In particular, a number of tetra-*ortho*-substituted biaryl compounds were synthesized in excellent

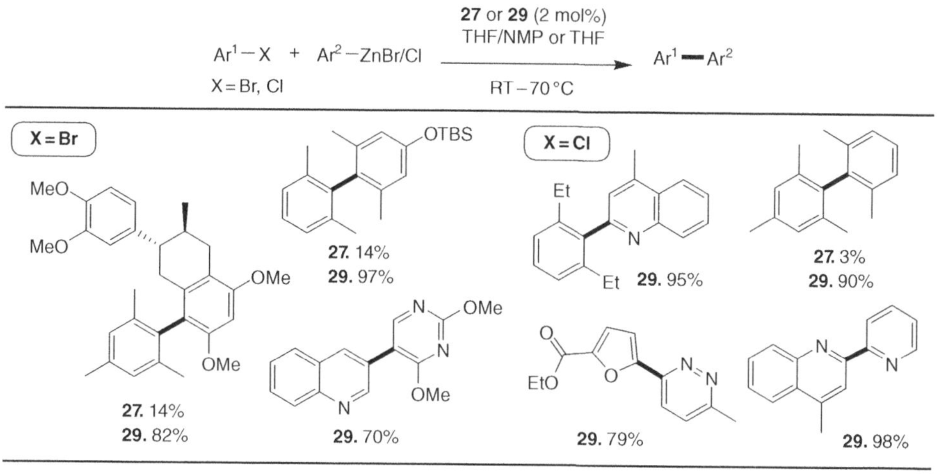

Scheme 29 *Reactivity comparison of Pd-NHC complexes **27** and **29** in the Negishi coupling*

yield and under very mild reaction conditions. Coinciding with results observed in the Suzuki–Miyaura cross-coupling, Organ and co-workers concluded that "flexible steric bulk"[18b] around the Pd center was essential for efficient catalysis in these more challenging cross-coupling reactions.

During the course of their studies in alkyl–alkyl Negishi coupling with **27**, Organ and co-workers discovered a unique solvent polarity "trigger" that allowed for the chemoselective coupling of C_{sp3}—Br bonds in the presence of C_{sp3}—Cl bonds. This encouraged the Organ group to further investigate this unique "trigger" and in 2011 they reported the first the one-pot orthogonal alkyl–alkyl Negishi cross-coupling of bifunctional and unactivated bromochloroalkanes.[77] The selective sequential coupling was made possible by the leaving group ability of the halide (C—I > C—Br >> C—Cl) and using the solvent polarity as a "trigger" to permit the chemoselective reaction. First, the C_{sp3}—Br bond was coupled in the presence of the C_{sp3}—Cl bond using a 1:2 DMI/THF solvent mixture. Then, by simply increasing the proportion of DMI in the reaction medium from 1:2 to 2:1 DMI/THF, the C_{sp3}—Cl bond was coupled with the second alkylzinc halide. Taking advantage of this solvent polarity "trigger" and the high reactivity of **27** in alkyl–alkyl coupling, a variety of functionalized alkanes were prepared in moderate to excellent yields at room temperature (Scheme 30).

Later that same year, Larrosa *et al.*[78] published a report highlighting the unusual reactivity observed with **27** in the Negishi reaction. During their studies towards the alkylation of 1,4-dibromobenzene with 1 equiv. of *n*-butylzinc bromide, they noticed that **27** selectively yielded the double alkylation product (>99.5%) whereas $Pd(PPh_3)_4$ gave mostly the monoalkylated product (94%) (Scheme 31). Further, the unusual behavior observed with **27** was not limited to the Negishi reaction. When 1,4-dibromobenzene was reacted with 1 equiv. of an organoboron or organomagnesium reagent under Suzuki–Miyaura or KTC reaction conditions, respectively, difunctionalized

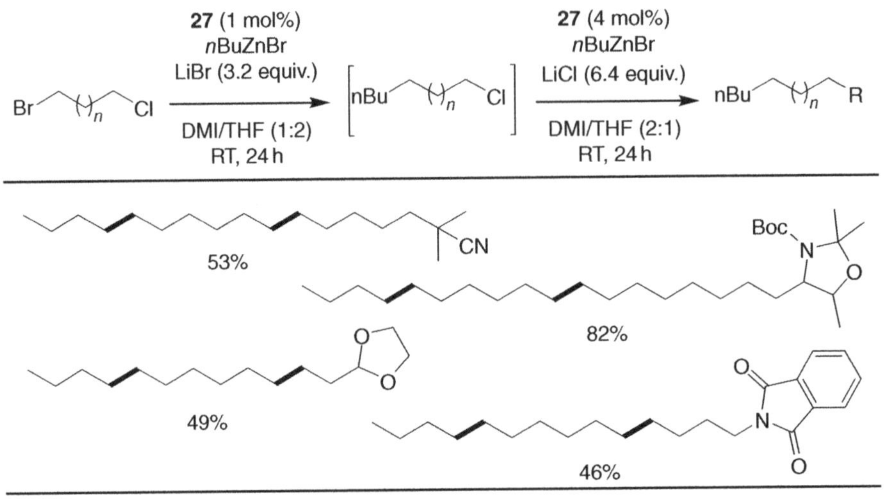

Scheme 30 *One-pot orthogonal alkyl–alkyl Negishi coupling catalyzed by **27***

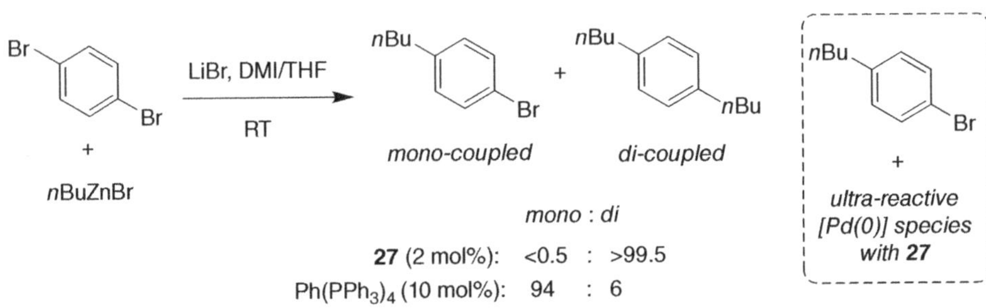

Scheme 31 *Unusual reactivity of **27** in the Negishi coupling of polybromo aromatic compounds*

products were obtained. The authors propose that the selective polyfunctionalization of polybromo aromatic compounds observed with Pd-PEPPSI-IPr is the result of an ultra-reactive Pd(0) species, which is generated after the initial reductive elimination step in the catalytic cycle, being in close proximity to the second oxidative addition site.

6.7.1 Mechanistic studies: investigating the role of additives and the nature of the active transmetalating species

Until recently, little was known about the mechanism of the Negishi reaction mediated by NHC-Pd complexes. Over the past decade, the Organ group has studied the Negishi reaction intensely using its Pd-PEPPSI complexes and many important details about the mechanism of transmetalation as well as the role of additives and polar co-solvents have been discovered.

In 2009, Organ and co-workers studied the mechanism of the alkyl–alkyl Negishi reaction and, through DFT computations, revealed transmetalation to be the rate-limiting step and not oxidative addition.[79] A key Pd–Zn interaction was also identified in the mechanism to persist beyond reductive elimination. This interaction, in combination with the NHC ligand was shown to help facilitate reductive elimination by creating a highly sterically crowded environment in the Pd coordination sphere.

In 2010, the Organ group studied the role of salt additives in alkyl–alkyl Negishi cross-coupling.[80] From previous work the authors noticed a difference in reaction outcome when different sources of zinc reagents were used. After conducting a comprehensive survey of salt additives, bromide salts were found to be the most effective promoters while the cation was found to be mechanistically benign. Double titration studies revealed that until at least 1 equiv. of MX_n (e.g., M=Li or Mg, X=Cl or Br) has been added to a salt-free alkylzinc reagent, no coupling is observed and that the reaction becomes optimal when ≥1.5 equiv. is employed. The titration study also revealed that the addition of $ZnBr_2$ inhibits formation of the active transmetalator. These results led the Organ group to propose that the active transmetalating agent is not the mono-anionic zincate (e.g., $RZnX_2^-$), as has been suggested for many years, but actually a higher-order zincate (i.e., $RZnX_3^{2-}$) since the reaction appears to become catalytic in halide ion after 1 equiv. of MX_n (relative to RZnX) has been added. Mass spectrometry and NMR spectroscopy provided further evidence consistent with the existence of higher-order zincates ($RZnX_3^{2-}$). The identity of variously charged zincates formed on mixing LiBr and *n*-BuZnBr were analyzed in mixtures of THF, DMI, and NMP and high dielectric solvents (DMI, NMP) were found to be essential for stabilizing the higher-order zincate.[81]

In 2012, Organ and co-workers were finally able to confirm the presence of high-order zincates in the alkyl–alkyl Negishi reaction by synthesizing the $RZnBr_3^{2-}$ zincates and then subjecting them to Negishi reaction conditions.[82] A variety of cross-coupled products were obtained in just 2 h and in excellent yield (Scheme 32).

Later, in 2014 McCann and Organ investigated the role of halide additives in the Negishi reaction catalyzed by **29** involving sp^2-hybridized zinc reagents.[83] Diarylzinc compounds were found to couple in low dielectric solvents (i.e., THF) with zero salt present whereas arylzinc halides failed to couple in THF alone, requiring the addition of salt. Further, unlike alkyl–alkyl coupling there was no evidence to support the

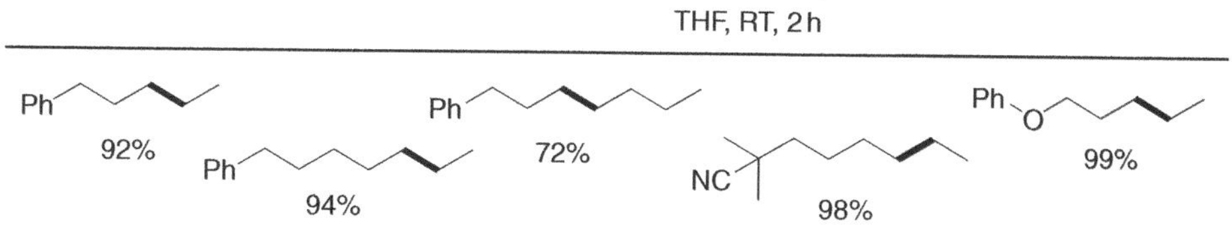

Scheme 32 *High-order zincates in alkyl–alkyl Negishi coupling catalyzed by* ***29***

involvement of high-order zincates in the coupling of aryl zinc reagents. These results are in stark contrast with sp^3-hybridized zinc reagents, which require salt, a high dielectric solvent (DMI, NMP), and the formation of high-order zincates for coupling to occur.

6.7.2 Selective cross-coupling of secondary organozinc reagents

One of the remaining challenges in the Negishi reaction is the reaction between C_{sp2} and secondary C_{sp3} centers to yield secondary alkyl aromatic substituents, common motifs found in the structure of many important active pharmaceutical ingredients.[84] Although a number of groups have developed methods to couple secondary alkyl halides with sp^2–hybridized aryl/alkenyl nucleophiles using Pd catalysis, there have been few studies published on the cross-coupling of secondary organometallic reagents with aryl halides, especially those involving secondary organozinc reagents.[68] One of the major difficulties in this transformation is controlling the regiochemical outcome of the reaction. After transmetalation, an undesired BHE/migratory insertion pathway competes with reductive elimination, leading to isomeric cross-coupled products (Scheme 33). One way for imparting better regiocontrol has been through ligand design. Indeed, the use of sterically bulky ligands has been shown to suppress isomerization by increasing the rate of reductive elimination relative to BHE.[85]

In 2011, Çalimsiz and Organ were the first to report the use of NHC ligands in this reaction.[86] Of the Pd-PEPPSI pre-catalysts screened, Pd-PEPPSI-IPent (**29**) was found to be the most active, coupling a wide variety of aryl/heteroaryl halides with secondary alkylzinc reagents with high selectivity for the non-isomeric, branched product (Scheme 33). The high regioselectivity imparted by **29** was attributed to increased ligand bulk in the metal coordination sphere, which reduces the rate of BHE relative to reductive elimination once transmetalation has occurred. However, oxidative addition partners possessing *ortho*-substituents and electron-rich substituents generally saw a decrease in regioselectivity.

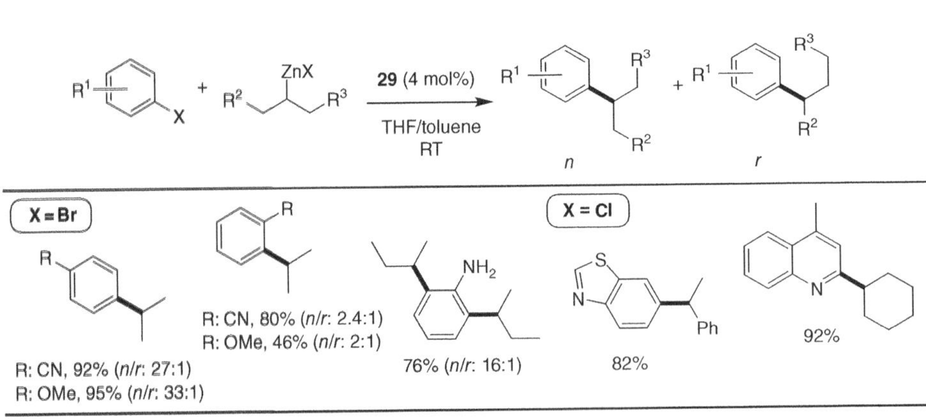

Scheme 33 *Reactivity of* ***29*** *in secondary alkyl Negishi coupling*

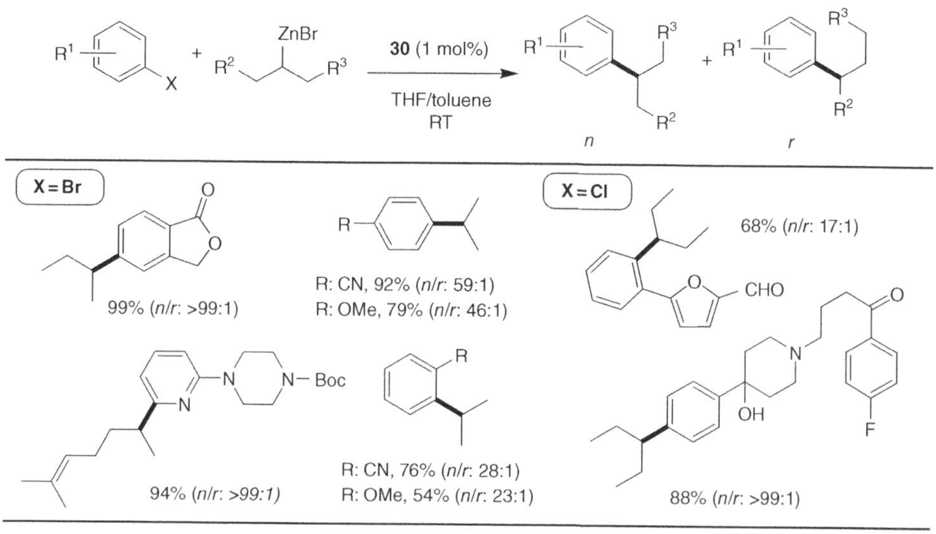

Scheme 34 *Reactivity of **30** in secondary alkyl Negishi coupling*

A year later, a series of new Pd-PEPPSI complexes containing NHCs with backbone substitution were reported by Pompeo *et al.*[87] and evaluated in the Negishi cross-coupling of secondary alkylzinc reagents. High regioselectivities were observed when using NHC ligands substituted with either electron-withdrawing or electron-releasing groups, leading the authors to conclude that the effect imparted by the backbone substituents is primarily steric in origin. Computational studies were also conducted and were found to be in accordance with the observed regioselectivities. These studies suggest that placing substituents on the backbone of the NHC ligand pushes the N-aryl groups closer to the metal coordination sphere, which in turn increases the rate of reductive elimination relative to BHE. Pd-PEPPSI-IPent^{Cl} (**30**) was identified as the most active and selective pre-catalyst, leading to virtually one desired isomer (i.e., the branched product) in the cross-coupling of secondary alkylzincs with a wide variety of electrophiles containing aldehydes, ketones, esters, amides, and even free amines, alcohols, and carboxylic acids (Scheme 34). More importantly, sterically hindered and electron-rich electrophiles could now be coupled with high selectivity for the desired branch product. However, whereas six-membered heterocyclic halides coupled with high selectivity with a variety of alkylzinc reagents, five-membered ring heterocycles were found to be stubborn substrates, yielding cross-coupled products with poor regioselectivity.

More recently, Atwater *et al.*[88] reported two new increasingly bulky Pd-NHC pre-catalysts, Pd-PEPPSI-IHept (**31**) and Pd-PEPPSI-IHept^{Cl} (**32**) and evaluated their activity in the selective cross-coupling of secondary alkylzinc reagents to five-membered ring heterocycles. Pre-catalyst **32** demonstrated the highest levels of regioselection, coupling secondary alkylzinc reagents with a wide variety of electrophiles including furans, thiophenes, benzothiophenes, benzofurans, and indoles substituted at both the 2- and

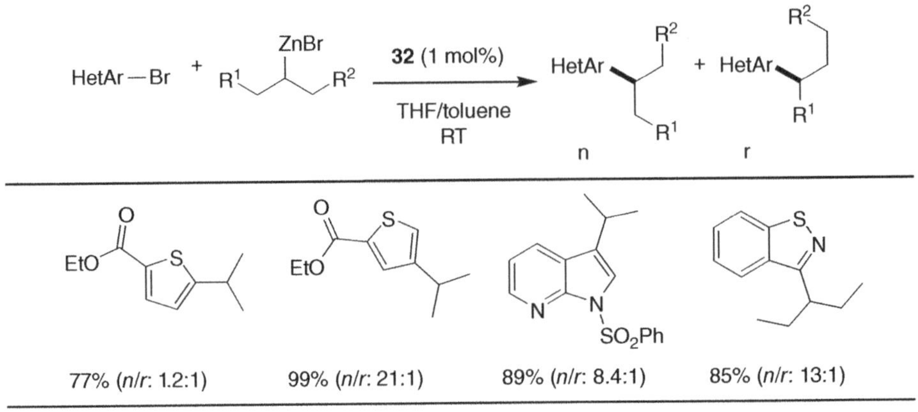

Scheme 35 Preparation of alkyl heterocycles using **32**

3-position. Moreover, in most instances, non-rearranged cross-coupled products were obtained almost exclusively demonstrating the high regioselectivity of **32** (Scheme 35).

6.8 Conclusion

The last decade has seen impressive strides being made in the cross-coupling of organozinc (Negishi), organoboron (Suzuki–Miyaura), and organomagnesium (KTC) reagents. Especially notable has been the rise of NHC ligands from chemical curiosities to robust ligands for Pd, where vast improvements in catalytic reactivity have been observed. It is now routine to be able to couple impressively sterically and/or electronically deactivated oxidative addition and organometallic starting materials. As a consequence, the number of applications of Pd-NHC catalyzed C—C cross-coupling reactions has exploded in the areas of medicinal chemistry, materials science, and natural products synthesis.

Based on results to date, it appears as though the most generally reactive Pd-NHC ligand complexes for C—C cross-coupling are those adorned with bulky groups at the 2- and 6-position of the N-aryl substituents off the imidazole-based carbene core. The notable outliers to this general trend are the spirocyclic pentacycles developed by Glorius' group. However, what is common to these quite different looking systems is that the significant bulk surrounding the Pd center is not rigidly held in place. This so-called "flexible steric bulk" is key in that it can be moved in toward the metal center to help promote reductive elimination, yet moved away so as not to negatively impact on the sterically sensitive steps of oxidative addition and transmetalation.

The relative electron richness of metals bound to NHC ligands when compared with the corresponding phosphine complexes means that oxidative addition will invariably be an easy process. Indeed, even aryl chlorides can undergo oxidative addition well below 0 °C. Transmetalation is promoted by an electron-poor metal center, establishing the first major challenge for these electron-rich Pd-NHC catalysts in the catalytic cycle. Reductive elimination is also promoted by an electron-poor Pd center, but driven forward by the pronounced steric bulk of the NHC N-aryl rings and their substituents that project into the metal's space. Thus, with bulky NHCs, it can be generally expected that transmetalation is apt to be the rate-limiting step across all members of the C—C cross-coupling family of reactions.

As cross-coupling continues its development, one thing that will be necessary is improved metrics to assess NHC ligand electronic and steric parameters.[89] Only in this way will further, systematic approaches be developed to improve catalyst performance in these mainstream chemical transformations. For example, the transmetalation of organozincs and organomagnesium reagents is facile, and in fact highly exothermic in some instances concerning the latter reagents. So, attempting to draw simplified generalizations to apply, for example to organoboron reagents that suffer from a much more difficult transmetalation, will be very difficult in the absence of more sophisticated tools to characterize ligand–metal physical parameters. The endpoint of such an exercise will be the development of a systematic, possibly mathematical approach to match ligand parameters with a particular cross-coupling reaction.

References

[1] (a) M. S. Kharasch, E. K. Fields *J. Am. Chem. Soc.* **1941**, *63*, 2316–2320; (b) M. S. Kharasch, C. F. Fuchs *J. Am. Chem. Soc.* **1943**, *65*, 504–507.

[2] The Nobel Prize in Chemistry **2010**. Nobelprize.org. 5 Aug 2015. http://www.nobelprize.org/nobel_prizes/chemistry/laureates/2010/ (accessed April 6, 2016).

[3] For a review of the Buchwald family of dialkylbiaryl phosphines, see: R. Martin, S. M. Buchwald *Acc. Chem. Res.* **2008**, *41*, 1461–1473.

[4] K. D. Hesp, M. Stradiotto *J. Am. Chem. Soc.* **2010**, *132*, 18026–18029.

[5] (a) H. W. Wanzlick *Angew. Chem. Int. Ed.* **1962**, *1*, 75–80; (b) H. W. Wanzlick, H. J. Schönherr *Angew. Chem. Int. Ed.* **1968**, *7*, 141–142.

[6] (a) K. Öfele *J. Organomet. Chem.* **1968**, *12*, 42; (b) K. Öfele *J. Organomet. Chem.* **1968**, *12*(3), P42–P43.

[7] A. J. Arduengo III, R. L. Harlow, M. Kline *J. Am. Chem. Soc.* **1991**, *113*, 361–363.

[8] W. A. Herrmann, M. Elison, J. Fischer, C. Köcher, G. R. J. Artus *Angew. Chem. Int. Ed.* **1995**, *34*, 2371–2374.

[9] A. J. Arduengo III, R. Krafczyk, R. Schmutzler *Tetrahedron* **1999**, *55*, 14523–14534.

[10] For selected seminal references, see: (a) W. A. Herrmann, C. P. Reisinger, M. Spiegler *J. Organomet. Chem.* **1998**, *557*, 93–96; (b) C. Zhang, J. Huang, M. L. Trudell, S. P. Nolan *J. Org. Chem.* **1999**, *64*, 3804–3805; (c) H. M. Lee, S. P. Nolan *Org. Lett.* **2000**, *2*, 2053–2055; (d) J. Huang, S. P. Nolan *J. Am. Chem. Soc.* **1999**, *121*, 9889–9890; (e) D. S. McGuinness, M. J. Green, K. J. Cavell, B. W. Skelton, A. H. White *J. Organomet. Chem.* **1998**, *565*, 165–178; (f) G. A. Grasa, S. P. Nolan *Org. Lett.* **2001**, *3*, 119–122; (g) J. Huang, G. Grasa, S. P. Nolan *Org. Lett.* **1999**, *1*, 1307–1309; (h) S. R. Stauffer, S. Lee, J. P. Stambuli, S. I. Hauck, J. F. Hartwig *Org. Lett.* **2000**, *2*, 1423–1426; (i) M. S. Viciu, R. F. Germaneau, S. P. Nolan *Org. Lett.* **2002**, *4*, 4053–4056; (j) C. Yang, S. P. Nolan *Organometallics* **2002**, *21*, 1020–1022.

[11] C. A. Tolman *Chem. Rev.* **1977**, *77*, 313–348.

[12] (a) A. C. Hillier, W. J. Sommer, B. S. Yong, J. L. Peterson, L. Cavallo, S. P. Nolan *Organometallics* **2003**, *22*, 4322–4326; For an online application that can be used to compute %VBur, see: (b) A. Poater, B. Cosenza, A. Correa, S. Giudice, F. Ragone, V. Scarano, L. Cavallo *Eur. J. Inorg. Chem.* **2009**, 1759–1766.

[13] R. Dorta, E. D. Stevens, N. M. Scott, C. Costabile, L. Cavallo, C. D. Hoff, S. P. Nolan *J. Am. Chem. Soc.* **2005**, *127*, 2485–2495.

[14] (a) H. Clavier, S. P. Nolan *Chem. Commun.* **2010**, *46*, 841–861; (b) O. Diebolt, G. C. Fortman, H. Clavier, A. M. Z. Slawin, E. C. Escudero-Adán, J. Benet-Buchholz, S. P. Nolan *Organometallics* **2011**, *30*, 1668–1676.

[15] For a review on the TEP and other methods for the measurement of net donor properties of phosphines, NHCs, and related ligands, see: D. G. Gusev *Organometallics* **2009**, *28*, 763–770.

[16] A. R. Chianese, X. Li, M. C. Janzen, J. W. Faller, R. H. Crabtree *Organometallics* **2003**, *22*, 1663–1667.

[17] A. R. Chianese, A. Kovacevic, B. M. Zeglis, J. W. Faller, R. H. Crabtree *Organometallics* **2004**, *23*, 2461–2468.

[18] (a) G. Altenhoff, R. Goddard, C. W. Lehmann, F. Glorius *J. Am. Chem. Soc.* **2004**, *126*, 15195–15201; (b) G. Altenhoff, R. Goddard, C. W. Lehmann, F. Glorius *Angew. Chem. Int. Ed.*, **2003**, *42*, 3690- 3693.

[19] G. D. Frey, C. F. Rentzshe, D. von Preysing, T. Scherg, M. Muehlhofer, E. Herdtweck, W. A. Herrmann *J. Organomet. Chem.* **2006**, *691*, 5725–5738.

[20] R. A. Kelly III, H. Clavier, S. Giudice, N. M. Scott, E. D. Stevens, J. Bordner, I. Samardjiev, C. D. Hoff, L. Cavallo, S. P. Nolan *Organometallics* **2008**, *27*, 202–210.

[21] For a recent review of the electronic properties of NHC ligands, see: D. J. Nelson, S. P. Nolan *Chem. Soc. Rev.* **2013**, *42*, 6723–6753.

[22] D. M. Khramov, V. M. Lynch, C. W. Biewlawski *Organometallics* **2007**, *26*, 6042–6049.

[23] S. Leuthäußer, D. Schwarz, H. Plenio *Chem. Eur. J.* **2007**, *13*, 7195–7203.

[24] J. F. Hartwig *The Organometallic Chemistry of the Transition Metals: From Bonding to Catalysis*, University Science, Sausalito, CA, **2010**.

[25] Y.-R. Lou *Comprehensive Handbook of Chemical Bond Energies*, CRC Press, Boca Raton, FL, **2007**.

[26] A. de Meijere, F. Diederich, *Metal-Catalyzed Cross-Coupling Reactions*, 2nd edn, Wiley-VCH, Weinheim, **2004**.

[27] C. C. C. Johansson Seechurn, M. O. Kitching, T. J. Colacot, V. Snieckus *Angew. Chem Int. Ed.* **2012**, *51*, 5062–5085.

[28] K. Tamao, K. Sumitani, M. Kumada *J. Am. Chem. Soc.* **1972**, *94*, 4374–4376.

[29] R. J. P. Corriu, J. P. Masse *J. Chem. Soc., Chem. Commun.* **1972**, *3*, 144a–144a.

[30] M. Yamamura, I. Moritani, S. Murahashi *J. Organomet. Chem.* **1975**, *91*, C39–C42.

[31] (a) J. Clayden, N. Greeves, S. Warren, P. Wothers *Organic Chemistry*, Oxford University Press, New York, **2001**; (b) M. B. Smith, J. March *March's Advanced Organic Chemistry: Reactions Mechanisms, and Structures*, 5th edn, John Wiley & Sons, Ltd, Chichester, **2000**.

[32] A. J. Arduengo III, H. V. R. Dias, J. C. Calabrese, F. J. Davidson *J. Am. Chem. Soc.* **1992**, *116*, 4391–4393.

[33] (a) J. Huang, E. D. Stevens, S. P. Nolan, J. L. Petersen *J. Am. Chem. Soc.* **1999**, *121*, 2674–2678; (b) J. Huang, H.-J. Schanz, E. D. Stevens, S. P. Nolan *Organometallics* **1999**, *18*, 2370–2375.

[34] A. C. Frisch, F. Rataboul, A. Zapf, M. Beller *J. Organomet. Chem.* **2003**, *687*, 403–409.

[35] M. G. Organ, M. Abdel-Hadi, S. Avola, N. Hadei, J. Nasielski, C. J. O'Brien, C. Valente *Chem. Eur. J.* **2007**, *13*, 150–157.
[36] C. E. Hartmann, S. P. Nolan, C. S. Cazin *J. Organometallics* **2009**, *28*, 2915–2919.
[37] J. Nasielski, N. Hadei, G. Achonduh, E. A. B. Kantchev, C. J. O'Brien, A. Lough, M. G. Organ *Chem.–Eur. J.* **2010**, *16*, 10844–10853.
[38] G. Ren, X. Cui, Y. Wu *Eur. J. Org. Chem.* **2010**, 2372–2378.
[39] Z. Jin, X.-P. Gu, L.-L. Qiu, G.-P. Wu, H.-B. Song, J.-X. Fang *J. Organomet. Chem.* **2011**, *696*, 859–863.
[40] Y.-H. Chang, Z.-Y. Liu, Y.-H. Liu, S.-M. Peng, J.-T. Chen, S.-T. Liu *Dalton Trans.* **2011**, *40*, 489–494.
[41] H. Türkmen, I. Kani *Appl. Organometal. Chem.* **2013**, *27*, 489–493.
[42] G. Bastug, S. P. Nolan *Organometallics* **2014**, *33*, 1253–1258.
[43] (a) N. Miyaura, K. Yamada, A. Suzuki *Tetrahedron Lett* **1979**, *20*, 3437–3440; (b) N. Miyaura, A. Suzuki *J. Chem. Soc. Chem. Commun.* **1979**, *866–867.*
[44] M. Suzuki *J. Organomet. Chem.* **2002**, *653*, 83–90.
[45] (a) N. Miyaura, A. Suzuki *Chem. Rev.* **1995**, *95*, 2457; (b) N. Miyaura *Top. Curr. Chem.* **2002**, *219*, 11–59; (c) N. Marion, S. P. Nolan *Acc. Chem. Res.* **2008**, *41*, 1440–1449; (d) E. A. B. Kantchev, C. J. O'Brien, M. G. Organ *Angew. Chem., Int. Ed.* **2007**, *46*, 2768–2813; (e) H. Doucet *Eur. J. Org. Chem.* **2008**, 2013–2030; (f) G. C. Fortman, S. P. Nolan *Chem. Soc. Rev.* **2011**, *40*, 5151–5169.
[46] J. Hassan, M. Svignon, C. Gozzi, E. Schulz, M. Lemaire *Chem. Rev.* **2002**, *102*, 1359–1469.
[47] M. G. Johnson, R. J. Foglesong *Tetrahedron Lett.* **1997**, *38*, 7001–7002.
[48] (a) J. Yin, M. P. Rainka, X. X. Zhang, S. L. Buchwald, *J. Am. Chem. Soc.* **2002**, *124*, 1162–1163; (b) S. D. Walker, T. E. Barder, J. R. Martinelli, S. L. Buchwald *Angew. Chem. Int. Ed.* **2004**, *43*, 1871–1876; (c) T. E. Barder, S. D. Walker, J. R. Martinelli, S. L. Buchwald *J. Am. Chem. Soc.* **2005**, *127*, 4685–4696.
[49] M. G. Organ, S. Çalimsiz, M. Sayah, K. H. Hoi, A. L. Lough *Angew. Chem. Int. Ed.* **2009**, *48*, 2383–2387.
[50] A. Schmidt, A. Rahimi *Chem. Commun.* **2010**, *46*, 2995–2997.
[51] L. Wu, E. Drinkel, F. Gaggia, S. Capolicchio, A. Linden, L. Falivene, L. Cavallo, R. Dorta *Chem. Eur. J.* **2011**, *17*, 12886–12890.
[52] A. Chartoire, M. Lesieur, L. Falivene, A. M. Z. Slawin, L. Cavallo, C. S. J. Cazin, S. P. Nolan *Chem. Eur. J.* **2012**, *18*, 4517–4521.
[53] T. Tu, Z. Sun, W. Fang, M. Xu, Y. Zhou *Org. Lett.* **2012**, *14*, 4250–4253.
[54] N. Debono, A. Labande, E. Manoury, J.-C. Daran, R. Poli *Organometallics* **2010**, *29*, 1879–1882.
[55] Y. Li, J. Tang, J. Gu, Q. Wang, P. Sun, D. Zhang *Organometallics* **2014**, *33*, 876–884.
[56] L. Benhamou, C. Besnard, E. P. Kündig *Organometallics* **2014**, *33*, 260 – 266.
[57] Y. Takeda, Y. Ikeda, A. Kuroda, S. Tanaka, S. Minakata *J. Am. Chem. Soc.* **2014**, *136*, 8544 – 8547.
[58] R. Jana, T. P. Pathak, M. S. Sigman *Chem. Rev.* **2011**, *111*, 1417–1492; and references therein.
[59] (a) J. H. Kirchhoff, M. R. Netherton, I. D. Hills, G. C. Fu *J.Am. Chem. Soc.* **2002**, *124*, 13662–13663; (b) J. H. Kirchhoff, C. Dai, G. C. Fu *Angew. Chem., Int. Ed.* **2002**, *41*, 1945–1947; (c) M. R. Netherton, G. C. Fu *Angew. Chem., Int. Ed.* **2002**, *41*, 3910–3912; (d) M. R. Netherton, C. Dai, K. Neuschütz, G. C. Fu *J. Am.Chem. Soc.* **2001**, *123*, 10099–10100.

[60] K. Arentsen, S. Caddick, F. G. N. Cloke, A. P. Herring, P. B. Hitchcock *Tetrahedron Lett.* **2004**, *45*, 3511–3515.
[61] C. Valente, S. Baglione, D. Candito, C. J. O'Brien, M. G. Organ *Chem. Commun.* **2008**, 735–737.
[62] G.-R. Peh, E. A. B. Kantchev, J.-C. Er, J. Y. Ying *Chem. Eur. J.* **2010**, *16*, 4010–4017.
[63] J. L. Farmer, H. N. Hunter, M. G. Organ *J. Am. Chem. Soc.* **2012**, *134*, 17470–17473.
[64] Y. Zhang, M.-T. Feng, J.-M. Lu *Org. Biomol. Chem.*, **2013**, *11*, 2266–2272.
[65] S. Choppin, P. Gros, Y. Fort *Org. Lett.* **2000**, *2*, 803–805.
[66] C. Song, Y. Ma, Q. Chai, C. Ma, W. Jiang, M. B. Andrus *Tetrahedron*, **2005**, *61*, 7438–7446.
[67] C. J. O'Brien, E. A. B. Kantchev, C. Valente, N. Hadei, G. A. Chass, A. L. Lough, A. C. Hopkinson, M. G. Organ *Chem. Eur. J.* **2006**, *12*, 4743–4748.
[68] M. R. Kumar, K. Park, S. Lee *Adv. Synth. Catal.* **2010**, *325*, 3525–3266.
[69] T. Liu, X. Zhao, Q. Shen, L. Lu *Tetrahedron* **2012**, *68*, 6535–6547.
[70] U. Yilmaz, S. Deniz, H. Küçükbay, N. Şireci *Molecules* **2013**, *18*, 3712–3724.
[71] M. Kuriyama, S. Matsuo, M. Shinozawa, O. Onomura *Org. Lett.* **2013**, *15*, 2716–2719.
[72] (a) E.-I. Negishi in *Handbook of Organopalladium Chemistry for Organic Synthesis, Vol. 1*, Wiley-Interscience, New York, **2002**, pp. 229–248; (b) M. Uchiyama, T. Furuyama, M. Kobayashi, Y. Matsumoto, K. Tanaka *J. Am. Chem. Soc.* **2006**, *128*, 8404–8405.
[73] E.-I. Negishi *Bull. Chem. Soc. Jpn.* **2007**, *80*, 233–257.
[74] (a) N. Hadei, E. A. B; Kantchev, C. J. O'Brien, M. G. Organ *Org. Lett.* **2005**, *7*, 3805–3807; (b) N. Hadei, E. A. B Kantchev, C. J. O'Brien, M. G. Organ *J. Org. Chem.* **2005**, *70*, 8503–8507.
[75] M. G. Organ, S. Avola, I. Dubovyk, N. Hadei, E. A. B. Kantchev, C. J. O'Brien, C. Valente *Chem. Eur. J.* **2006**, *12*, 4749–4755
[76] (a) S. Çalimsiz, M. Sayah, D. Mallik, M. G. Organ *Angew. Chem. Int. Ed.* **2010**, *49*, 2014–2017; (b) C. Valente, M. E. Belowich, N Hadei, M. G. Organ *Eur. J. Org. Chem.* **2010**, 4343–4345.
[77] N. Hadei, G. T. Achonduh, C. Valente, C. J. O'Brien, M. G. Organ *Angew. Chem. Int. Ed.* **2011**, *50*, 3896–3899.
[78] I. Larrosa, C. Somoza, A. Banquy, S. M. Goldup *Org. Lett.* **2011**, *13*, 146–149.
[79] G. A. Chass, C. J. O'Brien, N. Hadei, E. A. B. Kantchev, W.-H. Mu, D.-C. Fang, A. C. Hopkinson, I. G. Csizmadia, M. G. Organ *Chem. Eur. J.* **2009**, *15*, 4281–4288.
[80] G. T. Achonduh, N. Hadei, C. Valente, S. Avola, C. J. O'Brien, M. G. Organ *Chem. Commun.* **2010**, *46*, 4109–4111.
[81] H. N. Hunter, N. Hadei, V. Blagojevic, P. Patschinski, G. T. Achonduh, S. Avola, D. K. Bohme, M. G. Organ *Chem. Eur. J.* **2011**, *17*, 7845–7851.
[82] L. C. McCann, H. N. Hunter, J. A. C. Clyburne, M. G. Organ *Angew. Chem. Int. Ed.* **2012**, *51*, 7024–7024.
[83] L. C. McCann, M. G. Organ *Angew. Chem. Int. Ed.* **2014**, *53*, 4386 –4389.
[84] For a list of drugs containing secondary alkyl groups, see: Drug Information Online. Top 200 Drugs of 2012. http://www.drugs.com/stats/top100/2012/sales (accessed April 6, 2016).
[85] aFor examples of secondary alkyl coupling employing bulky phosphine ligands, see: (a) T. Hayashi, M. Konishi, Y. Kobori, M. Kumada, T. Higuchi, K. Hirotsy *J. Am. Chem. Soc.*, **1984**, *106*, 158–163; (b) S. D. Dreher, P. G. Dormer, D. L. Sandrock, G. A. Molander *J. Am. Chem. Soc.* **2008**, *130*, 9257–9259; (c) A. van den Hoogenband, J. H. M. Lange, J. M. Terpstra, M. Koch, G. M. Visser, M. Visser, T. J. Korstanje, J. T. B. H. Jastrzebski *Tetrahedron Lett.* **2008**, *49*, 4122–4124; (d) C. Han, S. L. Buchwald *J. Am. Chem. Soc.* **2009**, *131*, 7532–7533.

[86] S. Çalimsiz, M. G. Organ *Chem. Commun.* **2011**, *47*, 5181–5183.
[87] M. Pompeo, R. D. J. Froese, N. Hadei, M. G. Organ *Angew. Chem. Int. Ed.* **2012**, *51*, 11354–11357.
[88] B. Atwater, N. Chandrasoma, D. Mitchell, M. J. Rodriguez, M. Pompeo, R. D. J. Froese, A. J. Lough, M. G. Organ *Angew. Chem. Int. Ed.* **2015**, *54*, 9502–9506.
[89] S. V. C. Vummaleti, D. J. Nelson, A. Poater, A. Gómez-Suárez, D. B. Cordes, A. M. Z. Slawin, S. P. Nolan, L. Cavallo *Chem. Sci.*, **2015**, *6*, 1895–1904; and references therein.

7

Redox Non-innocent Ligands: Reactivity and Catalysis

Bas de Bruin, Pauline Gualco, and Nanda D. Paul[1]

Homogeneous and Supramolecular Catalysis Group, Van't Hoff Institute for Molecular Science (HIMS), University of Amsterdam (UvA), Science Park 904, 1098, XH Amsterdam, The Netherlands

[1]*Present address: Department of Chemistry, Indian Institute of Engineering Science and Technology, Shibpur (IIEST-Shibpur), Botanic Garden, Howrah 711103, India*

7.1 Introduction

The properties of a metal complex as a whole are the results of the interaction of the metal center and its surrounding ligands. To prepare new coordination complexes having useful applications, the first and most important aspect is to design new multifunctional ligands. In addition to the ligand functioning as a scaffold to bind a metal ion and to control the steric and electronic properties at the active metal site, ligands can also be more actively involved in a catalytic reaction. "Actor ligands" have one or more additional functions to control the reactivity of coordination complexes. For example, (i) proton responsive ligands are capable of undergoing a change of properties on gaining or losing one or more protons; (ii) ligands having hydrogen bonding donating or accepting functionalities can imply partial proton transfer to or from a suitable partner; (iii) redox non-innocent ligands are capable of undergoing a change of properties on gaining or losing one or more electrons; (iv) photo-responsive ligands are capable of undergoing a

Ligand Design in Metal Chemistry: Reactivity and Catalysis, First Edition. Edited by Mark Stradiotto and Rylan J. Lundgren.

change of properties on irradiation; and (v) ligands having a molecular recognition function can recognize certain substrates, metal ions or functional groups selectively.

The main concept is that the metal and the ligand can cooperate in a synergistic manner, and their interplay modifies the overall properties of the system and improves its applicability. Hence, from the application point of view (in particular, in the field of catalysis), development of new cooperative ligands for catalyst development is one of the most important issues of current catalysis research. Multi-electron reactions tend to be catalyzed most efficiently by scarce, expensive noble metals as they often tend to undergo two-electron oxidation state changes. However, with the continuous increase in price of these metals, cheap and easily accessible base metals are urgently needed as an alternative. Base metals often tend to undergo one-electron oxidation changes, and in order to confer nobility to base metal new concepts and approaches need to be developed. In this respect, "redox non-innocent" ligands are of growing interest. At the same time, base metals with redox active ligands may enable new reaction pathways, not available to other metal complexes.

The term "non-innocent" implies an uncertainty and refers to the ambivalence of oxidation state assignments in complexes bearing both redox active ligands and redox active metals. Over the past few years, assignment of the correct oxidation state of this class of ligands when bound to metal ions has been an actively pursued research topic [1]. More recently, inspired by intriguing radical-type reactions displayed by metallo-enzymes containing "ligand radicals", the research focus shifted towards the application domain; mainly in bringing about different useful chemical transformations (catalysis). In conventional approaches, the steric and electronic properties of the ligands are used to control the performance of the catalyst in which the ligands play a "spectator" role – the reactivity takes place at the metal center. Recent new approaches deviate from this concept, and make use of more reactive ("actor") ligands, that can play a much more prominent role in the elementary bond activation steps in a catalytic cycle [2]. This class of ligands has a unique ability to impart novel reactivity to the adjacent metal complexes by controlling the loss or gain of protons and electrons, thus permitting the overall framework to adopt a different electronic structure which is key to achieve new types of reactivity with high selectivity.

Redox transformations of coordination compounds have long drawn considerable attention in the chemical community, which is to a large extent due to (catalytic) reduction and oxidation reactions mediated by transition metal complexes. Usually, when a transition metal complex undergoes electron transfer, the redox event mainly involves the central metal ion, leaving the coordinated ligand unaffected. In some cases, the metallo-enzyme active site in particular, the reactivity relies on metal–ligand synergy wherein the ligands also participate in the redox and/or bond activation processes. The enzymes galactose oxidase, hydrogenases, and cytochrome P450 are the best described examples of such systems. In these cases, the ligand is referred to as being "redox non-innocent" or "redox active" [3]. In synthetic organometallic chemistry, the concept of using "reactive redox non-innocent ligands" is still dominated by stoichiometric examples. However, it is swiftly penetrating into the realm of catalytic processes to enhance reactivity and steer selectivity of transition catalysts.

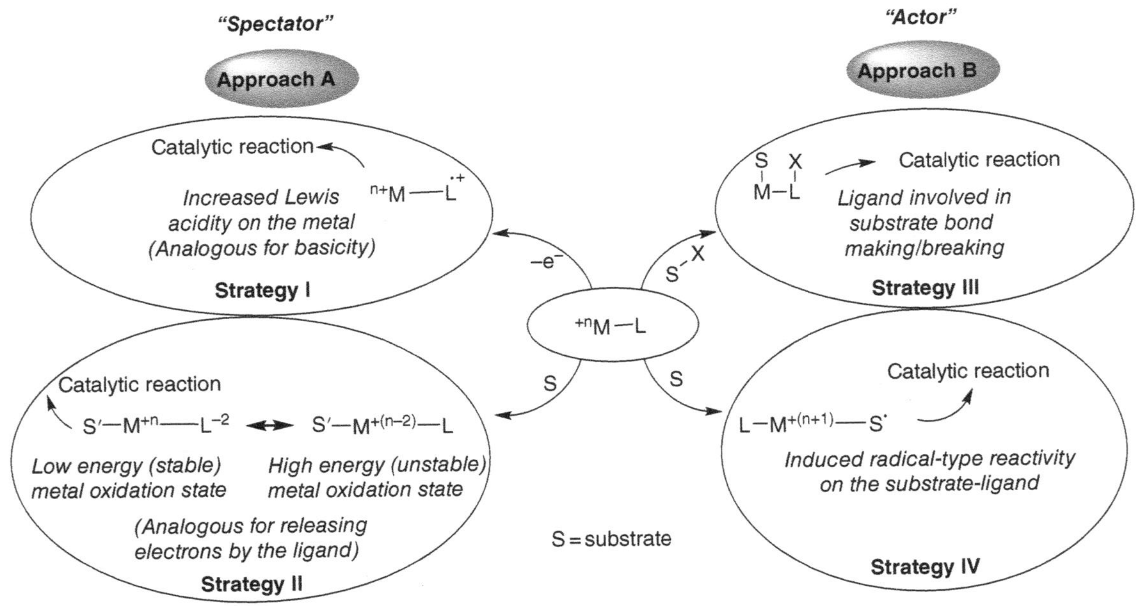

Figure 1 *Strategies of using redox non-innocent ligands in catalysis*

Redox active ligands, both in "spectator" or "actor" mode, can influence the overall reactivity and the selectivity of the catalyst. Four main strategies can be distinguished in this context (Figure 1): (**I**) As an electronic mediator, the redox active "spectator" active ligand can modify and control the Lewis acidity of the metal ions by selective ligand reduction/oxidation, thereby influencing the substrate affinity of the metal as well as the energy profile of subsequent follow-up reactions. (**II**) The redox active ligand can act as a discrete "electron reservoir", storing electrons from the metal on the ligand in elementary steps generating excessive electron density, and releasing electrons to the metal in elementary steps generating deficiencies, in all cases avoiding uncommon oxidation states of the metal. (**III**) Redox active ligands can act as a cooperative "actor" ligand, generating reactive ligand radicals during catalytic turnover which actively participate in the making and breaking of chemical bonds. Cooperative substrate activation by the redox non-innocent ligand and the metal allows reactions that are difficult to achieve otherwise. (**IV**) The last strategy involves (radical-type) activation of the substrates or modification of the substrate reactivity in cases where the substrate itself acts as a redox non-innocent ligand. Speaking more generally, redox non-innocent ligands can either participate in the catalytic cycle by accepting/releasing electrons ("spectator ligand", strategies **I** and **II**) or by forming/breaking chemical bonds of the substrate ("actor ligand", strategies **III** and **IV**).

This chapter gives an overview of different types of "redox active" ligands and the above mentioned four different application strategies. The number of publications on this concept is growing rapidly, especially over the past few years. Some excellent reviews and essays on this subject have appeared recently; a Forum issue of *Inorganic Chemistry* [4], a Special issue of the *European Journal of Inorganic Chemistry* [5], a tutorial review article in *ACS Catalysis* [6] and some other reviews [7] have been used in this chapter as a basis for much of the older work on this subject. This chapter also includes relevant new examples in the field of reactive redox active ligands.

7.2 Strategy I. Redox non-innocent ligands used to modify the Lewis acid–base properties of the metal

The Lewis acidity or basicity of metal ions plays a crucial role in catalysis. During catalysis of an organic reaction, Lewis acidic metal ions act as electron pair acceptors to increase the reactivity of the substrate. Enhanced Lewis acidity of a transition metal catalyst enables the metal ion (catalyst) to form an adduct with a lone pair bearing an electronegative atom in the substrate, such as oxygen (both sp^2 or sp^3), nitrogen, sulfur, and halogens. This leads to partial charge-transfer and makes the lone-pair donor effectively more electronegative, and thereby effectively activates the substrate towards nucleophilic attack, heterolytic bond cleavage, or cycloaddition with 1,3-dienes and 1,3-dipoles. Similarly, enhancing the Lewis basicity of a (catalytically active) metal complex can be beneficial in controlling its (catalytic) reactivity.

In coordination chemistry and catalysis, the surrounding ligands have a strong influence on the overall electronics and catalytic reactivity of the complex.

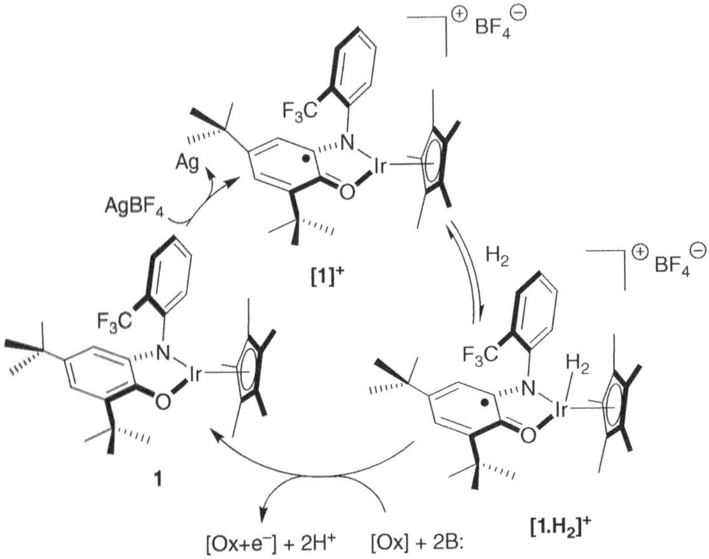

Scheme 1 Ligand oxidation leading to increased Lewis acidity of the metal: oxidation of H_2 by a Ir(III) catalyst containing a redox non-innocent ligand

In conventional approaches, such modifications are typically achieved by introducing electron-withdrawing or donating substituents at the ligand, which often requires laborious synthetic procedures. A more straightforward approach involves the use of redox active ligands. By controlling the redox events of the coordinated redox non-innocent ligands the Lewis acidic properties of the metal can be tuned in one step without altering the steric environment of the complex. This concept was recently applied in the oxidation of H_2 (Scheme 1) [8].

The complex **1** undergoes ligand centered oxidation by silver tetrafluoroborate forming compound [**1**]$^+$ containing a one-electron oxidized ligand radical. This in turn makes the metal a stronger Lewis acid than in the non-oxidized form **1**. Therefore [**1**]$^+$ reacts with H_2 to produce the adduct [**1·H**$_2$]$^+$, which undergoes deprotonation by non-coordinating base (2,6-di-^{t}Bu-pyridine; TBP). Further oxidation by [**1**]$^+$ or Ag^+, followed by yet another deprotonation step completes the catalytic redox process leading effectively to oxidation of H_2 by Ag^+ (Scheme 1). Other than just increasing the Lewis acidity of the metal ion, the ligand also acts as an electron reservoir; the electrons of H_2 actually reduce the oxidized form of the redox non-innocent ligand back to its neutral form. In the catalytic experiment, **1** (1 equiv.) was found to be able to oxidize 3 equiv. of H_2 in the presence of $AgBF_4$ (6 equiv.) and TBP (6 equiv.) within 1.5 h.

Other catalytic reactions may require an electron-rich metal to facilitate the rate-determining (slowest) step of the catalytic cycle – Lewis basicity of the catalyst then plays a key role in substrate activation. For example, oxidative addition of H_2 is frequently rate determining in olefin hydrogenation reactions. The rate of alkene hydrogenation catalyzed by Rh(I)-diphosphinoferrocene complexes of type **2a** (Figure 2) are known to increase the basicity of the phosphine moieties. In conventional approaches this can be achieved by

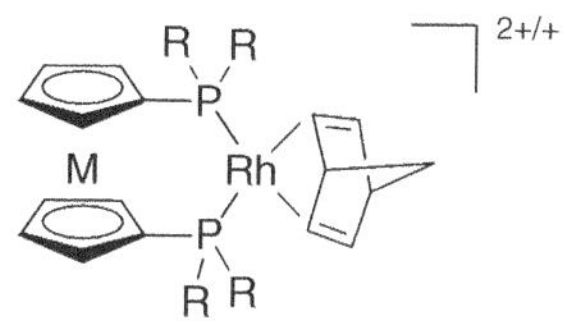

2a, M = Fe, R = Ph, *t*Bu, charge = +1
2b, M = Co, R = Ph, charge = +1,+2

Figure 2 *Redox non-innocent ligand supported Rh(I) catalysts employed in olefin hydrogenation*

introducing alkyl substituents (R=^{t}Bu) instead of phenyl substituents (R=Ph) on the phosphorus atom. The greater electron-donating ability of the alkyl substituents (R=^{t}Bu) enhances electron density on rhodium, which facilitates the oxidative addition step [9]. In isostructural complexes **2b**, a similar rate-enhancing effect on the hydrogenation of cyclohexene can also be achieved by changing the charge of the redox-active cobaltocene-based ligand moiety [10]. Both mono- and di-cationic complexes **2b** are found to be active in hydrogenation. Interestingly, in the electron-rich reduced form (**2b**, charge=+1) (enhanced basicity), hydrogenation proceeds 16 times faster than with the electron-deficient di-cationic form of the complex (**2b**, charge=+2). Some other examples employing related concepts have been reviewed by Allgeier and Mirkin [11].

7.3 Strategy II. Redox non-innocent ligands as electron reservoirs

The most direct application of redox non-innocent ligands in catalysis is their ability to function as electron reservoirs [12]. In homogeneous catalysis, many important transformations involve multi-electron (most typically two) transfer between the metal and the (activated) substrate (e.g., reductive elimination, oxidation addition). These multi-electron reactions tend to be catalyzed most efficiently by precious metals (Pd, Pt, Rh, Ir, etc.), since they typically can undergo multiple oxidation state changes. However, it is more difficult to achieve with cheaper and more abundant first-row transition metals (Fe, Co, etc.) which often prefer one-electron redox events. If additional electrons can be temporarily stored on (or released from) a redox active ligand, the complex as a whole can mediate multi-electron transformations avoiding uncommon oxidation states. In other words, redox active ligands in their role as an electron reservoir may be able to confer nobility on base metals by combining a one-electron ligand centered redox event with a one-electron redox change at the metal for an overall two-electron change. In recent years, several useful chemical transformations have been achieved using this concept of a spectator redox active ligand as an electron sink or reservoir.

Water oxidation is a thermodynamically unfavorable process which involves the transfer of four electrons. Several catalysts have been developed and simultaneously several mechanisms for these types of chemical transformations have been proposed [13]. Binuclear ruthenium complexes have drawn considerable attention in the context of water oxidation. The anthracene-bridged binuclear Ru bis-hydroxide bis-quinone complex **3** shows activity in (electro)catalytic water oxidation, and ligand centered redox events play a key role in the elementary reaction steps involved [14]. A bridged peroxo intermediate **4** is formed via a double deprotonation of the coordinated OH groups with simultaneous reduction of the two coordinated quinine ligands to semiquinone. The intermediate **4** then releases one electron each to produce the intermediate **5** where the coordinated semiquinone ligands return to the initial bis-quinone state without affecting the peroxo bridge. Further two-electron oxidation and evolution of O_2 with loss of two protons closes the catalytic cycle (Scheme 2).

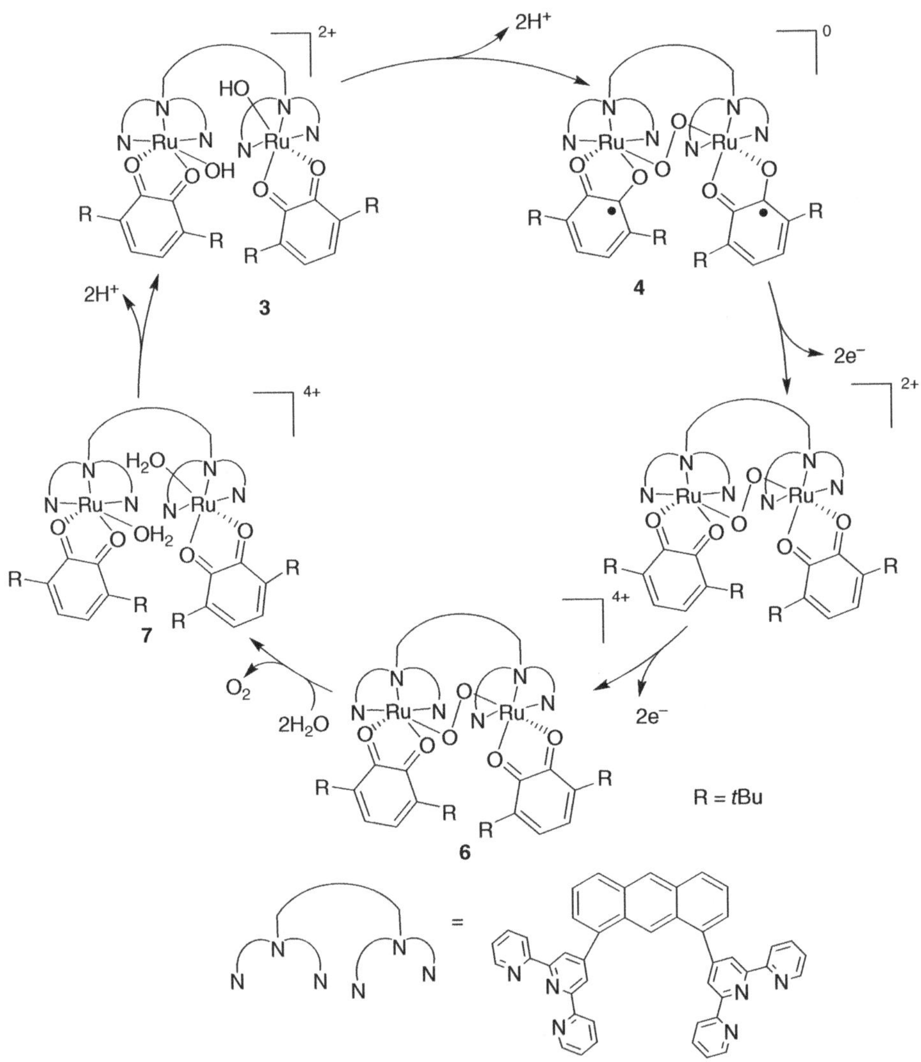

Scheme 2 *Redox active ligand participation in catalytic water oxidation by a binuclear Ru catalyst containing ortho-quinone/semiquinone ligands*

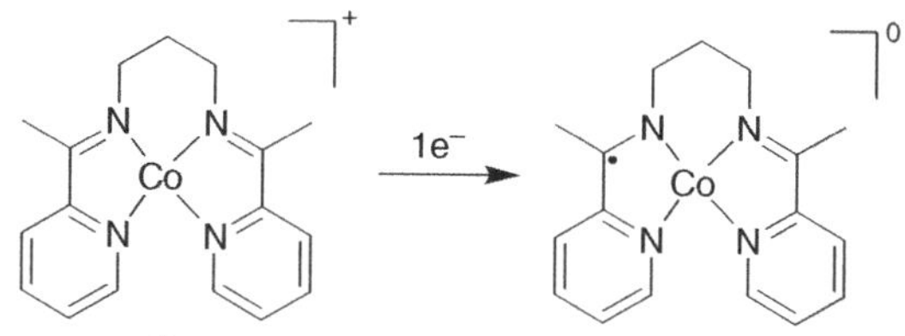

Scheme 3 *Ligand-centered redox event in a tetradentate bis-iminopyridine Co(II) complex in catalytic H_2 evolution*

The O–O bond formation step was proposed to involve two Ru(II)/Ru(III) redox steps and two ligand-centered quinone/seminquinone couples. However, careful investigations by other research groups suggested that all four oxidation steps may actually be entirely quinoid ligand based, and hence a cycle involving exclusively Ru(II) is energetically plausible [12].

The counterpart of catalytic water oxidation to O_2, catalytic H_2 evolution via water reduction, is of equal importance in the context of water splitting (artificial photosynthesis) [15]. In nature, hydrogenase enzymes show excellent catalytic rates and efficiencies and can catalyze both proton reduction and H_2 oxidation. Consequently, several biomimetic complexes have been developed which show excellent catalytic activity towards electrochemical H_2 production from water.

A cobalt complex **8** containing a redox active tetradentate bis-iminopyridine framework has been reported to support a water reduction catalyst, with activities of observed rate constant, k_{obs} of $10^7\,M^{-1}\,s^{-1}$ derived from voltammetry measurements (Scheme 3) [16]. Ligand-centered reduction of the coordinated imine function has been proposed as the first electrocatalytic step followed by protonation. Notably, this compound was shown to operate even under basic conditions at pH 8 (buffer) to give 10 liter of H_2 (mol catalyst^{-1} h^{-1}) albeit with a modest Faradaic efficiency of only 60%.

Similarly, the nickel complex **10** based on the 2,6-diiminepyridine ligand (Scheme 4) also shows electrochemical reduction of water to H_2, in which ligand-centered redox events play a key role [17]. A computational proposal for this process indicates the energetic accessibility of a proton-coupled electron transfer (PCET) step, which mainly involves ligand-centered redox events without altering the divalent oxidation state of the Ni center. The low-lying ligand-centered acceptor orbital (LUMO) of the complex can be much more easily accessed by the incoming electron than any of the Ni(II) orbitals which has been experimentally observed in the low over-potential. Complex **10** undergoes one-electron reduction in the first step to produce the monocationic intermediate **11** containing a one-electron reduced radical ligand. The radical intermediate **11** converts to nickel(II) hydride complex **12** via a PCET process. The H_2 release step was proposed to involve protonation of a nickel(II) hydride in coordinating solvents to regenerate the resting state **10** (Scheme 4).

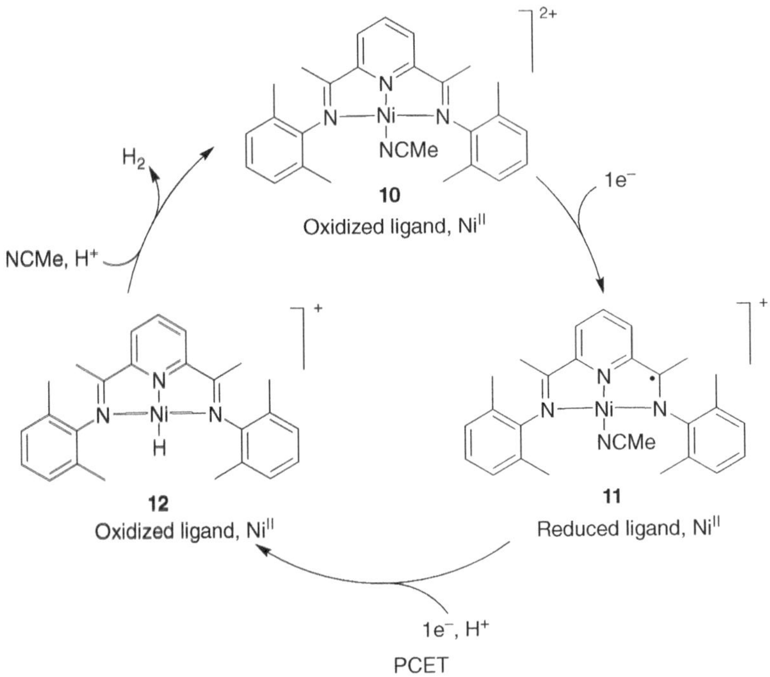

Scheme 4 *Ligand-centered reduction in electrocatalytic proton reduction mediated by a nickel(II) bisaryliminopyridine electrocatalyst*

Similar to catalytic O_2 or H_2 evolution, catalytic reduction of CO_2 into higher energy products is one of the most challenging topics of research to the chemist in the 21st century. The cobalt complex **13**, $[Co(III)N_4H(Br)_2]^+$, containing a redox active N_4H macrocyclic ligand (N_4H=2,12-dimethyl-3,7,11,17-tetraazabicyclo-[11.3.1]-heptadeca-1(7),2,11,13,15-pentane), has been used for electrocatalytic CO_2 reduction in wet MeCN with a glassy carbon working electrode (Scheme 5) [18]. Even in presence of large concentrations of water the catalyst preferentially reduces CO_2 over H^+ near the Co(I)/Co(0) redox couple [$E_{1/2}=-1.88$ *vs* $Fe(I)Cp_2/Fe(0)Cp_2$]. Systematic characterization of the whole catalytic cycle as well as that of the possible intermediates reveal that the formal Co(I) complex **14**, $[CoN_4H(MeCN)]^+$, is the actual precatalyst for CO_2 reduction. Moreover, broken symmetry density functional theory (DFT) calculations of **14** and **15** are suggestive of a more complicated electronic structure; an open-shell singlet configuration in which a low-spin Co(II) ion is antiferromagnetically coupled to a ligand-based anion radical ($N_4H^{\bullet-}$ and $N_4^{\bullet-}$) [7d, 19]. The stability of the reduced $N_4H^{\bullet-}$ ligand radical anion as well as the ability of the redox non-innocent ligand backbone to accommodate a second redox equivalent has been argued as the main reason for preferential CO_2 reduction even in wet MeCN.

Significant progress in the application of this concept has also been made in catalytic C—C bond formation reactions. Chirik and co-workers recently reported an interesting case involving [2π +2π] cycloaddition of dienes and enynes using the bis-dinitrogen

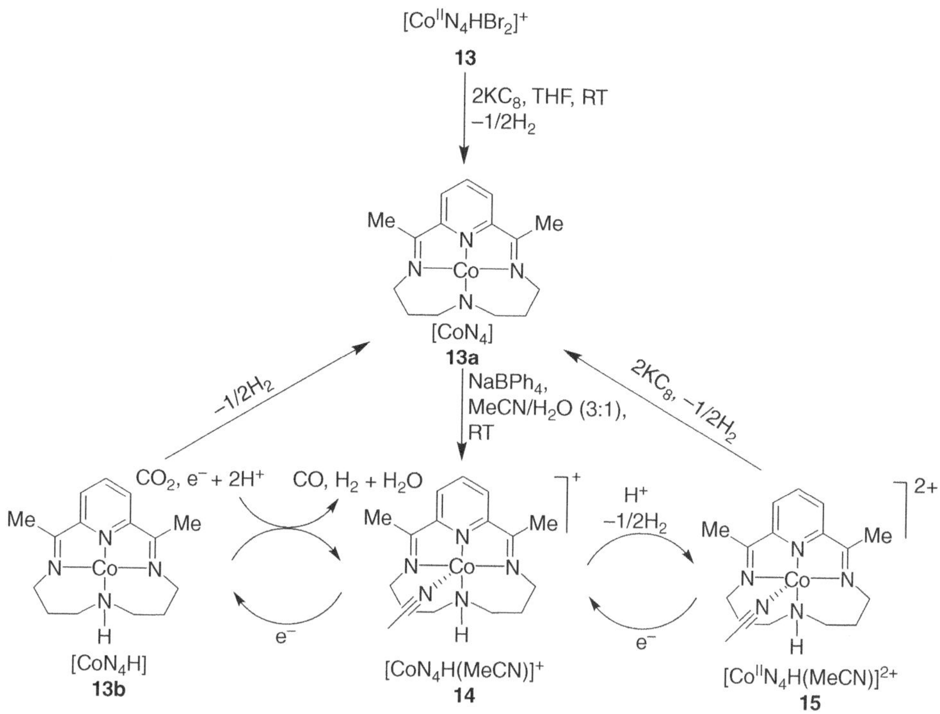

Scheme 5 *Preferential CO_2 reduction by complex **13** in wet MeCN*

iron complex **16** as the catalyst (Scheme 6) [20]. The catalyst (**16**, L=N_2) reacts with a diene substrate to form the π-complex **14** with loss of two dinitrogen ligands. Both **16** and **17** contain a dianionic tridendate N_3-ligand, which is formally the two-electron reduced form of the applied redox-active 2,6-diiminepyridine ligand [21]. According to the authors, complex **17** is in equilibrium with compound **18**. The subsequent C—C coupling step is formally a two-electron oxidative addition process which involves the oxidation of the reduced ligands in **16** and **17.** The electrons required for these transformations do not stem from the metal but from the dianionic, two-electron reduced 2,6-diiminepyridine ligand, thus allowing iron to maintain the energetically favorable Fe(II) oxidation state [avoiding the less favorable Fe(IV) oxidation state]. Subsequently, intermediate **18** undergoes a formal two-electron reductive elimination reaction to liberate the product. This process is again ligand based, leading to regeneration of complex **16** containing the two-electron reduced form of the 2,6-diiminepyridine ligand. The electron storage ability of the ligand allows the metal to maintain its stable Fe(II) oxidation state instead of a high-energy (unstable) Fe(0) oxidation state. The yield of the reaction depends on the nature of X (Scheme 6), reaching 95% for X=*N*-^{t}Bu with turnover frequency >250 h^{-1}. Related Fe(II) complexes are capable of catalyzing enyne cyclizations [22], intermolecular [2+2] cycloadditions of alkenes to butadienes [23] and olefin polymerization [24]. The role of the redox non-innocent 2,6-diiminepyridine ligand in the latter processes is however less clear and still debatable.

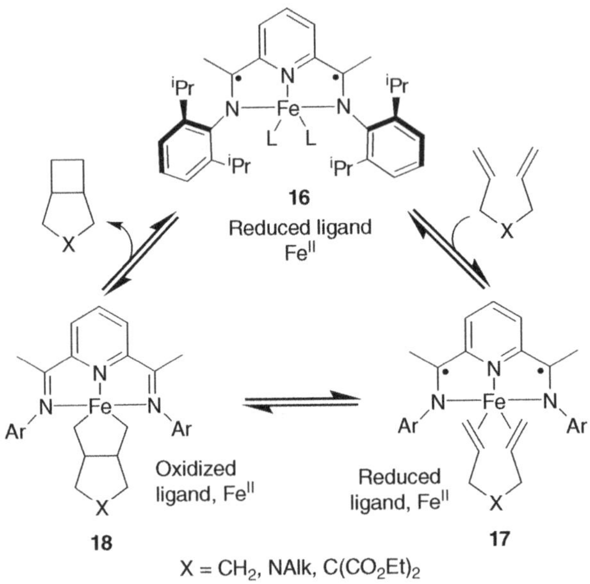

Scheme 6 *The redox non-innocent 2,6-diiminepyridine ligand acts as an electron reservoir in Fe-catalyzed [2 + 2] cycloaddition ring-closure reactions*

Using the same strategy, in a seminal example, Smith *et al.* [25] reported a C—C coupling between simple alkyl halides and aryl or alkyl zinc bromides using a relatively inexpensive Co(III) center supported by bis-iminophenolate ligands [7b, 7c, 25]. The square planar triplet cobalt(III) bis-iminophenolate complex **19**, upon reaction with alkyl halide, forms a five-coordinate square pyramidal Co(III) species **20** via pseudo-oxidative addition of an alkyl fragment to the Co(III) center without altering its oxidation state. The species **20** contains two one-electron oxidized radical ligands which are reported to remain antiferromagnetically coupled. This step can alternatively be rationalized as ligand-induced nucleophilic abstraction of R^+ from the alkyl halide by the metal. These species then undergo a formal R^+ group transfer either to aryl or alkyl zinc bromides to yield the C—C coupled products (Scheme 7).

Similar to C—C bond formation, C—C bond cleavage, aromatic C—C bond cleavage in particular, is a crucial step in the biodegradation of organic compounds such as catechols, aminophenols, hydroquinones, salicylic acid, and gentisic acid by aerobic microorganisms [26]. In biological systems, different iron oxygenases such as catechol dioxygenases, 2-aminophenol dioxygenases, and so on are found to be involved in catalytic C—C bond cleavage of aromatic compounds. Notably, the electron reservoir ability of the coordinated catecholate or 2-aminophenolate ligands plays a key role in catalytic turnover.

Interestingly, the electron reservoir properties of redox active ligands are also found to be useful to impose one-electron transformation on late transition metals. Rhenium complexes are known to be powerful oxo-transfer reagents [27]. However, closed-shell

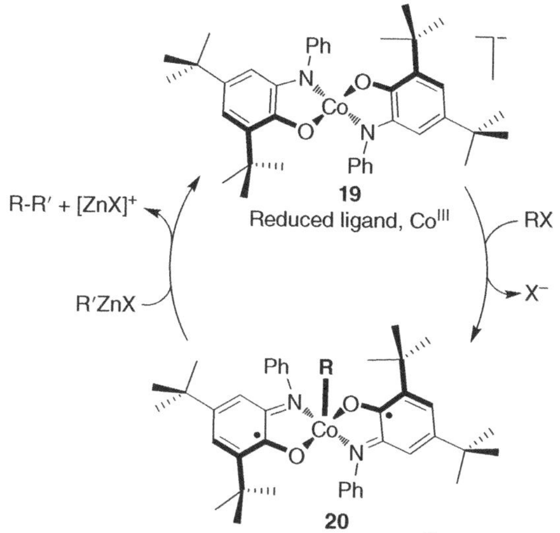

Scheme 7 *Redox non-innocent bis-iminophenolate ligands as electron reservoir in catalytic C—C bond formation by a Co(III) complex supported by bis-iminophenolate ligands*

rhenium complexes show very little affinity towards (triplet) O_2 to generate oxo complexes, which may be attributed to the "spin forbidden" nature of this transformation. The reactions of metal complexes with O_2 (even with closed-shell species) tend to proceed by a series of sequential, metal-mediated one-electron steps, via superoxo $[O_2^{\bullet}]^-$ and peroxo $[O_2]^{2-}$ complex intermediates [28]. Subsequent homolysis of the oxygen–oxygen bond then gives access to oxo species. Closed-shell d^2 Re(V) complexes have a general tendency to react via two-electron processes rather than one-electron redox reactions. Therefore, the complex **21** (Scheme 8) has a low probability to form an $[(L^{2-})_2(O)Re^{VI}–O–O\bullet]^-$ intermediate. However, the above type of superoxo intermediate is possible with a system capable of undergoing one-electron oxidation at the ligand center without altering the metal oxidation state. The presence of two redox non-innocent catecholate ligands bound to the metal allow rhenium to maintain its +5 oxidation state and give access to the intermediate $[(L^{2-})(L\bullet^-)(O)Re^V–O–O\bullet]^-$ species **22**, containing a one-electron oxidized catechol ligand (semiquinone). The species **22** produces the bis-oxo d^0 Re(VII) complex **24** via a bi-metallic intermediate **23**. Thus, the electron reservoir property of the coordinated catecholate ligands plays a decisive role in the formally "spin forbidden" reaction (Scheme 8). The overall reaction (**21** → **24**) requires half an equivalent of oxygen for each complex **21**, which shows that both oxygen atoms are incorporated in the product **24** [29].

The oxo-transfer abilities of complex **24** were also tested by oxidation of Ph_3P to $Ph_3P{=}O$ [30]. Strong binding affinity of PPh_3 towards complex **21** (to form **24·PPh$_3$**) inhibits true one-pot catalytic turnover in oxidation of Ph_3P to $Ph_3P{=}O$. Catalytic turnover however can be achieved with less coordinating substrates. Complex **21** catalyzes

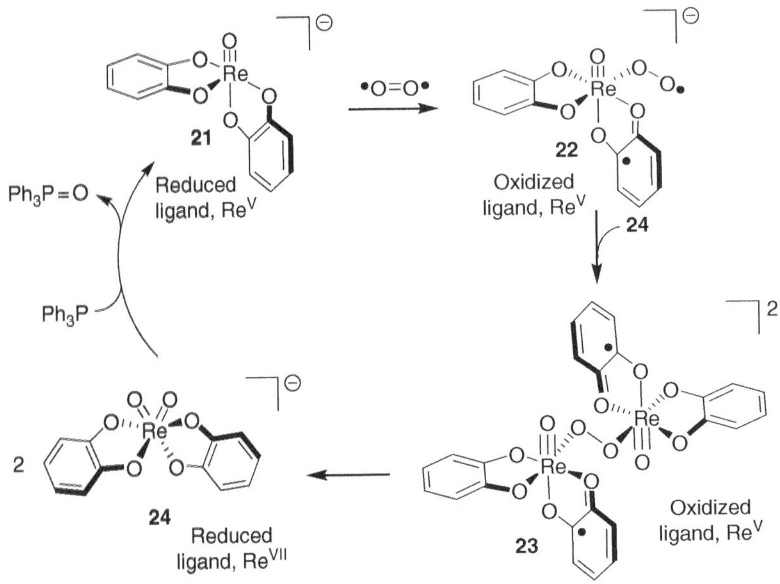

Scheme 8 *Electron-reservoir properties of redox active ligands assisting in O_2 activation*

aerobic oxidation of benzyl alcohol to benzaldehyde [31]. However, the coordinated hydroquinolate ligands are prone to oxidation which is a serious flaw of this catalytic system. By fine tuning the electronic property of the redox active ligands easy oxidation of the catalyst can be controlled. Modification of the ligand backbone with electron-withdrawing groups is one of the possible solutions to this problem which will make the ligand-centered oxidation difficult. In fact, the analogous complex of **21** containing the bis(tetrabromocatecholate) as ligands increases the catalytic activity and shows seven catalytic turnover cycles in oxidation of benzyl alcohol (neat). The interesting feature of the sequence of reactions shown in Scheme 8 is that the redox non-innocent ligands do not only act as electron reservoirs, but also seem to play an important role in lowering the exchange interactions through delocalization of the unpaired electron generated in the first step of the reaction, thus facilitating spin-crossover in the formally "spin forbidden" reaction between triplet oxygen and the closed-shell (singlet) d^2 Re(V) complex **21**.

The ability of redox active ligands to store and release electrons, when required, indeed enables redox transformations (oxidative addition or reductive elimination) to proceed even with metal complexes (catalyst) having d^0 electronic configuration. Unlike the system described above, this process can also involve electron pairs instead of single unpaired electrons and thus helps the metal from adopting uncommon oxidation states, and even allows redox transformations that are simply impossible otherwise.

For example, oxidative addition reaction becomes possible even with d^0 Zr(IV) complexes containing redox active ligands capable of releasing electron density [32]. The *ortho*-diamido Zr(IV) complex **25** catalyzes the disproportionation of diphenyl

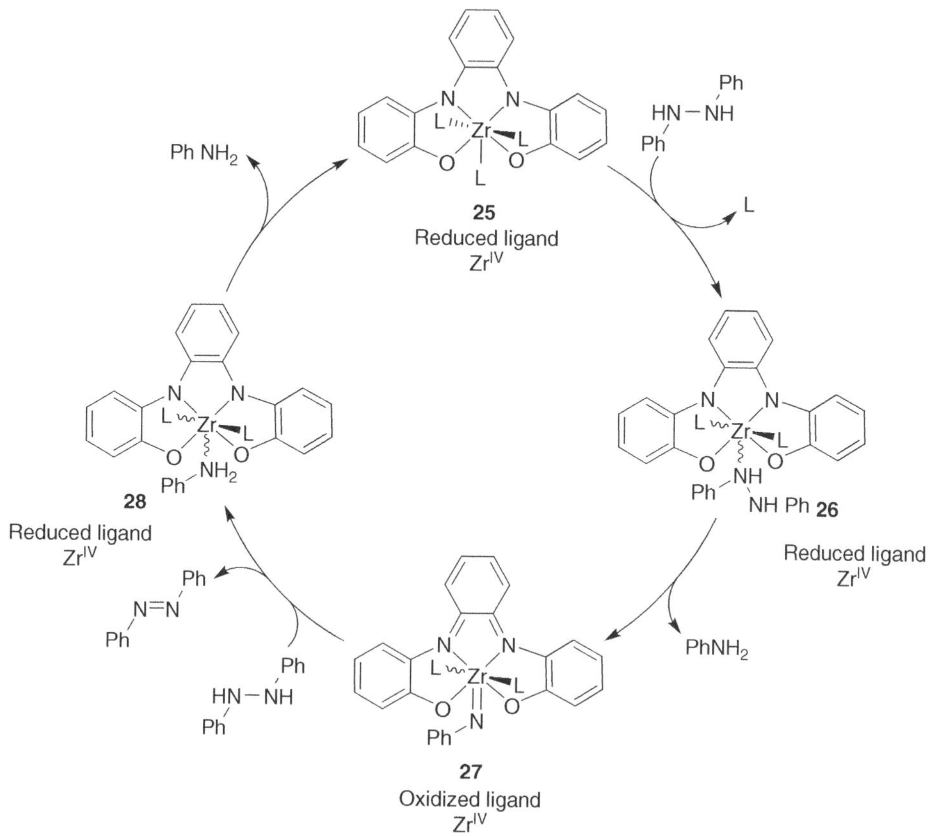

Scheme 9 *"Oxidative addition" reactions at d⁰ Zr(IV) complexes by virtue of the electron-releasing properties of redox active ligands*

hydrazine (PhNH—NHPh) to aniline ($PhNH_2$) and azobenzene (PhN=NPh) [30]. In the first step of the catalytic cycle diphenyl hydrazine coordinates to zirconium to produce intermediate complex **26**. The next step involves elimination of aniline, which is a two-electron "oxidative addition" process, thus requiring two electrons. Since the Zr(IV) has a d^0 electron configuration, the electrons required for the oxidative addition step stem from the redox active ligand. The ligand donates two electrons (via the metal, thus allowing the observed net oxidative addition at the metal site), and hence the complex converts to the *ortho*-diimine species **27**. In the next step a second molecule of diphenyl hydrazine reacts with complex **30** to form the aniline adduct **28** with simultaneous elimination of azobenzene. The oxidized form of the ligand (*ortho*-diimine form) accepts the two electrons that are released in this step of the catalytic cycle, and the ligand thereby gets reduced back to the *ortho*-diamide form in species **28**. The +4 oxidation state of zirconium thus remains unaltered throughout the catalytic cycle (Scheme 9). Elimination of aniline closes the catalytic cycle. Full conversion of diphenyl hydrazine into aniline and azobenzene (2:1 ratio) has been achieved with

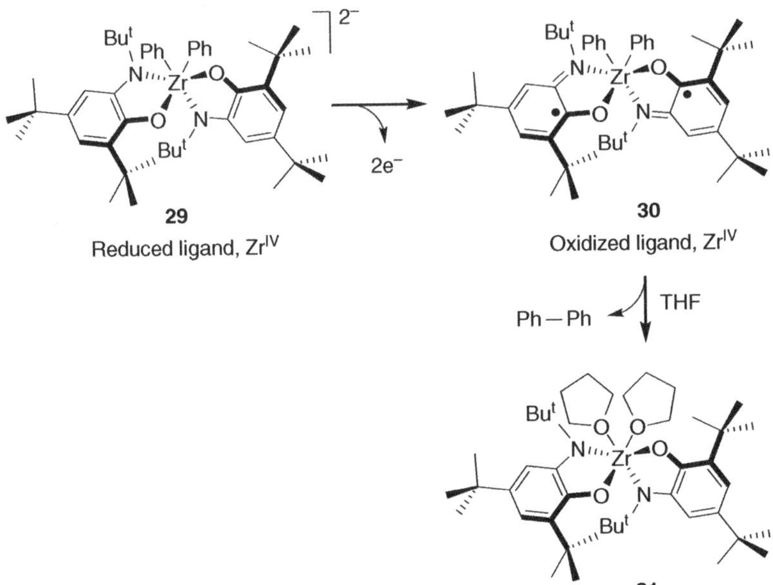

29 Reduced ligand, Zr^IV

30 Oxidized ligand, Zr^IV

31 Reduced ligand, Zr^IV

Scheme 10 *Ligand-centered redox events triggering C—C bond reductive elimination*

10 mol% catalyst loading within 1 day whereas for a lower catalyst loading of 1 mol% the reaction requires 6 days (yielding the same product ratio).

Similarly, Zr(IV) complex **29** containing redox active bis-aminophenoleto ligands undergoes ligand-centered redox events in an elementary C—C bond forming reductive elimination of a biaryl species (Scheme 10) [33]. In the first step of the catalytic reaction the Zr(IV) complex **30** is formed via two-electron reductive activation of the complex **29**. The two aminophenolato (ap) ligands undergo one-electron oxidation each. The next step involves reductive elimination of the biaryl species where the oxidized ligands in complex **31** accept one electron each, transforming the ligand back to the initial two-electron reduced catecholate form. Oxidations of the analogous dimethyl derivative, $[Li(OEt_2)]_2[Zr^{IV}Me_2(ap)_2]$, however, do not lead to clean ethane reductive elimination due to competing methyl radical expulsion pathways. The entire redox events for the formal two-electron reductive elimination process are shown to be supported entirely by the redox active ligand with no participation of the metal center.

Related approaches were recently employed in mediating (stoichiometric) "oxidative addition" and "reductive elimination" reactions at cobalt centers bearing redox active ligands [7b, 7c, 25, 34]. These reactions also proceed without a change in the d-electron configuration of the metal.

Another related zirconium complex **32** containing redox active tridentate NNN pincer ligand catalyzes nitrene transfer reactions from organic azides to *tert*-butyl isocyanide to form non-symmetrical carbodiimides where the metal maintains the Zr(IV)

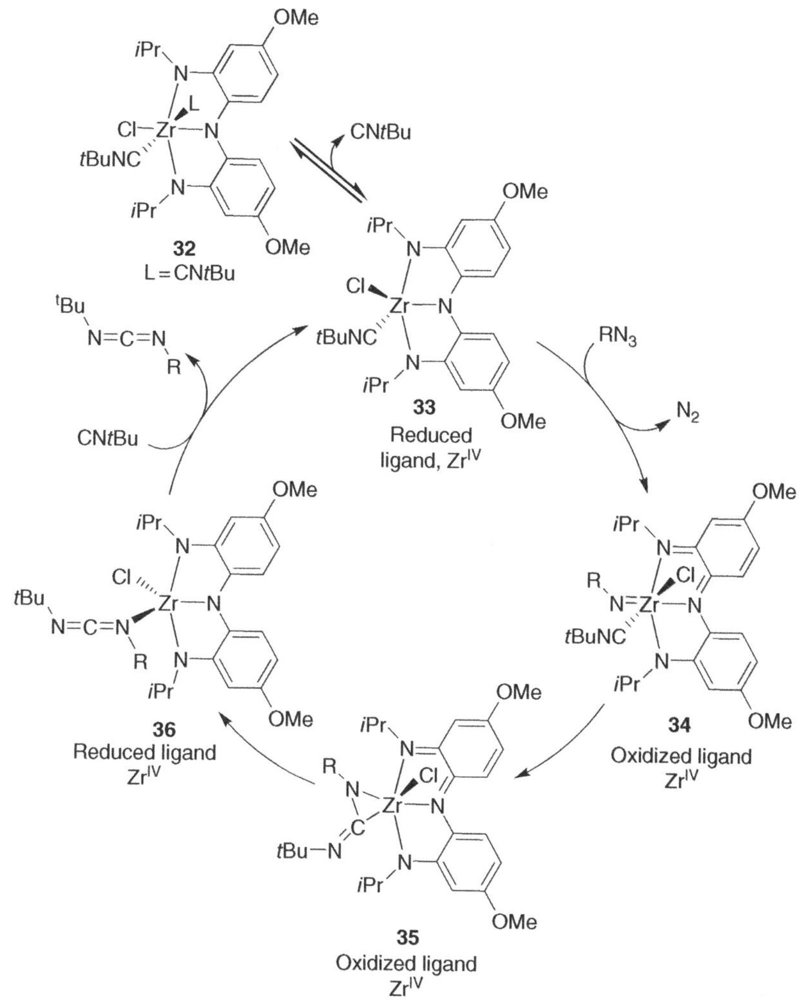

Scheme 11 *Redox active ligand participation in catalytic formation of carbodiimides from isocyanides and organic azides*

state throughout, while the pincer ligand interconverts between mono- and tri-anionic forms (Scheme 11) [35]. The organic azide then coordinates to produce the imido intermediate **34** via N_2 elimination (stepwise or simultaneously). Kinetic studies revealed that nitrogen elimination is rate-limiting. During the process of formation of the imido intermediate (similar to the example in Scheme 9) the ligand gets oxidized, since the d^0 Zr(IV) itself has no electrons available to enable this process. In the next step compound **35** is formed via an intramolecular nucleophilic attack of the imido nitrogen onto the isocyanide triple bond. Subsequent "reductive elimination" to form the carbodiimide complex **36** and dissociation of the product closes the catalytic cycle. During reductive elimination from **35** to **36**, the oxidized ligand formed in the initial imido complex **34** is reduced back to the initial state. Hence, the reductive elimination is again based on the redox non-innocence of the ligand, and does not lead to a change of the metal d-electron count. Complete conversion to the corresponding carbodiimide was achieved from the reaction of ^{t}BuNC with adamantyl and *tert*-butyl azides in 2 h, using 10 mol% of the catalyst [36].

7.4 Strategy III. Cooperative ligand-centered reactivity based on redox active ligands

In the approaches described above (strategies **I** and **II**) the reactivity is centered at the metal. The redox non-innocent ligands act as electron reservoirs enabling first-row transition metals to undergo multi-electron transformations, but they are otherwise spectator ligands. However, other cases are possible in which the redox non-innocent ligand plays a more direct and active role, participating directly in the making and breaking of bonds. Such "actor" ligands play an important role in several enzymatic processes, especially for metalloenzymes operating via radical-type reactions. Considerable amounts of related catalytic processes inspired by nature have been reported.

As a mimic of the well-studied galactose oxidase [37], a copper(II) thiophenol complex catalyzes the oxidation of primary alcohols to aldehydes in the presence of O_2 (Scheme 12) [38]. The latter also promotes the oxidation of secondary alcohols to diols (Scheme 12). The catalytic cycle starts with the oxidation of copper by O_2, leading to a biradical species. The intermediate **39** is produced from **38** by coordination of two alkoxide substrates. The rate-limiting step is the formation of **40** from **39** by a hydrogen atom transfer from the secondary alcoholate to the oxygen-centered radicals of the aminophenols ligands. The cycle is then closed by radical dimerization which leads the formation of the diol [39].

An iridium (instead of copper) complex surrounded by nitrogen (instead of oxygen)-centered ligand radicals, also inspired by galactose oxidase, showed activity in alcohol oxidation (Scheme 13) [40]. In the first stage of the catalytic cycle, **41** is deprotonated to form the anionic intermediate **42**. One-electron oxidation by benzoquinone (BQ) leads to the radical intermediate **43**. As previously described for the formation radical intermediates in the galactose oxidase cycle, **43** reacts with the alcoholate substrate. To facilitate the hydrogen atom transfer from the alcoholate substrate to the nitrogen-centered ligand radical, the system contains dibenzotropylamino substituents which play the same role as the tyrosine moiety in galactose oxidase. The aldehyde is then obtained by a second one-electron oxidation step involving semiquinonate ($SQ^{\bullet -}$) as the oxidant, thus regenerating the catalyst.

Redox non-innocent ligands have also been employed in other kinds of processes. For example, a nickel-based system has been used in the purification of ethene gas streams [41]. The two forms (reduced and oxidized) of the dithiolene complex have different affinities for olefin, leading to separation of ethene from gas mixtures (Scheme 14). Intermediate **47** is obtained after electrochemical oxidation of the anionic nickel complex **46**. The oxidized complex **47** reacts selectively with ethylene to form the adduct **48**, thus the non-olefinic contamination of the multi-component stream

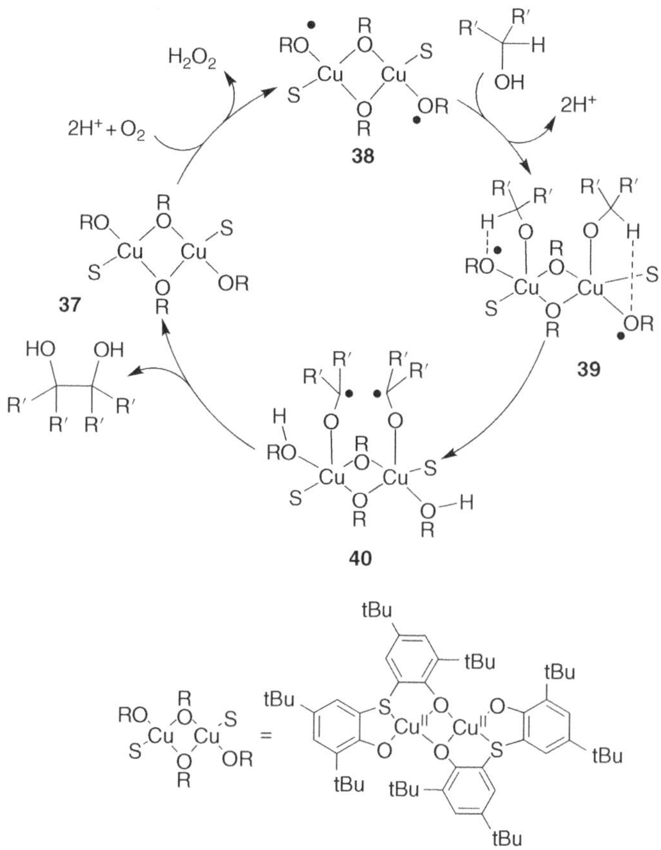

Scheme 12 *Catalytic cycle of Cu(II)-catalyzed dimerization of secondary alcohols (dashed lines represent hydrogen bonding)*

(MCS) remains uncoordinated. It is important to note that on the contrary to what is typically observed with most other complexes, the olefin does not form an adduct with the metal center, but reacts directly with the sulfur-based ligand. After reduction of complex **48** the olefin can be released and complex **46** recovered, thus closing the cycle. This mechanism of the trapping of ethylene probably involves thiyl-radical ligands as intermediates.

Another system based on rhenium and thiophosphine ligand is also capable of reversely binding alkene moieties (Scheme 15) [42]. After electrochemical oxidation of complex **49**, the ethylene is bound by the sulfur atoms of the ligand. As in complex **47**, the olefin has a stronger affinity for the oxidized form **49** (i.e., $\mathbf{49}^{+}$ and $\mathbf{49}^{2+}$) than for the reduced complex ($K_3 > K_2 > K_1$). In the neutral species (non-oxidized forms) the equilibrium is in favor of **49** over **50** even in a large excess of ethylene. In the case of the two-electron oxidized species, the equilibrium is in full favor of $\mathbf{50}^{2+}$, with no spectroscopic evidence of the presence of $\mathbf{49}^{2+}$.Whenever complexes are one-electron oxidized both mono-cationic forms ($\mathbf{49}^{+}$ and $\mathbf{50}^{+}$) are present, the ratio depending on the applied pressure of ethylene.

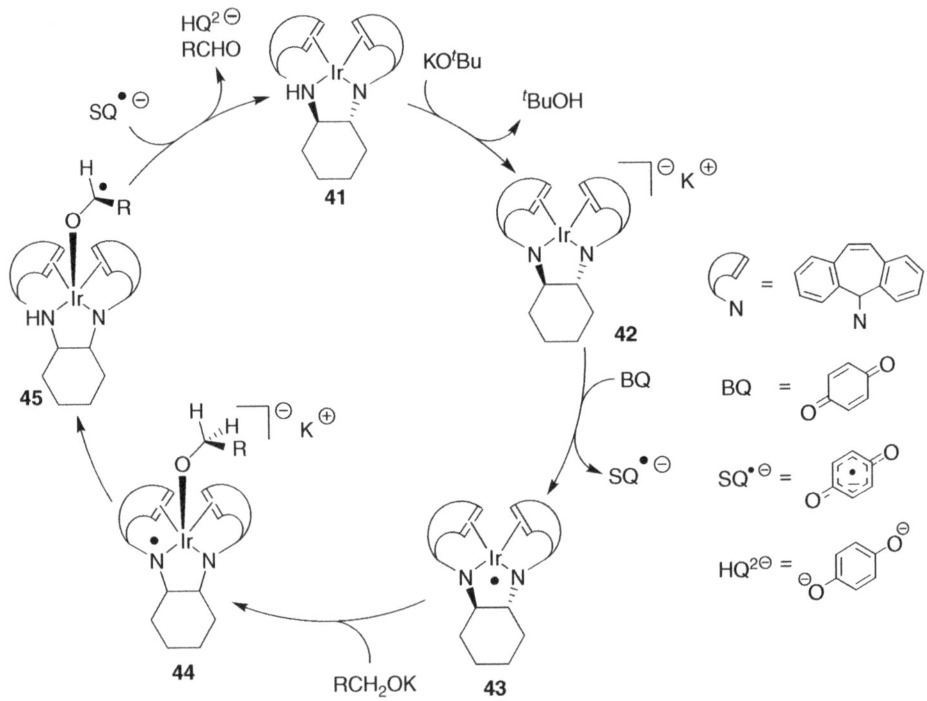

Scheme 13 Catalytic cycle of Ir-catalyzed oxidation of primary alcohols

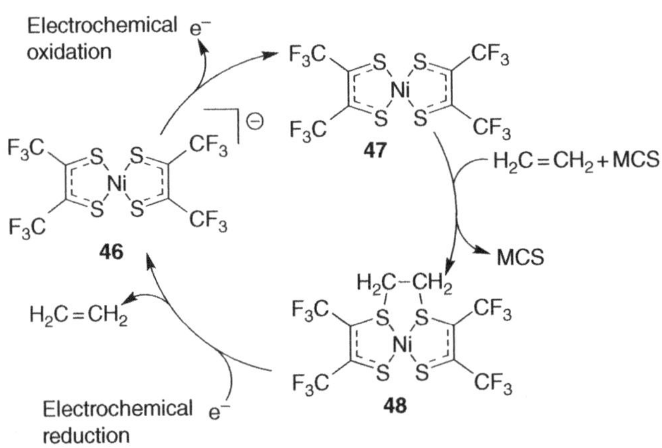

Scheme 14 Redox non-innocent ligands applied in ethene purification

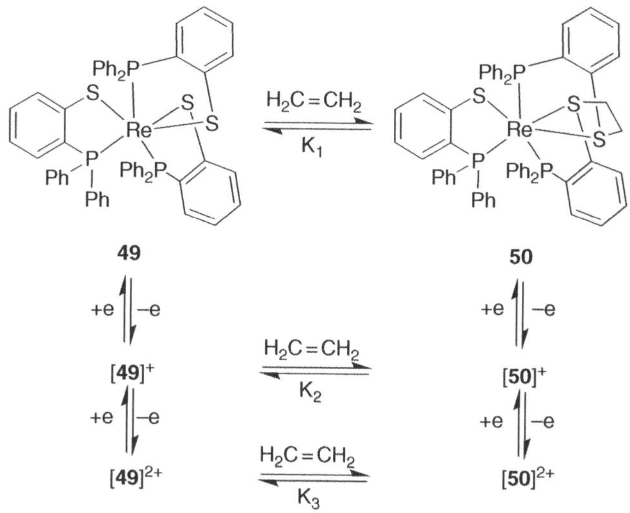

Scheme 15 Ethylene capture at the redox non-innocent thiyl radical of a rhenium complex

7.5 Strategy IV. Cooperative substrate-centered radical-type reactivity based on redox non-innocent substrates

Among the different classes of redox non-innocent ligands, the one using intrinsic redox features of the substrate is perhaps most interesting in view of catalytic turnover. Upon coordination to a redox active metal, redox active substrates can get reduced or oxidized by the metal center leading to an organic ligand-centered radical. This behavior has been recently applied in a variety of organometallic catalytic transformations. The new properties conferred to the substrate allowed organic transformations which were not achievable with closed-shell systems. Stoichiometric and catalytic reactions using those features have been already discussed [6, 7a, 7e, 43]. In this chapter we will focus on catalytic reactions where metal-activated substrates bear radical-type spin density, thus leading to an open-shell intermediate with new type of substrate–ligand–radical reactivity.

This idea was demonstrated by using open-shell porphyrin Co(II) complexes bearing a carbene or a nitrene (Scheme 16). Kinetics, spectroscopic and computational studies were used to determine the mechanism of the reaction, showing that the formation of the carbene (or nitrene) radical via loss of N_2 is the rate-determining step. The coordinated substrate undergoes one-electron reduction by the metal center leading to a coordinated organic radical-type moiety and a Co(III) center **52**. The radical intermediate can then react further with different organic substrates to produce intermediates of type **53** (Scheme 16), allowing radical-type cyclization reactions to form cyclopropanes (**54, X=C**) [44], or aziridines (**54, X=N**) [45]. This reaction is facilitated by the presence of a relatively weak axial Co—C bond [46], which allowed the regeneration of the catalyst by homolytic Co—C bond splitting simultaneous with radical-type C—C or C—N bond formation. Moreover, on account of the use of chiral

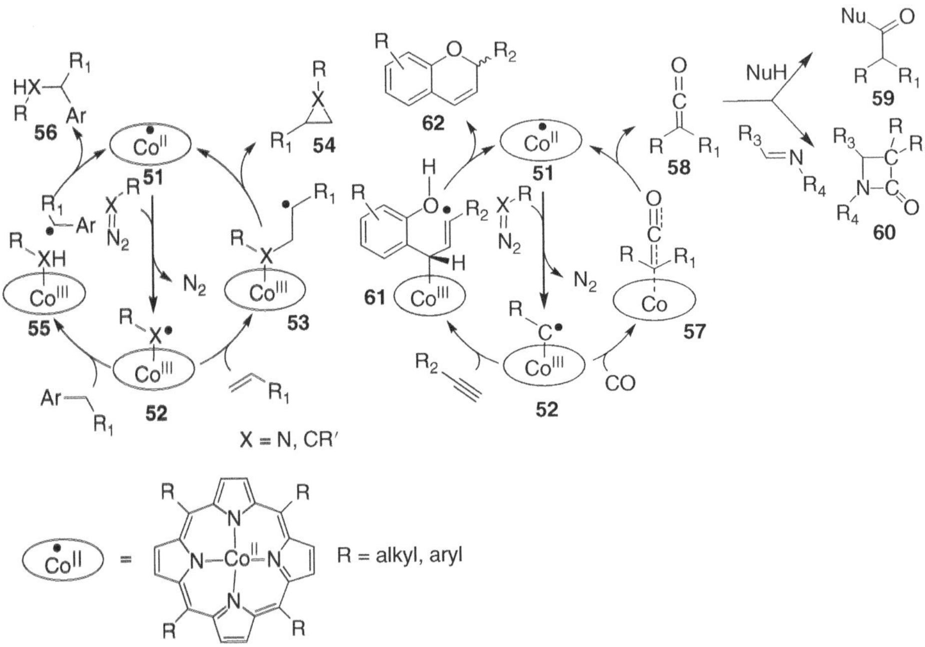

Scheme 16 *Formation of reactive substrate-based radical ligands in catalytic nitrene or carbene transfer reactions proceeding via controlled radical-type transformations (dashed lines represent partial bonds)*

cobalt porphyrin complexes high enantioselectivities have been achieved in the cyclopropanation and aziridination reaction [45b, 45e, 47]. Nitrenoid intermediates can also insert in the C—H bond of allylic or benzylic substrates to produce amine products **56** (via intermediates of type **55**).

This strategy has been recently employed in the catalytic synthesis of reactive ketenes **58**, involving carbene carbonylation (Scheme 16). The reactive ketenes could be trapped in situ by different nucleophiles and imines to form esters/imines **59** and β-lactams **60** [48]. Moreover, during this study *N*-tosylhydrazones (which upon deprotonation convert to diazo compounds) were found to be compatible with cobalt catalysis. This opened the way to a new class of reactions, allowing the use of a broader range of diazo compounds in catalytic reactions involving cobalt-bound "carbene radicals". For instance, chromenes **62** were efficiently synthesized in cobalt-catalyzed radical-type ring-closure reactions involving "carbene radical" and "vinyl radical" intermediates (Scheme 16) [49].

Interestingly, cobalt porphyrin catalysts tend to prevent carbene dimerization reactions, and allow cyclopropanation reactions with electron-deficient alkenes. This feature illustrates the more nucleophilic behavior of the carbenoid species formed as compared to typical electrophilic Fischer carbenes. The enhanced nucleophilic character of the carbene reduces its tendency to dimerize and allows reactions with more electron-deficient olefins.

In order to improve activity and robustness of the catalyst, cobalt-based porphyrin complexes have been encapsulated in a supramolecular host (Scheme 17) [50]. The so-called "molecular flask" (**63-Zn@64**, **63-Co@64**, **63-Zn@65,** and **63-Co@65**) prevents catalyst aggregation (ligand-bearing coordinating groups) and deactivation reactions (e.g., radical–radical coupling or catalyst deactivation via hydrogen atom transfer). Indeed, the encapsulated catalyst shows higher turnover number than the free analog in cyclopropanation reaction and *Z*-selective olefin synthesis using a diazo compound as the substrate. Moreover, cage modifications allowed cyclopropanation reactions in aqueous media, and the encapsulated catalysts show size selectivity. Smaller substrates are converted faster than larger substrates, which penetrate less easily through the pores of the flask.

A non-heme Fe(II) complex active in amination and aziridination reactions was isolated and characterized using adipyrromethene ligand (Scheme 18) [51]. Mixing **66** with a bulky organic azide (adamantylazide) in toluene reveals interesting reactivity involving nitrene insertion in the benzylic position of toluene. The optimized reaction conditions yielded 94% of benzyladamantylamine for a total of 10–12 turnovers at 60 °C. The proposed mechanism involves coordination of the organic azide with loss of N_2 leading to the formation of complex **68**. This intermediate has been isolated and characterized, and was proposed to be a high-spin Fe(III) (d^5, $S=5/2$) center coupled to a nitrene radical ($S=1/2$). In this way the intermediate **68** has a total spin of two with six unpaired electrons. Density functional theory calculations showed a substantial amount of spin density at the nitrene moiety, which hence adopts "nitrene radical" character [43a, 43b]. Complex **68** can abstract a hydrogen atom from toluene. Subsequent radical recombination leads to the formation of the aminated organic compound **72** and regeneration of catalyst. In a recent paper, the reaction was expanded to benzylic and allylic amination, as well as aziridination reactions, with more functional group tolerance [52]. The mechanism was confirmed to proceed via radical intermediates.

More recently, the same catalyst was used to produce cyclic amines with retention of stereochemistry from a simple linear aliphatic azide [53]. Treatment of a substituted aliphatic azide by complex **66** afforded the cyclized compound **75**, by insertion of the nitrene moiety in allylic, benzylic, and even in the less reactive tertiary C—H bonds. The catalyst is inhibited by coordination of the product to the metal center. However, that can be avoided by using an in situ protecting agent (Boc_2O is preferred over Fmoc-OSuc which leads to catalyst decomposition).

Very recently, a new redox-active N*NO* pincer ligand bound to a Pd(II) center permitted an intramolecular ligand-to-substrate electron shuttling, and then reacted with unactivated azide via an unusual "nitrene-substrate radical" (Scheme 19) [54]. Complex **76** has been synthetized in two steps from the neutral ligand and $PdCl_2(NCMe)_2$ precursor, followed by stirring under aerobic conditions in the presence of NEt_3. The latter, in the solid state, is air- and moisture- stable. X-ray analysis, magnetic susceptibility measurements (indicating an $S=1/2$ ground state), and EPR measurements correlated to DFT calculations indicate that complex **76** has an open-shell N*NO*^{ISQ} ligand radical electronic structure in which a mono-anionic iminobenzosemiquinonato ligand radical is coordinated to a square planar Pd(II) center. Chemical reduction of **76** with cobaltocene

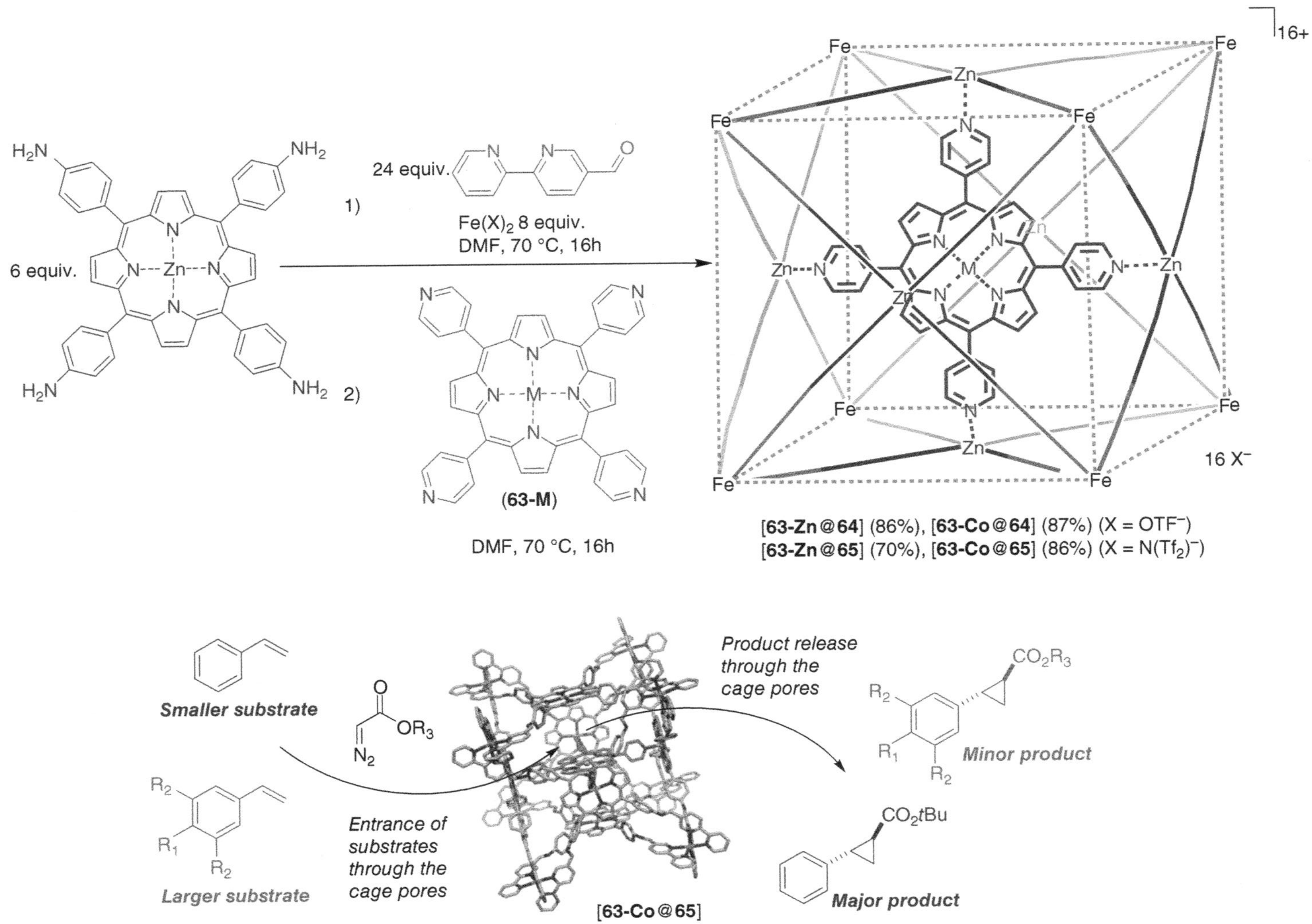

Scheme 17 *Molecular flask synthesis and application in size selective catalysis. (See insert for color/color representation of this figure)*

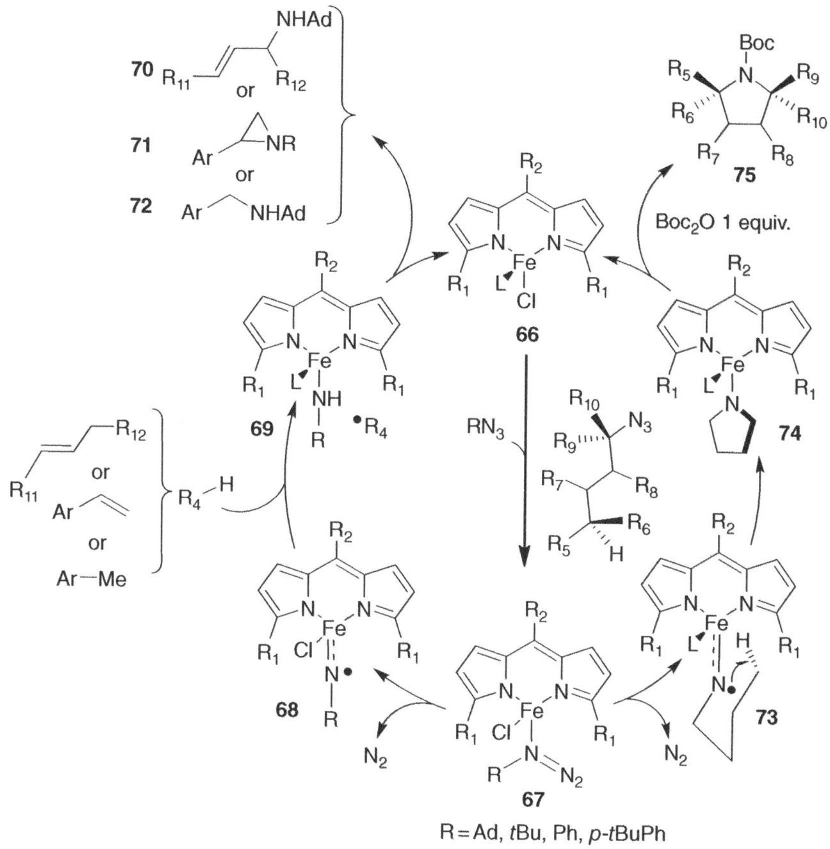

Scheme 18 *Catalytic cycle of nitrene insertion into the C—H bond of organic substrates proceeding via "nitrene radicals"*

afforded the highly air- and moisture-sensitive diamagnetic complex **77**. The N*NO*^{AP} (amidophenolato dianion) scaffold is able to donate a single electron, and on reaction with an aliphatic azide N_2 loss is associated with one-electron transfer from the redox active N*NO*^{AP} ligand to the thus formed nitrene substrate, initiating N-centered radical-type reactivity. The in situ reduction of **76** to **77** with cobaltocene in the presence of unactivated (4-azidobutyl)benzene and Boc_2O leads to stoichiometric formation of the Boc-protected pyrrolidine **81** (~1 equiv. with respect to **77**). Hence, intramolecular one-electron transfer from the redox-active ligand to the substrate produces an unusual "nitrene-substrate radical, ligand-radical" Pd(II) intermediate. Similar to the reactions reported above for the Co(porphyrin) and Fe(dipyrromethene) systems, subsequent hydrogen-atom abstraction and radical rebound steps lead to C—H bond amination. Reaction of intermediate **80** with Boc_2O was proposed to generate the pyrrolidine compound **81**. Hence, the use of a redox-active N*NO* ligand enables one-electron reaction pathways for Pd(II), a metal that more commonly reacts via two-electron (e.g., oxidative addition, reductive elimination) pathways.

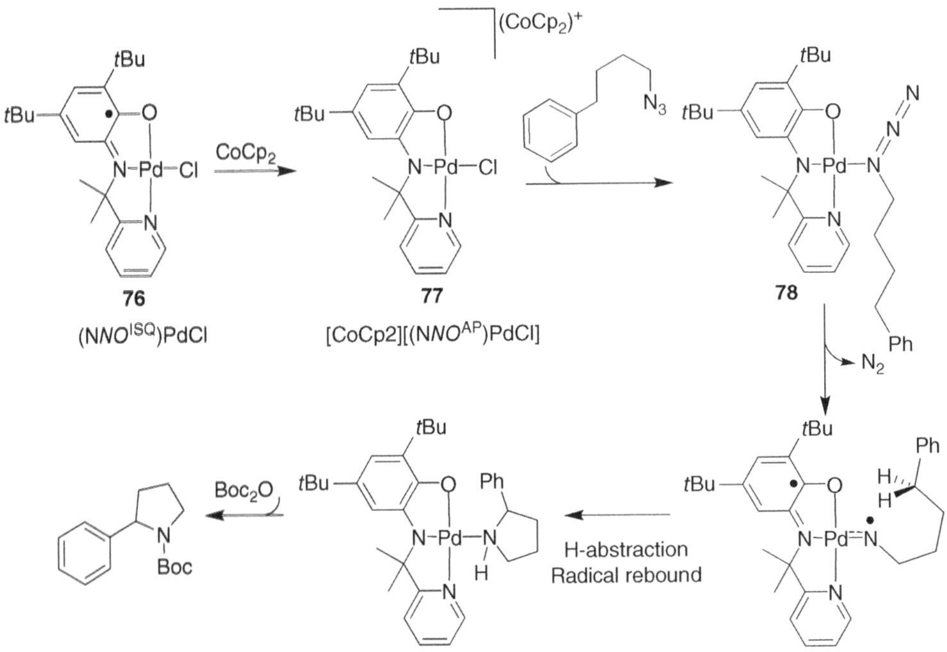

Scheme 19 *Reactivity of a Pd(II) center bound to a redox active NNO pincer ligand in sp^3 C—H amination from unactivated azide via ligand-to-substrate one-electron transfer*

7.6 Conclusion

To ensure a sustainable society, highly selective and efficient but cheap new catalysts need to be developed. Using "redox non-innocent ligands" the activity as well as selectivity of a transition metal catalyst can be controlled in such a way that several difficult chemical transformations can be facilitated. Redox active ligands can be used as tunable "spectator" ligands in transition metal coordination complexes ("the catalyst"), allowing redox control of the Lewis acidity/basicity of the metal, hence controlling the substrate affinity/selectivity of the reactive metal center. As a "spectator", redox active ligands can also influence the overall catalytic process by acting as electron reservoirs without any covalent interaction with the substrates. Redox active ligands can also play a more active role in catalysis, allowing cooperative bond making/breaking during catalytic turnover. As an electron reservoir, redox active ligands can prevent the metal ions adopting unfavorable oxidation states throughout the whole catalytic reaction. The ability of this class of ligands to store and release electrons when required, allows multi-electron transformations (usually favored for noble, late second- and third-row transition metals) with cheaper first-row transition metal ions (which commonly prefer one-electron transformations). Hence, nobility may be conferred to base metals [20a, 55]. This unique ability of redox non-innocent ligands also permits oxidative addition processes to take place during the catalytic cycle even with catalysts possessing d^0 metal centers. In "actor" mode, the substrate itself can act as a redox active ligand, forming open-shell ligand-centered radicals during catalysis. Many of such ligand-centered radical-type reactions proceed in a controlled and selective manner, most frequently leading to metal–carbon and carbon–carbon radical couplings. For the more sterically hindered systems selective ligand-centered hydrogen-atom abstractions are a common alternative reaction pathway. These selective ligand-centered radical-type reactions are synthetically useful in organometallic synthesis. Especially carbene- and nitrene-substrate-based radical ligands have shown their ability towards selective radical-type transformations. Completely different reactivity and chemoselectivities were shown by the one-electron reduced open-shell forms of the carbenes and nitrenes compared with their closed-shell analogs. Finally, the role of redox non-innocent ligands in different metabolic/enzymatic reactions in biological systems is now well documented. Their application is now expanding in the field of synthetic chemistry, expanding the toolbox of synthetic chemists. Many new discoveries can be expected from future research in this area, especially in the field of catalysis.

References

[1] (a) W. Kaim, A. Klein, M. Glöckle, *Acc. Chem. Res.* **2000**, *33*, 755–763; (b) M. D. Ward, J. A. McCleverty, *J. Chem. Soc., Dalton Trans.* **2002**, 275–288; (c) C. C. Lu, E. Bill, T. Weyhermüller, E. Bothe, K. Wieghardt, *J. Am. Chem. Soc.* **2008**, *130*, 3181–3197.

[2] (a) H. Grützmacher, *Angew. Chem. Int. Ed.* **2008**, *47*, 1814–1818; (b) H. Grützmacher, *Angew. Chem.* **2008**, *120*, 1838–1842; (c) J. I. van der Vlugt, J. N. Reek, *Angew. Chem. Int. Ed.* **2009**, *48*, 8832–8846; (d) J. I. van der Vlugt, J. N. Reek, *Angew. Chem.* **2009**, *121*, 8990–9004.

[3] (a) W. Kaim, B. Schwederski, *Coord. Chem. Rev.* **2010**, *254*, 1580–1588; (b) W. Kaim, *Coord. Chem. Rev.* **1987**,*76*, 187–235.

[4] P. J. Chirik, *Inorg. Chem.* **2011**, *50*, 9737–9740.

[5] K. Hindson, B. de Bruin, *Eur. J. Inorg. Chem.* **2012**, 340–342.

[6] V. Lyaskovskyy, B. de Bruin, *ACS Catal.* **2012**, *2*, 270–279.

[7] (a) J. I. van der Vlugt, *Eur. J. Inorg. Chem.* **2012**, 363–375; (b) W. I. Dzik, J. I. van der Vlugt, J. N. Reek, B. de Bruin, *Angew. Chem. Int. Ed.* **2011**, *50*, 3356–3358; (c) W. I. Dzik, J. I. van der Vlugt, J. N. Reek, B. de Bruin, *Angew. Chem.* **2011**, *123*, 3416–3418; (d) D. Zhu, I. Thapa, I. Korobkov, S. Gambarotta, P. H. Budzelaar, *Inorg. Chem.* **2011**, *50*, 9879–9887; (e) W. I. Dzik, B. de Bruin, *Organomet. Chem.* **2011**,*37*, 46–78.

[8] (a) M. R. Ringenberg, S. L. Kokatam, Z. M. Heiden, T. B. Rauchfuss, *J. Am. Chem. Soc.* **2008**, *130*, 788–789; (b) M. R. Ringenberg, T. B. Rauchfuss, *Eur. J. Inorg. Chem.* **2012**, 490–495.

[9] W. R. Cullen, T. J. Kim, F. W. B. Einstein, T. Jones, *Organometallics* **1983**, *2*, 714–719.

[10] I. M. Lorkovic, R. R. Duff, M. S. Wrighton, *J. Am. Chem. Soc.* **1995**, *117*, 3617–3618.

[11] (a) A. M. Allgeier, C. A. Mirkin, *Angew. Chem. Int. Ed.* **1998**, *37*, 894–908; (b) A. M. Allgeier, C. A. Mirkin, *Angew. Chem.***1998**, *110*, 936–952.

[12] (a) J. L. Boyer, J. Rochford, M. Tsai, J. T. Muckerman, E. Fujita, *Coord. Chem. Rev.* **2010**, *254*, 309–330; (b) J. T. Muckerman, D. E. Polyansky, T. Wada, K. Tanaka, E. Fujita, *Inorg. Chem.* **2008**, *47*, 1787–1802.

[13] (a) R. A. Binstead, C. W. Chronister, J. Ni, C. M. Hartshorn, T. J. Meyer, *J. Am. Chem. Soc.* **2000**, *122*, 8464–8473; (b) J. J. Concepcion, J. W. Jurss, M. K. Brennaman, P. G. Hoertz, A. O. T. Patrocinio, N. Y. Murakami Iha, J. L. Templeton, T. J. Meyer, *Acc. Chem. Res.* **2009**, *42*, 1954–1965; (c) F. Liu, J. J. Concepcion, J. W. Jurss, T. Cardolaccia, J. L. Templeton, T. J. Meyer, *Inorg. Chem.* **2008**, *47*, 1727–1752.

[14] T. Wada, K. Tsuge, K. Tanaka, *Inorg. Chem.* **2001**, *40*, 329–337.

[15] (a) V. Artero, M. Fontecave, *Coord. Chem. Rev.* **2005**, *249*, 1518–1535; (b) V. Artero, M. Chavarot-Kerlidou, M. Fontecave, *Angew. Chem. Int. Ed.* **2011**, *50*, 7238–7266; (c) V. Artero, M. Chavarot-Kerlidou, M. Fontecave, *Angew. Chem.* **2011**, *123*, 7376–7405; (d) C. W. Cady, R. H. Crabtree, G. W. Brudvig, *Coord. Chem. Rev.* **2008**, *252*, 444–455; (e) M. Yagi, M. Kaneko, *Chem. Rev.* **2001**, *101*, 21–36.

[16] B. D. Stubbert, J. C. Peters, H. B. Gray, *J. Am. Chem. Soc.* **2011**, *133*, 18070–18073.

[17] O. R. Luca, S. J. Konezny, J. D. Blakemore, D. M. Colosi, S. Saha, G. W. Brudvig, V. S. Batista, R. H. Crabtree, *New J. Chem.* **2012**, *36*, 1149–1152.

[18] D. C. Lacy, C. C. L. McCrory, J. C. Peters, *Inorg. Chem.* **2014**, *53*, 4980–4988.

[19] (a) D. Sieh, M. Schlimm, L. Andernach, F. Angersbach, S. Nückel, J. Schöffel, N. Šušnjar, P. Burger, *Eur. J. Inorg. Chem.* **2012**, 444–462; (b) M. Ghosh, T. Weyhermüller, K. Wieghardt, *Dalton Trans.* **2010**, *39*, 1996–2007; (c) C. C. Lu, T. Weyhermüller, E. Bill, K. Wieghardt, *Inorg. Chem.* **2009**, *48*, 6055–6064; (d) Q. Knijnenburg, S. Gambarotta, P. H. Budzelaar, *Dalton Trans.* **2006**, *46*, 5442–5448; (e) Q. Knijnenburg, D. Hetterscheid, T. M. Kooistra, P. H. M. Budzelaar, *Eur. J. Inorg. Chem.* **2004**, 1204–1211; (f) A. M. Tait, F. V. Lovecchio, D. H. Busch, *Inorg. Chem.* **1977**, *16*, 2206–2212.

[20] (a) P. J. Chirik, K. Wieghardt, *Science* **2010**, *327*, 794–795; (b) M. W. Bouwkamp, A. C. Bowman, E. Lobkovsky, P. J. Chirik, *J. Am. Chem. Soc.* **2006**, *128*, 13340–13341.

[21] (a) B. de Bruin, E. Bill, E. Bothe, T. Weyhermuller, K. Wieghardt, *Inorg. Chem.* **2000**, *39*, 2936–2947; (b) P. H. Budzelaar, B. de Bruin, A. W. Gal, K. Wieghardt, J. H. van Lenthe, *Inorg. Chem.* **2001**, *40*, 4649–4655.

[22] K. T. Sylvester, P. J. Chirik, *J. Am. Chem. Soc.* **2009**, *131*, 8772–8774.

[23] S. K. Russell, E. Lobkovsky, P. J. Chirik, *J. Am. Chem. Soc.* **2011**, *133*, 8858–8861.

[24] (a) A. M. Tondreau, C. Milsmann, A. D. Patrick, H. M. Hoyt, E. Lobkovsky, K. Wieghardt, P. J. Chirik, *J. Am. Chem. Soc.* **2010**, *132*, 15046–15059; (b) M. W. Bouwkamp, E. Lobkovsky, P. J. Chirik, *Inorg. Chem.* **2006**, *45*, 2–4.

[25] A. L. Smith, K. I. Hardcastle, J. D. Soper, *J. Am. Chem. Soc.* **2010**, *132*, 14358–14360.

[26] (a) S. Fetzner, *Appl. Environ. Microbiol.* **2012**, *78*, 2505–2514; (b) G. Fuchs, M. Boll, J. Heider, *Nat. Rev. Microbiol.* **2011**, *9*, 803–816; (c) S. Fetzner, B. Tshisuaka, F. Lingens, R. Kappl, J. Hüttermann, *Angew. Chem. Int. Ed.* **1998**, *37*, 576–597; (d) S. Fetzner, B. Tshisuaka, F. Lingens, R. Kappl, J. Hüttermann, *Angew. Chem.* **1998**, *110*, 596–617.

[27] (a) G. Owens, *Catal. Today* **2000**, *55*, 317–363; (b) R. H. Holm, *Chem. Rev.* **1987**, *87*, 1401–1449.

[28] (a) C. R. Landis, C. M. Morales, S. S. Stahl, *J. Am. Chem. Soc.* **2004**, *126*, 16302–16303; (b) B. V. Popp, J. E. Wendlandt, C. R. Landis, S. S. Stahl, *Angew. Chem. Int. Ed.* **2007**, *46*, 601–604; (c) B. V. Popp, J. E. Wendlandt, C. R. Landis, S. S. Stahl, *Angew. Chem.* **2007**, *119*, 607–610; (d) L. Häller, L. Jonas, E. Mas-Marzá, A. Moreno, J. P. Lowe, S. A. Macgregor, M. F. Mahon, P. S. Pregosin, M. K. Whittlesey, *J. Am. Chem. Soc.* **2009**, *131*, 9618–9619.

[29] C. A. Lippert, S. A. Arnstein, C. D. Sherrill, J. D. Soper, *J. Am. Chem. Soc.* **2010**, *132*, 3879–3892.

[30] K. J. Blackmore, N. Lal, J. W. Ziller, A. F. Heyduk, *J. Am. Chem. Soc.* **2008**, *130*, 2728–2729.

[31] C. A. Lippert, K. Riener, J. D. Soper, *Eur. J. Inorg. Chem.* **2012**, 554–561.
[32] A. F. Heyduk, R. A. Zarkesh, A. I. Nguyen, *Inorg. Chem.* **2011**, *50*, 9849–9863.
[33] M. R. Haneline, A. F. Heyduk, *J. Am. Chem. Soc.* **2006**, *128*, 8410–8411.
[34] (a) M. van der Meer, Y. Rechkemmer, I. Peremykin, S. Hohloch, J. van Slageren, B. Sarkar, *Chem. Commun.* **2014**, DOI: 10.1039/C4CC03309D; (b) A. L. Smith, L. A. Clapp, K. I. Hardcastle, J. D. Soper, *Polyhedron* **2010**, *29*, 164–169.
[35] A. I. Nguyen, R. A. Zarkesh, D. C. Lacy, M. K. Thorson, A. F. Heyduk, *Chem. Sci.* **2011**, *2*, 166–169.
[36] F. Lu, R. A. Zarkesh, A. F. Heyduk, *Eur. J. Inorg. Chem.* **2012**, 467–470.
[37] (a) L. Que, W. B. Tolman, *Nature* **2008**, *455*, 333–340; (b) J. W. Whittaker, *Chem. Rev.* **2003**, *103*, 2347–2363.
[38] (a) P. Chaudhuri, M. Hess, U. Flörke, K. Wieghardt, *Angew. Chem. Int. Ed.* **1998**, *37*, 2217–2220; (b) P. Chaudhuri, M. Hess, U. Flörke, K. Wieghardt, *Angew. Chem.* **1998**, *110*, 2340–2343.
[39] P. Chaudhuri, K. Wieghardt, T. Weyhermüller, T. K. Paine, S. Mukherjee, C. Mukherjee, *Biol. Chem.* **2005**, *386*, 1023–1033.
[40] (a) M. Königsmann, N. Donati, D. Stein, H. Schönberg, J. Harmer, A. Sreekanth, H. Grützmacher, *Angew. Chem. Int. Ed.* **2007**, *46*, 3567–3570; (b) M. Königsmann, N. Donati, D. Stein, H. Schönberg, J. Harmer, A. Sreekanth, H. Grützmacher, *Angew. Chem.* **2007**, *119*, 3637–3640.
[41] (a) K. Wang, E. I. Stiefel, *Science* **2001**, *291*, 106–109; (b) D. J. Harrison, N. Nguyen, A. J. Lough, U. Fekl, *J. Am. Chem. Soc.*, **2006**, *128*, 11026–11027; (c) L. Dang, M. F. Shibl, X. Yang, A. Alak, D. J. Harrison, U. Fekl, E. N. Brothers, M. B. Hall, *J. Am. Chem. Soc.*, **2012**, *134*, 4481–4484.
[42] (a) C. Grapperhaus, K. Ouch, M. S. Mashuta, *J. Am. Chem. Soc.* **2009**, *131*, 64–65; (b) K. Ouch, M. S. Mashuta, C. A. Grapperhaus, *Inorg. Chem.* **2011**, *50*, 9904–9914.
[43] (a) A. I. O. Suarez, V. Lyaskovskyy, J. N. H. Reek, J. I. van der Vlugt, B. de Bruin, *Angew. Chem. Int. Ed.* **2013**, *52*, 12510–12529; (b) A. I. O. Suarez, V. Lyaskovskyy, J. N. H. Reek, J. I. van der Vlugt, B. de Bruin, *Angew. Chem.* **2013**, *125*, 12740–12760; (c) W. I. Dzik, X. P. Zhang, B. de Bruin, *Inorg. Chem.* **2011**, *50*, 9896–9903.
[44] (a) H. Lu, W. I. Dzik, X. Xu, L. Wojtas, B. de Bruin, X. P. Zhang, *J. Am. Chem. Soc.* **2011**, *133*, 8518–8521; (b) D. Intrieri, A. Caselli, E. Gallo, *Eur. J. Inorg. Chem.* **2011**, 5071–5081; (c) J. L. Belof, C. R. Cioce, X. Xu, X. P. Zhang, B. Space, H. L. Woodcock, *Organometallics* **2011**, *30*, 2739–2746; (d) W. I. Dzik, X. Xu, X. P. Zhang, J. N. H. Reek, B. de Bruin, *J. Am. Chem. Soc.* **2010**, *132*, 10891–10902; (e) S. Zhu, X. Xu, J. A. Perman, X. P. Zhang, *J. Am. Chem. Soc.* **2010**, *132*, 12796–12799; (f) S. Zhu, J. V. Ruppel, H. Lu, L. Wojtas, X. P. Zhang, *J. Am. Chem. Soc.* **2008**, *130*, 5042–5043; (g) Y. Chen, J. V. Ruppel, X. P. Zhang, *J. Am. Chem. Soc.* **2007**, *129*, 12074–12075.
[45] (a) A. I. Olivos Suarez, H. Jiang, X. P. Zhang, B. de Bruin, *Dalton Trans.* **2011**, *40*, 5697–5705; (b) V. Subbarayan, J. V. Ruppel, S. Zhu, J. A. Perman, X. P. Zhang, *Chem. Commun.* **2009**, 4266–4268; (c) J. E. Jones, J. V. Ruppel, G. Gao, T. M. Moore, X. P. Zhang, *J. Org. Chem.* **2008**, *73*, 7260–7265; (d) J. V. Ruppel, J. E. Jones, C. A. Huff, R. M. Kamble, Y. Chen, X. P. Zhang, *Org. Lett.* **2008**, *10*, 1995–1998; (e) G. Gao, J. E. Jones, R. Vyas, J. D. Harden, X. P. Zhang, *J. Org. Chem.* **2006**, *71*, 6655–6658.
[46] M. F. Summers, P. J. Toscano, N. Bresciani-Pahor, G. Nardin, L. Randaccio, L. G. Marzilli, *J. Am. Chem. Soc.* **1983**, *105*, 6259–6263.
[47] (a) X. Xu, S. Zhu, X. Cui, L. Wojtas, X. P. Zhang, *Angew. Chem. Int. Ed.* **2013**, *52*, 11857–11861; (b) X. Xu, S. Zhu, X. Cui, L. Wojtas, X. P. Zhang, *Angew. Chem.* **2013**,

125, 12073–12077; (c) L. Jin, X. Xu, H. Lu, X. Cui, L. Wojtas, X. P. Zhang, *Angew. Chem. Int. Ed.* **2013**, *52*, 5309–5313; (d) L. Jin, X. Xu, H. Lu, X. Cui, L. Wojtas, X. P. Zhang, *Angew. Chem.* **2013**, *125*, 5417–5421; (e) S. Zhu, X. Cui, X. P. Zhang, *Eur. J. Inorg. Chem.* **2012**, 430–434; (f) X. Cui, X. Xu, L. Wojtas, M. M. Kim, X. P. Zhang, *J. Am. Chem. Soc.* **2012**, *134*, 19981–19984; (g) H. Lu, Y. Hu, H. Jiang, L. Wojtas, X. P. Zhang, *Org. Lett.* **2012**, *14*, 5158–5161; (h) X. Cui, X. Xu, H. Lu, S. Zhu, L. Wojtas, X. P. Zhang, *J. Am. Chem. Soc.* **2011**, *133*, 3304–3307; (i) X. Xu, H. Lu, J. V. Ruppel, X. Cui, S. Lopez de Mesa, L. Wojtas, X. P. Zhang, *J. Am. Chem. Soc.* **2011**, *133*, 15292–15295; (j) H. Lu, H. Jiang, L. Wojtas, X. P. Zhang, *Angew. Chem. Int. Ed.* **2010**, *49*, 10192–10196; (k) H. Lu, H. Jiang, L. Wojtas, X. P. Zhang, *Angew. Chem.* **2010**, *122*, 10390–10394.

[48] N. D. Paul, A. Chirila, H. Lu, X. P. Zhang, B. de Bruin, *Chem. Eur. J.* **2013**, *19*, 12953–12958.

[49] N. D. Paul, S. Mandal, M. Otte, X. Cui, X. P. Zhang, B. de Bruin, *J. Am. Chem. Soc.* **2014**, *136*, 1090–1096.

[50] (a) M. Otte, P. F. Kuijpers, O. Troeppner, I. Ivanovic-Burmazovic, J. N. H. Reek, B. de Bruin, *Chem. Eur. J.* **2013**, *19*, 10170–10178; (b) M. Otte, P. F. Kuijpers, O. Troeppner, I. Ivanovic-Burmazovic, J. N. H. Reek, B. de Bruin, *Chem. Eur. J.* **2014**, *20*, 4880–4884.

[51] E. R. King, E. T. Hennessy, T. A. Betley, *J. Am. Chem. Soc.* **2011**, *133*, 4917–4923.

[52] E. T. Hennessy, R. Y. Liu, D. A. Iovan, R. A. Duncan, T. A. Betley, *Chem. Sci.* **2014**, *5*, 1526–1532.

[53] E. T. Hennessy, T. A. Betley, *Science* **2013**, *340*, 591–595.

[54] D. L. J. Broere, B. de Bruin, J. N. H. Reek, M. Lutz, S. Dechert, J. I. van der Vlugt, *J. Am. Chem. Soc.* **2014**, DOI: 10.1021/ja502164f.

[55] O. R. Luca, R. H. Crabtree, *Chem. Soc. Rev.* **2013**, *42*, 1440–1459.

8

Ligands for Iron-based Homogeneous Catalysts for the Asymmetric Hydrogenation of Ketones and Imines

Demyan E. Prokopchuk, Samantha A. M. Smith, and Robert H. Morris

Department of Chemistry, University of Toronto, 80 St. George Street, Toronto, Ontario, Canada M5S 3H6

8.1 Introduction: from ligands for ruthenium to ligands for iron

8.1.1 Ligand design elements in precious metal homogeneous catalysts for asymmetric direct hydrogenation and asymmetric transfer hydrogenation

The asymmetric direct hydrogenation (ADH) of ketones and imines is performed in the pharmaceutical, agrochemical and fragrance industries to produce valuable enantiopure alcohols and amines. For example, specially designed ruthenium catalysts containing elaborate enantiodirecting diphosphine and diamine ligands are important for catalyzing the addition of hydrogen gas to inexpensive arylketones in order to obtain (*S*)- or (*R*)-alcohol products for perfumes or pharmaceutical intermediates. Scheme 1 shows the ADH of 3-(dimethylamino)propiophenone to the (*R*)-enantioenriched alcohol using a ruthenium complex developed by Ohkuma *et al.*[1] This alcohol is used to prepare (*R*)-fluoxetine, a potent serotonin-uptake inhibitor used for treating depression.

Ligand Design in Metal Chemistry: Reactivity and Catalysis, First Edition. Edited by Mark Stradiotto and Rylan J. Lundgren.

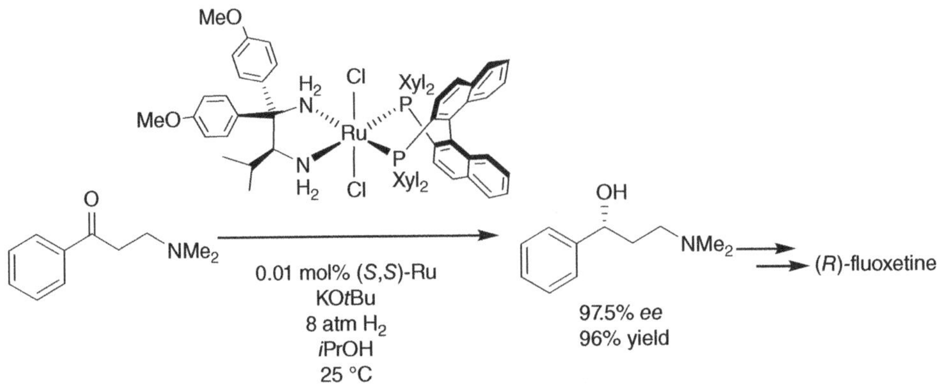

Scheme 1 *The ADH of a prochiral phenylalkylketone using (S,S)-trans-[RuCl$_2${(S)-xylbinap}{(S)-daipen}] (where Xyl = 3,5-Me$_2$C$_6$H$_3$–)*

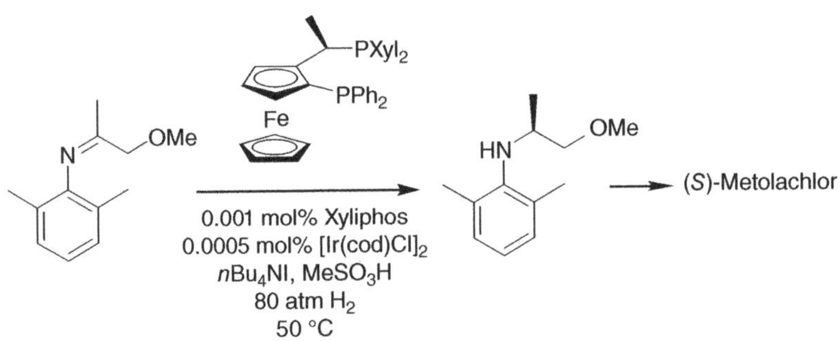

Scheme 2 *The ADH of a prochiral imine performed on a large scale in industry*

Similarly, an iridium catalyst with a specific ferrocenylphosphine ligand (Scheme 2) is used in the production of the chiral amine (*S*)-metolachlor, a herbicide, by the asymmetric direct hydrogenation of a prochiral imine.[2]

Another useful reduction process is asymmetric transfer hydrogenation (ATH) where the hydrogen is transferred from the solvent, often isopropanol, to the ketone or imine function to produce the enantiopure alcohol or amine. For example, Baratta *et al.*[3] made ruthenium complexes containing the (*R,S*)-Xyliphos ligand to reduce a simple ketone to (*S*)-1-(3-trifluoromethylphenyl)ethanol, used in the synthesis of the fungicide (*S*)-MA20565 (Scheme 3).

In these examples the ligands have been carefully crafted for optimum turnover frequency (TOF) and turnover number (TON). Several design elements for effective catalysis have emerged in the development of such catalysts.

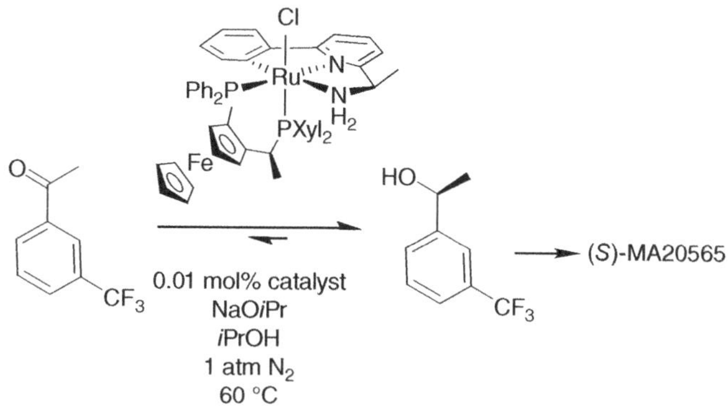

Scheme 3 *The ATH of methyl(3-trifluoromethylphenyl)ketone using a ruthenium Xyliphos complex in basic isopropanol (the source of hydrogen)*

8.1.1.1 The hydride–metal–nitrogen–proton motif and chiral ligand elements for enantioselectivity

In the foregoing examples a hydride forms on the ruthenium or iridium by the action of hydrogen gas (in ADH) or solvent and base (in ATH) and then attacks the carbon of the polar bond. A primary or secondary amino group is often present in the ligand structure to accelerate the hydrogenation by means of the "NH effect."[4–8] Here the proton from the nitrogen hydrogen bonds with the oxygen of the ketone and orients it to attack by the metal hydride; usually it serves to transfer a proton to the oxygen when the metal transfers the hydride to the carbon in a low energy barrier transition state. This is referred to as bifunctional catalysis (Figure 1a).[9, 10]

The chiral diamine and diphosphine structures on Noyori-type ruthenium catalysts (e.g., Scheme 1) each contribute to the chiral environment around the catalyst that leads to enantioselectivity. Figure 1b illustrates the proposed role of each in the *trans*-dihydride complex that attacks the carbonyl of the ketone in the enantio-determining step.[9] This model has been supported by recent computational studies.[11, 12] The substituents on the diamine lock the five-membered N-Ru-N ring so that there is one axial N—H in line with each Ru—H bond. The carbonyl of the incoming ketone is then aligned for attack. The binaphthyl group locks the P-Ru-P ring so that one aryl group on each phosphorus, stacked in a rigid fashion against a naphthyl ring, is held in an axial position. The ketone approaches the H-Ru-N-H motif so that its bulky aryl group is away from this axial group. As Noyori writes "the enantiofaces of prochiral ketones are differentiated on the molecular surface of the coordinatively saturated RuH intermediate. This notion is in contrast to the conventional mechanism for hydrogenation of unsaturated bonds that requires the metal-substrate complexation."[13] (*S*)-chirality is induced in the product alcohol when the (*R*)-binaphthyl and (*R,R*)-diamine are present in the catalyst.[9] The chirality of the diphosphine and diamine must be matched (*R*) with (*R,R*) for high enantioselectivity in the resulting

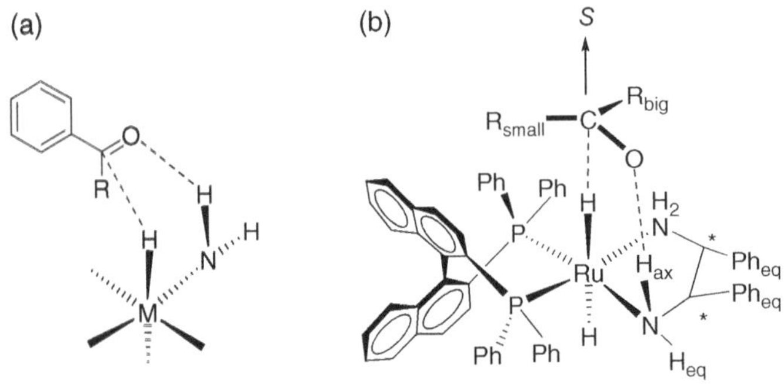

Figure 1 *The hydride–metal–nitrogen–proton motif in the catalyst attacking a ketone in a bifunctional manner (a) and the source of enantioselectivity in the ADH of ketones catalyzed by trans-RuH$_2$((R)-binap)(diamine) complexes (b)*[9, 10]

(*S*)-alcohol. The (*R*) with (*S,S*) combination does not have the required chiral array of aryl groups in the complex and so has poor selectivity.[5]

8.1.1.2 A nucleophilic hydride and weakly acidic N—H bond in the H-M-N-H motif

The carbon in a ketone or imine functional group is weakly electrophilic and so a hydride group must be strongly nucleophilic. A ligand *trans* to the hydride with high *trans* influence[14] weakens the metal–hydrogen bond thus facilitating hydride transfer to the ketone. For ketone hydrogenation catalysts this *trans* ligand has a donor of low electronegativity, either hydrogen (hydride), carbon (e.g., carbonyl, alkyl) or phosphorus (e.g., phosphine).[15–17] For example, *trans*-RuHCl(binap)(diamine) with electronegative chloride *trans* to hydride does not react with ketones at room temperature while *trans*-Ru(H)$_2$((*R*)-binap)(diamine) does (Figure 1b).[9]

A neutral or negative charge on the complex also makes a hydride more nucleophilic than a comparable cationic complex.[15, 18] For example, the energy barrier for the attack on acetophenone by the cationic complex [Ru(H)(NHC-NH$_2$)(benzene)]$^+$ is calculated to be 16 kcal/mol higher in energy than that of the isoelectronic neutral complex Ru(H)(NHC-NH$_2$)(C$_5$H$_5$) which is a very active ketone direct hydrogenation (DH) catalyst (Figure 2, NHC-NH$_2$ is an *N*-heterocyclic carbene ligand with a primary amino group tethered to it).[18]

Charge has a large effect on the acidity of metal hydride complexes[19] although acidity and hydricity are not necessarily coupled.[20] Of the complexes [M(P-N-P)H]$^+$ where M = Ni, Pd, Pt, the palladium complex is the most hydridic and one of the strongest acids of the triad.[20] However, in general, phosphorus and nitrogen donors make hydrides less acidic and more hydridic and nucleophilic than π-accepting ligands such as carbonyls.[15, 19]

The amine group on ruthenium involved in bifunctional catalysis should have aliphatic substituents as in Schemes 1 and 3 and Figures 1 and 2. When saturated carbon(s) is (are) attached to the nitrogen of a primary or secondary amine, respectively,

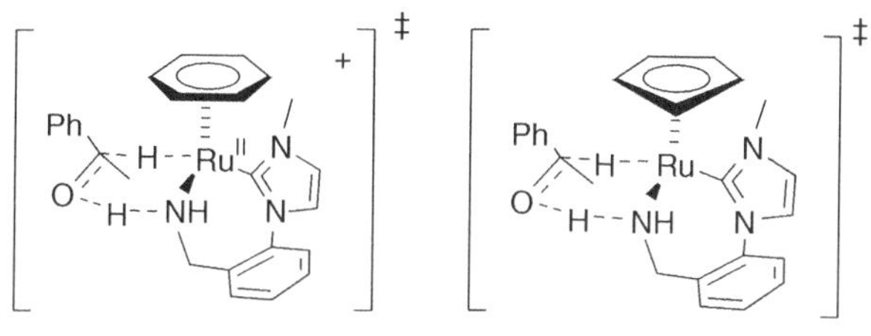

Figure 2 Comparing free energy barriers to hydride transfer from ruthenium to acetophenone for neutral and cationic N-heterocyclic carbene ruthenium catalysts

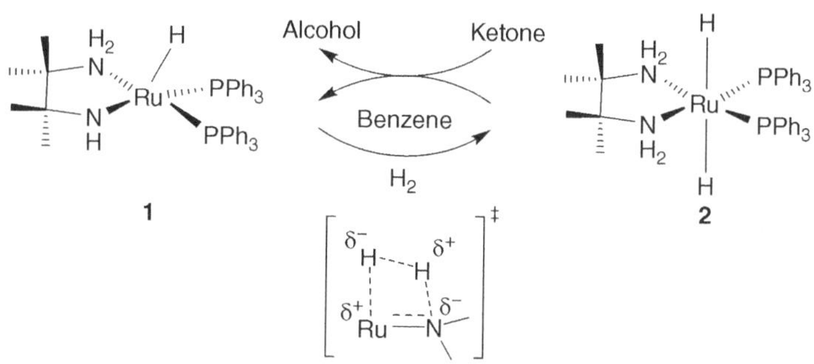

Scheme 4 Well defined amido complex **1** and trans-dihydride complex **2** catalyze the direct hydrogenation of ketones

the deprotonated form of the amine, the amido form, is often basic enough to facilitate the heterolytic splitting of dihydrogen at ruthenium. Dihydrogen is an extremely weak acid, but when coordinated in a neutral metal complex with the correct ancillary ligands it becomes acidic enough to be deprotonated by a strong base. For example, the well-defined amido complex **1** in Scheme 4 is known to catalyze the efficient DH of ketones via the heterolytic splitting of dihydrogen at the ruthenium–amido bond and subsequent bifunctional attack of the hydride and proton on the ketone. When diamines such as *ortho*-phenylenediamine or binaphthyldiamine are coordinated in ruthenium phosphine complexes, the nitrogen of the resulting amido complex is usually not basic enough to deprotonate coordinated dihydrogen because of the delocalization of the amido electrons into the aromatic ring. Thus, such diamines have been referred to as poisons to ruthenium-catalyzed ketone DH.[21]

8.1.1.3 Iminophosphine ligands and early catalyst design

A wide range of ruthenium DH ketone catalysts have been made with phosphorus and nitrogen ligands.[15, 22] The Schiff base condensation of phosphine aldehydes and amines to generate iminophosphines is an important reaction for providing a range of

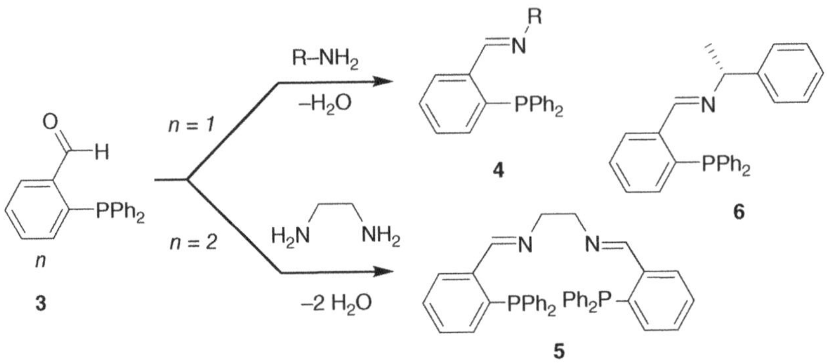

Scheme 5 Synthesis of iminophosphine ligands by Schiff base condensation

imino- and aminophosphine P-N-N-P donor ligands for asymmetric transition metal catalyzed processes. First discovered in 1973, air stable *ortho*-(diphenylphosphino) benzaldehyde (**3**, Scheme 5)[23, 24] could be reacted with various amines and diamines to afford free iminophosphine ligands, a process first exploited by Rauchfuss in 1978 (**4**, Scheme 5).[25, 26] If 2 equiv. of phosphine aldehyde are reacted in the presence of ethylenediamine, the symmetrical P-N-N-P ligand is generated (**5**).[27] It is interesting to note that in 1984 the first application of this Schiff base condensation for ADH was with iminophosphine **6** and its reduced aminophosphine derivative; poor activity and enantioselectivity was observed.[28] The diverse reactivity of **3** in Schiff base condensation reactions with various transition metals has been reviewed elsewhere;[29] moreover, chiral bidentate PN iminophosphine chelates with the general formula **4** are especially useful for palladium-catalyzed allylic alkylation and have been reviewed elsewhere.[30]

Several years later, it was reported that $Ru(OAc)_2(PPh_3)_2$ could be reacted with **5** to generate the structurally characterized complexes *trans*-$RuCl_2$(P-N-N-P) or *trans*-$Ru(OAc)_2$(P-N-N-P), depending on the reaction conditions.[31, 32] Subsequently, the chiral diamine (*S,S*)-diaminocyclohexane and 2 equiv. of **3** were reacted to generate the chiral P-N-N-P ligand **7** and reduction with $NaBH_4$ yielded the diamine product P-NH-NH-P (**8**).[33] At about the same time, Gao, Ikariya, and Noyori went on to synthesize (*S,S*)-*trans*-$RuCl_2$(P-N-N-P) (**9**) and (*S,S*)-*trans*-$RuCl_2$(P-NH-NH-P) (**10**), which are both precatalysts for the ATH of aromatic ketones in basic isopropanol (Scheme 6).[34, 35] However, complex **9** was poorly active/selective for the ATH of acetophenone (3% yield, 18% *ee* after 48 h where *ee* is enantiomeric excess) while **10** had excellent activity and selectivity (91% yield, 97% *ee* after 25 h). Several other chiral derivatives can be easily synthesized by varying the chiral diamine used in the condensation reaction, followed by reduction with $NaBH_4$.[35] After this discovery, using *ortho*-(diphenylphosphino)benzaldehyde (**3**) as a ligand synthon became recognized as a facile way to generate chiral tridentate[36–44] and tetradentate[45–49] aminophosphine ligands for the ruthenium-catalyzed hydrogenation of polar bonds.

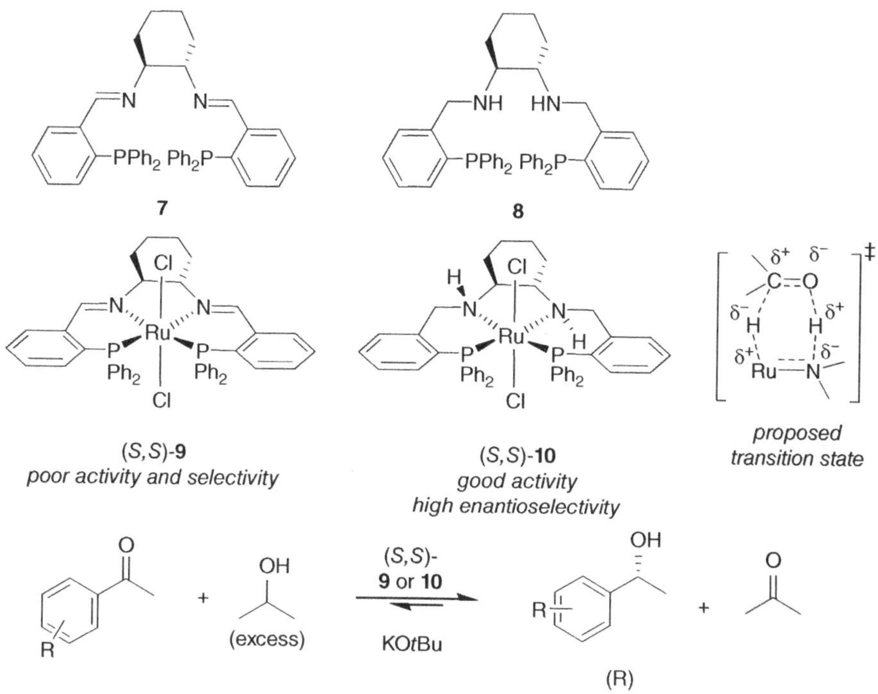

Scheme 6 *Asymmetric transfer hydrogenation of acetophenone derivatives by Ru precatalysts (S,S)-**9** and **10** with tetradentate ligands*[34]

The dramatic difference in activity for diamine precatalysts versus diimines can be rationalized by the NH effect (Section 8.1.1.1).[4–8] Base was required during catalysis and it was postulated that its role was to deprotonate the amine (NH) functionality and generate a reactive metal-amido moiety. This reacts with dihydrogen and then transfers a proton and hydride equivalent in the outer coordination sphere to ketones in a bifunctional manner, going through a six-membered pericyclic transition state (Scheme 6, right). However, rigorous mechanistic investigations in recent years have shown that the nature of proton and hydride transfer is likely to be a two-step process as opposed to a concerted one in many cases, and this is a subject of ongoing experimental and computational investigation.[50–56]

The ruthenium-catalyzed reduction of polar bonds using H_2 gas, rather than a sacrificial reductant such as isopropanol, is an atom-economical reaction that has been thoroughly explored.[15] A key discovery by our laboratory was that neutral, structurally characterized metal-amido complexes such as **11** (Scheme 7) could cleave H_2 heterolytically to yield the *trans*-dihydride complex **12**, and that these are crucial intermediates during catalysis.[47, 57, 58] Once the H-N-Ru-H moiety is in place, proton and hydride can then be transferred to the substrate. Having methyl groups instead of hydrogens on carbons alpha to the amido group (beta to the ruthenium) in this case and in the case of **1** (Scheme 4) was important to allow the isolation of an amido

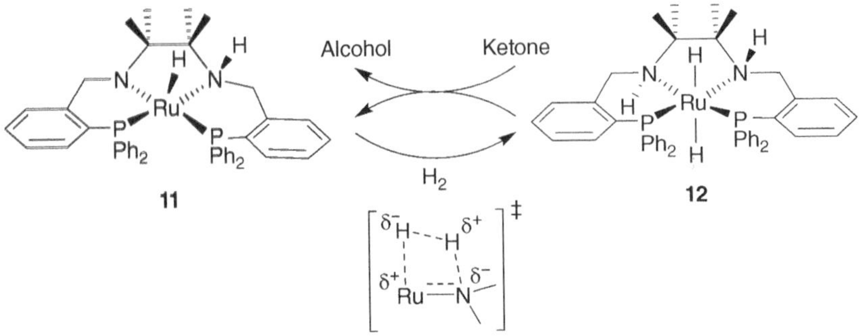

*Scheme 7 Reaction of ruthenium hydrido-amido complex **11** with hydrogen to give the dihydride **12** during ketone hydrogenation*

species. If hydrogens are present, the ruthenium amido complexes tend to undergo β-hydride elimination to form catalytically inactive imine complexes.

We also found that the chiral complex *trans*-RuH(Cl)(P-NH-NH-P) [where P-NH-NH-P is (*S,S*)-**8**] when treated with base was highly active for both ATH and ADH; however, the activity and enantioselectivity were not the same.[46] This suggested that different catalysts or mixtures of catalysts were forming in solution depending on the reaction conditions, specifically the amount of added base and the nature of the reductant.

With an understanding of the general mechanism of ketone hydrogenation using diamine-diphosphine and tetradentate ruthenium P-N-N-P complexes, we were motivated to explore iron-based systems. Attempts by our group to synthesize iron diamine-diphosphine analogs akin to dihydrides **1** and **12** were unsuccessful, although their potential as hydrogenation catalysts has been investigated using density functional theory (DFT).[59] Our rationale was that by using a higher denticity P-N-N-P scaffold, the ligand might remain more securely anchored to the 3d metal, preventing ligand dissociation or catalyst decomposition/deactivation. Furthermore, the low cost of iron, its high abundance, and its reduced toxicity are all desirable catalyst features for the pharmaceutical, fragrance and agrochemical industries.

8.1.2 Effective ligands for iron-catalyzed ketone and imine reduction

The development of active iron catalysts for ATH and ADH processes followed from the ligand design principles enumerated above and further principles uncovered from 2008 to the present. Before that time there were reports of poorly active catalysts based on iron carbonyl cluster complexes for transfer hydrogenation (TH)[60–62], ATH [63] or iron dicarbonyls with a hydroxyl-functionalized cyclopentadienyl group for DH.[64, 65] A breakthrough came in 2008 when our group discovered that monocarbonyl iron complexes with tetradentate P-N-N-P ligands, when treated with base in isopropanol, were quite active for ATH.[66] Since that time we have uncovered further ligand design principles that have led to successively more active and enantioselective catalysts. The progress can be classified into three generations of precatalyst types as shown in Figure 3.

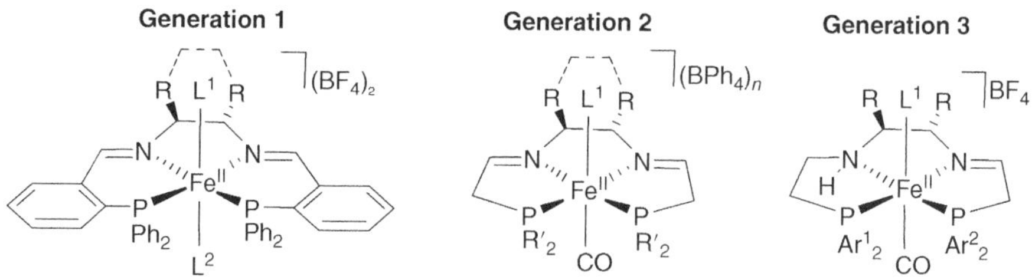

R = H, alkyl, aryl, or part of cyclohexyldiamine ring

L^1 = MeCN, L^2 = CO
or L^1 = MeCN, L^2 = CN*t*Bu

R = H, alkyl, aryl, or part of cyclohexyldiamine ring
R′ = Et, Aryl

L^1 = MeCN (n = 2),
or L^1 = Br^- (n = 1)

R = H, aryl

L1 = Cl^- or Br^-

Figure 3 *General structures of three generations of successively more active iron ATH catalysts*

The synthesis and properties of these catalysts will be discussed in more detail below but first some aspects of the structures will be discussed to emphasize important ligand design features for iron-catalyzed hydrogenation processes.

8.1.3 Ligand design elements for iron catalysts

8.1.3.1 Strong field ligands on iron(II) keep it diamagnetic

Common to the structures shown in Figure 3 and to those of several other hydrogenation and hydrosilylation catalysts based on iron complexes[65, 67–82] are the strong ligand field carbonyl and phosphine ligands. Diamagnetic octahedral iron(II) complexes tend to be substitution inert and less reactive compared with paramagnetic iron(II) complexes which can undergo rapid substitution.[83] In particular, substitution of the chiral groups in the molecule is undesirable for enantioselective catalysis, which depends on structurally well-defined, single sites for catalysis. Therefore, paramagnetic complexes are usually not as effective for these applications.

A further advantage of diamagnetic complexes is that the reaction solutions can be studied by NMR methods in order to obtain mechanistic information. Low coordinate iron(II) complexes with nitrogen ligands as shown in Figure 4 are active catalysts for olefin (**13**)[84] and ketone hydrosilylation (**14**)[85] as well as asymmetric hydrosilylation (**15**).[86] However, these highly reactive complexes are paramagnetic and difficult to study, leaving the nature of the active catalytic species undefined.

8.1.3.2 Strong field ligands promote hydride formation

Hundreds of iron hydride complexes with phosphine ligands have been isolated and characterized. These complexes are almost always diamagnetic. For example, a spectacularly active electrocatalyst for dihydrogen oxidation is the cyclopentadienyl iron

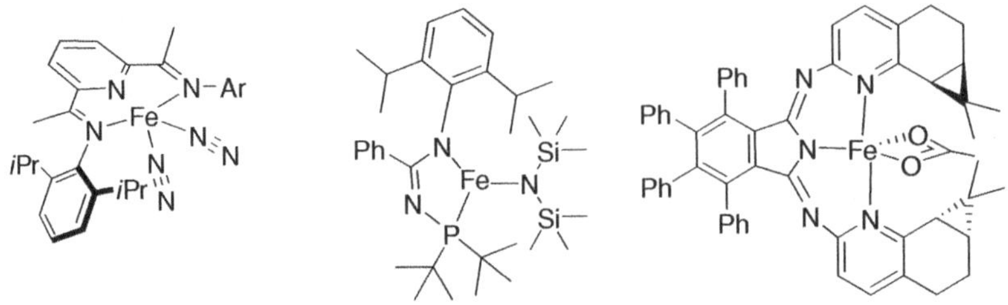

13 14 15

Figure 4 *Paramagnetic iron complexes **13–15** are catalysts for hydrosilylation*

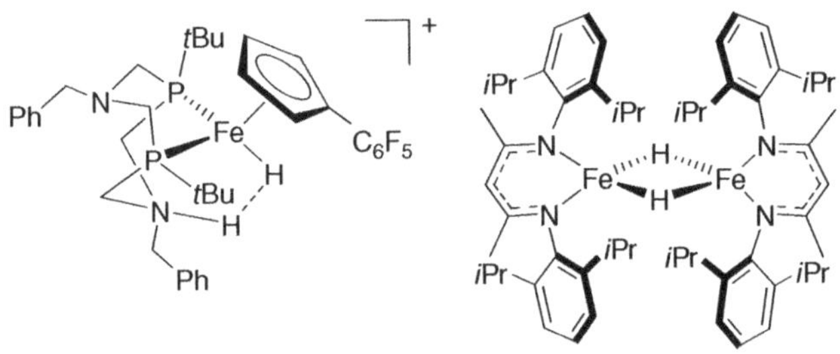

16 **17**

Figure 5 *Examples of diamagnetic (**16**) and paramagnetic (**17**) iron(II) hydride complexes*

phosphine hydride complex **16**[87] shown in Figure 5. Both the chelating phosphine ligand and the η^5-cyclopentadienyl ligands are strong field ligands. This hydride complex and catalytic intermediate is formed by the extremely rapid reaction of the iron(II) with dihydrogen gas followed by rapid heterolytic splitting of the H–H bond to produce the complex shown where the proton and hydride are still interacting in a dihydrogen bond.[88–90] Such complexes have been characterized by many techniques including NMR and crystallography.[87, 91] In contrast, paramagnetic hydride complexes of iron such as **17**[92] are rare and highly reactive.

8.1.3.3 Privileged polydentate phosphorus and nitrogen donor ligands

Tridentate P-N-P and tetradentate P-N-N-P are more effective and selective in iron-based catalytic hydrogenation than bidentate or monodentate ligands. There have been attempts by our research group and others to prepare iron analogs to the ruthenium diphosphine diamine complexes of Noyori shown in Scheme 1. These have so far proven unsuccessful. Similarly, the complexes $RuCl_2(P\text{-}NH_2)_2$ and $RuH(Cl)(P\text{-}NH_2)_2$, where $P\text{-}NH_2$ is a chelating ligand with phosphine and primary amine donors, are efficient hydrogenation catalysts when treated with base under hydrogen for the reduction of ketones to alcohols,[10] imines to amines,[93] and esters to alcohols,[94] while the iron complexes are unknown. In contrast, extremely active iron catalysts based on tetradentate P-N-N-P ligands are known as will be discussed extensively below.

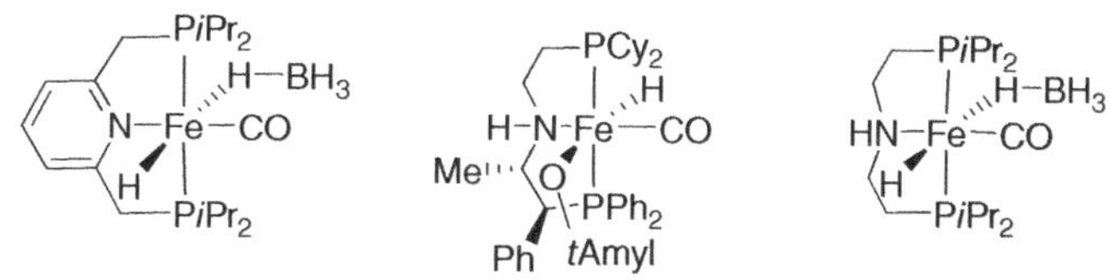

18 19 20

Figure 6 *Iron precatalysts with P-N-P ligands for DH (**18, 20**) and ADH (**19**)*

Tridentate phosphorus–nitrogen donors have recently been shown to provide active iron catalysts for DH and ADH. Figure 6 shows three types of iron hydride complexes that are active catalysts for ketone, imine and ester hydrogenation. Complex **18** is a DH catalyst for ketones at 4.1 atm H_2, achieving TONs for acetophenone hydrogenation of up to 1720 with a TOF 300 h^{-1} at 40 °C.[68] Complex **19** is a precatalyst and the first example of an effective iron-based system for the ADH of ketones to (*S*)-alcohols; when activated with base under 5 atm H_2 at 50 °C, this system produces alcohols with *ee* up to 85% at TON of up to 5000 and TOF up to 2000 h^{-1}.[67] Complex **20**[95] and related derivatives are active catalysts for ester hydrogenation[96, 97] and imine hydrogenation[98] as well as other dehydrogenation processes.

8.1.3.4 *Optimum chelating ring size*

Certain tetradentate P-N-N-P and P-N-NH-P ligands have proven very effective for ATH with iron (Figure 3). As will be discussed in more detail below, the complexes with five-membered Fe-P-N- and Fe-N-N- rings are more active than those with six-membered rings. This has been attributed to flexibility of the six-membered ring allowing for the formation of iron(0), leading to catalyst decomposition.[99] The complexes with five-membered rings are more rigid and provide higher enantioselectivity to (*S*) or (*R*) alcohols than the ones with six-membered rings.

8.1.3.5 *Versatile template synthesis of P-N-P and P-N-N-P ligands*

A beautiful aspect of the chemistry of cationic iron(II) complexes is that the metal center acts as a template and promotes the Schiff base condensation reaction of amine and phosphine-aldehyde components so that a wide range of tridentate and tetradentate ligands can be readily assembled on the metal. The metal template effect was first described by Busch and co-workers[100, 101] and has been used extensively in the synthesis of polydentate ligands directly on the metal when their synthesis off the metal would be tedious or impossible.[102, 103] For example, the ligand for complex **19** of Figure 6 was created by reacting the precursor of the phosphine-aldehyde PCy_2CH_2CHO (see below for more details) with the enantiopure compound (*S,S*)-$PPh_2CHPhCHMeNH_2$ in the presence of $FeBr_2$ and CO(g) followed by reduction with $LiAH_4$ and quenching with alcohol to convert the imine to an amine.[67] This is a powerful ligand synthesis strategy, allowing the ready exploration of catalyst structure to find the one that yields the highest activity and selectivity for a particular substrate.

8.2 First generation iron catalysts with symmetrical [6.5.6]-P-N-N-P ligands

At about the same time that Gao *et al.*[34] published their seminal work on the asymmetric hydrogenation of ketones with tetradentate Ru-P-N-N-P complexes, the achiral diimine complex *trans*-$[Fe(NCMe)_2(P\text{-}N\text{-}N\text{-}P)][ClO_4]_2$ (**21**) and diamine complex *trans*-$[Fe(NCMe)_2(P\text{-}NH\text{-}NH\text{-}P)][ClO_4]_2$ (**22**) were synthesized by refluxing free ligand **5** (Scheme 5) or its reduced version with $[Fe(H_2O)_6][ClO_4]_2$ in acetonitrile; however, structural characterization and catalytic activity were not reported.[104] It was also known that chiral diaminodiphosphine P-NH-NH-P ligands could be combined with $[HNEt_3][Fe_3H(CO)_{11}]$ to catalyze the asymmetric transfer hydrogenation of ketones, but the nature of the active species was not well known.[63] Thus, we sought to synthesize a well-defined system for this application.

Chiral Fe(II) P-N-N-P complex **23** (Scheme 8) can be synthesized by mixing the free P-N-N-P ligand (*R,R*)-7 [(*S,S*)-7 could also be used] with $[Fe(H_2O)_6][BF_4]_2$ in refluxing acetonitrile or by mixing with $FeCl_2$ and then performing a salt metathesis with $NaBF_4$.[66] Complex **23** was structurally characterized using single crystal X-ray diffraction, confirming that the acetonitrile ligands are in a *trans* orientation and the P-N-N-P ligand is coordinating in a tetradentate fashion. This iron complex was tested for the ADH of acetophenone in basic isopropanol solvent (25 atm H_2, 50 °C, substrate:catalyst:base ratio of 225:1:15) and was found to be somewhat active [TOF = 5 h^{-1}, 40% conversion, 27% *ee* (*S*)]; nonetheless, it was the first well-defined ADH catalyst based on iron for the reduction of polar bonds.[66] However, it was inactive as a catalyst for the ATH of ketones.

Next, π-acidic carbonyl and isonitrile ligands were incorporated into the coordination sphere either by exposing **23** to CO gas (1 atm) or *tert*-butyl isonitrile (2 equiv.) to obtain complexes **24** and **25**, respectively, in quantitative yields (Scheme 8).

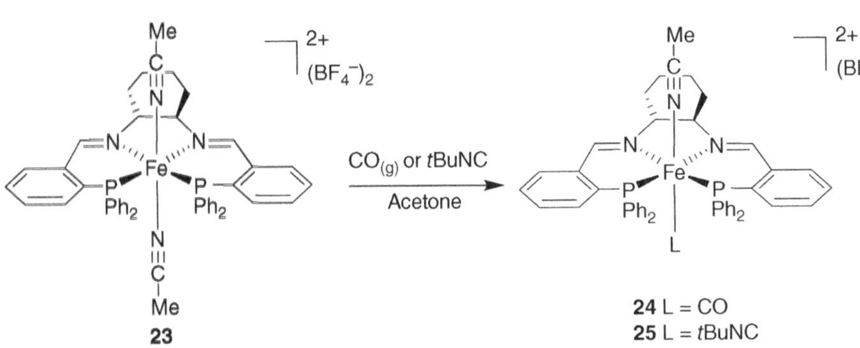

Scheme 8 *The first well-defined iron complexes for the ADH (**23**) and ATH (**24, 25**) of ketones*

We rationalized that the presence of such π-acceptors, much like those found in the active site of Fe-hydrogenase,[105–108] might also enhance the catalytic activity of these systems. Furthermore, the π-acceptor ligand might also encourage the 3d metal to remain in a low spin configuration, preventing catalyst degradation or decomposition. Interestingly, complex **24** was found to be *inactive* for DH but *active* for the ATH of ketones as well as TH of aldehydes and an aldimine in basic isopropanol at room temperature with TOFs as high as 1000 h^{-1}, *ee* of up to 61% (*S*), and conversions of up >99% for various ketones.[66] Complex **25** was found to be much less active but more selective for the ATH of acetophenone under the same conditions, with a TOF of 28 h^{-1}, *ee* of 76% (*S*), and conversion of 34%.

8.2.1 Synthetic routes to ADH and ATH iron catalysts

Various chiral and achiral derivatives of the *trans*-acetonitrile bis(iminophosphine) [6.5.6]-P-N-N-P iron complexes can be synthesized by a one-pot condensation/template reaction involving $[Fe(H_2O)_6][BF_4]_2$, **3** (2 equiv.), and a diamine (**26–28**) to yield complexes (**29–31**) (Scheme 9).[109, 110] These new compounds were tested for catalytic activity in the DH of acetophenone to 1-phenylethanol. Low to moderate conversions were observed in most cases at room temperature and 25 atm H_2 pressure, with precatalyst **29** being able to achieve up to 95% conversion at 50 °C in 18 h. Only complex (*R,R*)-**31** was enantioselective [*ee* of 61% (*S*)], but with poor conversion (4%) at elevated temperatures.

In addition, the tetrafluoroborate salt of P-NH-NH-P complex **22** showed slightly better activity (95% conversion) than the diimine precatalyst **29** under identical conditions. This

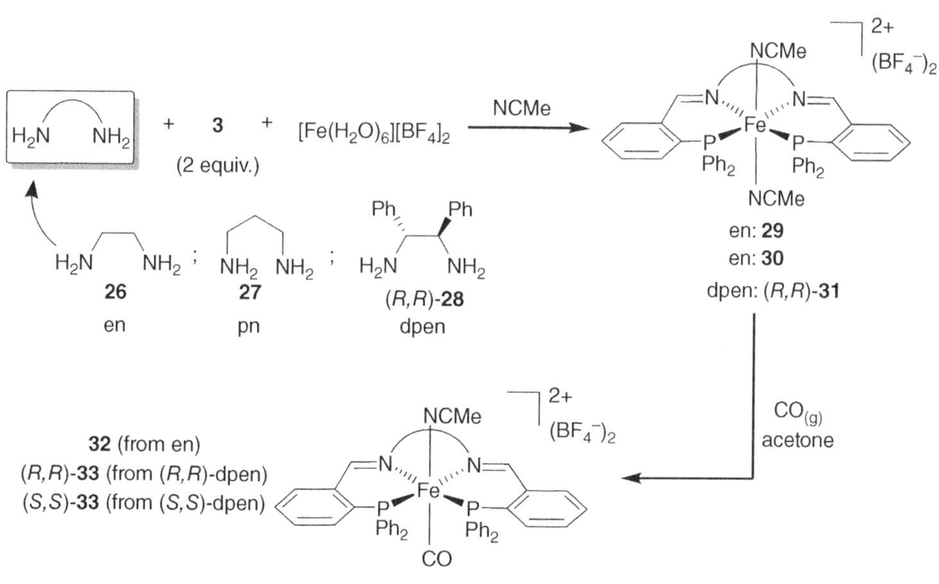

Scheme 9 *Synthesis of iron(II) [6.5.6]-P-N-N-P DH catalysts **29–31** by a one-pot Fe(II) template reaction followed by CO exchange to generate ATH catalysts **32** and **33***

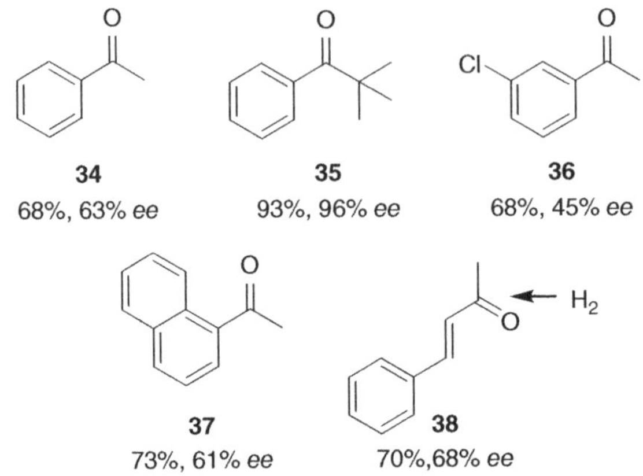

Figure 7 *The conversion of a variety of ketones to their alcohols with the ee (S) via ATH catalyzed by (R,R)-**33***

suggested that the NH functionality was important for catalytic activity and that the imine group(s) of **29** were being reduced during catalysis. None of the bis-acetonitrile iron complexes in Scheme 9 were found to be active for the TH of acetophenone in basic isopropanol.

Stirring the *trans*-acetonitrile complexes **29**, (*S*,*S*)-**31** and (*R*,*R*)-**31** under an atmosphere of carbon monoxide in acetone cleanly generates the new monosubstituted carbonyl complexes **32**, (*S*,*S*)-**33** and (*R*,*R*)-**33**.[111] Structural characterization of **32** and (*S*,*S*)-**33** confirm that the P-N-N-P ligand is in a planar arrangement, with the carbonyl and acetonitrile ligands occupying the apical positions. These complexes are active for the ATH of ketones, with the highest TOF and *ee* observed using (*S*,*S*)-**33** or (*R*,*R*)-**33**, reaching a maximum TOF of 2600 h^{-1}, maximum *ee* of 96%, and maximum conversion of 93%. A few examples that illustrate the substrate scope using catalyst (*R*,*R*)-**33** are shown in Figure 7.

8.2.2 Catalyst properties and mechanism of reaction

The identity of active catalytic species for the TH of ketones with our iron carbonyl [6.5.6]-P-N-N-P complexes was still unclear. Did the imine or imines on the ligand get reduced in situ, allowing catalysis to occur through a bifunctional outer sphere mechanism, as seen with the analogous ruthenium systems? This question drove us to further investigate the mechanism of transfer hydrogenation with our first generation [6.5.6]-P-N-N-P systems.

All the iron [6.5.6]-P-N-N-P hydrogenation catalysts require activation by an external base. The catalytic profile for achiral complex **32** showed that there is an induction period before rapid catalysis takes place for the TH of acetophenone in basic isopropanol. Unexpectedly, reacting **32** with sodium isopropoxide in either benzene or isopropanol yields the pentadentate ferraaziridine complex **39**, which was isolated and structurally characterized (Scheme 10).[99] This unusual complex is diamagnetic, with

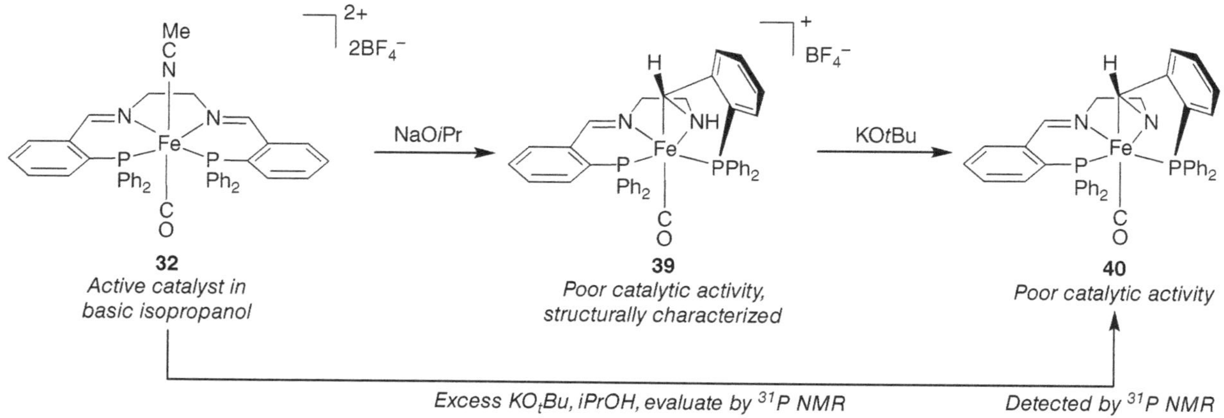

Scheme 10 *The conversion of complex **32** to the ferraaziridine complex **39** and then to electron-rich complex **40***

a tetrahedral alkyl substituent in the apical position, which is singly bonded to iron and a secondary amine, forming a three membered Fe-C-N metallacycle. The amine ligand can be deprotonated in the presence of base to form the neutral ferraziridinide complex **40**, whose P–P coupling constant is spectroscopically distinguishable from **39** by ^{31}P NMR spectroscopy.

Both complexes **39** and **40** are poor catalysts for the TH of acetophenone in basic isopropanol, suggesting that the protonated or deprotonated form is not within the catalytic cycle. When a catalytic mixture with **32** is evaluated by ^{31}P NMR, the major species detected in solution is **40** along with free ligand and oxidized free ligand. Mass balance experiments, as evaluated by ^{31}P NMR, showed that a maximum of 75% of the iron in solution is NMR inactive, suggesting that the active species is either NMR inactive or NMR active and found in extremely low concentrations. Furthermore, an IR spectrum of a crude solution of **40** revealed two distinct carbonyl stretching frequencies at 1862 and 1870 cm^{-1}, which are indicative of electron-rich iron carbonyl complexes.

With the support of DFT calculations, we proposed that the two observed carbonyl stretches below 1900 cm^{-1} belong to ferraziridinide **40** and a reduced Fe(0) complex, which could serve as a pathway towards NMR inactive iron species (Scheme 10).[99] The possibility of an Fe(0) species being responsible for the absence of 75% of iron-containing compound in our mass balance NMR experiments pointed to the presence of iron nanoparticles. This led us to discover moderately efficient nanoparticle-catalyzed ATH as described elsewhere.[110]

8.3 Second generation iron catalysts with symmetrical [5.5.5]-P-N-N-P ligands

8.3.1 Synthesis of second generation ATH catalysts

8.3.1.1 Synthesis of phosphonium dimers: "protected" phosphine aldehydes

In order to design a ligand system more suitable for iron, we rationalized that the larger six-membered M-P-N rings containing the *orthophenylene* moiety needed to be reduced in size. In 1996, Matt *et al.*[112] found that a stable phosphonium dimer could be isolated under acidic conditions. The use of these phosphonium dimers, which could be converted to phosphine aldehydes in the presence of base, was a promising route to achieve this goal. Our group improved the synthesis of the dimers and made a wide variety of alkyl (Scheme 11, route **A**) and aryl variants (routes **B** and **C**), which are air and moisture stable.[113, 114] Phosphines with electron-withdrawing groups on the aryl substituents are not electrophilic enough to undergo dimerization; the diarylphosphinoaldehyde diethyl acetals were synthesized and used in further reactions (route **D**).[115]

8.3.1.2 Template assisted synthesis of iron P-N-N and P-N-N-P complexes

Our group developed a convenient multi-component template synthesis for iron complexes with ligands that form five-membered M-P-N- metallacycles. It has been used to make P-N-N, P-N-S, P-N-P mono- and bis-tridentate iron complexes,

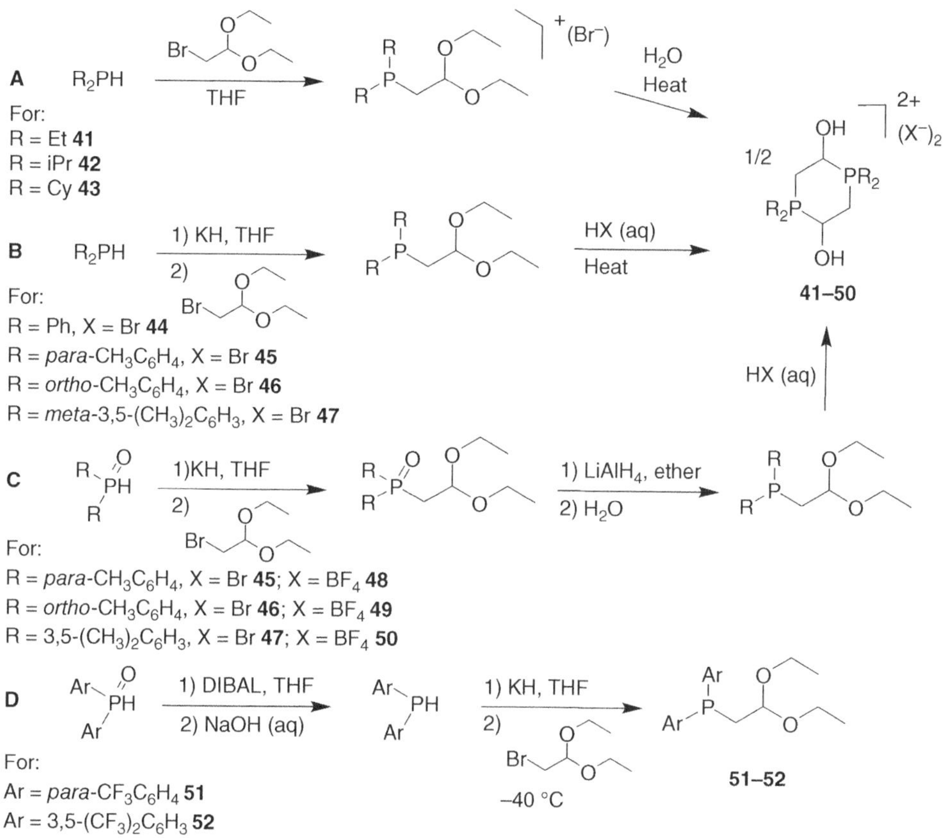

Scheme 11 *Syntheses of phosphonium dimers and α-phosphinoacetaldehyde diethylacetals (DIBAL = diisobutylaluminum hydride)*

and tetradentate P-N-N-P iron complexes.[113, 115–118] In contrast to using *ortho*-diphenylphosphinobenzaldehyde, a template reaction is required for the synthesis of the tetradentate P-N-N-P ligands with alkyl-linked phosphinoaldehydes because the diamine and the phosphinoaldehyde do not condense cleanly without a metal. Selected examples using (*R,R*)-diamines are shown in Scheme 12 and the following discussion will focus on the use of the (*R,R*)-dpen (**28**) as the diamine source during template synthesis.

When the phosphonium dimers were reacted with base in the presence of Fe^{2+}, there was evidence for the formation of new intermediate iron complexes, which were characterized as bis-tridentate iron complexes when using smaller PR_2 groups (R = Et, Ph) on phosphorus (Scheme 12). These complexes were isolated as the BPh_4^- salts after salt exchange of BF_4^- and $FeBr_4^-$ by the addition of $NaBPh_4$. When these kinetic products were left in refluxing acetonitrile for several days, the desired tetradentate complexes were formed.[116]

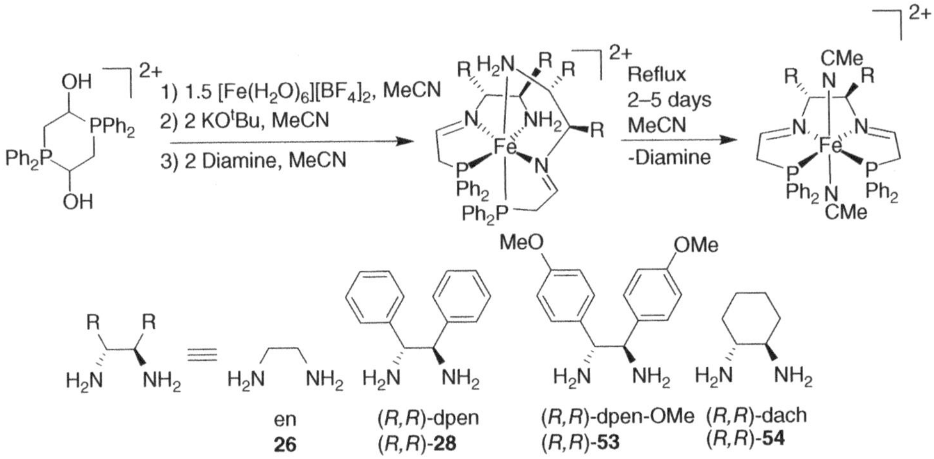

Scheme 12 *Template synthesis of the first-formed bis(P-N-NH$_2$) complex and thermodynamic P-N-N-P product complex*

This synthesis was later optimized by using an acetonitrile/methanol mixture as the solvent at room temperature (Scheme 13). Both (*R,R*)- and (*S,S*)-dpen (**28**) have been used.[113, 115, 119] These complexes can be isolated by use of $NaBPh_4$ but are usually generated in situ for reaction with CO to generate the ATH catalyst precursors. The diarylphosphinoaldehyde diethyl acetals **51** and **52** are stable under basic conditions, so an acid promoted template synthesis was developed to accommodate these phosphinoacetals (Scheme 13).[115]

For the synthesis of the second generation ATH catalysts (see Figure 3 for the general structure), the bis(acetonitrile) complexes were dissolved in acetone and placed under CO atmosphere in the presence of KBr to give, after salt metathesis, complexes **64–72** (Scheme 13).[113, 115, 117] Alternatively, one acetonitrile can be replaced with CO by following the same procedure in the absence of KBr to obtain complex **73**, which was tested for ATH with a variety of ketones.[120]

8.3.2 Asymmetric transfer hydrogenation catalytic properties and mechanism

The bis-acetonitrile complexes **55–63** were not active for either TH or DH of acetophenone in *i*PrOH/KO*t*Bu. However, the monocarbonyl complexes **67, 68, 70** and **73** with the correct steric and electronic properties at phosphorus[115,121] and nitrogen,[117, 121, 122] as defined in Section 8.3.2.1 were active for TH but not DH as long as they were first activated with KO*t*Bu in *i*PrOH. The complexes **67** and **73** with bromide and acetonitrile *trans* to carbonyl provided the same ATH activity and selectivity. The bromide complexes were easiest to prepare and were used for most of the catalyst testing.

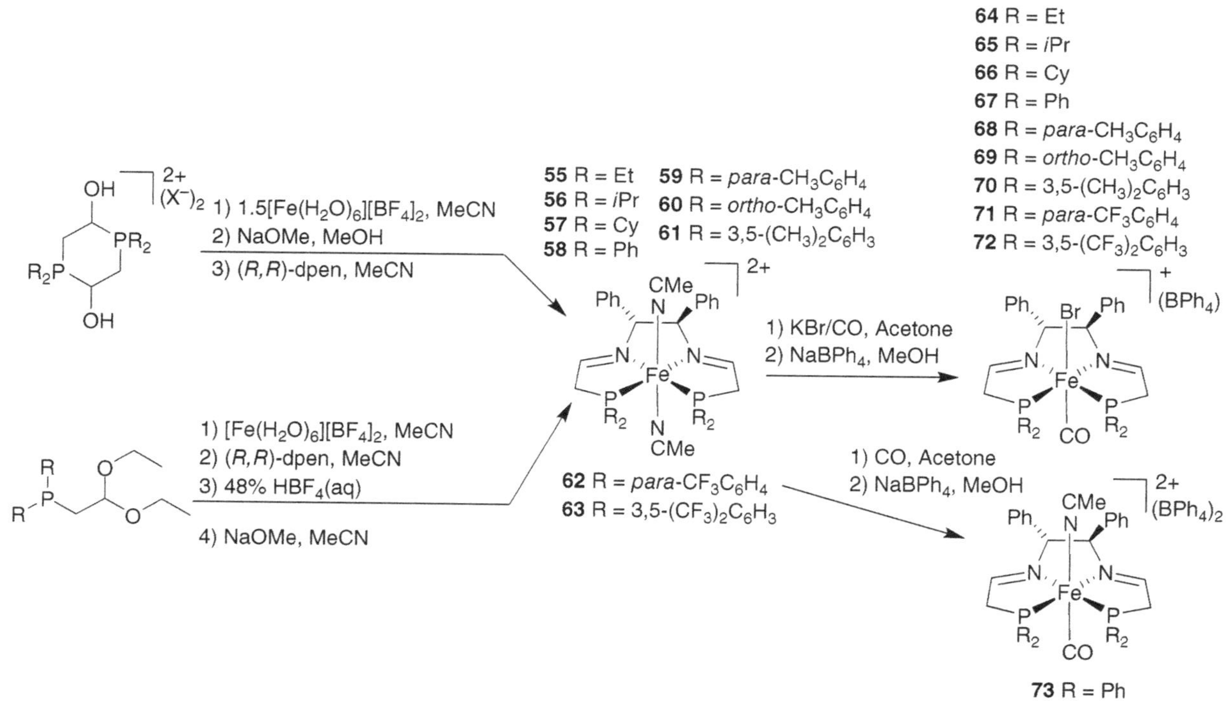

Scheme 13 *Template synthesis of [5.5.5]-P-N-N-P bis(acetonitrile) iron complexes (**55–73**)*

8.3.2.1 The effect of changes of the phosphine on ATH activity

A series of monocarbonyl complexes were tested with phenyl groups on the diamine and various substituents on the phosphorus atoms (complexes **64–72**). The electronic effects of the ligand can be quantified by use of the carbonyl stretching wavenumber of the complex while the steric effects can be expressed by the Tolman cone angle for groups around the phosphorus atoms.[115] Figure 8 shows a plot of these two parameters versus the TOF for the ATH of acetophenone in isopropanol at 30 °C catalyzed by complexes **64–72** (0.02 mol%) which have been activated by KO*t*Bu (0.16 mol%).

The plot shows that there is a small range of substituent sizes defined by cone angles from 132 (PEt_2) to 150° (P*m*Xyl_2) with a small region of electron density as reflected by CO stretching frequencies in the IR ranging between 1951 cm^{-1} (PEt_2) and 1975 cm^{-1} (PPh_2) that define active ATH catalysts. The catalyst of maximum activity is **68** with $P(tol)_2$ groups and second is **67** containing PPh_2 groups, which both furnish (*R*)-1-phenylethanol with an *ee* of 82–84%. The third most active is **70** with *meta*-xylyl groups, but it is the most enantioselective [90% *ee* (*R*)]. The complex with PEt_2 groups (**64**) had very low activity at 30 °C while arylphosphine systems with electron-

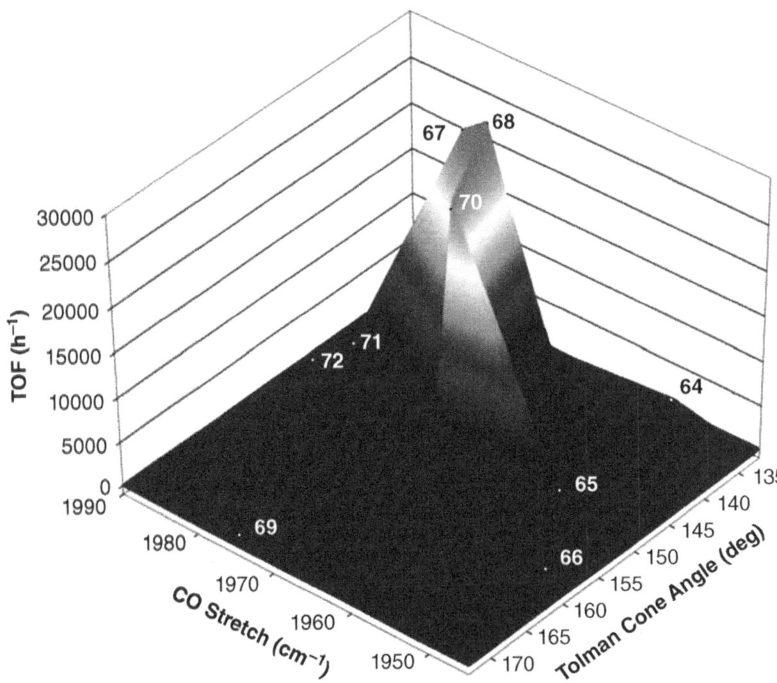

Figure 8 *The plot of catalyst activity for the ATH of acetophenone (TOF, h⁻¹) catalyzed by complexes* ***64–72*** *(0.02 mol%; 0.2 mol% KOtBu) at 30 °C versus CO stretching frequency, as an electronic parameter, and Tolman cone angle, as a steric parameter. Reproduced from [115] with permission from the American Chemical Society. (See insert for color/color representation of this figure)*

withdrawing CF_3 substituents (**71** and **72**) or bulky *ortho*-CH_3 groups (**69**) result in no activity. Similarly, complexes with bulky *i*Pr and Cy groups are inactive.

8.3.2.2 The effect of changes of the diamine on ATH activity

A series of monocarbonyl complexes analogous to **67** with PPh_2 groups and various diamines were tested for ATH (Figure 9). Complexes with aryl substituents on the diamines [using (*R,R*)-**53** or (*R,R*)-**54**] provided very active catalysts for the ATH of acetophenone at 30 °C, with a TOF of $2.0 \times 10^4\ h^{-1}$ at 0.02% catalyst loading and an *ee* of 82% (*S*). The *S* configuration is consistent with a transition state involving an H-Fe-N-H attack (see below) that looks very similar to that of the RuH_2((*R,R*)-dpen)(binap) (Figure 1).

The complex prepared using (*R,R*)-diaminocyclohexane (**76**) was four times less active and produced 1-phenylethanol in only 60% *ee* (*S*). The use of ethylenediamine (**75**) was even less active and decomposed after 40% conversion, possibly due to β-hydride elimination from a metal amide intermediate (Section 1.1.4). The diaminocyclohexane moiety is quite flat and the ethylenediamine backbone is small, both leading to less steric interaction with the incoming ketone substrate. The rates for catalysts **77** and **78** with diethylphosphino substituents are lower than those with diarylphosphino substituents. These may form less active catalysts with less acidic NH groups (cf. Scheme 6). The active forms of complexes **67** and **74** appear to have optimum hydridicity and NH acidity resulting in high ATH activity.

The diethylphosphino-substituted complexes **77** and **78** are less active for the ATH of acetophenone and so reactions were conducted at 50 °C (Figure 9).[67] The catalysts decomposed before the reaction equilibrium could be reached. Again, the aryl-substituted diamine provided the more active catalyst (**77** versus **78**).[113]

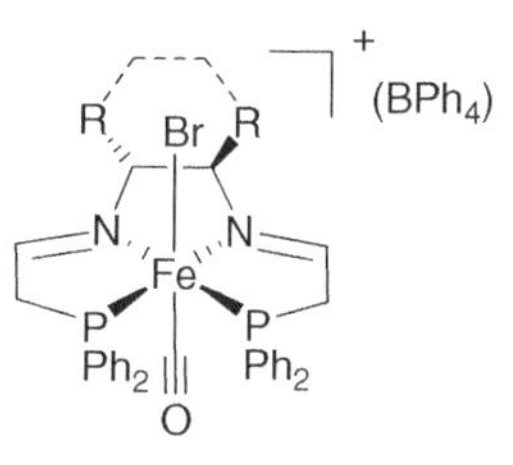

67 R = Ph, diamine (*R,R*)-**28**
TOF = $2.0 \times 10^4\ h^{-1}$ at 28 °C %*ee* = 81 (*S*)
74 R = 4-MeOC$_6$H$_4$, diamine (*R,R*)-**53**
TOF = $2.0 \times 10^4\ h^{-1}$ at 28 °C, %*ee* = 82 (*S*)
75 R = H, diamine **26**
TOF = $2.1 \times 10^3\ h^{-1}$ at 28 °C
76 R = –CH$_2$CH$_2$CH$_2$CH$_2$–, diamine (*R,R*)-**54**
TOF = $4.9 \times 10^3\ h^{-1}$ at 28 °C, %*ee* = 60 (*S*)

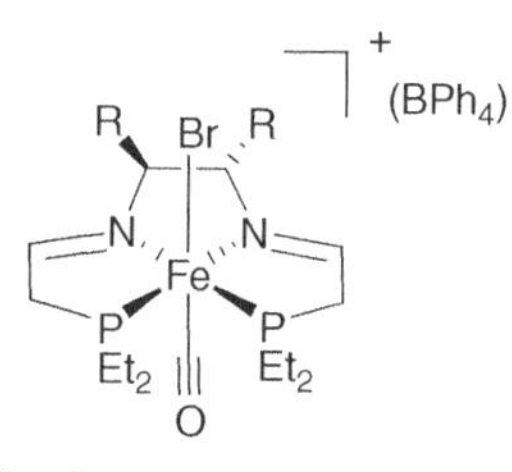

77 R = Ph, diamine (*S,S*)-**28**
TOF = $2.4 \times 10^3\ h^{-1}$ at 50 °C, %*ee* = 57 (*R*)
78 R = H, diamine **26**
TOF = $735\ h^{-1}$ at 50 °C

Figure 9 *Complexes **67** and **74–76** with PPh_2 groups[117] and **77** and **78** with PEt_2 groups[121] prepared using various diamines*

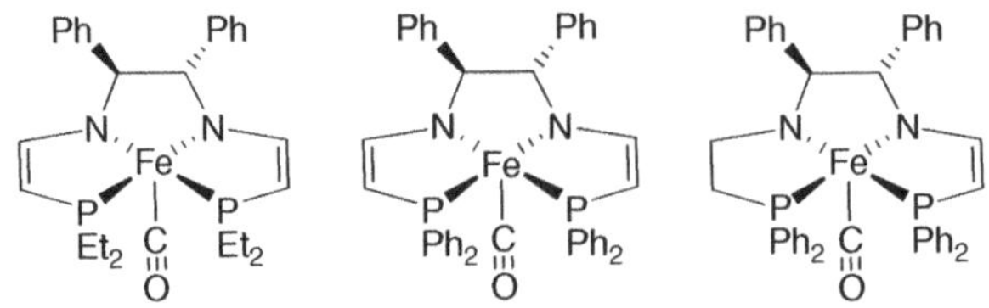

79 **80** **81**

Figure 10 *Bis-ene-amido complexes **79** and **80** and the catalytic intermediate amido(ene-amido) complex **81***

The reaction of complex (*S,S*)-**67** or (*S,S*)-**77** prepared with excess base in benzene leads to the bis-ene-amido complexes **79**[122] and **80**[123] (Figure 10). These are active for the ATH of acetophenone without the addition of extra base. However, there is still an induction period in the catalytic reaction profile, which is associated with the reduction of one of the double bonds in the ligand backbone to produce the active catalyst, an amido(ene-amido) complex **81** (see below).[123]

8.3.2.3 *Mechanism*

Kinetic[123] and DFT studies[124] provided a mechanism for the ATH of acetophenone catalyzed by (*S,S*)-**67** and base in isopropanol (Scheme 14). The slowest step is activation of the catalyst by ligand reduction via hydride transfer from coordinated isopropoxide to afford the amido(ene-amido) complex **81** (Scheme 14, top left). The catalytic cycle is very similar to that of ruthenium hydrogenation catalysts with H-Ru-N-H motifs as discussed above (e.g., Scheme 6 and Scheme 7) where an H-Fe-N-H moiety attacks the ketone in the outer coordination sphere. H-transfer is a stepwise process, with rapid proton transfer to nitrogen followed by hydride transfer from a stabilized isopropoxide ion in the outer coordination sphere to the metal, forming the amino-hydrido complex **82**. The turnover-limiting step is hydride transfer from iron to the substrate, followed by proton transfer from the NH group to the coordinated alkoxide to regenerate **81**. More recently, complexes of the type (*S,S*)-**83** with one imine group reduced were shown to rapidly react with base to directly enter the catalytic cycle, making the overall catalytic process much faster.[118] In addition the amido(ene-amido) complex **81** and the hydride complex **82** have been fully characterized by NMR spectroscopy.[118]

The bis(amine) complex [Fe(Br)(CO)(P-NH-NH-P)]BPh_4 was also synthesized and tested for catalysis, and showed a much lower conversion of only 10% after 2 h.[123] Thus, the ene-amido portion of the active catalyst is critical for superior activity.

8.3.3 Substrate scope

Complex (*R,R*)-**67** is an excellent ATH catalyst, superior in activity to known precious metal catalysts under the same conditions. The TOF approaches 30 000 h^{-1} at 30 °C. Its enantioselectivity in the ATH of a variety of ketones to the (*S*) alcohols is fair to excellent as indicated by the examples in Figure 11.[125] The enantioselectivity of (*S,S*)-**67** in the

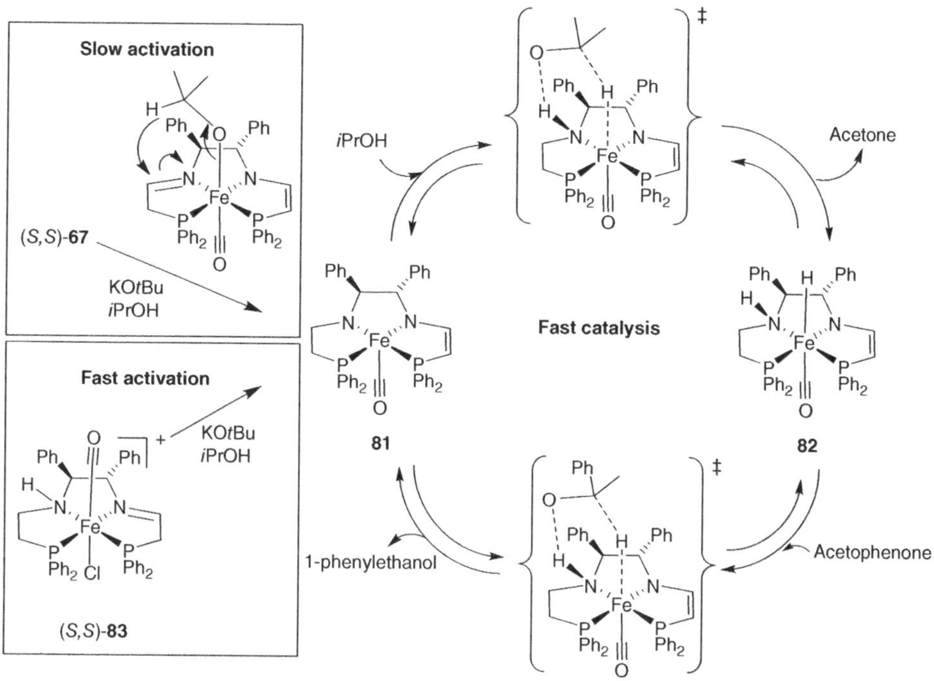

Scheme 14 *The proposed catalytic cycle for the ATH of acetophenone catalyzed by (S,S)-**67** and KOtBu, or more efficiently by (S,S)-**83** and KOtBu in iPrOH and the structures of the amido(ene-amido) complex **81** and amine(hydride) complex **82***

ATH of a range of ketimines is almost perfect. Thus the combination of PPh_2 groups and NCHPh groups on the ligand provides an effective ATH catalyst.

8.4 Third generation iron catalysts with unsymmetrical [5.5.5]-P-NH-N-P′ ligands

The previous mechanistic and experimental studies of our second generation catalysts for ATH suggested that one of the imine moieties of complex **67** was reduced to an amido moiety by a hydride from isopropoxide, producing the amido(ene-amido) complex **81**. Thus we decided to target this complex using a new synthetic pathway.

8.4.1 Synthesis of bis(tridentate)iron complexes and P-NH-NH_2 ligands

New enantiopure P-NH-NH_2 ligands required for the synthesis of complexes leading to **81** can by prepared in an iron-template procedure (Scheme 15) that makes use of the phosphonium dimers described above (Scheme 11).[118] After the formation of bis(tridentate) complexes, they can be released from their template and isolated to afford enantiopure P-NH-NH_2 synthons.

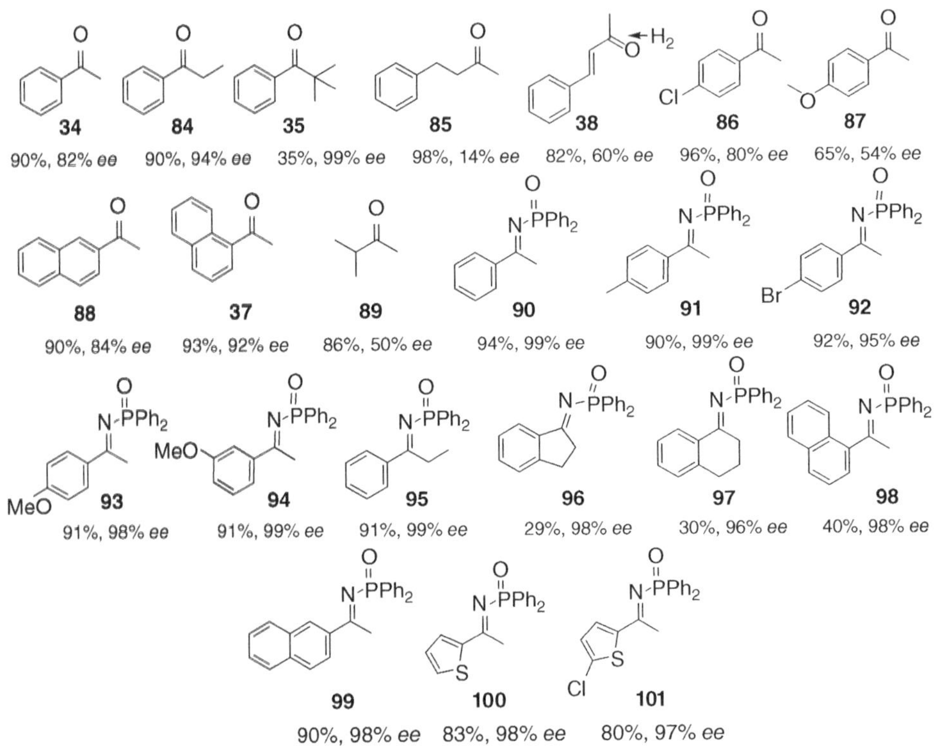

Figure 11 *The ATH of ketones and ketimines catalyzed by 0.1–4 mol% (R,R)-**67**, 0.8 mol% KOtBu in iPrOH at 30 °C; the yield and ee of the resulting (S) alcohols and amines are provided*[125]

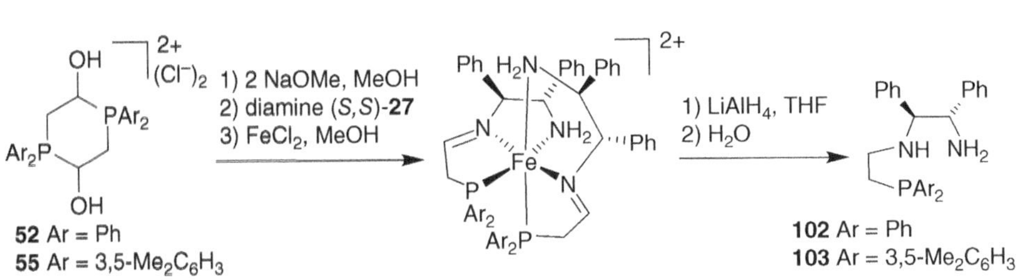

Scheme 15 *Template synthesis of enantiopure P-NH-NH$_2$ ligands (S,S)-**102** and (S,S)-**103***

8.4.2 Template-assisted synthesis of iron P-NH-N-P′ complexes

The P-NH-NH$_2$ ligands are then condensed with another 0.5 equiv. of phosphonium dimer in the presence of base and $[Fe(H_2O)_6][BF_4]_2$ to give partially unsaturated P-NH-N-P frameworks on iron (II) (Scheme 16). The resulting iron complexes are then reacted with NaCl under a CO atmosphere to give the iron complexes (*S,S*)-**104** and

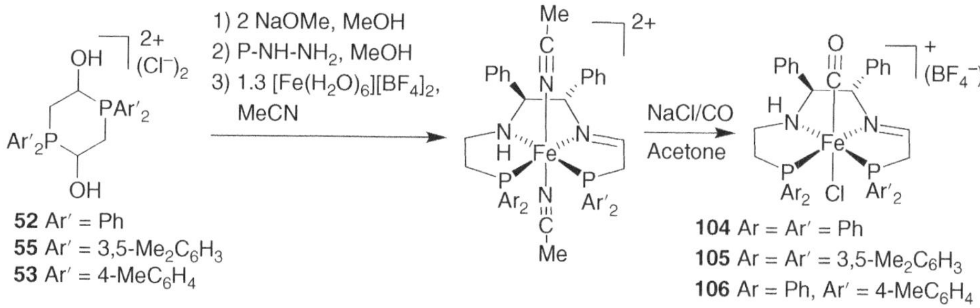

Scheme 16 *Template synthesis of the third generation precatalysts (S,S)-**104**, (S,S)-**105**, and (S,S)-**106***

(*S,S*)-**105.** A new dimension of ligand variation can now be incorporated by using a different combination of phosphonium dimer and P-NH-NH_2 ligand, as illustrated by the synthesis of (*S,S*)-**106**. This new pathway allows the synthesis of a wide range of iron complexes with unsymmetrical P-NH-N-P′ ligands.[118]

8.4.3 Selected catalytic properties

This new generation of iron precatalysts was tested for the reduction of ketones and imines, in both ATH and, most recently, ADH.[118, 126] In both cases, no induction period was observed and high activity was obtained, more so for ATH than DH. To date, these are the most active ATH catalysts known on the basis of their exceptional TOF for acetophenone to (*R*)-1-phenylethanol at 30°C.[118]

8.4.3.1 Asymmetric transfer hydrogenation

Catalysis with **104-106** resulted in the conversion of ketones to the corresponding alcohols significantly faster than our previous generations of catalysts, even at 28 °C. Unprecedented TOF of over 200 s^{-1} at 50% conversion and TONs of up to 6100 were displayed using **104** for a variety of ketones with *ee* up to 98% (Figure 12). The catalyst was tolerant towards pyridyl and furyl functional groups and reduced **38** selectively to the allyl alcohol. The tetradentate ligand system appears to prevent the pyridyl group in particular from coordinating and poisoning catalysis. Under standard conditions, the complex that gave the highest TOF was **106** while the highest *ee* was found with the use of **105**, consistent with substituent effects previously observed with second generation iron catalysts.[118] Acetophenone (**34**) was reduced to 1-phenylethanol with an *ee* of 88% using **104** and 92% using **105.** Similarly, ketone **108** was reduced with a final *ee* of 91% using **104** and 98% using **105**, with the product alcohol serving as a precursor to the antiemetic drug Aprepitant (Emend). These iron-based catalysts are not only faster than the most efficient ruthenium-based catalysts, but approach enzymatic activity. They also provide an *ee* of >99% for certain imines.

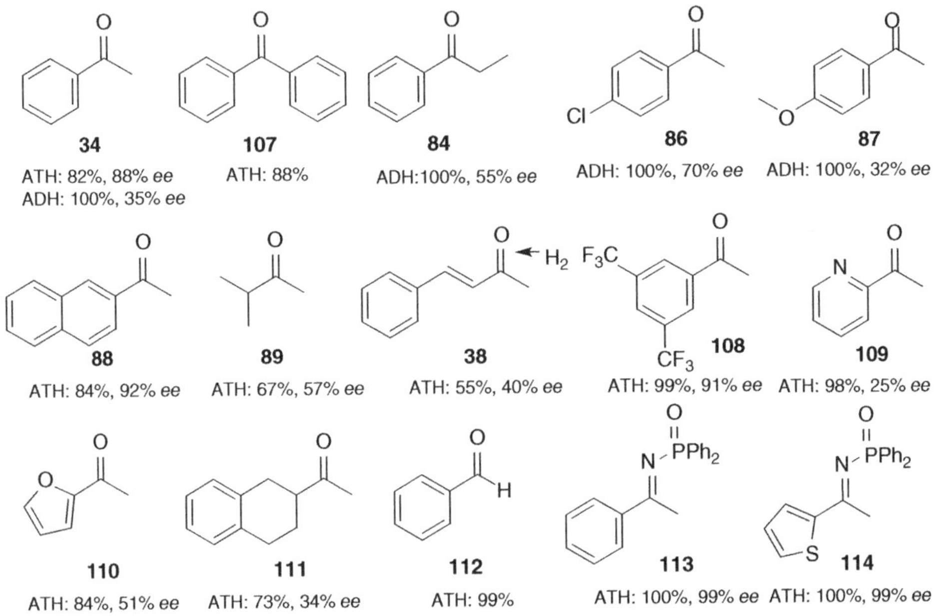

Figure 12 *The ATH and ADH of selected ketones and imines catalyzed by the third generation catalysts (S,S)-**104** and (S,S)-**105**, respectively. ATH: (0.02 mol%) with KOtBu (0.16 mol%) in isopropanol at 28 °C; ADH: 50 °C with 20 atm H_2 in THF*[126]

8.4.3.2 *Asymmetric direct hydrogenation*

Complexes **104** and **105**, when treated with 2 equiv. of KO*t*Bu, were found to be active for the DH of ketones with TOF up to 80 h^{-1}.[126] Complex **104** with PPh_2 groups yielded racemic alcohols under these conditions while *(S,S)*-**105** with P*m*Xyl_2 groups gave (*R*) alcohols with moderate *ee* values (Figure 12); in general, the *ee* of the alcohol products decreased over time. Side reactions at the higher temperature of ADH (50 °C) versus ATH (30 °C) seem to cause a loss of enantioselectivity.

8.4.4 Mechanism

When complex **104** is treated with KO*t*Bu in THF, the amido(ene-amido) complex **81** of Scheme 14 can be characterized by NMR.[118] When this solution is subjected to a hydrogen source, either 2-propanol or H_2 gas, the active amine iron hydride species **82** (Scheme 14) was detected. Experimental results as well as DFT calculations suggest that the mechanisms of ATH and ADH for this system are very similar, the major difference being the reaction of the amido(ene-amido) complex **81** with H_2 instead of isopropanol to generate the hydride complex **82**. The heterolytic splitting of dihydrogen by **81** is the turn over limiting step (TLS) in ADH[126] just as the H^+/H^- transfer from isopropanol to **81** is the TLS in ATH, with the calculated and experimental energy barrier to ADH being higher than those to ATH.

8.5 Conclusions

Three generations of successively more active iron catalysts for the ATH of ketones and imines were developed by optimizing the structure of P-N-N-P and then P-NH-N-P tetradentate ligands. The procedure for ligand modifications was simplified by the condensation of amine and aldehyde components to form imines by using iron(II) as a template, yielding in most cases only one diastereomer. A wide range of enantiopure diamines are commercially available to set the chirality of the ligand, while four different routes were introduced to make new α-diarylphosphinoacetaldehyde and α-dialkylphosphinoacetaldehyde components. An iron-assisted synthesis of enantiopure P-NH-NH_2 ligands is also possible, allowing further ligand variation by introducing unsymmetrical phosphine donors. The flexibility of the synthetic methods means that catalyst discovery for a particular reduction can be accelerated by the availability of a wide range of catalyst structures. Tetradentate ligands that form three five-membered rings ([5.5.5]) with iron(II) are preferable to ones that form more flexible six-membered rings ([6.5.6]) found in first generation catalysts.

There are similarities and differences between the ligand structures that activate iron(II) and ruthenium(II) for homogeneous hydrogenation processes. A very important common element for ketone and imine reduction is an amine group on the ligand that, along with a *cis*-hydride in the H-M-N-H motif, efficiently transfers proton and hydride equivalents to the polar bond in the outer coordination sphere. Providing an amine function in the tetradentate ligand in the third generation $[Fe(Cl)(CO)(P\text{-}NH\text{-}N\text{-}P)]^+$ precatalysts resulted in much higher activity than in the second generation structures where an imine had to be converted to an amine function during the inefficient activation of the precatalyst. The combination of phosphorus and nitrogen coordinated to a neutral M(II) center along with a hydride (for Ru) or carbonyl (for Fe) *trans* to a hydride makes the hydride a potent nucleophile. However, electron-withdrawing substituents on the phosphines deactivate the iron catalyst. The steric environment may be more restricted for Fe versus Ru, explaining why only a narrow range of cone angles at phosphorus yields active catalysts for Fe. The strong field P, H and CO donors appear to keep the complexes in the catalytic cycle in a low spin electron configuration, even the five-coordinate iron(II) amido(ene-amide) intermediate Fe(CO)(P-N-NCH=CHP) **81**. Often, low coordinate iron complexes are high spin (see Figure 5). As expected, the amido intermediate Fe(CO)(P-N-NCH=CHP) reacts with dihydrogen or isopropanol in a fashion similar to ruthenium amido complexes, allowing the regeneration of the hydride amine complex Fe(H)(CO)(P-NH-NCH = CHP) after the hydrogenation of the substrate. Thus, these iron complexes can serve as ATH or ADH catalysts. The iron catalyst appears to be tolerant to nitrogen-donor functional groups on the substrates because the tetradentate ligand prevents these groups from coordinating and poisoning the catalyst.

The chiral array of aryl groups presented to the incoming substrate seems very similar for $Ru(H)_2$(diamine)(diphosphine) and Fe(H)(CO)(P-NH-N-P) complexes. In each case, the complex with a structure derived from the (*R,R*)-diamine produces the (*S*)-alcohol or (*S*)-amine from prochiral aryl ketones or imines. The origin for this selectivity is readily seen by inspecting how the incoming ketone hydrogen bonds with one NH group which is locked parallel to the M—H bond by the chiral diamine. The larger substituent on the ketone is kept away from the bulky groups on the phosphorus donor, biasing the interaction to the pro-(*S*) face of the ketone (Figure 1).

These and other design principles will continue to be exploited in the on-going effort to replace chemical processes that are currently catalyzed by precious metal complexes with more abundant, sustainable and less toxic metals.

Acknowledgments

R.H.M. thanks NSERC Canada for a Discovery grant and a graduate scholarship to D.E.P., the Canada Foundation for Innovation and Ontario Ministry of Research for equipment, and all of the students in the research group involved in this project and mentioned in the references.

References

[1] T. Ohkuma, D. Ishii, H. Takeno, R. Noyori, *J. Am. Chem. Soc.* **2000**, *122*, 6510–6511.
[2] H. U. Blaser, B. Pugin, F. Spindler, M. Thommen, *Acc. Chem. Res.* **2007**, *40*, 1240–1250.
[3] W. Baratta, G. Chelucci, S. Magnolia, K. Siega, P. Rigo, *Chem. Eur. J.* **2009**, *15*, 726–732.
[4] R. Noyori, S. Hashiguchi, *Acc. Chem. Res.* **1997**, *30*, 97–102.
[5] R. Noyori, T. Ohkuma, *Angew. Chem. Int. Ed.* **2001**, *40*, 40–73.
[6] T. Ikariya, K. Murata, R. Noyori, *Org. Biomolec. Chem.* **2006**, *4*, 393–406.
[7] B. Zhao, Z. Han, K. Ding, *Angew. Chem. Int. Ed.* **2013**, *52*, 4744–4788.
[8] R. Noyori, M. Yamakawa, S. Hashiguchi, *J. Org. Chem.* **2001**, *66*, 7931–7944.
[9] K. Abdur-Rashid, M. Faatz, A. J. Lough, R. H. Morris, *J. Am. Chem. Soc.* **2001**, *123*, 7473–7474.
[10] R. Guo, A. J. Lough, R. H. Morris, D. Song, *Organometallics* **2004**, *23*, 5524–5529.
[11] R. Feng, A. Xiao, X. Zhang, Y. Tang, M. Lei, *Dalton Trans.* **2013**, *42*, 2130–2145.
[12] H. Y. T. Chen, D. Di Tommaso, G. Hogarth, C. R. A. Catlow, *Dalton Trans.* **2012**, *41*, 1867–1877.
[13] R. Noyori, *Angew. Chem. Int. Ed.* **2002**, *41*, 2008–2022.
[14] T. G. Appleton, H. C. Clark, L. E. Manzer, *Coord. Chem. Rev.* **1973**, *10*, 335–422.
[15] S. E. Clapham, A. Hadzovic, R. H. Morris, *Coord. Chem. Rev.* **2004**, *248*, 2201–2237.
[16] R. Abbel, K. Abdur-Rashid, M. Faatz, A. Hadzovic, A. J. Lough, R. H. Morris, *J. Am. Chem. Soc.* **2005**, *127*, 1870–1882.
[17] S. E. Clapham, R. Guo, M. Zimmer-De Iuliis, N. Rasool, A. J. Lough, R. H. Morris, *Organometallics* **2006**, *25*, 5477–5486.
[18] W. W. N. O, A. J. Lough, R. H. Morris, *Organometallics* **2012**, *31*, 2137–2151.
[19] R. H. Morris, *J. Am. Chem. Soc.* **2014**, *136*, 1948–1959.
[20] C. J. Curtis, A. Miedaner, J. W. Raebiger, D. L. DuBois, *Organometallics* **2004**, *23*, 511–516.
[21] K. Mikami, T. Korenaga, Y. Yusa, M. Yamanaka, *Adv. Synth. Catal.* **2003**, *345*, 246–254.
[22] W. H. Zhang, S. W. Chien, T. S. A. Hor, *Coord. Chem. Rev.* **2011**, *255*, 1991–2024.

[23] G. N. P. Schiemenz, H. Kaack, *Justus Lieb. Ann. Chem.* **1973**, *1973*, 1480–1493.
[24] J. E. Hoots, T. B. Rauchfuss, D. A. Wrobleski, H. C. Knachel, in *Inorganic Syntheses, Vol. 21* (ed. J. P. Fackler Jr), John Wiley & Sons, Inc., New York, **1982**, pp. 175–179.
[25] T. B. Rauchfuss, *J. Organomet. Chem.* **1978**, *162*, C19-C22.
[26] J. E. Hoots, T. B. Rauchfuss, S. P. Schmidt, J. C. Jeffery, P. A. Tucker, in *Catalytic Aspects of Metal Phosphine Complexes* (eds E. C. Alyea, D. W. Meek), American Chemical Society, Washington, DC, **1982**, pp. 303–311.
[27] J. C. Jeffery, T. B. Rauchfuss, P. A. Tucker, *Inorg. Chem.* **1980**, *19*, 3306–3315.
[28] H. Brunner, A. F. M. Mokhlesur Rahman, *Chem. Ber.* **1984**, *117*, 710–724.
[29] M. A. Garralda, *C. R. Chim.* **2005**, *8*, 1413–1420.
[30] P. J. Guiry, C. P. Saunders, *Adv. Synth. Catal.* **2004**, *346*, 497–537.
[31] W.-K. Wong, J.-X. Gao, Z.-Y. Zhou, T. C. W. Mak, *Polyhedron* **1993**, *12*, 1415–1417.
[32] W.-K. Wong, K.-K. Lai, M.-S. Tse, M.-C. Tse, J.-X. Gao, W.-T. Wong, S. Chan, *Polyhedron* **1994**, *13*, 2751–2762.
[33] W.-K. Wong, T.-W. Chik, K.-N. Hui, I. Williams, X. Feng, T. C. W. Mak, C.-M. Che, *Polyhedron* **1996**, *15*, 4447–4460.
[34] J.-X. Gao, T. Ikariya, R. Noyori, *Organometallics* **1996**, *15*, 1087–1089.
[35] J.-X. Gao, H. Zhang, X.-D. Yi, P.-P. Xu, C.-L. Tang, H.-L. Wan, K.-R. Tsai, T. Ikariya, *Chirality* **2000**, *12*, 383–388.
[36] S. Laue, L. Greiner, J. Wöltinger, A. Liese, *Adv. Synth. Catal.* **2001**, *343*, 711–720.
[37] C. Z. Flores-López, L. a. Z. Flores-López, G. Aguirre, L. H. Hellberg, M. Parra-Hake, R. Somanathan, *J. Mol. Cat. A: Chem.* **2004**, *215*, 73–79.
[38] M. L. Clarke, M. B. Díaz-Valenzuela, A. M. Z. Slawin, *Organometallics* **2007**, *26*, 16–19.
[39] J. A. Fuentes, M. L. Clarke, A. M. Z. Slawin, *New J. Chem.* **2008**, *32*, 689–693.
[40] M. B. Díaz-Valenzuela, S. D. Phillips, M. B. France, M. E. Gunn, M. L. Clarke, *Chem. – Eur. J.* **2009**, *15*, 1227–1232.
[41] S. D. Phillips, J. A. Fuentes, M. L. Clarke, *Chem. – Eur. J.* **2010**, *16*, 8002-8005.
[42] I. Carpenter, S. C. Eckelmann, M. T. Kuntz, J. A. Fuentes, M. B. France, M. L. Clarke, *Dalton Trans.* **2012**, *41*, 10136–10140.
[43] J.Fuentes,S.Phillips,M.Clarke,*Chem.Cent.J.***2012**,*6*,DOI:10.1186/1752-1153X-1186-1151.
[44] A. Del Zotto, W. Baratta, M. Ballico, E. Herdtweck, P. Rigo, *Organometallics* **2007**, *26*, 5636–5642.
[45] A. Mezzetti, C. Bonaccorsi, *Curr. Org. Chem.* **2006**, *10*, 225–240.
[46] V. Rautenstrauch, X. Hoang-Cong, R. Churlaud, K. Abdur-Rashid, R. H. Morris, *Chem. Eur. J.* **2003**, *9*, 4954–4967.
[47] T. Li, R. Churlaud, A. J. Lough, K. Abdur-Rashid, R. H. Morris, *Organometallics* **2004**, *23*, 6239–6247.
[48] T. Li, I. Bergner, F. N. Haque, M. Zimmer-De Iuliis, D. Song, R. H. Morris, *Organometallics* **2007**, *26*, 5940–5949.
[49] F. N. Haque, A. J. Lough, R. H. Morris, *Inorg. Chim. Acta* **2008**, *361*, 3149–3158.
[50] R. J. Hamilton, S. H. Bergens, *J. Am. Chem. Soc.* **2006**, *128*, 13700–13701.
[51] R. J. Hamilton, S. H. Bergens, *J. Am. Chem. Soc.* **2008**, *130*, 11979–11987.
[52] S. Takebayashi, N. Dabral, M. Miskolzie, S. H. Bergens, *J. Am. Chem. Soc.* **2011**, *133*, 9666–9669.
[53] F. Hasanayn, R. H. Morris, *Inorg. Chem.* **2012**, *51*, 10808–10818.
[54] P. A. Dub, T. Ikariya, *J. Am. Chem. Soc.* **2013**, *135*, 2604–2619.

[55] J. M. John, S. Takebayashi, N. Dabral, M. Miskolzie, S. H. Bergens, *J. Am. Chem. Soc.* **2013**, *135*, 8578–8584.
[56] P. A. Dub, N. J. Henson, R. L. Martin, J. C. Gordon, *J. Am. Chem. Soc.* **2014**, *136*, 3505–3521.
[57] K. Abdur-Rashid, S. E. Clapham, A. Hadzovic, J. N. Harvey, A. J. Lough, R. H. Morris, *J. Am. Chem. Soc.* **2002**, *124*, 15104–15118.
[58] M. Zimmer-De Iuliis, R. H. Morris, *J. Am. Chem. Soc.* **2009**, *131*, 11263–11269.
[59] H.-Y. T. Chen, D. Di Tommaso, G. Hogarth, C. R. A. Catlow, *Dalton Trans.* **2011**, *40*, 402–412.
[60] S. Enthaler, G. Erre, M. K. Tse, K. Junge, M. Beller, *Tetrahedron Lett.* **2006**, *47*, 8095–8099.
[61] S. Enthaler, K. Junge, M. Beller, *Angew. Chem. Int. Ed.* **2008**, *47*, 3317–3321.
[62] S. Enthaler, B. Spilker, G. Erre, K. Junge, M. K. Tse, M. Beller, *Tetrahedron* **2008**, *64*, 3867–3876.
[63] J.-S. Chen, L.-L. Chen, Y. Xing, G. Chen, W.-Y. Shen, Z.-R. Dong, Y.-Y. Li, J.-X. Gao, *Acta Chim. Sinica (Huaxue Xuebao)* **2004**, *62*, 1745–1750.
[64] C. P. Casey, H. Guan, *J. Am. Chem. Soc.* **2007**, *129*, 5816–5817.
[65] C. P. Casey, H. R. Guan, *J. Am. Chem. Soc.* **2009**, *131*, 2499–2507.
[66] C. Sui-Seng, F. Freutel, A. J. Lough, R. H. Morris, *Angew. Chem. Int. Ed. Engl.* **2008**, *47*, 940–943.
[67] P. O. Lagaditis, P. E. Sues, J. F. Sonnenberg, K. Y. Wan, A. J. Lough, R. H. Morris, *J. Am. Chem. Soc.* **2014**, *136*, 1367–1380.
[68] R. Langer, M. A. Iron, L. Konstantinovski, Y. Diskin-Posner, G. Leitus, Y. Ben-David, D. Milstein, *Chem. – Eur. J.* **2012**, *18*, 7196–7209.
[69] S. Warratz, L. Postigo, B. Royo, *Organometallics* **2013**, *32*, 893-897.
[70] Y. Li, S. Yu, X. Wu, J. Xiao, W. Shen, Z. Dong, J. Gao, *J. Am. Chem. Soc.* **2014**, *136*, 4031–4039.
[71] Y. Sunada, H. Tsutsumi, K. Shigeta, R. Yoshida, T. Hashimoto, H. Nagashima, *Dalton Trans.* **2013**, *42*, 16687–16692.
[72] L. C. M. Castro, H. Q. Li, J. B. Sortais, C. Darcel, *Chem. Commun.* **2012**, *48*, 10514–10516.
[73] S. Moulin, H. Dentel, A. Pagnoux-Ozherelyeva, S. Gaillard, A. Poater, L. Cavallo, J. F. Lohier, J. L. Renaud, *Chem. - Eur. J.* **2013**, *19*, 17881–17890.
[74] D. Bezier, G. T. Venkanna, L. C. M. Castro, J. Zheng, J.-B. Sortais, C. Darcel, *Adv. Synth. Catal.* **2012**, *354*, 1879.
[75] S. Das, Y. H. Li, K. Junge, M. Beller, *Chem. Commun.* **2012**, *48*, 10742–10744.
[76] T. N. Plank, J. L. Drake, D. K. Kim, T. W. Funk, *Adv. Syn. Catal.* **2012**, *354*, 597–601.
[77] J. P. Hopewell, J. E. D. Martins, T. C. Johnson, J. Godfrey, M. Wills, *Org. Biomol. Chem.* **2012**, *10*, 134–145.
[78] K. Junge, K. Schroder, M. Beller, *Chem. Commun.* **2011**, *47*, 4849–4859.
[79] T. C. Johnson, G. J. Clarkson, M. Wills, *Organometallics* **2011**, *30*, 1859–1868.
[80] S. Chakraborty, H. R. Guan, *Dalton Trans.* **2010**, *39*, 7427–7436.
[81] A. Berkessel, S. Reichau, A. von der Hoh, N. Leconte, J. M. Neudorfl, *Organometallics* **2011**, *30*, 3880–3887.
[82] E. Buitrago, F. Tinnis, H. Adolfsson, *Adv. Syn. Catal.* **2012**, *354*, 217–222.
[83] C. E. Housecroft, A. G. Sharpe, *Inorganic Chemistry*, Pearson Education Ltd, London, **2012**.
[84] A. M. Tondreau, C. C. H. Atienza, K. J. Weller, S. A. Nye, K. M. Lewis, J. G. P. Delis, P. J. Chirik, *Science* **2012**, *335*, 567–570.
[85] A. J. Ruddy, C. M. Kelly, S. M. Crawford, C. A. Wheaton, O. L. Sydora, B. L. Small, M. Stradiotto, L. Turculet, *Organometallics* **2013**, *32*, 5581–5588.

[86] B. K. Langlotz, H. Wadepohl, L. H. Gade, *Angew. Chem. Int. Ed.* **2008**, *47*, 4670–4674.
[87] T. Liu, D. L. Dubois, R. M. Bullock, *Nature Chem.* **2013**, *5*, 228–233.
[88] R. H. Crabtree, P. E. M. Siegbahn, O. Eisenstein, A. L. Rheingold, T. F. Koetzle, *Acc. Chem. Res.* **1996**, *29*, 348–354.
[89] A. J. Lough, S. Park, R. Ramachandran, R. H. Morris, *J. Am. Chem. Soc.* **1994**, *116*, 8356–8357.
[90] J. C. Lee Jr, A. L. Rheingold, B. Muller, P. S. Pregosin, R. H. Crabtree, *J. Chem. Soc., Chem. Commun.* **1994**, 1021–1022.
[91] T. Liu, X. Wang, C. Hoffmann, D. L. Dubois, R. M. Bullock, *Angew. Chem. Int. Ed.* **2014**, *53*, 5300–5304.
[92] T. R. Dugan, E. Bill, K. C. Macleod, W. W. Brennessel, P. L. Holland, *Inorg. Chem.* **2014**, *53*, 2370–2380.
[93] K. Abdur-Rashid, R. Guo, A. J. Lough, R. H. Morris, D. Song, *Adv. Syn. Catal.* **2005**, *347*, 571–579.
[94] L. A. Saudan, C. M. Saudan, C. Debieux, P. Wyss, *Angew. Chem. Int. Ed.* **2007**, *46*, 7473–7476.
[95] I. Koehne, T. J. Schmeier, E. A. Bielinski, C. J. Pan, P. O. Lagaditis, W. H. Bernskoetter, M. K. Takase, C. Würtele, N. Hazari, S. Schneider, *Inorg. Chem.* **2014**, *53*, 2133–2143.
[96] S. Chakraborty, H. Dai, P. Bhattacharya, N. T. Fairweather, M. S. Gibson, J. A. Krause, H. Guan, *J. Am. Chem. Soc.* **2014**, *136*, 7869–7872.
[97] S. Werkmeister, K. Junge, B. Wendt, E. Alberico, H. Jiao, W. Baumann, H. Junge, F. Gallou, M. Beller, *Angew. Chem. Int. Ed.* **2014**, *53*, 8722–8726.
[98] S. Chakraborty, W. W. Brennessel, W. D. Jones, *J. Am. Chem. Soc.* **2014**, *136*, 8564–8567.
[99] D. E. Prokopchuk, J. F. Sonnenberg, N. Meyer, M. Zimmer-De Iuliis, A. J. Lough, R. H. Morris, *Organometallics* **2012**, *31*, 3056–3064.
[100] M. C. Thompson, D. H. Busch, *J. Am. Chem. Soc.* **1964**, *86*, 213–217.
[101] D. H. Busch, D. C. Jicha, M. C. Thompson, J. W. Wrathall, E. L. Blinn, *J. Am. Chem. Soc.* **1964**, *86*, 3642–3650.
[102] C. R. K. Glasson, G. V. Meehan, C. A. Motti, J. K. Clegg, M. S. Davies, L. F. Lindoy, *Aust. J. Chem.* **2012**, *65*, 1371–1376.
[103] A. Flores-Figueroa, T. Pape, K.-O. Feldmann, F. E. Hahn, *Chem. Commun.* **2010**, *46*, 324–326.
[104] J.-X. Gao, H.-L. Wan, W.-K. Wong, M.-C. Tse, W.-T. Wong, *Polyhedron* **1996**, *15*, 1241–1251.
[105] R. H. Morris, in *Concepts and Models in Bioinorganic Chemistry* (eds B. Kraatz, N. Metzler-Nolte), Wiley-VCH, Weinheim, **2006**, pp. 329–360.
[106] J. C. Fontecilla-Camps, A. Volbeda, C. Cavazza, Y. Nicolet, *Chem. Rev.* **2007**, *107*, 4273–4303.
[107] C. Tard, C. J. Pickett, *Chem. Rev.* **2009**, *109*, 2245–2274.
[108] S. Dey, P. K. Das, A. Dey, *Coord. Chem. Rev.* **2013**, *257*, 42–63.
[109] C. Sui-Seng, F. N. Haque, A. Hadzovic, A.-M. Pütz, V. Reuss, N. Meyer, A. J. Lough, M. Zimmer-De Iuliis, R. H. Morris, *Inorg. Chem.* **2009**, *48*, 735–743.
[110] J. F. Sonnenberg, N. Coombs, P. A. Dube, R. H. Morris, *J. Am. Chem. Soc.* **2012**, *134*, 5893–5899.
[111] N. Meyer, A. J. Lough, R. H. Morris, *Chem. Eur. J.* **2009**, *15*, 5605–5610.
[112] D. Matt, R. Ziessel, A. De Cian, J. Fischer, *New J. Chem.* **1996**, *20*, 1257–1263.

[113] P. O. Lagaditis, A. A. Mikhailine, A. J. Lough, R. H. Morris, *Inorg. Chem.* **2010**, *49*, 1094–1102.
[114] A. A. Mikhailine, P. O. Lagaditis, P. Sues, A. J. Lough, R. H. Morris, *J. Organomet. Chem.* **2010**, *695*, 1824–1830.
[115] P. E. Sues, A. J. Lough, R. H. Morris, *Organometallics* **2011**, *30*, 4418–4431.
[116] A. A. Mikhailine, E. Kim, C. Dingels, A. J. Lough, R. H. Morris, *Inorg. Chem.* **2008**, *47*, 6587–6589.
[117] A. A. Mikhailine, R. H. Morris, *Inorg. Chem.* **2010**, *49*, 11039–11044.
[118] W. Zuo, A. J. Lough, Y. Li, R. H. Morris, *Science* **2013**, *342*, 1080–1083.
[119] A. A. Mikhailine, A. J. Lough, R. H. Morris, *J. Am. Chem. Soc* **2009**, *131*, 1394–1395.
[120] R. H. Morris, *Chem. Soc. Rev.* **2009**, *38*, 2282–2291.
[121] P. O. Lagaditis, A. J. Lough, R. H. Morris, *Inorg. Chem.* **2010**, *49*, 10057–10066.
[122] P. O. Lagaditis, A. J. Lough, R. H. Morris, *J. Am. Chem. Soc.* **2011**, *133*, 9662–9665.
[123] A. A. Mikhailine, M. I. Maishan, A. J. Lough, R. H. Morris, *J. Am. Chem. Soc.* **2012**, *134*, 12266–12280.
[124] D. E. Prokopchuk, R. H. Morris, *Organometallics* **2012**, *31*, 7375–7385.
[125] A. A. Mikhailine, M. I. Maishan, R. H. Morris, *Org. Lett.* **2012**, *14*, 4638–4641.
[126] W. Zuo, S. Tauer, D. E. Prokopchuk, R. H. Morris, *Organometallics* **2014**, *33*, 5791–5801.

9

Ambiphilic Ligands: Unusual Coordination and Reactivity Arising from Lewis Acid Moieties

Ghenwa Bouhadir and Didier Bourissou

CNRS, Université Paul Sabatier, France

9.1 Introduction

The success of homogeneous catalysis can largely be attributed to the development of a diverse range of ligand frameworks. This has led to not only highly active and selective catalysts through judicious choice of metal and ligand, but also to the discovery of new transition metal mediated transformations. Most commonly, ligands are based on donor moieties that naturally bind to metals as σ-donors. Lewis acids are also frequently used in transition metal chemistry, either as co-catalysts or additives, but they have rarely been incorporated within ligands. Yet, unusual bonding situations and reactivity patterns may be envisioned with Lewis acids. To study and develop such systems, it is desirable to control the way Lewis acids are introduced in the coordination sphere of transition metals and participate in coordination as well as reactivity. To this end, ambiphilic ligands combining donor and acceptor sites are particularly appealing and fascinating.[1] Pioneering contributions in this field were reported in the early 1960s and 1980s,

Ligand Design in Metal Chemistry: Reactivity and Catalysis, First Edition. Edited by Mark Stradiotto and Rylan J. Lundgren.

but the last decade has witnessed a spectacular upsurge of interest. This was especially stimulated by the novel and versatile coordination properties evidenced for phosphine-boranes, the prototypical ambiphilic ligands, as well as the parallel development of Lewis pair activation/functionalization of small molecules, so-called frustrated Lewis pair (FLP) chemistry.[2]

In this chapter, the design and structure of ambiphilic ligands are discussed. The different ways Lewis acids can participate in coordination are then presented as well as their impact on the electronic, geometric and optical properties of metal fragments. The reactivity of complexes deriving from ambiphilic ligands is also described, with special emphasis on the influence of the Lewis acid moiety on the stoichiometric/catalytic transformations of interest. Attention has been given to well-defined complexes deriving from preformed ambiphilic ligands.

9.2 Design and structure of ambiphilic ligands

By design, ambiphilic ligands combine donor and acceptor sites on the same skeleton (Figure 1). The basic idea is to use donor sites as anchors to introduce Lewis acids in the coordination sphere of transition metals. Fine tuning of the structure of the ambiphilic ligands (coordination sites and linker) gives the possibility to control the position of the Lewis acid moiety and the way it participates in bonding and/or reactivity.

Within this general scheme, many variations are conceivable. Representative structures are given in Figure 2 to showcase the most studied systems and illustrate their broad diversity.

Hereafter are discussed the most important structural modulations achieved to date.

The number of donor and acceptor sites. Ambiphilic ligands featuring a Lewis acid and one, two and three Lewis bases have been investigated. The number of donor buttresses has been used to modulate the constraint associated with chelation assistance from relatively flexible systems (one donor anchor) to highly rigidified structures in which the Lewis acid is maintained tightly in the coordination sphere of the metal (cf. cage complexes deriving from triphosphine-boranes and related ligands **8**).[3]

Donor sites D. Phosphines have been largely privileged here due to their high affinity for mid and late transition metals with which Lewis acids have been mainly associated so far. Practical reasons (easy synthesis and handling, ^{31}P NMR probe)

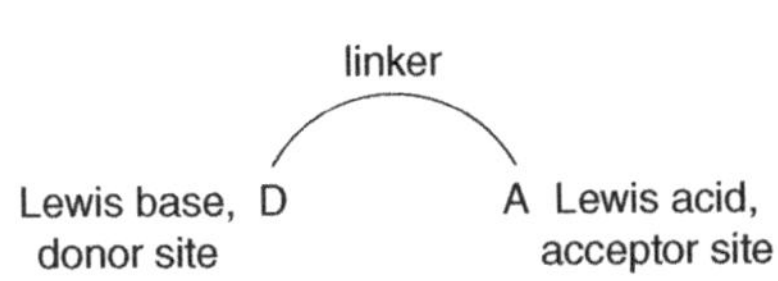

Figure 1 *General structure of ambiphilic ligands*

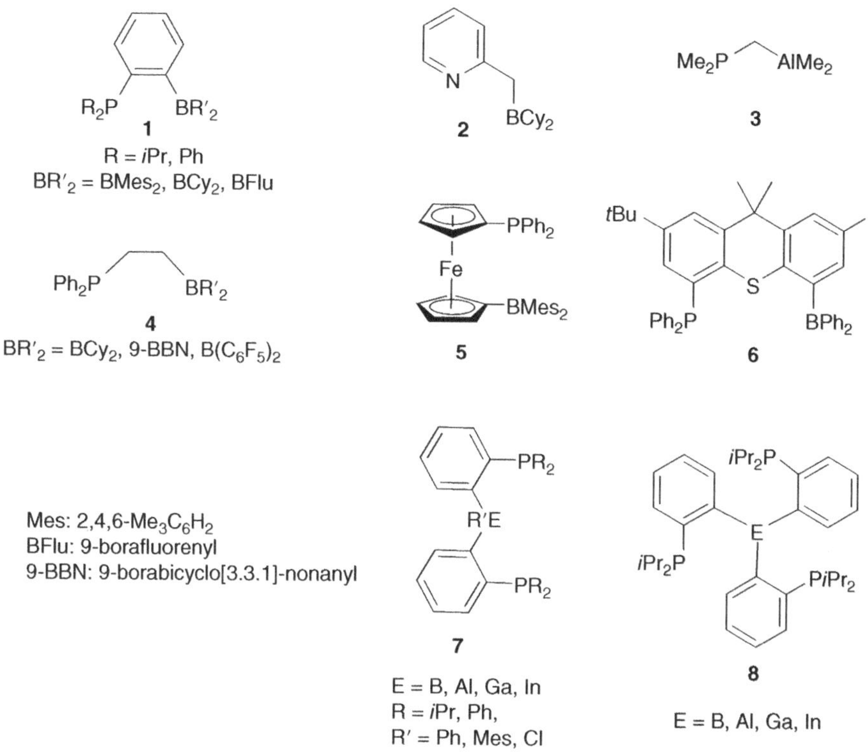

Figure 2 *Representative structures of ambiphilic ligands*

have also clearly played a role. Nitrogen-based systems, such as the pyridine-borane **2**,[4] have also been occasionally studied and it is likely that other donor groups, such as *N*-heterocyclic carbenes or phosphazenes, will be employed in future work.

Acceptor sites A. Here, the archetypal Lewis acid is boron and a large majority of ambiphilic ligands feature a borane moiety, whose behavior can be probed efficiently by using ^{11}B NMR. Synthetically, boron can be easily introduced by ionic coupling from halo-boranes and this methodology has been applied broadly, in particular for the preparation of rigid C_2-bridged phosphine-boranes **1**,[5] bisphosphine-boranes **7**[6] and triphosphine-boranes **8**.[6b] Hydroboration also provides an efficient entry to ambiphilic ligands, and flexible C_2-bridged phosphine-boranes **4**, have been prepared from vinyl-phosphines. Here, the ligand can be assembled before coordination or in the coordination sphere, depending on the sequence of hydroboration/coordination.[7] Varying the substituents at boron provides a means of modulating its behavior (by tuning Lewis acidity and steric shielding), although the incorporation of highly electrophilic boranes remains limited and this certainly represents a challenge for future work. So far, ambiphilic ligands featuring electron-withdrawing C_6F_5 groups at boron are limited to rare examples (such as C_2-bridged phosphine-borane **4**).[7d] Another way to increase Lewis acidity at boron is to incorporate it in a formally anti-aromatic borafluorene fragment (BFlu), and this strategy has been applied to an *ortho*-phenylene bridged phosphine-borane **1**.[5a] Replacement of boron for heavier group 13 elements, in particular aluminum, has drawn significant attention. Alanes are generally highly Lewis acidic, even when substituted by simple alkyl/aryl fragments, and are thus expected to interact more strongly with metal fragments. Because of synthetic limitations, the variety of corresponding ambiphilic ligands is still relatively limited, but significant contributions have been reported with the C_1-bridged phosphine-alane **3**[8] as well as *ortho*-phenylene bridged di- and tri-phosphine alanes of types **7**/**8**.[9] Besides unsaturated fragments of the group 13 elements, heavier group 14–16 elements can also act as Lewis acids due to their ability to form hypervalent structures.[10]

Linker L. Ambiphilic ligands have been built up on different types of organic linkers. The key features here are the length and flexibility of the backbone that directly influence the relative position of the donor and acceptor sites, and thus their coordination behavior. Essentially short linkers (C_1 and C_2) have been used, with the aim to introduce the Lewis acid moiety in the first coordination sphere of the metals. In this regard, the nature of the linker and in particular its rigidity play a key role, and the *ortho*-phenylene linker proved particularly well-suited. The corresponding ligands are preorganized so that the donor and acceptor sites point in the same direction, and at the same time, they retain some flexibility so that the Lewis acid can accommodate its environment. A few extended systems have also been investigated. Here, the Lewis acid usually remains pendant and it may be used to anchor substrates in the second coordination sphere of the metals (see Section 3.1).

Donor→acceptor (D→A) interactions. The combination of donor and acceptor sites raises the possibility of D→A interactions (intramolecular and intermolecular). In most systems, such D→A interactions are prevented by the steric demand of the substituents (steric frustration) and/or the nature of the organic linker (geometric frustration). Noticeable exceptions are illustrated in Scheme 1: (i) the C_1-bridged phosphine-alane **3** adopts a head-to-tail dimeric structure and is usually reacted in the presence of an additional Lewis base to displace the P→Al interactions;[11] (ii) the di- and tri-phosphine boranes **7** and **8** equilibrate in solution between *open* and *closed* forms which were unambiguously identified in the solid state by X-ray diffraction analyses;[6b] (iii) the pyridine-borane **2** exists in solution as a mixture of *closed* monomeric form and head-to-tail dimeric structure, but N→B interactions are readily cleaved upon coordination to Ru.[4]

Most typically, the formation of D→A interactions is disfavored by rigid C_2-linkers (it would result in strained four-membered rings). The corresponding *open* forms (free of D→A interactions) are usually the ground-state structures or lie only slightly higher in energy than the corresponding *closed* forms. Conversely, some linkers are incompatible with ambiphilic behavior due to enforced D→A interactions. This is the case of 1-phosphino 2-boryl naphthalenes **9** which feature strong P→B interactions even in the presence of very bulky substituents at phosphorus and boron (Figure 3).[12]

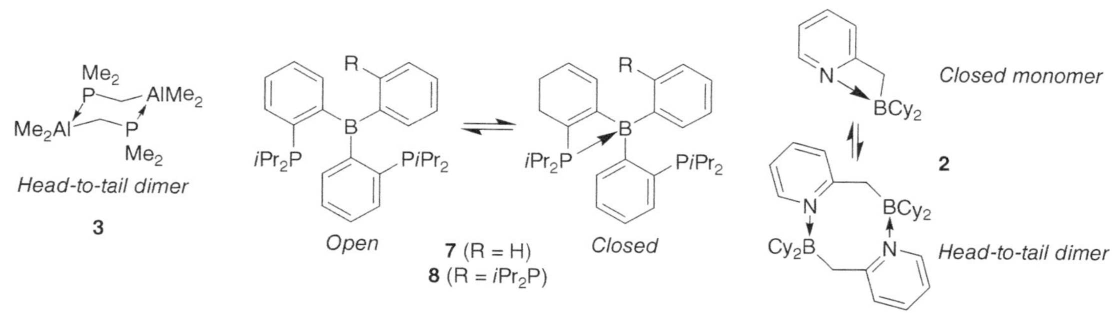

Scheme 1 *Intermolecular and intramolecular donor → acceptor interactions*

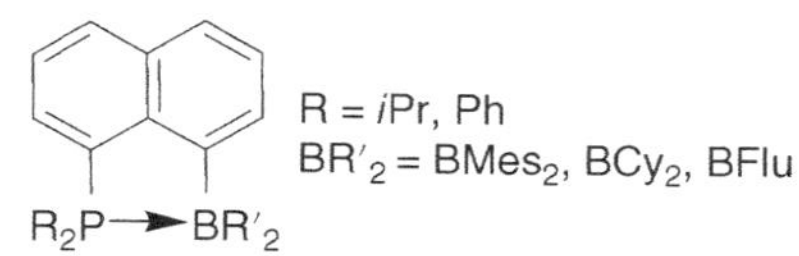

9

Figure 3 *Representative structure of 1-phosphino 2-boryl naphthalenes*

9.3 Coordination of ambiphilic ligands

The coordination properties of ambiphilic ligands have been thoroughly investigated. Depending on the role of the Lewis acid, four different situations can be distinguished and they have all been authenticated experimentally (Figure 4). Representative examples will be discussed in the following sections. Special attention will be devoted to the three coordination modes in which the Lewis acid site actively participates.

9.3.1 Complexes featuring a pendant Lewis acid

Early contributions on ambiphilic ligands relate to this coordination mode. Kagan and Jacobsen explored in parallel boraDIOP ligands and soon after, Landis investigated ferrocenyl bisphosphines featuring a benzoxaborolidine moiety.[13] In both cases, the idea was to introduce a remote borane moiety in the second coordination sphere of transition metals. This pendant functionality was expected to act as an anchor for incoming substrates, seeking to facilitate their coordination and control their orientation. Corresponding Rh, Pd and Pt complexes (Figure 5) were prepared, spectroscopically characterized and evaluated in a few catalytic transformations. Unfortunately, no clear improvement was observed compared with the corresponding boron-free systems, probably due to the weak Lewis acidity of the involved borane moieties.

More recently, complexes featuring pendant Lewis acids were prepared from C_2-bridged phosphine-boranes (Figure 6). As mentioned above, the *ortho*-phenylene linker usually favors the participation of the Lewis acid in coordination, but in the Pd complex **13**,[5c] this effect is counterbalanced by the bulky Mes which prevents interaction of the boron center with the metal fragment. Despite smaller substituents at boron, the Ru complex **14** adopts the same coordination mode.[7c] The flexibility of the CH_2CH_2 linker probably comes into play here (the phosphine-borane ligand adopts antiperiplanar conformation).

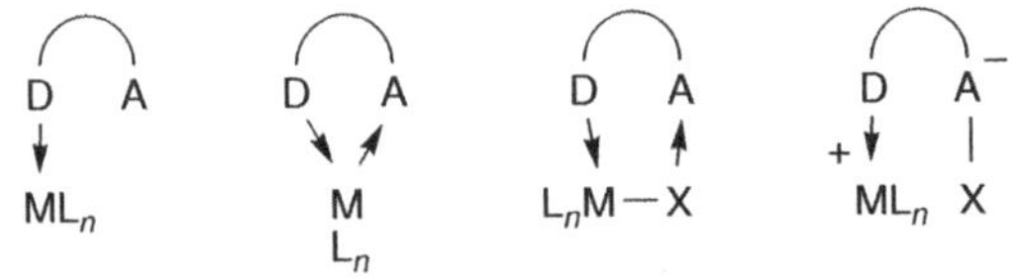

Figure 4 Coordination modes of ambiphilic ligands

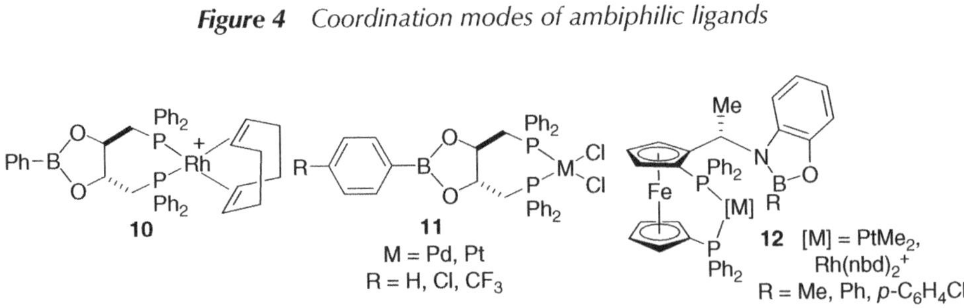

Figure 5 Rh, Pd and Pt complexes featuring pendant Lewis acids

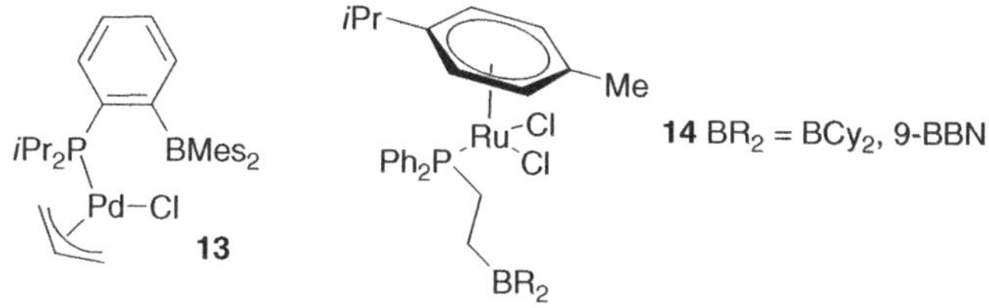

Figure 6 Pd and Ru complexes featuring pendant Lewis acids

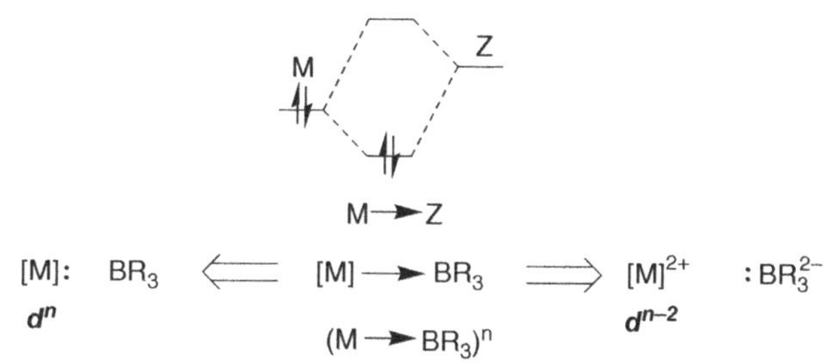

Figure 7 Schematic representations of two-center, two-electron M→B interaction

To date, the ability of external Lewis bases to bind to the pendant Lewis acid (anchoring effect) has not been structurally substantiated, but $^{1}H/^{11}B$ NMR evidence has been reported in a few instances using *N*-based donors.[13b,14]

9.3.2 Bridging coordination involving M→Lewis acid interactions

9.3.2.1 Concept of σ-acceptor, Z-type ligand

The ability of Lewis acids to bind to transition metals as σ-acceptor ligands was recognized early on[15] and the common classification of ligands as L (two-electron donor) and X (one-electron donor) was extended to Z (two-electron acceptor).[16] In the latter case, the metal behaves as a Lewis base and engages in dative M→Z interactions. Figure 7 provides a simplified orbital diagram for the corresponding two-center, two-electron interaction.[17]

For a long time, complexes involving M→LA interactions remained chemical curiosities, but chelating assistance has enabled significant progress over the last decade. In particular, borane complexes, long considered as putative species, were unambiguously authenticated and better knowledge has been gained on the nature and magnitude of M→B interactions.[1,18] The advent of phosphine-boranes has clearly played a major role here, providing straightforward access to M→B interactions (in addition to the formation of borane complexes upon coordination of hydrido-borates, as pioneered by Hill).[19] Progressively, the concept of σ-acceptor ligands has been developed and M→LA interactions have been exemplified with a great variety of metal fragments (group 8–11 metals, with coordination numbers from 2 to 5, and electron counts from

14 to 18). Lewis acids can also be varied broadly, from group 13 elements to heavier group 14–16 elements. It is not the purpose of this section to describe all the complexes featuring M→LA interactions. Only representative examples will be presented and the key features of M→LA interactions will be discussed.

9.3.2.2 Bridging coordination of phosphine-boranes

Gold complexes **15** nicely illustrate the ability of ambiphilic ligands to support M→Lewis acid interactions (Figure 8).[5a] Two phosphine-boranes differing in the Lewis acidity of the boron were studied and the Au→B interaction was found to strengthen from BCy_2 to BFlu. The first indication came from ^{11}B NMR, with a high-field shift of about 10 ppm of the BFlu signal upon formation of complex **15b**. Geometric data, as determined by X-ray crystallography, provide further insight. Both complexes adopt T-shaped geometry. The PAuCl skeleton is almost linear (which prevents interaction of B with Cl instead of Au) and the boron atom comes close to gold, all the more so as the Lewis acidity of the borane increases. The two AuB distances fall well within the sum of the van der Waals radii (3.7 Å). A convenient and normalized descriptor is the ratio *r* between the M/Lewis acid distance and the sum of the corresponding covalent radii. The pyramidalization of the boron environment (as estimated from the sum of CBC bond angles, referred to as ΣB_α) is another useful indicator for M→B interactions. The values of *r* (1.2–1.3) and ΣB_α (356–358°) indicate the presence of relatively weak Au→B interactions in complexes **15**. Valuable information can also be gained from density functional theory (DFT) calculations. Second-order perturbation theory analyses (NBO calculations) are particularly suited for the description of D→A interactions. Accordingly, donation from an occupied d orbital at Au to the vacant 2p orbital at B was found. The corresponding delocalization energy increases with the Lewis acidity of boron (ΔE_{NBO} from 5 kcal/mol in **15a** to 13 kcal/mol for **15b**). The bonding interaction between Au and B is clearly apparent from the associated natural localized molecular orbital, a bonding combination of d(Au) and vacant 2p(B). Since M→B interactions are associated with a transfer of electron density from the gold center to the borane fragment, one can also refer to atomic charges. Accordingly, the charge at gold was found to increase by 0.06 to 0.16 e^- upon coordination to BCy_2/BFlu, using the corresponding boron-free complex as a reference.

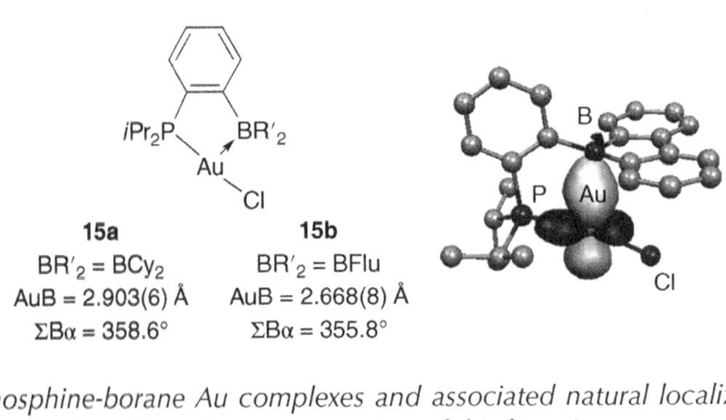

Figure 8 *Phosphine-borane Au complexes and associated natural localized molecular orbital. (See insert for color/color representation of this figure)*

9.3.2.3 M→LA interactions supported by bisphosphine-boranes and related ambiphilic ligands

A variety of d^8 fragments have been shown to engage in M→B interactions and form square-pyramidal complexes. The Rh complex **16a** (Figure 9) offered the first representative of such complexes and its bonding situation was analyzed thoroughly.[6a] The ^{11}B NMR signal is shifted upfield by ca. 30 ppm upon coordination. The boron atom occupies an apical position, with a short RhB distance (r = 1.02) and pronounced pyramidalization (ΣB_α = 339°). Density functional theory calculations support the presence of a relatively strong D→A interaction between the occupied d_z2 orbital at rhodium and the vacant 2p orbital at boron (ΔE_{NBO} 86 kcal/mol).

Consistently, Lewis bases such as DMAP or CO do not interact with the boron center of **16a** but rather split the chloro bridge to give mononuclear complexes (*cis* complex **16b** and *trans* complexes **16c**).[6a,20] Due to the π-accepting character of CO, the Rh center of **16c** is less electron-rich and thus forms a slightly weaker Rh→B interaction. In turn, the ν_{CO} stretching frequency is shifted to higher energy by 35 cm^{-1} upon coordination of Rh to boron, indicating that the borane moiety exerts a substantial withdrawing effect on the metal.

The corresponding Pd and Pt dichloro complexes adopt similar square-pyramidal structures, but the M→B interaction weakens significantly from Rh to Pt and Pd, in line with the decrease of the Lewis basicity of the metal.[20] Despite the presence of *ortho*-phenylene linkers, the bisphosphine-borane ligands are remarkably flexible. They can accommodate facial as well as meridional coordination and the two phosphine anchors support but do not impose M→B interactions.

Related Au complexes **17** (Figure 10) have attracted most interest.[21] The presence of a second phosphine sidearm was shown to significantly strengthen the Au→B interaction (δ ^{11}B ~ 26 ppm, r ~ 1.03, ΣB_α ~ 342°) compared with the corresponding monophosphine-borane complexes **5**. According to DFT calculations, complexes **17** feature strong Au→B interactions (ΔE_{NBO} ~ 55 kcal/mol), but the transfer of electron density is not large enough to consider that gold is oxidized from Au(I) to Au(III), as established unequivocally by ^{197}Au Mössbauer spectroscopy. The square-planar geometry of **17** is unprecedented for Au(I) complexes, demonstrating that the coordination of Lewis acids may challenge the basic rules dictating the geometry of complexes. The related

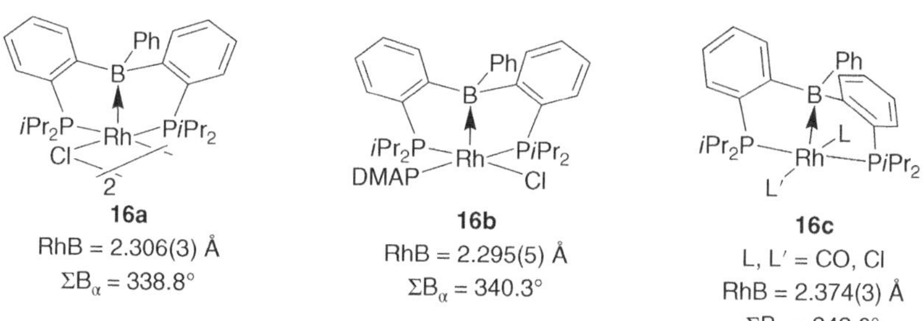

Figure 9 *Bisphosphine-borane Rh complexes*

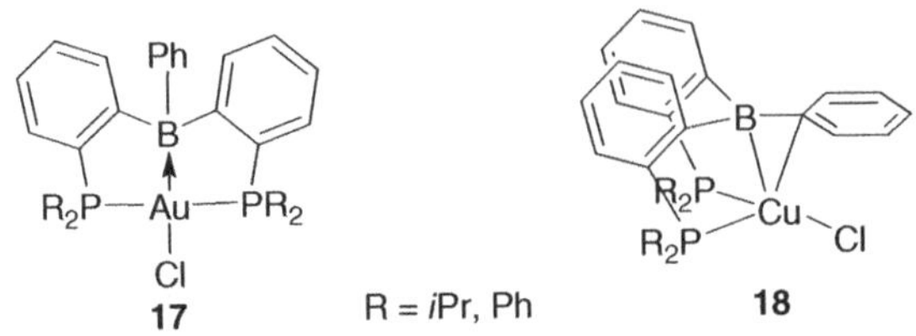

Figure 10 *Bisphosphine-borane Au and Cu complexes*

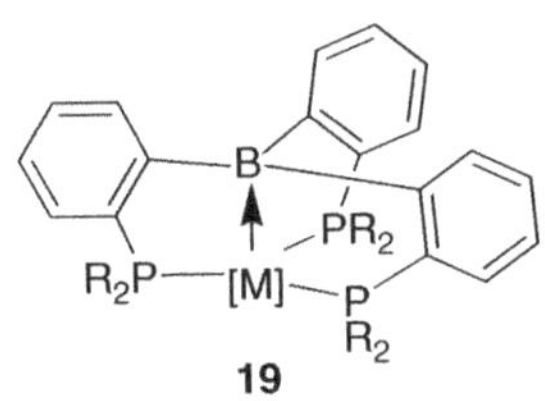

Figure 11 *Triphosphine-borane complexes featuring M→B interaction*

copper complexes **18** display completely different structures.[22] Because of the more electron-deficient character of Cu, the bisphosphine-borane is tilted and the central BPh moiety adopts a new η^2-BC coordination mode. The ability of aryl-boranes to engage into multi-center BC and BCC coordinations is quite general[23] and opens interesting perspectives in reactivity.

Gold complexes **17** were used as models to vary the nature of the Lewis acid and study its influence on M→LA interactions. Replacement of boron for heavier group 13 elements (Al and Ga) leads to zwitterionic complexes as the result of chloride abstraction from Au (see Section 3.4).[9a,24] Heavier group 14–16 elements are significantly less Lewis acidic, but nevertheless, they are prone to coordinate Lewis bases and form thereby hypervalent compounds. Such a situation with metals (Au, Pd, Pt) acting as Lewis bases was illustrated with Si and Sn,[10a,b] Sb and Bi,[10c,g] as well as Te.[10h]

9.3.2.4 M → LA interactions supported by triphosphine-boranes and related ambiphilic ligands

The presence of three donor buttresses increases further chelating assistance and results in cage complexes. In this respect, triphosphine-boranes have attracted most attention and a broad variety of metallaboratranes have been reported. First, a comprehensive study was carried out on complexes of the group 10 and 11 metals **19** (Figure 11).[3,25] The cage structure enforces in all cases transannular M→B bonding, but the system retains flexibility and the magnitude of this interaction varies significantly from one metal to the other. Accordingly, group 10 metals form stronger M→B interactions than group 11 metals, and M→B interactions strengthen from 3d to 5d

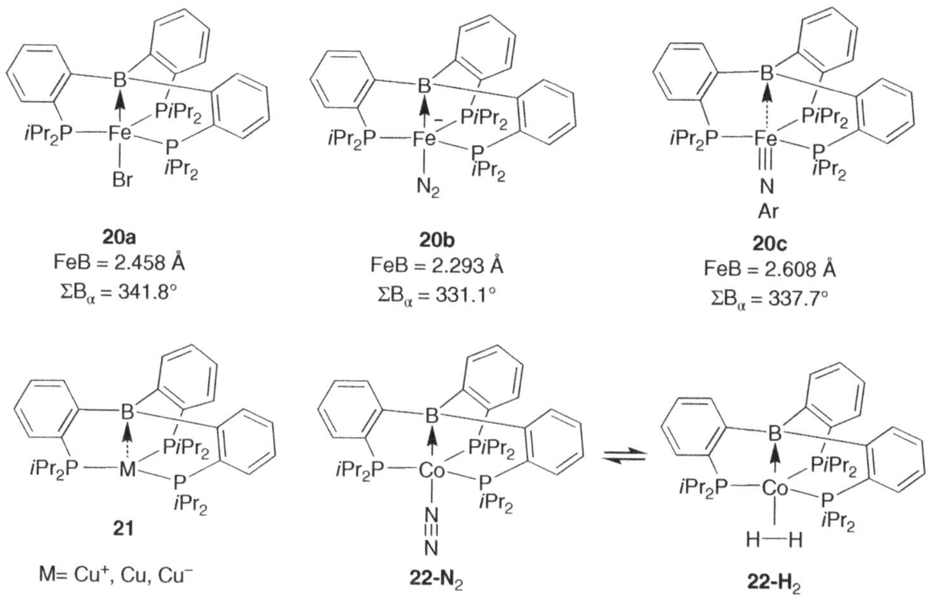

Figure 12 *Triphosphine-borane Fe, Cu and Co complexes*

elements, especially for the coinage metals.[25] The strongest interactions were met with Pt and Au, which are more Lewis basic than their lighter congeners due to relativistic effects (5d orbitals are raised in energy). This work revealed another interesting feature of M → B interactions concerning UV-vis properties. Indeed, coordination of the Lewis acid induces a significant bathochromic shift as the result of the introduction in the MO diagram of a low-lying vacant orbital (corresponding to the antibonding combination of the 2p orbital at B with the d_{z^2} orbital at M).

A series of Fe complexes **20** were then described by Moret and Peters,[26] illustrating the ability of the triphosphine-borane to accommodate different electronic configurations, from low-valent N_2 to high-valent imido species (Figure 12). All complexes adopt cage structures but the central Fe → B interaction showed high plasticity (the FeB distance varies from 2.293 to 2.608 Å). Similar observations were made recently for related Cu complexes **21**.[27] The triphosphine-borane was shown to accommodate three different oxidation states of the metal (Cu^+, Cu and Cu^-), with CuB distances varying from 2.198 to 2.495 Å. The neutral species is most remarkable as it features a 1-electron Cu → B interaction with spin density located mostly on B (57%). Paramagnetic Co complexes **22** were also reported and boron was found to facilitate exchange between end-on N_2 and side-on H_2 at Co.[28]

Nakazawa extended the variety of metallaboratranes to Rh/Ir and isolated neutral, cationic as well as anionic complexes **23** (Figure 13).[29] The neutral species **23b** undergoes unusually facile and reversible CO/phosphine exchange, illustrating strong *trans* influence and labilizing effect of the borane.

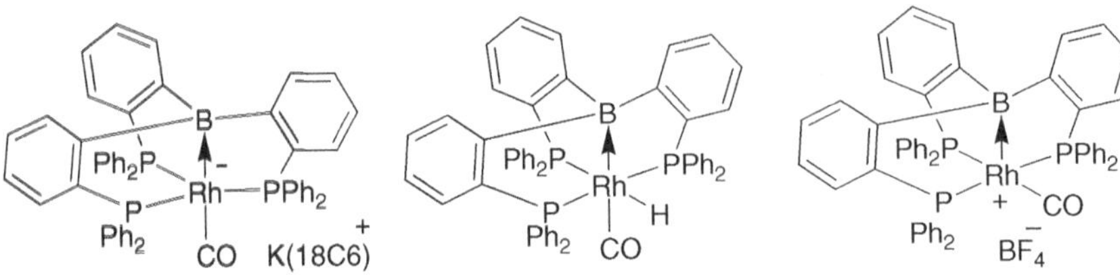

23a **23b** **23c**

Figure 13 *Anionic, neutral and cationic triphosphine-borane Rh complexes*

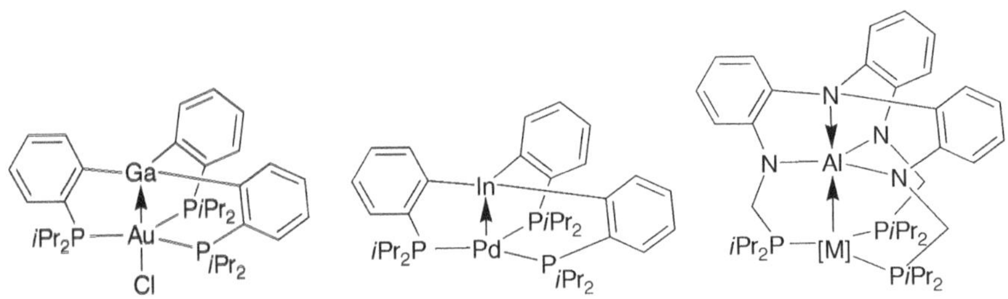

24a **24b** **25**

[M] = Ni, Co(N_2), Fe(N_2)

Figure 14 *Cage complexes with heavier group 13 elements*

A few cage complexes with heavier group 13 elements coordinated as σ-acceptor ligands have also been reported (Figure 14).[24,30] Despite the larger size of Ga and In, triphosphine gallane and indane were found to engage in strong Au → Ga and Pd → In interactions. In addition, Lu and co-workers developed a novel aluminatrane platform featuring three pendant phosphine arms and zerovalent complexes **25** featuring strong M → Al interactions (M = Ni, Co and Fe) were thereby obtained.[31]

9.3.3 Bridging coordination of M–X bonds

Interaction of the Lewis acid moiety with a co-ligand at the metal (typically a halogen or a hydride) gives rise to chelate coordination of the ambiphilic ligand. This coordination mode was found in particular with the iPr$_2$(o-C_6H_4)BCy$_2$ ligand. The corresponding chloro complexes of Pd and Rh **26** (Figure 15) were studied both experimentally and theoretically.[5a,c] Characteristic features for the Cl → B interactions are the high-field shift of the ^{11}B NMR resonance signal, short ClB distance and noticeable pyramidalization of the boron environment. The pyridine-borane Ru complex **27** provides another example of such M–Cl bridging coordination.[4]

Bridging coordination of M–X bonds (M = Rh, Pd, Pt; X = Cl, Br, I) with a tridentate phosphine-thioether-borane ligand has been extensively investigated by Emslie. Close inspection of the [Rh(CO)X] complexes **28** (Figure 16) indicated that the boron center interacts more strongly with Cl and Br than with I.[32]

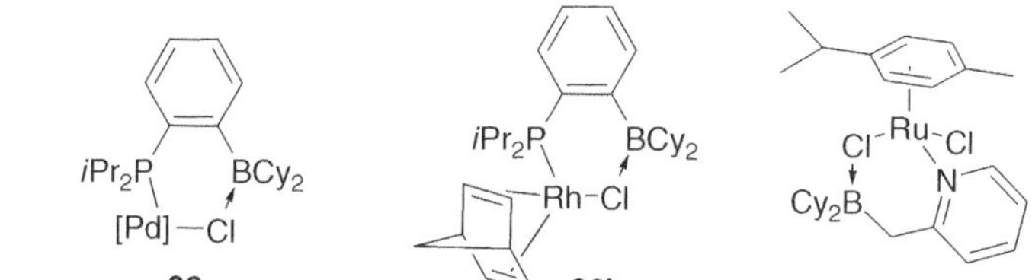

26a **26b** **27**

[Pd] = Pd(allyl), $PdClPPh_3$

Figure 15 *Bridging coordination of M—Cl bonds (M = Pd, Rh, Ru)*

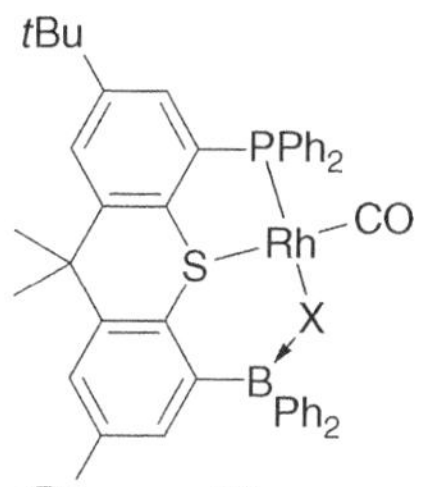

28

	X = Cl	Br	I
B-X (Å)	1.995	2.190	3.125
Rh-X (Å)	3.470	3.647	3.150
ΣBα (°)	340	342	357

Figure 16 *Representative Rh complexes featuring Rh—X—B interaction (X = Cl, Br, I)*

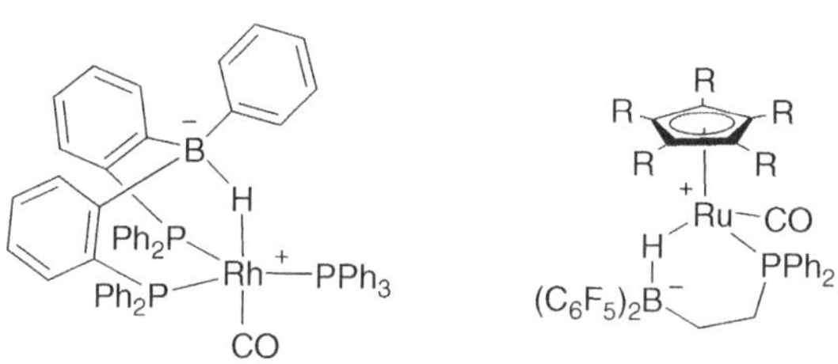

29 **30**

Figure 17 *Bridging coordination of M—H bonds (M = Rh, Ru)*

Such bridging coordination of M—X bonds with ambiphilic ligands is expected to be associated with some degree of bond activation. M—X bond lengths proved hardly informative in that respect, but computational studies, including comparison with related Lewis acid-free complexes, have clearly evidenced substantial weakening of M—Cl bonds upon interaction with boron.[32]

Recently, the bridging coordination of phosphine-boranes has been extended to metal hydrides (Figure 17). Kameo and Nakazawa[33] first described the bisphosphine-borane Rh hydride complex **29**. The presence of a Rh—H—B bridge was unambiguously established by using ^{11}B NMR and X-ray diffraction analyses. Comparing the ν_{CO} stretching frequency with that of the related boron-free complex $[RhH(CO)(PPh_3)_3]$ revealed a noticeable shift to higher energy (by 24 cm^{-1}), indicating that the borane moiety exerts a significant withdrawing effect on Rh upon coordination to the hydride.

In addition, several RuH complexes **30** deriving from phosphine-boranes of type **4** have been reported by Ostapowicz *et al.*[7d] Taking advantage of the possibility to assemble the CH_2CH_2-bridged ligands before or after coordination, the Lewis acidity of boron was varied. All compounds adopt bridging coordination in the solid state, but only the $B(C_6F_5)_2$ moiety forms strong Rh—H—B interactions in solution, as clearly apparent from ^1H and ^{11}B NMR.

Brigding coordination of M—X bonds by ambiphilic ligands opens interesting perspectives. It can contribute to facilitate the activation of σ-bonds and can participate in reversible abstraction/transfer of atoms.

9.3.4 Ionization of M—X bonds

The bridging coordination of M—X bonds represents a preliminary stage in their intramolecular activation and may ultimately lead to zwitterionic complexes upon heterolytic cleavage. Such a process has been observed and unambiguously authenticated with simple phosphine-boranes of type **4** as well as di- and tri-phosphine derivatives of the heavier group 13 elements **7** and **8**.

Fischbach *et al.*[7b] reported the zwitterionic complexes **31** (Scheme 2). Coordination of the phosphine to Ni is accompanied by the abstraction of a Me group at Ni by the borane moiety. This results in a formal insertion of the phosphine-borane ligand into a Ni—Me bond.

In the meantime, we studied the coordination of the bisphosphine-alane of type **7** to gold and obtained thereby the zwitterionic complex **32** (Scheme 3).[9a] The transfer of chloride from gold to aluminum was established crystallographically. The coordination mode of **32** markedly contrasts with that of the related bisphosphine-borane complexes **17** (neutral square-planar complexes with Au → B interactions). The difference between

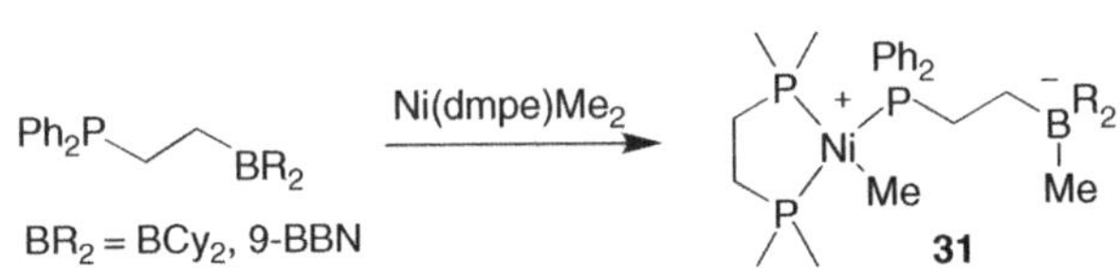

Scheme 2 *Formation of zwitterionic Ni complexes upon coordination of monophosphine-boranes*

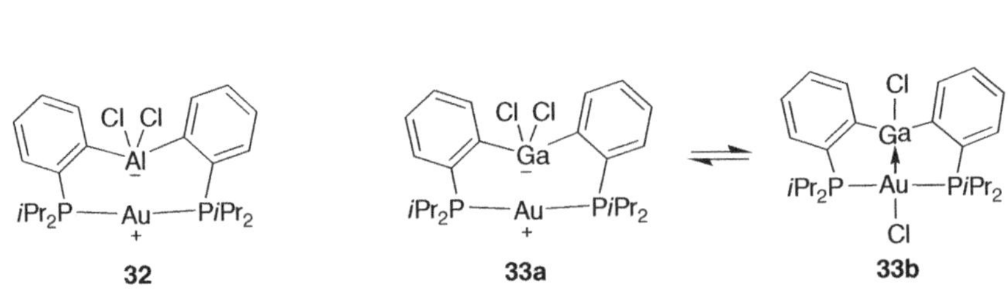

Scheme 3 *Coordination of bisphosphine-alane and galane to Au*

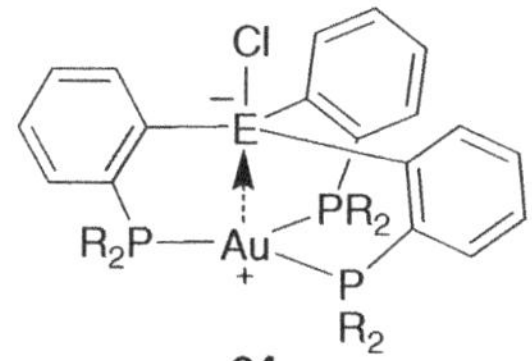

34
E = Al, In; R = *i*Pr

Figure 18 Zwitterionic Au complexes of triphosphine-alane and indane

Al and B was investigated by computational means. Accordingly, the zwitterionic structure is favored with aluminum due to stronger affinity for Cl^-, while the neutral form is favored with borane due to stronger Au → B interactions. Using the related bisphosphine-gallane ligand, it was possible to observe directly by NMR the chloride shift between the metal center and the Lewis acid.[24] Indeed, in this case, the zwitterionic and neutral forms **33a,b** coexist and slowly interconvert in solution. But the presence of a third phosphine sidearm inhibits the transfer of Cl^- between Au and Ga. Indeed, the two coordination isomers of the corresponding triphosphine-gallane complex were found to be separable and not to interconvert.

Zwitterionic complexes were also obtained upon coordination of triphosphine-alane and indane ligands **34** to gold (Figure 18).[9b,30] In addition to chloride abstraction, the aluminum and indium centers engage in weak Au^+ → Lewis acid interactions (enabled by the ability of heavier group 13 elements to form hypervalent structures). This is apparent geometrically (short AuAl and AuIn contacts; trigonal bipyramidal geometry around Al/In) and computationally (D → A interactions were identified by NBO analyses).

9.4 Reactivity of metallic complexes deriving from ambiphilic ligands

So far, the potential of ambiphilic ligands in reactivity and catalysis has only been scarcely investigated. The kinetic, thermodynamic and even the outcome of metal-mediated transformations can be influenced by the presence of a Lewis acid in the coordination sphere. As discussed hereafter, the first studies performed in this area are very promising and further developments are certainly to be expected.

9.4.1 Lewis acid enhancement effect in Si—Si and C—C coupling reactions

Ambiphilic ligands were investigated early on in Si—Si and C—C coupling with the aim to substantiate the beneficial influence the introduction of boranes or alanes may have. Although the exact role of the Lewis acid moiety during catalysis remains generally unclear, comparison with Lewis acid-free systems has often revealed significant improvments in terms of activity and/or selectivity.

9.4.1.1 Dehydrogenative coupling of silanes catalyzed by phosphine-alane Ni complexes

In the early 2000s, Fontaine and Zargarian[11] used the dimeric phosphine-alane $(Me_2PCH_2AlMe_2)_2$ **3** instead of the common MAO cocatalyst for the dehydrogenative coupling of phenylsilane (Scheme 4). The bifunctional PAl cocatalyst was found to increase the activity of the reference Ni complex [(1-MeInd)Ni(PPh_3)Me] **35** by factor of 50.

The phosphine-alane is expected to displace PPh_3 at Ni, but no reaction takes place in the absence of substrate due to strong intramolecular P→Al interactions. Lewis bases such a triethylamine are able to split the head-to-tail phosphine-alane dimer and consistently, further increase catalytic activity. The precise role of the Lewis acid moiety in the catalytic cycle remains unknown. It is supposed to interact with the methyl group at Ni and to facilitate methyl/silyl exchange. The key active species **36a** could not be characterized, but its Lewis base adduct **36b** was identified by NMR (Figure 19).

Subsequent studies by Fontaine and co-workers on related Rh complexes have provided further insight. Complex **37a** featuring a DMSO→Al interaction was characterized spectroscopically (Scheme 5).[34] Fast exchange between the Me groups at aluminum and rhodium was unambiguously evidenced by ^1H NMR upon addition of $AlMe_3$ to trap DMSO. This supports reversible abstraction of Me at Rh by the alane moiety, and the corresponding zwitterionic complex could be trapped by PMe_3 and ethylene.

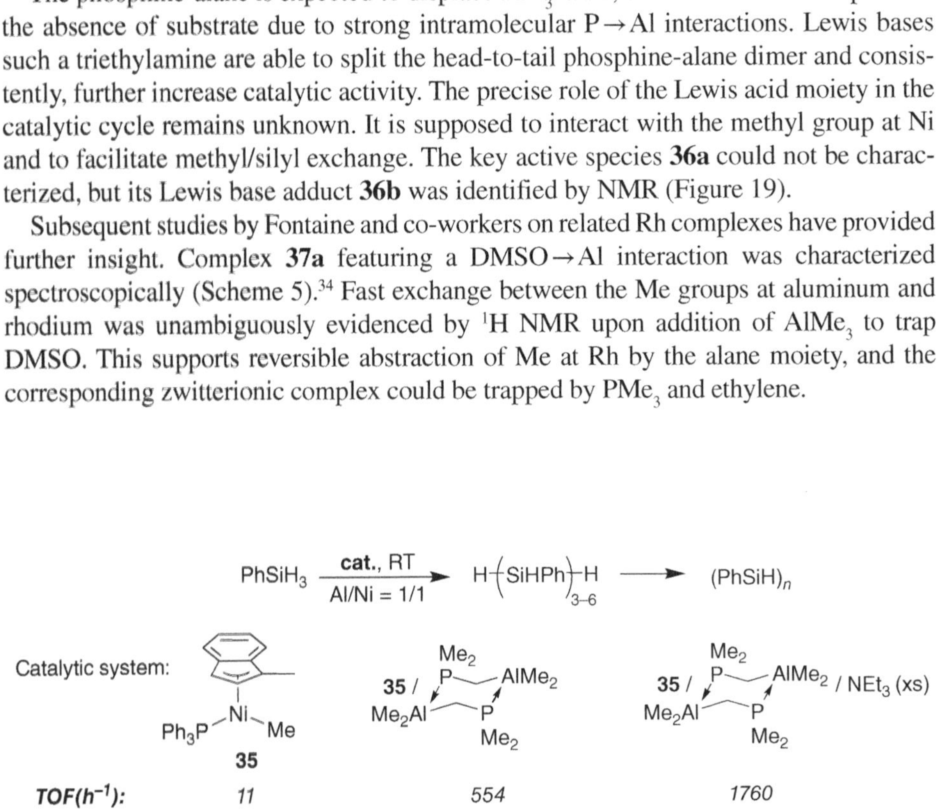

Scheme 4 Dehydrogenative coupling of silanes. TOF = turnover frequency

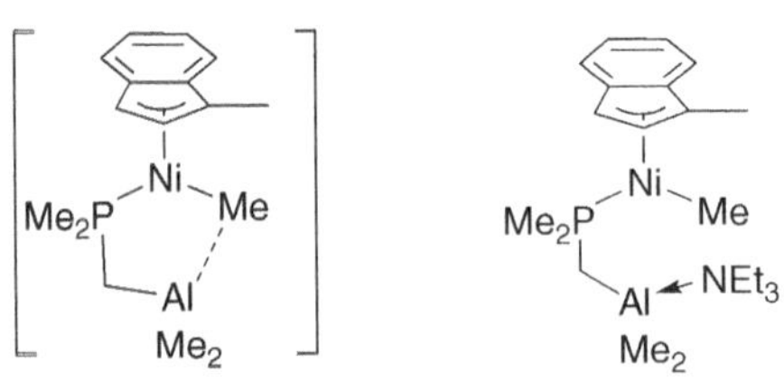

Figure 19 Active species proposed to account for dehydrogenative coupling of silanes and its Lewis base adduct

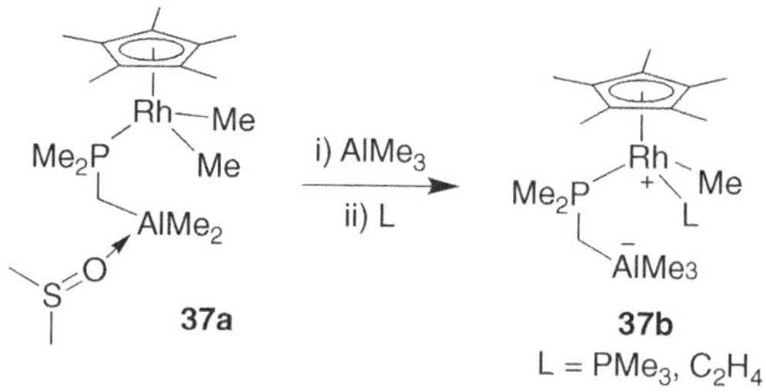

Scheme 5 *Exchange between the Me groups at aluminum and rhodium*

9.4.1.2 ortho-Boryl phosphines versus biaryl-phosphines in Pd-catalyzed Suzuki–Miyaura cross-coupling

As mentioned above, *ortho*-phenylene phosphine-boranes possess versatile coordination properties. The presence of a boryl moiety in *ortho*-position to phosphorus also opens interesting perspectives in catalysis. The prototypical ligand Ph_2P-(*o*-C_6H_4)-$BMes_2$ **1a** was initially evaluated in Rh-catalyzed hydroformylation of 1-octene and compared with the boron-free Buchwald-type biaryl-phosphine **38** (Figure 20).[35] Modest activities and branched:linear selectivities were obtained in both cases, with no beneficial influence of the $BMes_2$ group. But the similar catalytic properties unexpectedly observed between the two ligands stimulated further studies.

The air-stable phosphine-borane **1a**[36] was evaluated in the Pd-catalyzed Suzuki–Miyaura reaction.[37] Typically, 4-bromoanisole and phenyl boronic acid were efficiently coupled using 1 mol% of $Pd(OAc)_2$ and 2 mol% of ligand **1a** (Scheme 6). The $BMes_2$ moiety is compatible with the cross-coupling and it actually improves catalytic activity (under the same conditions, PPh_3 gives a notably lower yield). Interestingly, a catalytically competent Pd(0) complex **39** was isolated and fully characterized. The phosphine-borane adopts a new coordination mode: besides phosphorus, the Pd center is coordinated by one of the Mes groups at B (η^2-$C_{ipso}C_{ortho}$ interaction). This weak π interaction, which is reminiscent of those commonly encountered with biaryl-phosphines,[38] is likely to stabilize the electronically unsaturated Pd fragment and enhance its catalytic performance.

Given the importance of cross-coupling reactions for the preparation of heterobiaryl compounds, phosphine-borane Pd complexes were then applied to the reaction of chloro-pyridines and related *N*-heterocycles.[39] The Lewis acid site of the ligand may anchor and eventually activate these substrates via N → B interactions prior to oxidation addition of the C—Cl bond. For a large scope of substrates, the desired cross-coupled products were obtained in high yields (>90%) within 20 h at 100°C, using 1 mol% of $PdCl_2$(cod) (where cod = 1,5-cyclooctadiene) and 2 mol% of **1a** (Scheme 7). The reaction even tolerates free amino groups, and the phosphine-borane ligand actually gave better results than PPh_3 and the biaryl-phosphine **38** in the coupling of 4-amino-2-chloropyridine (92%, *versus* 11 and 79% yields, respectively).

The phosphine-borane/Pd catalytic system was also used to achieve sequential cross-coupling of a 2,6-chloropyridine. In contrast to the biaryl-phosphine **38**, mono-coupled

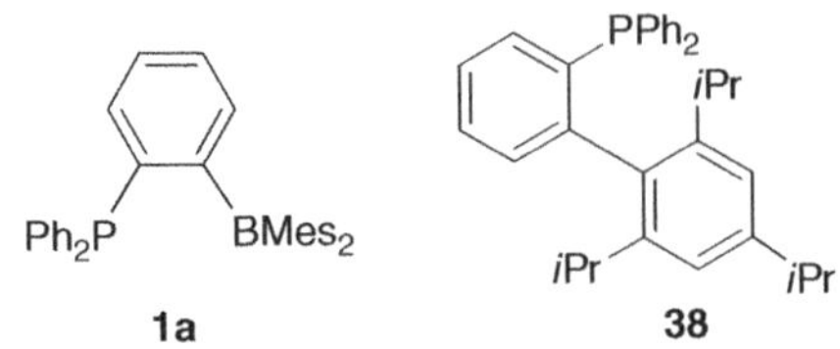

Figure 20 *Phosphine-borane **1a** and the related boron-free Buchwald-type ligand **38***

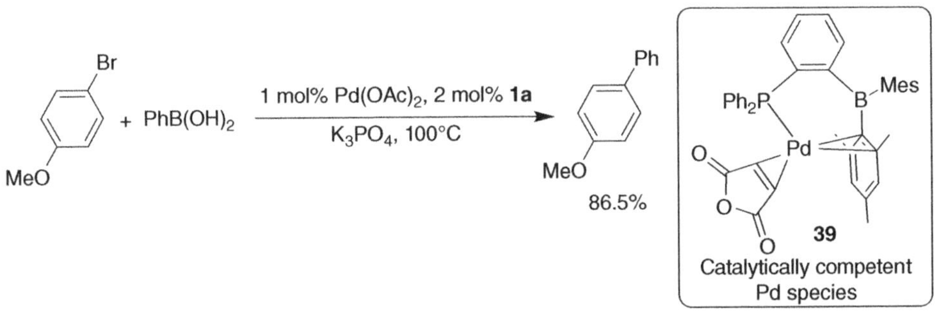

Scheme 6 *Suzuki–Miyaura cross-coupling mediated by phosphine-borane Pd complexes*

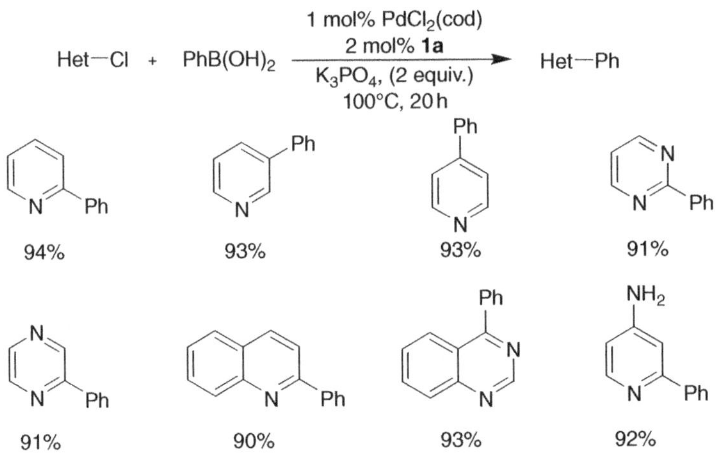

Scheme 7 *Synthesis of heterobiaryl compounds catalyzed by phosphine-borane Pd complexes*

products of 2,6-dichloro,3-nitropyridine were obtained with good selectivity (~70%) with **1a**. Accordingly, the two regio-isomeric unsymmetrical pyridines **40a** and **40b** could be readily prepared, using phenyl- and *ortho*-tolyl-boronic acids, simply changing their order of reaction (Scheme 8).

Recently, Liu extended further the variety of *ortho*-phenylene phosphine-boranes and Pd-catalyzed transformations.[40] The boron center was incorporated in a 1,4-azaborine

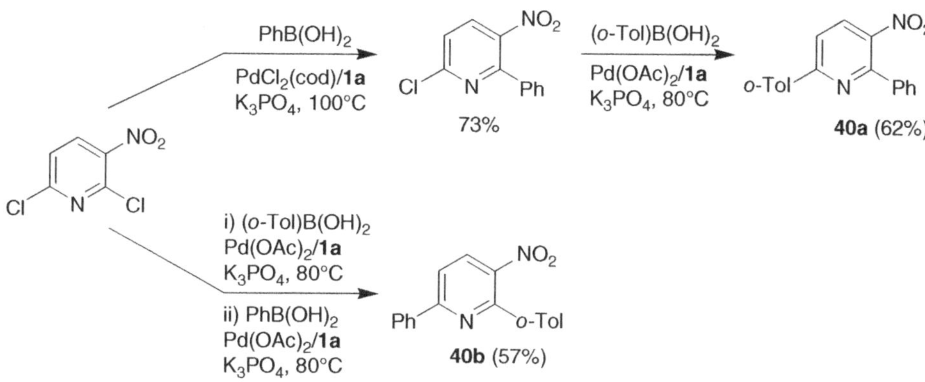

Scheme 8 *Preparation of unsymmetrical pyridines by sequential cross-coupling*

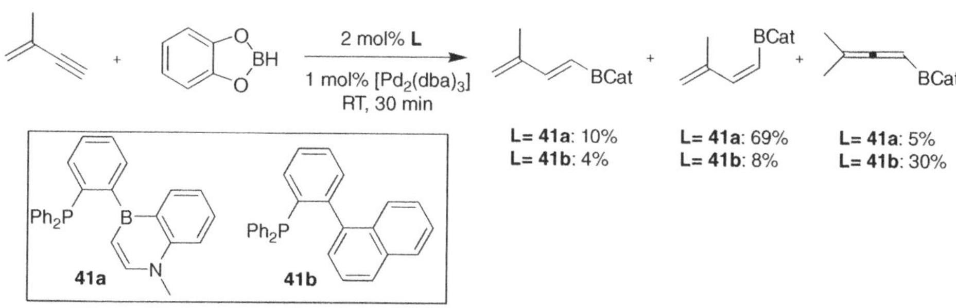

Scheme 9 *Hydroboration mediated by a phosphine-borane Pd complex*

moiety, and the ensuing ligand **41a** was evaluated in the catalytic hydroboration of enynes, using $Pd_2(dba)_3$ as precursor (Scheme 9). Beneficial influence of the Lewis acid group on activity and selectivity was substantiated here also by comparison with a related boron-free biaryl-phosphine **41b**.

9.4.2 Hydrogenation, hydrogen transfer and hydrosilylation reactions assisted by boranes

9.4.2.1 Boranes as hydride shuttles: hydride shift from boron to transition metals

It was recognized early on that boranes may act as hydride shuttles in the first coordination sphere of transition metals.[41] Such a behavior opens interesting possibilities in a broad range of catalytic transformations, starting from hydrogenation and hydrogen transfer. The associated elementary processes, namely hydride transfer between boron and transition metals on the one hand, and between boron and organic residues on the other hand, have drawn much interest. In this context, flexible scorpionate ligands proved particularly valuable. Crossley and Hill[42] investigated

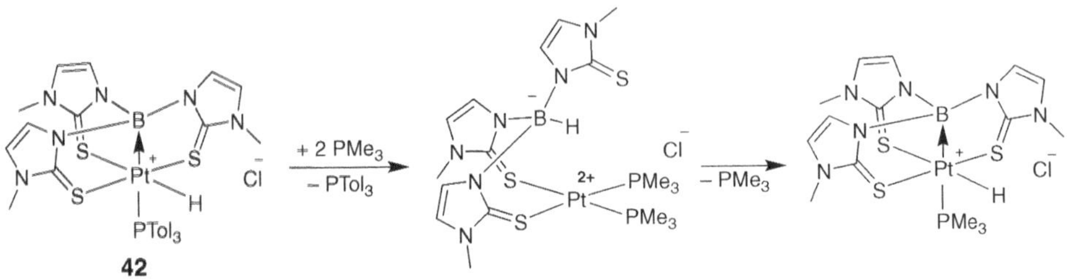

Scheme 10 Hydride shuttling process between Pt and B

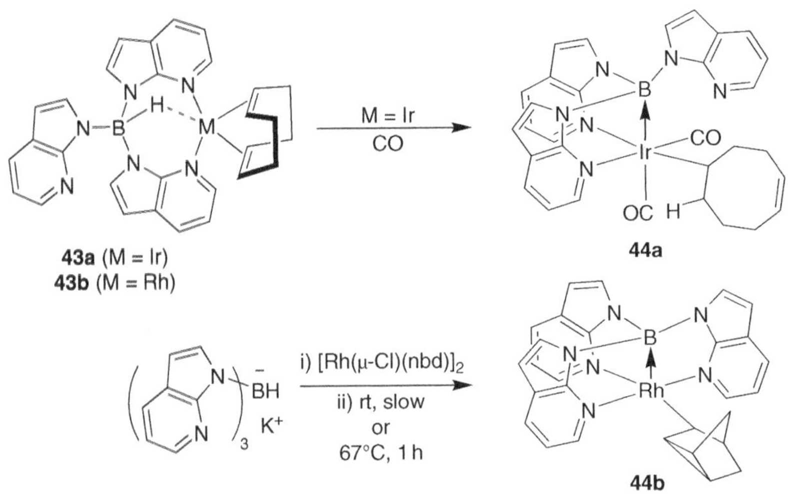

Scheme 11 Reactivity of Rh and Ir complexes featuring M—H—B interactions

platinaboratranes **42** derived from tri(methimazolyl)borohydrides and evidenced reversible locking/unlocking of the cage structure as the result of hydride transfer between B and Pt (Scheme 10).

In the meantime, Owen and co-workers studied the coordination of the tri(azaindolyl) borohydride to rhodium and iridium (Scheme 11). The presence of strong bridging M—H—B interactions in complexes **43a,b** was indicated by X-ray diffraction, IR and ^{1}H NMR data. However, hydride migration can be induced by modifying the coordination sphere of the metal. The Ir complex **43a** reacts with carbon monoxide to give the corresponding metallaboratrane **44a**. The reaction probably proceeds via dissociation of one of the double bonds of cod, followed by hydride migration from B to Ir and insertion of the coordinated olefin into the Ir—H bond.[43] A similar process was substantiated for a related Rh complex, with rearrangement of norbonadiene into nortricyclyl (as the result of a C—C coupling reaction).[44]

The catalytic properties of the Rh and Ir complexes **43a,b** towards transfer hydrogenation of ketones were evaluated using *i*PrOH as reducing agent (Scheme 12).[45] The two complexes exhibit similar profiles and require the presence of KOH. Although it is

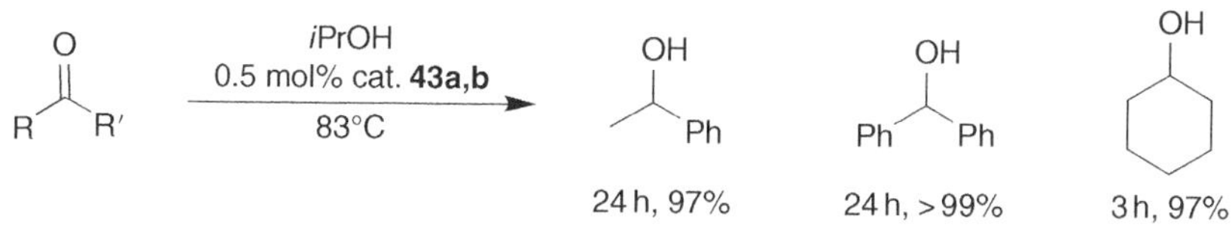

Scheme 12 *Catalytic transfer hydrogenation of ketones with complexes* ***43a,b***

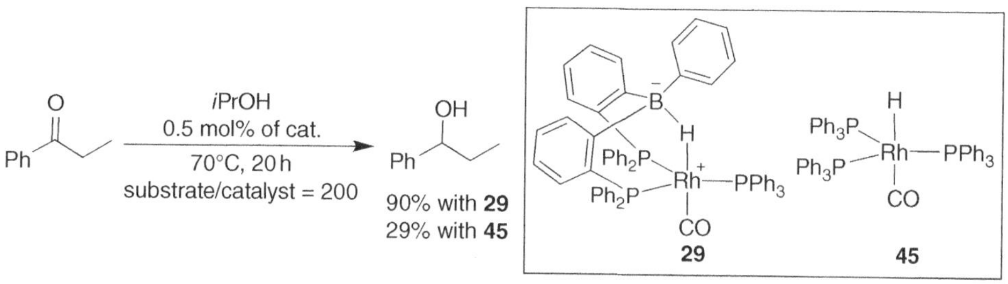

Scheme 13 *Transfer hydrogenation mediated by Rh complexes*

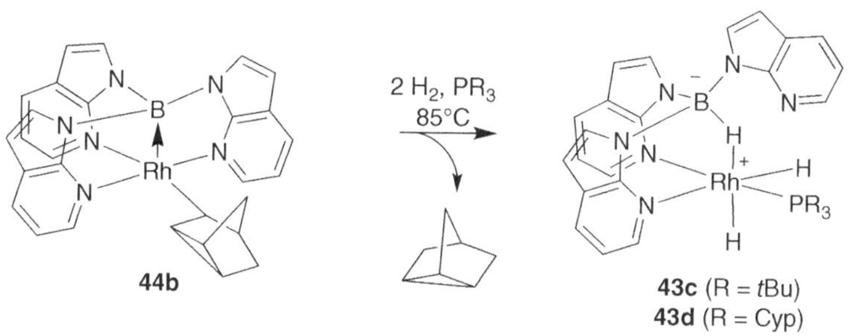

Scheme 14 *H_2 activation across Rh→B*

difficult to understand the role and influence of the boron center in these transformations, they highlight the potential application of such systems in catalysis.

Related catalytic studies were performed recently by Kameo and Nakazawa[33] on bisphosphine-borane Rh and Ir complexes (Scheme 13). Accordingly, transfer hydrogenation of ethyl phenyl ketone could be successfully achieved with the Rh complex **29** at 70°C in the absence of any base. Complexes featuring M→B interactions proved significantly less active. Comparison with the related boron-free complex **45** highlights the important and beneficial influence of M—H—B bridging interaction.

Besides acting as a hydride relay, the Lewis acid fragment can also participate in substrate activation. This possibility was first illustrated by Owen and co-workers while investigating the reactivity of the Rh complex **44b** with sterically demanding phosphines (Scheme 14).[44] No reaction is observed in the absence of H_2, but under 2.5 bar of H_2, rhodium hydride complexes **43c,d** are formed along with tricyclo[2.2.1.0^{2,6}]heptane. This transformation involves the addition of H_2, the cleavage of the Rh→B interaction and the formation of a C—H bond (presumably via reductive

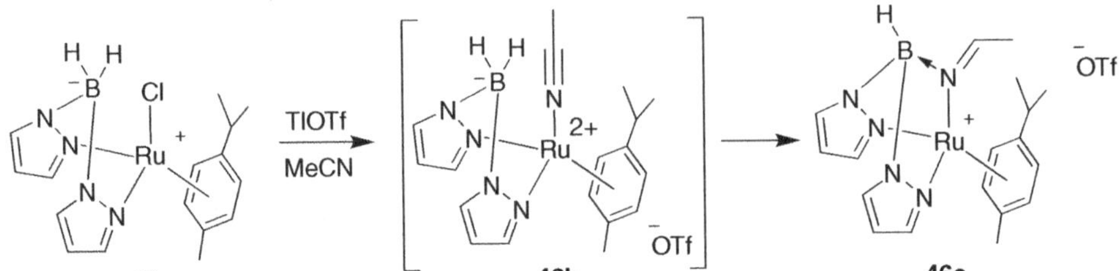

46a 46b 46c

Scheme 15 *Hydride transfer from B to acetonitrile in the coordination sphere of Ru*

Table 1 *Representative examples of nitrile reductions[a] catalyzed by* **46c**

Substrate	Product	*n* equiv. $NaBH_4$	Reflux (h)	Yield (%)
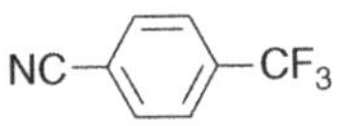	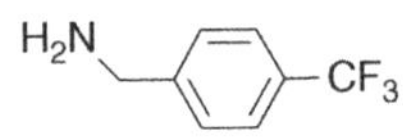	4	12	82
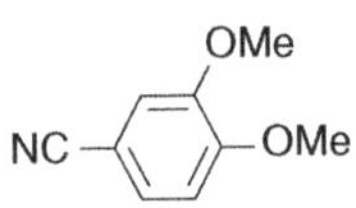	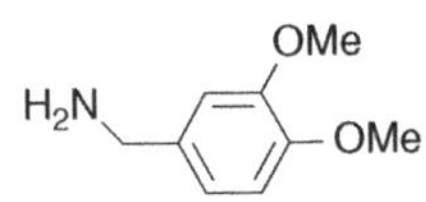	8	12	87
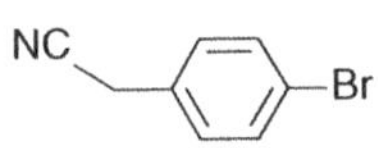	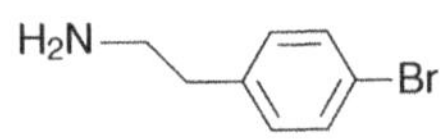	8	14	85
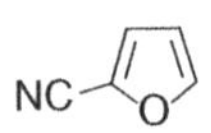	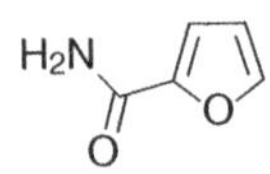	4	8	56

[a] 5 mol% **46c**, *n* equiv. $NaBH_4$, 1 equiv. NaO*t*Bu, MeOH, reflux.

elimination from an intermediate hydride nortricyclyl Rh complex). Such a behavior is appealing for olefin hydrogenation, and indeed, complex **44b** was found to efficiently catalyze the reduction of styrene and cyclooctene at 85°C and 2.5 bar. Using 1 mol% catalyst loading, conversions are complete after 18 h.

The ability of boranes to coordinate and activate an incoming substrate was also proposed recently by Lu and Williams.[46] The di(pyrazolyl)borohydride was first coordinated to ruthenium. Chloride abstraction in acetonitrile then afforded the imido complex **46c** as a result of intramolecular hydroboration of the CN triple bond (Scheme 15). The process is amenable to catalysis using excess of $NaBH_4$ and 1 equiv. of NaO*t*Bu (Table 1). A broad variety of electron-poor and electron rich-aromatic nitriles were thereby reduced into primary amines using 5 mol% of **46c**. With electron-rich heterocycles, hydration instead of hydrogenation is observed and amides are obtained.

9.4.2.2 Cooperative activation of H_2 and silanes across M→B and transfer to alkenes/ketones[47]

Further progress in the Lewis acid assisted activation of H_2 was achieved by Harman and Peters[48] with bisphosphine-borane Ni complexes. An ideal balance between accessibility, stability and reactivity was met in the diamagnetic Ni(0) complex **47a**. Obtained by

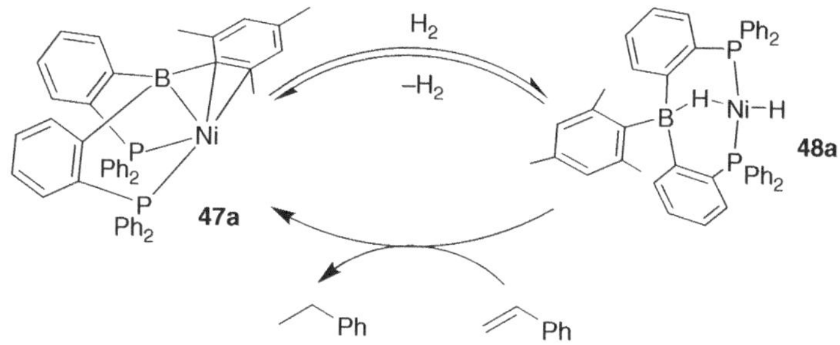

Scheme 16 *Reduction of styrene catalyzed by bisphosphine-borane Ni complex*

reduction of the corresponding Ni(I) bromide complex, it adopts a pyramidal geometry with η^3-BC$_{ipso}$C$_{ortho}$ coordination of the central BMes moiety and rapidly reacts with H_2 (Scheme 16). The reaction is equilibrated (K_{obs} ~ 5 under 1 atm at 25°C) and reversible. According to multi-nuclear NMR spectroscopy, the ensuing dihydrido nickel species **48a** adopts square-planar geometry and *trans* arrangement, with one hydride engaged in a Ni—H—B bridging interaction. The Mes group at boron proved to be critical. When it is replaced by a Ph ring, the Ni complex, which binds a THF molecule and features strong η^2-BC$_{ipso}$ coordination, no longer reacts with H_2 even after days at 60°C. Complex **48a** readily transfers H_2 and converts styrene into ethylbenzene. The process is amenable to catalysis and allows for efficient hydrogenation under mild conditions (complete reaction within 1 h with 1 mol% catalytic loading).

Soon after this discovery, Zeng and Sakaki[49] studied the whole catalytic cycle computationally (Scheme 17). Activation of H_2 was confirmed to give a *trans* dihydrido Ni complex stabilized by Ni—H—B bridging interaction. In the initial stage of the reaction, H_2 coordinates to Ni in a side-on fashion. The formation of the square-planar complex **48a** releases the strain associated with η^3-BCC coordination and this plays an important role in the exothermicity of H_2 activation ($\Delta G = -4.2$ kcal/mol from **47a**). Then styrene coordinates to Ni and inserts into the terminal Ni—H bond. The resulting alkyl complex **49a** is stabilized by β-agostic interaction and further evolves by reductive elimination, with cleavage of the Ni—H—B bridge and regeneration of complex **47a**. The active role played by the borane site was further emphasized by considering a related boron-free system (with a CHMes central moiety).

Further insight into the mechanism of H_2 activation and into the role of the borane was gained with the bisphosphine-borane featuring *i*Pr groups at P and a Ph ring at boron.[50] The corresponding Ni(0) complex **47b** was isolated as an end-on N_2 adduct with relatively strong η^2-B,C$_{ipso}$ interaction (Scheme 18). In this case, reaction with H_2 to give the *trans* hydrido-borohydrido Ni complex **48b** is quantitative but requires hours at room temperature. This enables NMR characterization of the side-on H_2 complex **50b** (the first σ-H_2 complex of Ni to be authenticated) as an intermediate towards complete H_2 activation. More insight into the mechanism of H_2 oxidative addition across Ni → B was obtained from kinetic and computational studies. Accordingly, cooperation between the nickel center and the borane moiety significantly lowers the activation barrier for H_2 cleavage (by 9 kcal/mol). This situation is reminiscent, albeit

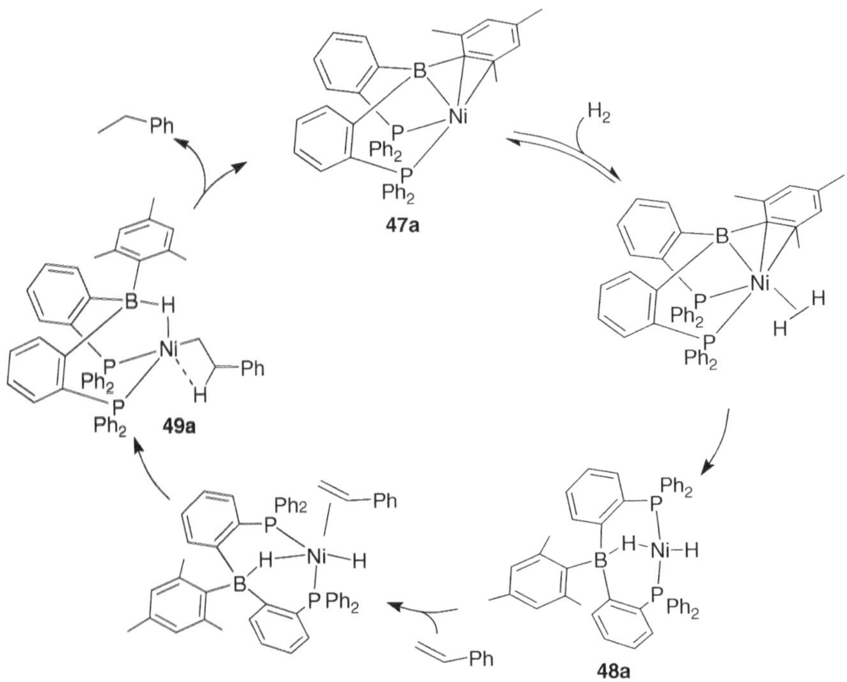

Scheme 17 *Catalytic cycle computed for the Ni–borane cooperative hydrogenation of styrene*

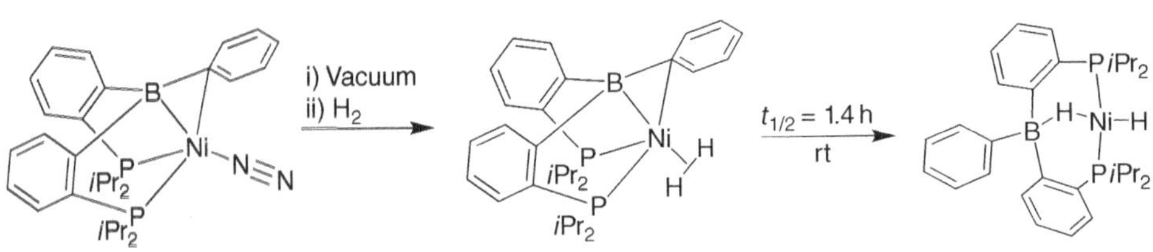

Scheme 18 *Preparation of trans hydrido-borohydrido Ni complex* ***48b***

electronically reverse, to that encountered upon activation of H_2 across M—X bonds (X being an element with a lone pair such as N and O).[51]

Such cooperativity between Ni and borane was then extended to the activation of Si—H bonds and applied to catalytic hydrosilylation (Scheme 19).[52] Complex **47a** rapidly reacts with phenylsilane and diphenylsilane at room temperature to give the borohydrido-silyl complexes analog to **48a**. Once activated, the silane can be transferred to aldehydes and a range of *para*-substituted benzaldehydes could be catalytically reduced into the corresponding silyl ethers under mild conditions (5 mol% of **47a**, room temperature). Although it is not possible to provide a comprehensive picture of the catalytic cycle at this stage, mechanistic studies support the intermediacy of a borohydrido-siloxyalkyl Ni complex **51** (resulting from the insertion of the aldehyde into the Ni–Si bond).

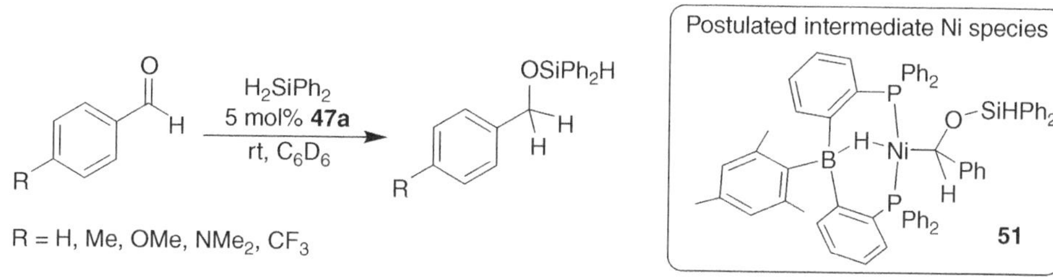

Scheme 19 *Catalytic hydrosilylation of ketones*

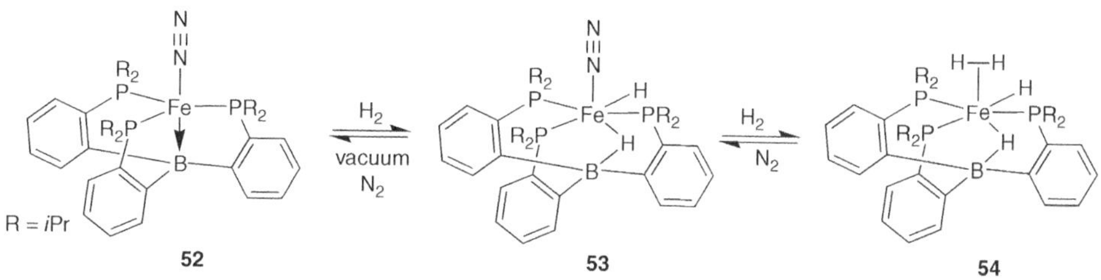

Scheme 20 *Reversible activation of H_2 across Fe→B*

Table 2 *Catalytic hydrogenation of alkenes and alkynes mediated by **52**[a]*

Substrate	Time (h)	TOF (h^{-1})
Ethylene	144	15
Styrene	2	0.27
Phenylacetylene	4	0.16

[a]Conditions: substrate/catalyst=30, room temperature, 1 atm H_2

Another example of bifunctional metal-borane catalysis was reported with triphosphine-borane iron complexes.[53] Despite the rigidity associated with the cage structure, activation of H_2 proceeds with cleavage of the Fe→B interaction and replacement with a Fe–H–B bridge (Scheme 20). Under excess H_2, the dinitrogen co-ligand at iron is displaced to give the corresponding dihydrogen complex **52**. The process is reversible and **54** gives back **53** and then **52** when exposed to vacuum and then N_2. The borane moiety not only stabilizes the dihydrido species but it also acts as a hydride shuttle and promotes the hydrogenation of alkenes and alkynes. Indeed, complexes **52–54** efficiently catalyze the conversion of ethylene into ethane and of styrene/phenylacetylene into ethylbenzene (Table 2).

9.4.2.3 Lewis acid enhancement upon π-coordination of arylboranes, application to the catalytic hydrogenation of imines

Besides cooperative activation of H_2 across M→B, bisphosphine-borane complexes have been used to mediate H_2 activation and transfer in a FLP-type manner.[54] Starting from $[RuCl_2(PPh_3)_3]$, the cationic ruthenium complex **55** was readily prepared by phosphine displacement and chloride abstraction. The phenyl group at boron is η^6-coordinated to Ru

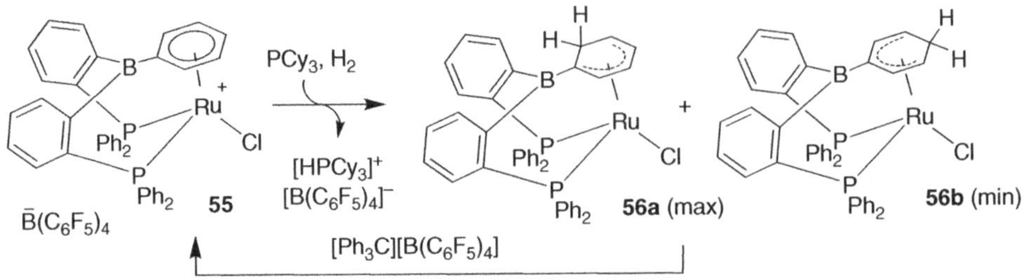

Scheme 21 *Heterolytic cleavage of H_2 upon reaction with bisphosphine-borane Ru complex **55***

and despite the presence of a free borane moiety, complex **55** behaves as a *C*-centered Lewis acid upon treatment with H_2 and bulky phosphines. A hydride is delivered at the *ortho* or *para* positions of the Ph ring to afford neutral complexes **56a,b** (Scheme 21). The cationic complex **55** can be recovered by hydride abstraction with the trityl cation and similarly to $B(C_6F_5)_3$, it was shown to catalyze the hydrogenation of imines. This requires high H_2 pressure (102 atm), but complete conversions could be achieved within a few hours at room temperature using 1–5 mol% catalyst loadings.

9.4.3 Activation/functionalization of N_2 and CO

9.4.3.1 From CO to C_2 fragments

Introduction of a Lewis acid in the second coordination sphere of transition metals was shown early on to facilitate migratory insertion of CO.[55] In the context of syngas conversion into useful organic compounds, pendant boranes were also envisioned to act as hydride acceptors to promote CO reduction at transition metals.[56] For this purpose, Labinger and Bercaw investigated the reaction of the bis(phosphine-borane) Rh complex **57** with hydride donors such as $NaBEt_3H$ or $[PtH(dmpe)_2]^+$ (Scheme 22). The Lewis acid moiety facilitates hydride transfer and promotes C—C coupling to give the anionic boroxy(boroxymethyl)carbene complex **58**. It is unlikely to make such a process catalytic, given the strength of B—O bonds, but these results nicely illustrate the ability of pendant boranes to act as hydride shuttles and to promote thereby reductive coupling of CO.

Another process involving reductive coupling of CO was discovered with bisphosphine-borane iron complexes (Scheme 23).[57] A series of carbonyl complexes were prepared, substantiating once again the ability of the ambiphilic ligand to accommodate different electronic configurations. The borane moiety is engaged in η^2-BC coordination in the neutral Fe(0) complex **59**, while the corresponding mono and dianionic complexes **60** feature two-center Fe→B interactions. Reduction of **60** with excess potassium following by treatment with trimethylsilyl triflate results in double *O*-silylation and affords the novel dicarbyne complex **61** in which one of the Fe≡C(OSiMe$_3$) carbyne moiety interacts with the boron center. C—C coupling does not occur spontaneously but is readily induced by addition of H_2 at room temperature to give the bis-silylated enol ether derivative **62** with complete Z selectivity. The overall process stands as a rare example of reductive coupling and functionalization of CO promoted by iron.

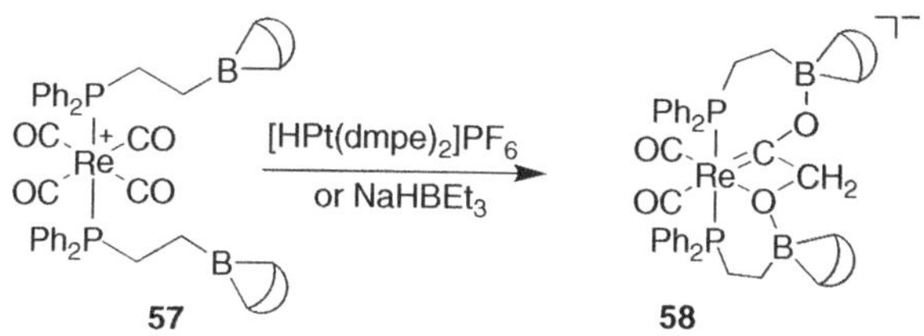

Scheme 22 *Reduction and coupling of CO assisted by a pendant borane at Re*

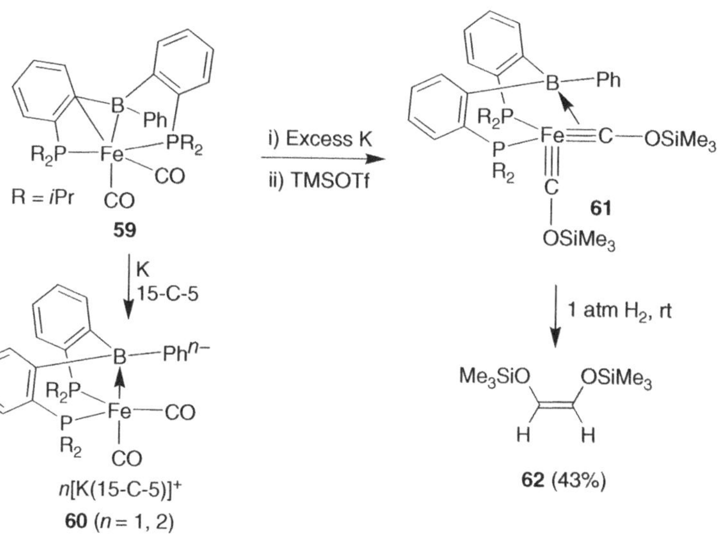

Scheme 23 *Reductive coupling and functionalization of CO at bisphosphine-borane Fe complexes*

9.4.3.2 *Activation/functionalization of N_2*

Low-valent Fe complexes of bisphosphine-boranes are also interesting as nitrogenase models.[58] Reduction of the Fe(I) bromide complex **63a** under N_2 atmosphere gives the binuclear N_2-bridged complex **63b** (Scheme 24). In the presence of 1,2-bis(chlorodimethylsilyl)ethane, the N_β atom undergoes bis-silylation to give the amino-imido complex **63c**. The central borane moiety participates in coordination via η^2-BC (**63a**) or η^3-BCC (**63b,c**) coordination. Subsequent reaction of **63c** with phenylsilane occurs via silylation of N_α and cleavage of the BPh/Fe interaction. Such Si–H bond activation parallels that observed across Ni→B and takes advantage of the flexibility of the borane moiety (the ensuing trisilyl-hydrazido complex **64** is stabilized by Fe–H–B bridging coordination). Triphosphine-borane Fe complexes were also studied in this context and mono as well as bis-silylation reactions of N_β were observed.[58b]

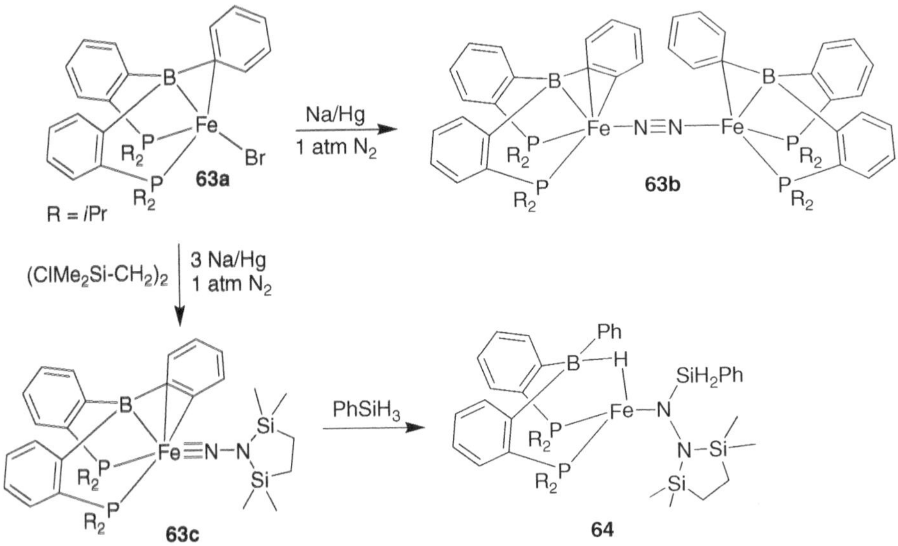

Scheme 24 *Functionalization of N_2 at bisphosphine-borane Fe complexes*

9.4.3.3 From N_2 to NH_3[59]

These reactivities encouraged the investigation of N_2 reduction into NH_3 and spectacular results were obtained with triphosphine-borane Fe complexes. Various *N*-bound Fe complexes **65** relevant to the catalytic reduction of N_2 into NH_3 were prepared.[60] Most significant are the protonation of the [Fe]–NH_2 complex into [Fe]–NH_3^+ and the reductive displacement of NH_3 by N_2, which are the final steps proposed to account for the fixation/reduction of N_2 at Fe (Scheme 25). This stoichiometric reaction sequence was adapted into a catalytic transformation.[61] Treatment of the anionic complex **65d** with excess H^+ and KC_8 at –78°C results in NH_3 production. Under these conditions, up to 7 NH_3 molecules were generated per Fe complex and more than 40% of the furnished protons were delivered to N_2. The borane moiety plays a major role in this process, being capable of accommodating the various [Fe(N_xH_y)] species involved in the catalytic cycle due to its versatile coordination properties. This is further supported by the inability of the related triphosphine-silyl complex to promote the reduction of N_2 under similar conditions.

9.5 Conclusions and outlook

The research on phosphine-boranes and related ambiphilic ligands is still in its infancy, but their coordination properties are particularly appealing and open interesting possibilities in reactivity. Depending on the role of the Lewis acid, four different situations can be distinguished:

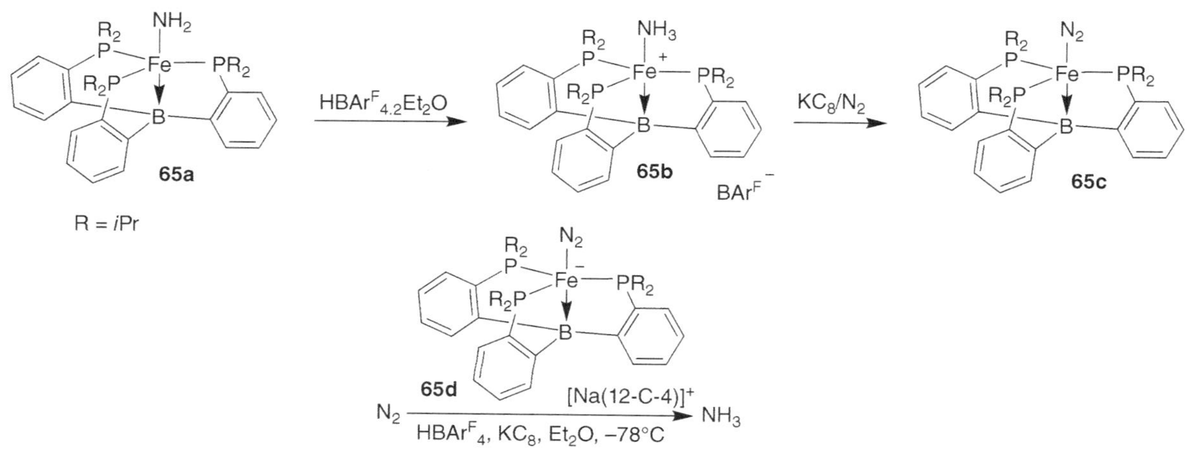

Scheme 25 *Catalytic reduction of N_2 into NH_3 at a triphosphine-borane Fe complex*

1. The Lewis acid may not participate in coordination and remain pendant. As such, it can play the role of an anchor for incoming substrates and promote their reaction with metal fragments.
2. Lewis acids may act as σ-acceptor ligands. This type of metal → ligand interaction no longer appears as a chemical curiosity but represents a new way to modulate the properties of metal fragments. Recent achievements have highlighted the flexibility and lability of such M → LA interactions. Most promising is the cooperative activation of small molecules that may occur across such M → LA interactions, as clearly evidenced experimentally and computationally with H_2 for example.
3. Lewis acid may engage in M—X → LA bridging coordination (X: co-ligand such as Cl, Br, H, ...). This behavior may activate metal fragments, as proposed in the case of Ni-catalyzed dehydrogenative coupling of phenylsilane.
4. Lewis acid may activate M—X bonds with complete abstraction of the X co-ligand, leading to zwitterionic complexes. This behavior may represent a valuable alternative to the use of external Lewis acids, as recently illustrated for PAl gold complexes.[62]

The structural modularity of ambiphilic ligands will undoubtedly be extended further in the future. New bonding situations and reactivity patterns involving appended Lewis acids will also certainly be discovered. In particular, important developments are expected in several directions of major catalytic relevance such as substrate anchoring, activation of inert bonds by metal–ligand cooperation, in situ generation of zwitter-ionic complexes or atom/group shuttling.

References

[1] (a) G. Bouhadir, A. Amgoune, D. Bourissou, *Adv. Organomet. Chem.* **2010**, *58*, 1–106; (b) A. Amgoune, G. Bouhadir, D. Bourissou, *Top. Curr. Chem.* **2013**, *334*, 281–311; (c) I. Kuzu, I. Krummenacher, J. Meyer, F. Armbruster, F. Breher, *Dalton Trans.* **2008**, 5836–5865; (d) F. G. Fontaine, J. Boudreau, M. H. Thibault, *Eur. J. Inorg. Chem.* **2008**, *2008*, 5439–5454; (e) H. Kameo, H. Nakazawa, *Chem. Asian J.* **2013**, *8*, 1720–1734.

[2] (a) D. W. Stephan, G. Erker, *Angew. Chem. Int. Ed.* **2010**, *49*, 46–76; (b) D.W. Stephan, G. Erker (eds), in *Frustrated Lewis Pairs I*, Springer, Berlin, **2013**, pp. 85–110.

[3] Cage Pt and Au complexes were described first, see: S. Bontemps, G. Bouhadir, W. Gu, M. Mercy, C. H. Chen, B. Foxman, L. Maron, O. Ozerov, D. Bourissou, *Angew. Chem. Int. Ed.* **2008**, *47*, 1481–1484.

[4] J. Vergnaud, T. Ayed, K. Hussein, L. Vendier, M. Grellier, G. Bouhadir, J. C. Barthelat, S. Sabo-Etienne, D. Bourissou, *Dalton Trans.* **2007**, 2370–2372.

[5] (a) S. Bontemps, G. Bouhadir, K. Miqueu, D. Bourissou, *J. Am. Chem. Soc.* **2006**, *128*, 12056–12057; (b) M. W. P. Bebbington, S. Bontemps, G. Bouhadir, D. Bourissou, *Angew. Chem., Int. Ed.* **2007**, *46*, 3333–3336; (c) S. Bontemps, G. Bouhadir, D. C. Apperley, P. W. Dyer, K. Miqueu, D. Bourissou, *Chem. Asian J.* **2009**, *4*, 428–435.

[6] (a) S. Bontemps, H. Gornitzka, G. Bouhadir, K. Miqueu, D. Bourissou, *Angew. Chem. Int. Ed.* **2006**, *45*, 1611–1614; (b) S. Bontemps, G. Bouhadir, P. W. Dyer, K. Miqueu, D. Bourissou, *Inorg. Chem.* **2007**, *46*, 5149–5151.

[7] (a) P. Spies, G. Erker, G. Kehr, K. Bergander, R. Froehlich, S. Grimme, D. W. Stephan, *Chem. Commun.* **2007**, 5072–5074; (b) A. Fischbach, P. R. Bazinet, R. Waterman, T. D. Tilley, *Organometallics* **2008**, *27*, 1135–1139; (c) J. Vergnaud, M. Grellier, G. Bouhadir, L. Vendier, S. Sabo-Etienne, D. Bourissou, *Organometallics* **2008**, *27*, 1140–1146; (d) T. G. Ostapowicz, C. Merkens, M. Hölscher, J. Klankermayer, W. Leitner, *J. Am. Chem. Soc.* **2013**, *135*, 2104–2107.

[8] H. H. Karsch, A. Appelt, F. H. Koehler, G. Mueller, *Organometallics* **1985**, *4*, 231–238.

[9] (a) M. Sircoglou, G. Bouhadir, N. Saffon, K. Miqueu, D. Bourissou, *Organometallics* **2008**, *27*, 1675–1678; (b) M. Sircoglou, N. Saffon, K. Miqueu, G. Bouhadir, D. Bourissou, *Organometallics* **2013**, *32*, 6780–6784.

[10] (a) P. Gualco, T. P. Lin, M. Sircoglou, M. Mercy, S. Ladeira, G. Bouhadir, L. M. Perez, A. Amgoune, L. Maron, F. P. Gabbai, D. Bourissou, *Angew. Chem. Int. Ed.* **2009**, *48*, 9892–9895; (b) P. Gualco, M. Mercy, S. Ladeira, Y. Coppel, L. Maron, A. Amgoune, D. Bourissou, *Chem. Eur. J.* **2010**, *16*, 10808–10817; (c) C. R. Wade, F. P. Gabbaï, *Angew. Chem. Int. Ed.* **2011**, *50*, 7369–7372; (d) I. S. Ke, F. P. Gabbaï, *Inorg. Chem* **2013**, *52*, 7145–7151; (e) I. S. Ke, J. S. Jones, F. P. Gabbaï, *Angew. Chem. Int. Ed.* **2014**, *53*, 2633–2637; (f) C. Tschersich, C. Limberg, S. Roggan, C. Herwig, N. Ernsting, S. Kovalenko, S. Mebs, *Angew. Chem. Int. Ed.* **2012**, *51*, 4989–4992; (g) T. P. Lin, I. S. Ke, F. P. Gabbaï, *Angew. Chem. Int. Ed.* **2012**, *51*, 4985–4988; (h) T. P. Lin, F. P. Gabbaï, *Angew. Chem. Int. Ed.* **2013**, *52*, 3864–3868.

[11] F. G. Fontaine, D. Zargarian, *J. Am. Chem. Soc.* **2004**, *126*, 8786–8794.

[12] (a) S. Bontemps, M. Devillard, S. Mallet-Ladeira, G. Bouhadir, K. Miqueu, D. Bourissou, *Inorg. Chem* **2013**, *52*, 4714–4720; (b) J. Beckmann, E. Hupf, E. Lork, S. Mebs, *Inorg. Chem* **2013**, *52*, 11881–11888; (c) Y. F. Li, Y. Kang, S. B. Ko, Y. Rao, F. Sauriol, S. Wang, *Organometallics* **2013**, *32*, 3063–3068.

[13] (a) A. Boerner, J. Ward, K. Kortus, H. Kagan, *Tetrahedron: Asymmetry* **1993**, *4*, 2219–2228; (b) L. B. Fields, E. N. Jacobsen, *Tetrahedron: Asymmetry* **1993**, *4*, 2229–2240; (c) B. F. M. Kimmich, C. R. Landis, D. R. Powell, *Organometallics* **1996**, *15*, 4141–4146.

[14] S. Chikkali, S. Magens, D. Gudat, M. Nieger, I. Hartenbach, T. Schleid, *Eur. J. Inorg. Chem.* **2008**, 2207–2213.

[15] J. M. Burlitch, M. E. Leonowicz, R. B. Petersen, R. E. Hughes, *Inorg. Chem.* **1979**, *18*, 1097–1105.

[16] (a) R. B. King, *Adv. Chem. Ser.* **1967**, *62*, 203–220; (b) M. L. H. Green, *J. Organomet. Chem.* **1995**, *500*, 127–148.

[17] (a) G. Parkin, *Organometallics* **2006**, *25*, 4744–4747; (b) A. F. Hill, *Organometallics* **2006**, *25*, 4741–4743.

[18] (a) A. Amgoune, D. Bourissou, *Chem. Commun.* **2011**, *47*, 859–871; (b) J. Bauer, H. Braunschweig, R. D. Dewhurst, *Chem. Rev.* **2012**, *112*, 4329–4346; (c) H. Braunschweig, R. D. Dewhurst, A. Schneider, *Chem. Rev.* **2010**, *110*, 3924–3957.

[19] A. F. Hill, G. R. Owen, A. J. P. White, D. J. Williams, *Angew. Chem., Int. Ed.* **1999**, *38*, 2759–2761.

[20] S. Bontemps, M. Sircoglou, G. Bouhadir, H. Puschmann, J. A. K. Howard, P. W. Dyer, K. Miqueu, D. Bourissou, *Chem. Eur. J.* **2008**, *14*, 731–740.

[21] M. Sircoglou, S. Bontemps, M. Mercy, N. Saffon, M. Takahashi, G. Bouhadir, L. Maron, D. Bourissou, *Angew. Chem., Int. Ed.* **2007**, *46*, 8583–8586.

[22] M. Sircoglou, S. Bontemps, M. Mercy, K. Miqueu, S. Ladeira, N. Saffon, L. Maron, G. Bouhadir, D. Bourissou, *Inorg Chem* **2010**, *49*, 3983–3990.

[23] D. J. H. Emslie, B. E. Cowie, K. B. Kolpin, *Dalton Trans.* **2012**, *41*, 1101–1117.

[24] M. Sircoglou, M. Mercy, N. Saffon, Y. Coppel, G. Bouhadir, L. Maron, D. Bourissou, *Angew. Chem., Int. Ed.* **2009**, *48*, 3454–3457.

[25] M. Sircoglou, S. Bontemps, G. Bouhadir, N. Saffon, K. Miqueu, W. Gu, M. Mercy, C. H. Chen, B. M. Foxman, L. Maron, O. V. Ozerov, D. Bourissou, *J. Am. Chem. Soc.* **2008**, *130*, 16729–16738.

[26] M. E. Moret, J. C. Peters, *Angew. Chem. Int. Ed.* **2011**, *50*, 2063–2067.

[27] M. E. Moret, L. Zhang, J. C. Peters, *J. Am. Chem. Soc.* **2013**, *135*, 3792–3795.

[28] D. L. M. Suess, C. Tsay, J. C. Peters, *J. Am. Chem. Soc.* **2012**, *134*, 14158–14164.

[29] (a) H. Kameo, Y. Hashimoto, H. Nakazawa, *Organometallics* **2012**, *31*, 4251–4258; (b) H. Kameo, Y. Hashimoto, H. Nakazawa, *Organometallics* **2012**, *31*, 3155–3162.

[30] E. J. Derrah, M. Sircoglou, M. Mercy, S. Ladeira, G. Bouhadir, K. Miqueu, L. Maron, D. Bourissou, *Organometallics* **2011**, *30*, 657–660.

[31] P. A. Rudd, S. Liu, L. Gagliardi, V. G. Young, C. C. Lu, *J. Am. Chem. Soc.* **2011**, *133*, 20724–20727.

[32] (a) D. J. H. Emslie, B. E. Cowie, S. R. Oakley, N. L. Huk, H. A. Jenkins, L. E. Harrington, J. F. Britten, *Dalton Trans.* **2012**, *41*, 3523–3535; (b) B. E. Cowie, D. J. H. Emslie, H. A. Jenkins, J. F. Britten, *Inorg. Chem* **2010**, *49*, 4060–4072.

[33] H. Kameo, H. Nakazawa, *Organometallics* **2012**, *31*, 7476–7484.

[34] (a) M. H. Thibault, J. Boudreau, S. Mathiotte, F. Drouin, O. Sigouin, A. Michaud, F. G. Fontaine, *Organometallics* **2007**, *26*, 3807–3815; (b) J. Boudreau, F. G. Fontaine, *Organometallics* **2011**, *30*, 511–519.

[35] M. W. P. Bebbington, S. Bontemps, G. Bouhadir, M. J. Hanton, R. P. Tooze, H. van Rensburg, D. Bourissou, *New J. Chem.* **2010**, *34*, 1556–1559.

[36] Another common feature of *o*-phenylene phosphine-boranes and biaryl-phosphines is their resistance to oxidation by O_2: (a) T. E. Barder, S. L. Buchwald, *J. Am. Chem. Soc.* **2007**, *129*, 5096–5101; (b) S. Porcel, G. Bouhadir, N. Saffon, L. Maron, D. Bourissou, *Angew. Chem., Int. Ed.* **2010**, *49*, 6186–6189.

[37] R. Malacea, N. Saffon, M. Gomez, D. Bourissou, *Chem. Commun.* **2011**, *47*, 8163–8165.

[38] T. E. Barder, S. D. Walker, J. R. Martinelli, S. L. Buchwald, *J. Am. Chem. Soc.* **2005**, *127*, 4685–4696.

[39] R. Malacea, F. Chahdoura, M. Devillard, N. Saffon, M. Gomez, D. Bourissou, *Adv. Synth. Catal.* **2013**, *355*, 2274–2284.

[40] S. Xu, F. Haeffner, B. Li, L. N. Zakharov, S. Y. Liu, *Angew. Chem. Int. Ed.* **2014**, *53*, 6795–6799.

[41] G. R. Owen, *Chem. Soc. Rev.* **2012**, *41*, 3535–3546.

[42] I. R. Crossley, A. F. Hill, *Dalton Trans.* **2008**, 201–203.

[43] N. Tsoureas, M. F. Haddow, A. Hamilton, G. R. Owen, *Chem. Commun.* **2009**, 2538–2540.

[44] (a) N. Tsoureas, T. Bevis, C. P. Butts, A. Hamilton, G. R. Owen, *Organometallics* **2009**, *28*, 5222–5232; (b) N. Tsoureas, Y.-Y. Kuo, M. F. Haddow, G. R. Owen, *Chem. Commun.* **2011**, *47*, 484–486.

[45] N. Tsoureas, G. R. Owen, A. Hamilton, A. G. Orpen, *Dalton Trans.* **2008**, 6039–6044.

[46] Z. Lu, T. J. Williams, *Chem. Commun.* **2014**, *50*, 5391–393.

[47] M. Devillard, G. Bouhadir, D. Bourissou, *Angew. Chem., Int. Ed.* **2015**, *54*, 730–732.

[48] W. H. Harman, J. C. Peters, *J. Am. Chem. Soc.* **2012**, *134*, 5080–5082.

[49] G. Zeng, S. Sakaki, *Inorg. Chem* **2013**, *52*, 2844–2853.

[50] W. H. Harman, T. P. Lin, J. C. Peters, *Angew. Chem. Int. Ed.* **2014**, *53*, 1081–1086.

[51] R. Noyori, T. Ohkuma, *Angew. Chem. Int. Ed.* **2001**, *40*, 40–73.

[52] S. N. MacMillan, W. Hill Harman, J. C. Peters, *Chem. Sci.* **2014**, *5*, 590–597.
[53] H. Fong, M. E. Moret, Y. Lee, J. C. Peters, *Organometallics* **2013**, *32*, 3053–3062.
[54] M. P. Boone, D. W. Stephan, *J. Am. Chem. Soc.* **2013**, *135*, 8508–8511.
[55] (a) J. A. Labinger, J. S. Miller, *J. Am. Chem. Soc.* **1982**, *104*, 6856–6858; (b) J. A. Labinger, J. N. Bonfiglio, D. L. Grimmett, S. T. Masuo, E. Shearin, J. S. Miller, *Organometallics* **1983**, *2*, 733–740; (c) D. L. Grimmett, J. A. Labinger, J. N. Bonfiglio, S. T. Masuo, E. Shearin, J. S. Miller, *J. Am. Chem. Soc.* **1982**, *104*, 6858–6859; (d) D. L. Grimmett, J. A. Labinger, J. N. Bonfiglio, S. T. Masuo, E. Shearin, J. S. Miller, *Organometallics* **1983**, *2*, 1325–1332.
[56] (a) A. J. M. Miller, J. A. Labinger, J. E. Bercaw, *J. Am. Chem. Soc.* **2008**, *130*, 11874–11875; (b) A. J. M. Miller, J. A. Labinger, J. E. Bercaw, *Organometallics* **2010**, *29*, 4499–4516; (c) A. J. M. Miller, J. A. Labinger, J. E. Bercaw, *Organometallics* **2011**, *30*, 4308–4314; (d) A. J. M. Miller, J. A. Labinger, J. E. Bercaw, *J. Am. Chem. Soc.* **2010**, *132*, 3301–3303; (e) N. M. West, A. J. M. Miller, J. A. Labinger, J. E. Bercaw, *Coord. Chem. Rev.* **2011**, *255*, 881–898.
[57] D. L. M. Suess, J. C. Peters, *J. Am. Chem. Soc.* **2013**, *135*, 12580–12583.
[58] (a) D. L. M. Suess, J. C. Peters, *J. Am. Chem. Soc.* **2013**, *135*, 4938–4941; (b) M. E. Moret, J. C. Peters, *J. Am. Chem. Soc.* **2011**, *133*, 18118–18121.
[59] H. Broda, F. Tuczek, *Angew. Chem. Int. Ed.* **2014**, *53*, 632–634.
[60] J. S. Anderson, M. E. Moret, J. C. Peters, *J. Am. Chem. Soc.* **2013**, *135*, 534–537.
[61] J. S. Anderson, J. Rittle, J. C. Peters, *Nature* **2013**, *501*, 84–87.
[62] M. Devillard, E. Nicolas, A. W. Ehlers, J. Backs, S. Mallet-Ladeira, G. Bouhadir, J. C. Slootweg, W. Uhl, D. Bourissou, *Chem. Eur. J.* **2015**, *21*, 74–79.

10

Ligand Design in Enantioselective Ring-opening Polymerization of Lactide

Kimberly M. Osten, Dinesh C. Aluthge, and Parisa Mehrkhodavandi

Department of Chemistry, The University of British Columbia, 2036 Main Mall, Vancouver, British Columbia, Canada V6T 1Z1

10.1 Introduction

The development of commercially useful polymers in the early 20th century ushered in an era where mass-produced, organo-polymeric materials have become a ubiquitous part of daily life.[1] Sixty years after the Nobel prize-winning discovery by Ziegler and Natta,[2] the scale of worldwide polyolefin production is massive. The current estimated annual global production of polyolefins is over 150 million metric tons.[3] However, the inherent chemical inertness of these substances causes them to persist in the environment centuries after they have been discarded.[4] The detrimental environmental impact of a man-made waste problem of this scale has generated an interest in commercially viable, biodegradable alternatives.[5]

The ring-opening polymerization (ROP) of cyclic esters can be used to generate biodegradable polyesters such as poly(caprolactone) (PCL), poly(hydroxybutyrate) (PHB), and poly(lactic acid) (PLA) in a controlled fashion. Of these polyesters PLA is the most widely used, for applications ranging from food packaging to automotive parts.[6]

Ligand Design in Metal Chemistry: Reactivity and Catalysis, First Edition. Edited by Mark Stradiotto and Rylan J. Lundgren.

The ROP of lactide affords high molecular weight PLA polymers with better control of the polymerization process relative to polycondensation. These advantages can be directly attributed to the fact that ROP can be a *living* polymerization process. Living polymerization is a chain-growth polymerization where chain termination is absent[7] and is characterized by a linear relationship between the monomer to initiator ratio and the experimental molecular weight, and narrow dispersity ($Đ_M$) [$Đ_M$ indicates the molecular weight distribution of the polymers and is the ratio between the weight average (M_w) and the number average molecular weights (M_n)].[8] With some systems in ROP, an exogenous chain transfer agent such as an alcohol can be added to grow multiple polymer chains – a process known as *immortal* polymerization.[9] This allows the use of lower catalyst loadings and can be used for polymer functionalization. These advantages make ROP a powerful technique in polymer synthesis.

10.1.1 Tacticity in PLA

Three possible stereoisomers of lactide (LA) exist; D-, L- and *meso*-lactide (Figure 1). A racemic mixture of D- and L-lactide is referred to as *rac*-lactide (*rac*-LA).[10] The stereochemistry of these monomers, when incorporated into a polymer chain, creates material with a certain stereocomplexity or tacticity.[11] The tacticity of a given PLA sample may be defined by two parameters: P_m [probability of forming adjacent stereocenters with the same chirality or a meso (m) linkage] and P_r [probability of

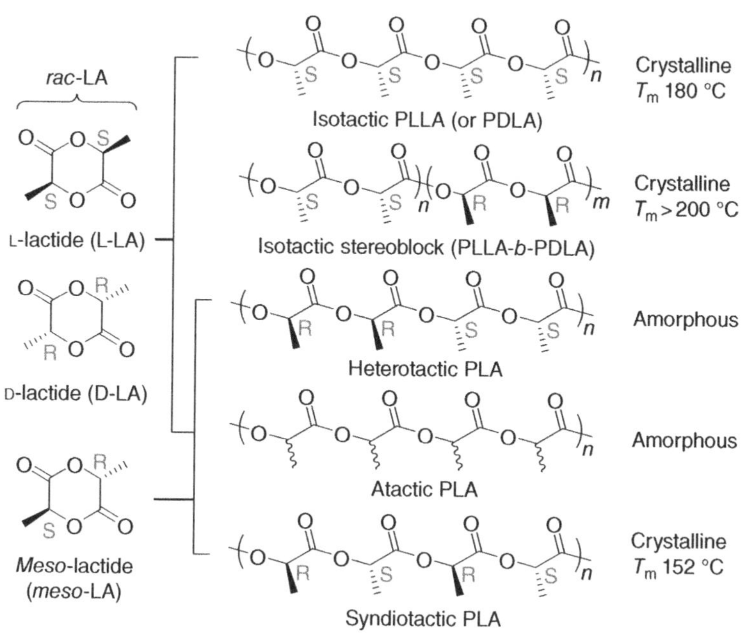

Figure 1 *Microstructures and properties of PLAs. (See insert for color/color representation of this figure)*

forming adjacent stereocenters with the opposite chirality or a racemic (r) linkage], which can be statistically calculated using ^{1}H and ^{13}C NMR spectroscopy. Hence, in the polymerization of *rac*-lactide a perfectly isotactic polymer sample will have $P_m = 1$ and a perfectly heterotactic sample will have $P_r = 1$. Alternatively, in the polymerization of *meso*-lactide a perfectly heterotactic polymer sample will have $P_m = 1$ and a perfectly syndiotactic polymer sample will have $P_r = 1$. To calculate the P_m and/or P_r value(s) of a PLA sample a homonuclear decoupled ^{1}H{^{1}H} NMR spectrum of the methine region of the material is obtained to remove the scalar coupling between the methyl and methine protons. For the polymerization of *rac*-lactide, five distinguishable resonances corresponding to mmm, mmr(rmm), rmm(mmr), mrm and rmr tetrad sequences (four adjacent stereocenters) can be observed. The relative integrations of these resonances are substituted into a set of equations derived from Bernoullian statistics to determine P_m and P_r values.[12]

The tacticity of the polymer chain has an enormous impact on the bulk properties of the resulting material.[13] This fact becomes apparent when taking into account the melting points of PLA polymers with differing tacticities (Figure 1). PLA with irregular microstructure, known as atactic PLA, is amorphous and is not very useful for most applications.[14] Heterotactic PLA, which is formed through the incorporation of alternating single monomer units of D-lactide and L-lactide, is also amorphous although the polymer chains contain clear stereoregularity.[15] Heterotactic PLA can also be generated from *meso*-lactide.[12] Isotactic (PLLA/PDLA) and syndiotactic PLA formed from D-, L-, or *meso*-lactide, respectively, are crystalline and have melting temperatures (T_m) of 180 and 152 °C, respectively.[12, 13]

Lactide derived from natural sources is composed of >90% L-lactide with small amounts of D-lactide and *meso*-lactide.[16] Polymerization of this mixture with a simple homoleptic catalyst such as tin(II) 2-ethylhexanoate, $Sn(Oct)_2$, or aluminum isopropoxide, $Al(OiPr)_3$, produces isotactic PLA with $T_m = 165–170$ °C depending on the exact composition of the monomer feedstock.[6, 17] These initiators do not impart stereoselectivity during polymerization. Stereoblock PLA (PLLA-*b*-PDLA), where a single polymer chain contains two distinct regions with opposite stereochemistry, can be generated through coupling of homochiral chains,[18] the sequential addition of enantiopure D- and L-lactide,[19] or via the polymerization of mixtures of D- and L-lactide with a highly stereoselective catalyst.[12] Due to stereocomplex formation[20] PLLA-*b*-PDLA microstructures give higher melting points (>200 °C) than the optically pure isotactic polymers (PDLA/PLLA) and better mechanical properties that are commercially desirable.[21] Hence, the ability to control tacticity during the polymerization of racemic lactide represents one of the most important and fundamental aspects of this field.

10.1.2 Metal catalysts for the ROP of lactide

The generally accepted mechanism for the metal-catalyzed ROP of lactide is a coordination–insertion mechanism (Figure 2), which was proposed by Dittrich and Schulz.[22] In this mechanism, the lactide is activated after coordination to a metal center through the carbonyl oxygen. Then an initiator, such as an alkoxide, attacks the

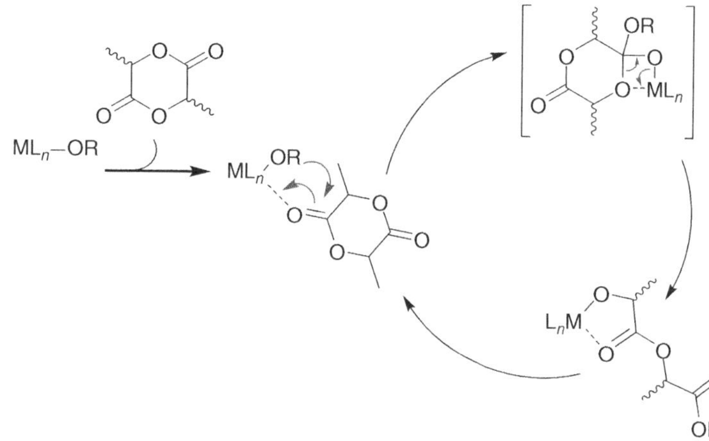

Figure 2 *Coordination–insertion mechanism for the polymerization of lactide by a metal complex with an alkoxide initiator*

carbonyl carbon and eventually leads to the ring-opening of the lactide ring and the formation of a new metal–alkoxide bond. This metal–alkoxide bond will act as the initiator for the incoming monomer to turn over the catalytic cycle.[15] The thermodynamic driving force for the polymerization reaction is the release of the −23 kJ/mol ring strain in the lactide molecule.[23]

Stereocontrol in polymerization can arise from either chain-end control, where the stereochemistry of the chain-end determines the configuration of the incoming monomer, or enantiomorphic site control, where the chirality of the ligand imparts stereoselectivity.[24] However, these two mechanisms are not mutually exclusive, and often act in concert in a given system.[25]

Numerous metal-based catalysts, bearing a diverse array of ligand architectures, as well as families of organocatalysts have been reported for the ROP of lactide.[15, 24c, 26] The simple homoleptic complex $Sn(Oct)_2$ catalyzes lactide polymerization both in solution and in the melt at temperatures >130 °C and is the most widely used catalyst for lactide polymerization industrially.[15] As previously mentioned, commercially available lactide with >90% L-LA polymerized with $Sn(Oct)_2$ will generate ~90% isotactic PLA. While this material has commercial applications, its thermal and mechanical properties are not suitable for many applications where polyolefins are typically used.[6, 27] One way to improve the polymer properties is by increasing the isotacticity or forming isotactic stereoblock PLA through stereoselective polymerization of L- and D-LA mixtures.[6] However, this cannot be achieved with $Sn(Oct)_2$ and other simple initiators.[15]

Since the bulk properties of PLA are highly dependent on the stereoregularity or the tacticity of the polymer,[27] the development of catalysts for *rac*-lactide (or *meso*-lactide) polymerization has been focused on achieving stereoselectivity.[24c, 26a, 28] Numerous organocatalysts such as *N*-heterocyclic carbenes (NHCs)[29] and phosphine-based compounds[30] have been investigated for the controlled ROP of lactide.

Organocatalysts are an attractive option due to their ability to generate PLA without residual metal contaminants; however, a highly stereoselective organocatalyst remains elusive.[8a] The vast majority of efforts on developing stereoselective catalysts have been focused on metal-based catalysts with diverse ligand designs.[15, 24c, 26]

10.1.3 Ligand design in the enantioselective polymerization of racemic lactide

The enantioselective metal-catalyzed polymerization of racemic lactide to form high melting stereoblock PLA remains a benchmark goal in the development of higher-value PLA material. A wide array of ligands with different donors and denticities to support metal-based initiators has been reported for the ROP of racemic lactide.[24c, 26a, 28] As commonly seen in asymmetric catalysis, diverse ligand architectures, with modular donor and stereoelectronic properties, allow for the development of a broad range of catalysts for a given reaction. Ligand modifications can alter the electrophilicity of the metal center as well as enhance or diminish the steric crowding within the catalyst. While examples of enantioselective catalysts for the ROP of lactide featuring bidentate ligands can be found in the literature, their control of tacticity is mainly based on chain-end control.[24b, 31] Tri- and tetradentate ligand supports are more widely used with success in metal-catalyzed ROP of lactide and these will be discussed in this chapter in some detail.[15, 24c, 26] This text is not a comprehensive review of the literature but instead seeks to highlight important contributions to the field of lactide polymerization within the broad classifications of tri- and tetradentate ligand architectures. The following discussion has been organized into sections by ligand type, starting with tridentate ligands (Sections 1.3.1–1.3.5) followed by tetradentate ligands (Sections 1.3.6–1.3.8) and finally ending with a more in depth discussion of important indium and zinc catalysts featuring tri- and tetradentate aminophenolate ligands developed by the Mehrkhodavandi group (Sections 2 and 3).

10.1.3.1 Tripodal homo- and heteroscorpionate ligands

Neutral and anionic tripodal homo- and heteroscorpionate ligands have been used successfully as supports for a variety of metals in the stereoselective ROP of *rac*-lactide. The homoscorpionate Mg complex, $[HB(3\text{-}^tBupz)_3]MgOEt$, featuring a bulky *tert*-butyl substituted tris(pyrazolyl)borate ligand can polymerize L-LA forming isotactic PLA without epimerization (Figure 3).[32] It undergoes the typical coordination insertion mechanism with acyl bond cleavage, as evidenced by the presence of ester end groups in the ^1H NMR spectra of the polymers. Controlled molecular weights and low $Đ_M$ (~1.2) are observed up to 1000 equiv. of monomer. Analogous Mg,[33] Zn[33] and Ca[34] complexes featuring both tris(pyrazolyl)borate and chiral tris(indazolyl)borate ligands with a variety of alkoxide and amide initiator groups have been reported (Figure 3). The selectivity of the tris(pyrazolyl)borate Ca complex (P_r ~ 0.9) is higher than the related Mg or Zn analogs (P_r ~ 0.5).[34] The selectivity is completely lost in complexes without sufficient steric bulk, and the lack of bulk leads to formation of homoleptic CaL_2 due to a rapid Schlenk equilibrium.[34b] These systems have been reviewed extensively.[35]

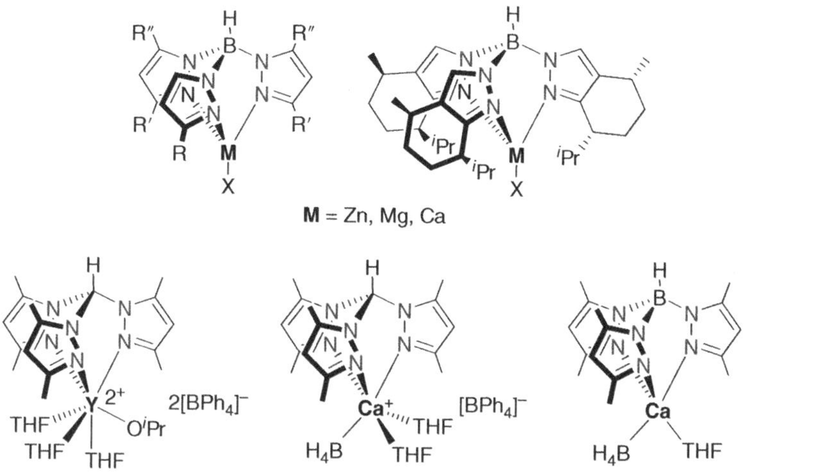

Figure 3 *Tris(pyrazolyl)borate, tris(indazolyl)borate and tris(pyrazolyl)methane complexes of Zn, Mg, Ca and Y*

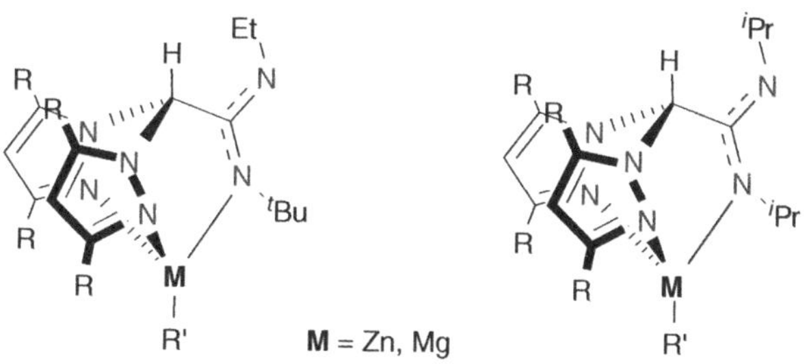

Figure 4 *Magnesium and zinc complexes bearing bis(pyrazolyl)methane ligands with amidinate pendant arms*

Yttrium[36] and Ca[37] cationic and zwitterionic complexes bearing tris(pyrazolyl) borate and the related tris(pyrazolyl)methane ligand systems have also been reported. The Y dicationic complex is much less active than the related neutral Mg and Ca complexes for the ROP of *rac*-LA and forms atactic PLA.[36] The formally zwitterionic Ca analog is heteroselective with a P_r value up to 0.80 at room temperature and up to 0.90 at lower temperatures (Figure 3).[37]

Heteroscorpionate ligands based on bis(pyrazolyl)methane systems with various pendant arms as supports for Mg,[38] Zn,[38b, 39] Al,[40] and rare-earth metals[41] have been reported. Mg and Zn complexes bearing amidinate-based ligands having either *iso*-propyl or *tert*-butyl/ethyl groups on the amidinate pendant arm and either methyl or *tert*-butyl pyrazolyl substituents have been reported (Figure 4).[38a, 38c, 39a, 42] Most of these complexes show no epimerization in the polymerization of L-LA, producing isotactic PLA with generally well-controlled molecular weights and low $Đ_M$s (1.04–1.2). There

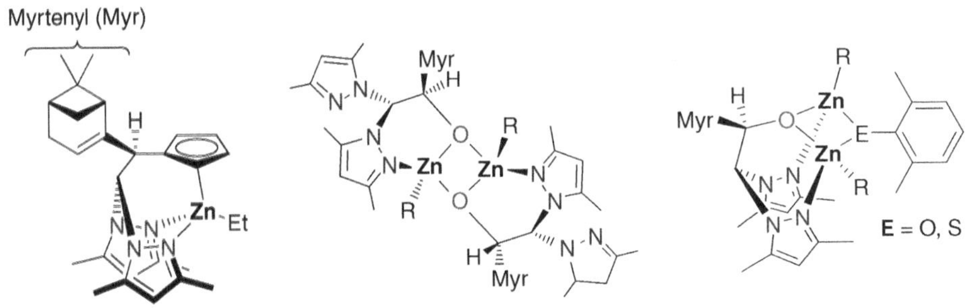

Figure 5 Enantiopure Zn complexes with bis(pyrazolyl)methane ligands bearing chiral myrtenyl-substituted cyclopentadienyl and alkoxy pendant arms

are stark contrasts in the activity of the systems based on the ligand substituents; the Mg analogs with the less bulky methyl substituted pyrazolyl groups are much less active than their bulkier analogs with *tert*-butyl substituents.[38a, 38c] The trends in the selectivity of these systems in the polymerization of *rac*-LA are related to their steric bulk, with all analogs featuring the bulkier *tert*-butyl pyrazolyl substituents showing a preference for the formation of heterotactic PLA (P_r=0.60–0.79) and all analogs with the less bulky methyl pyrazolyl substituents forming atactic PLA.[38a, 38c, 39a]

Other notable examples of catalysts bearing tripodal ligands containing cyclopentadienyl[38b, 39c] and alkoxy[39b, 39d] pendant groups are the enantiopure Zn complexes based on chiral myrtenyl-substituted arms, with the cyclopentadienyl version showing excellent isoselectivity in the polymerization of *rac*-LA (P_m=0.73 – 0.77) (Figure 5).[39c] Related dimeric Zn complexes featuring myrtenyl-substituted alkoxy pendant arms, with either alkyl or alkoxy/thioalkoxy initiators, show a shift in the selectivity from heterotactic (P_r=0.77) to isotactic (P_m=0.71–0.73).[39b, 39d] These are rare examples of Zn initiators capable of producing isotactic PLA from *rac*-LA.

10.1.3.2 Tridentate diamidoamino and related ligands

Variously substituted diamidoamino ligands have been used as supports in ROP catalysis for a range of metals in various oxidation states. A series of Al, Ga, and In compounds with tridentate diamidoamino ligands were reported in 1998 (Figure 6).[43] Of the large number of compounds reported, only the Al—CH_3 and Al—H compounds were active for the polymerization of *rac*-LA, converting 50 equiv. to 37 and 76% conversion in 5 and 7 days, respectively. Activation of the analogous cationic aluminum chloride complex with propylene oxide forms a reactive alkoxide in situ that can convert 46% of 50 equiv. of *rac*-LA to PLA in 5 days. The selectivity of these systems was not reported. Related Zn, Sm, and Sn compounds with these ligands are active for the copolymerization of lactide and glycolide.[44]

A series of Al,[45] In,[46] and Ti[47] complexes coordinated by a similar tridentate sulfonamide ligand system has been reported (Figure 7). The aluminum ethyl complex in the series is the least active and controlled, polymerizing 100 equiv. of *rac*-LA

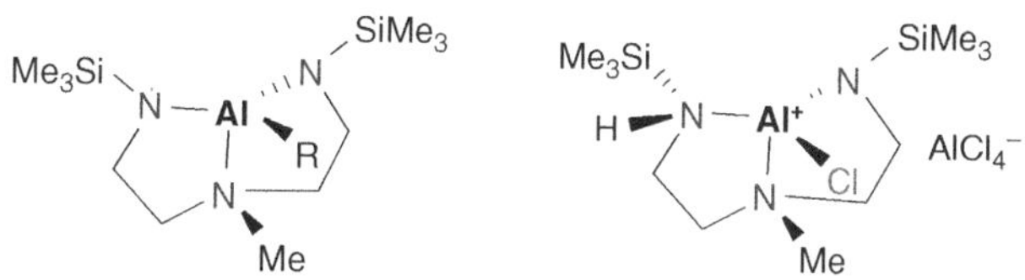

Figure 6 *Aluminum complexes featuring tridentate diamidoamino ligands.*[43]

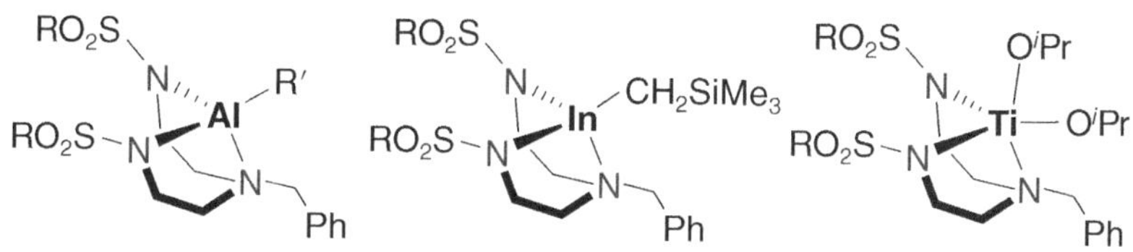

Figure 7 *Tridentate sulfonamide ligand-supported Al, In and Ti complexes*

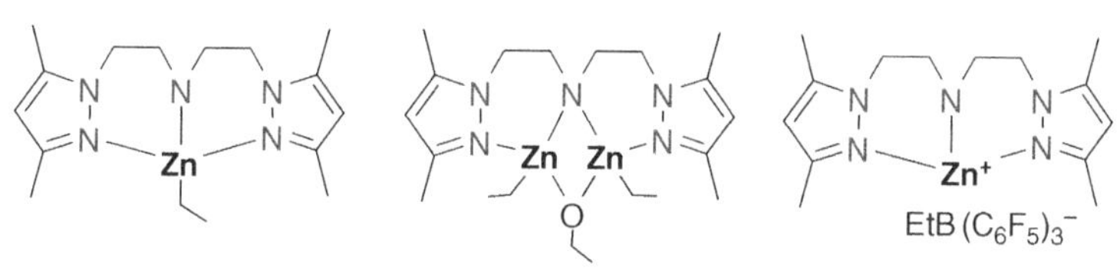

Figure 8 *Zinc complexes featuring tridentate bis(pyrazolyl)amine ligands*

(toluene, 70 °C) to only 20% conversion in 72 h, with higher than expected molecular weights and high $Đ_M$ (1.47). In comparison, the related aluminum alkoxide complexes reach >70% conversion in the same time frame with controlled molecular weights and low $Đ_M$ (1.12), although they produce only atactic PLA.[45] The related indium alkyl complex has a comparable rate under the same conditions.[46] The related titanium alkoxide complexes are more active than their Al or In counterparts; however, the catalysts become inactive at longer reaction times and high conversion of monomer cannot be reached.[47]

Related monometallic, dimetallic, and cationic Zn complexes supported by tridentate bis(pyrazolyl)amine ligands have been reported for the polymerization of *rac*-LA and methyl methacrylate (Figure 8).[48] All three complexes are active for the polymerization of *rac*-LA, reaching conversions of >90% in 30 h (unoptimized time) at room temperature. Whereas the ethoxide complex is slightly more active and better controlled than the ethyl and cationic complexes, it produces atactic PLA.

10.1.3.3 Tridentate diamidoether and linked bis(phenolate) ligands

Tridentate diamidoether (N,O,N) ligands have been used as supports for a series of Al and Ga complexes with amide and alkoxide initiators (Figure 9).[49] The Al "ate" complexes polymerize *rac*-LA with moderate heteroselectivity (P_r=0.65–0.67).[49a] In comparison, the related neutral alkoxy dimer requires more forcing conditions to reach high conversions and is isoselective (P_m=0.62).[49b] The related aluminum and gallium amido complexes are more active than the alkoxide and can generate PLA with P_m=0.70.[49b]

Several Ti complexes supported by bis(phenolate) ligands and bridged by chalcogen donors for use in the polymerization of several cyclic esters, including lactide, have been reported (Figure 10).[50] Among the reported complexes, the polymerization of lactide was investigated with only the Te- and S-bridged species. The Te-bridged complex was active for the polymerization of L-LA in toluene, anisole and dioxane, producing isotactic PLA (no epimerization) with controlled molecular weights and low $Đ_M$ (~1.1). The controlled nature of the reactivity of the complex also allowed for the copolymerization of L-LA and ε-caprolactone.

Tridentate bis(phenolate) ligands bridged by NHCs have been reported as supports for Ti and Zr complexes (Figure 10).[51] Both the Zr and Ti complexes show relatively controlled molecular weights with low $Đ_M$ (<1.1); however, the Ti complex produces atactic PLA whereas the Zr analog produces highly heterotactic PLA with a P_r>0.95.[51]

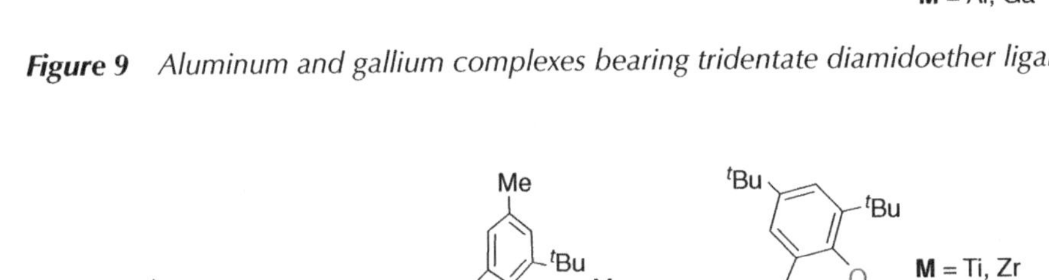

Figure 9 *Aluminum and gallium complexes bearing tridentate diamidoether ligands*

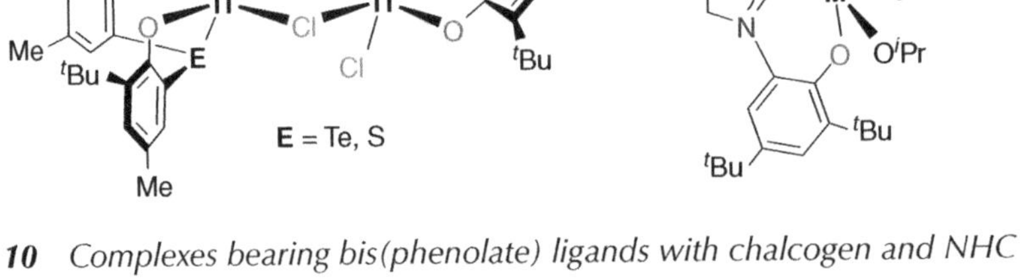

Figure 10 *Complexes bearing bis(phenolate) ligands with chalcogen and NHC bridges*

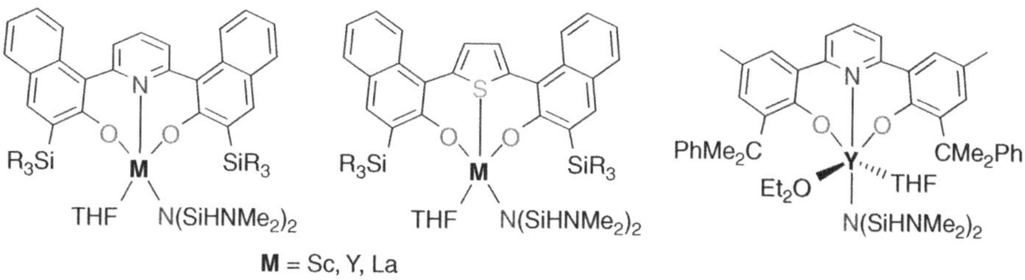

Figure 11 *Group 3 complexes bearing bis(naphtholate) and bis(phenolate) ligands bridged by pyridine or thiophene groups*

The Zr catalyst is versatile and able to polymerize unpurified commercial *rac*-LA, carry out melt polymerization of *rac*-LA, and catalyze the immortal ROP of LA with BnOH with little change in performance or selectivity.

Similar group 3 complexes bearing bis(naphtholate) and bis(phenolate) ligands bridged by either pyridine or thiophene groups are all highly active in the polymerization of *rac*-LA at room temperature and show some intriguing trends in reactivity and selectivity (Figure 11).[52] In the series with bis(naphtholate) ligands, all compounds polymerize *rac*-LA to produce heterotactic PLA with controlled molecular weights and reasonably low $Đ_M$ (1.32–1.90) in a few hours in either toluene or THF; however, the polymerization rate is faster in toluene. The selectivity is highest in THF; polymers formed in toluene are atactic while those formed in THF have varying degrees of heterotacticity. The thiophene-bridged complexes are less selective than the pyridine-bridged complexes. There is also a strong dependence on steric bulk: in complexes with Si^tBuMe_2 groups the selectivity increases with decreasing ionic radius of the metal center (P_r=0.93 for Sc, 0.84 for Y, and 0.50 for La), while in the related $SiPh_3$-substituted ligands the Sc complex is the least selective (P_r=0.65) and the Y and La complexes have similar selectivity (P_r~0.9).[52a] The related bis(phenolate), pyridine-bridged Y complex is more selective than the bis(naphtholate) analogs producing heterotactic PLA in toluene (P_r=0.55–0.60) and THF (P_r=0.94–0.96); the same trend of increased selectivity in THF over toluene is also seen with this complex.[52b]

10.1.3.4 Ketiminate ligands

Lin and co-workers have studied ketiminate ligands with a third amine donor arm for use in Mg, Zn, Al and Ca complexes for the polymerization of lactide (Figure 12).[53] Studies of dinuclear Mg complexes with simple aminoketiminate ligands show that the bulkiest complexes promote dissociation of the dimers to a greater extent, and therefore elicit higher reactivity. There is a contrasting electronic effect with electron-withdrawing CF_3 substituents: these promote decreased nucleophilicity of the OBn groups, thereby reducing their ability to initiate polymerization.[53a]

Lin and co-workers also investigated a related ligand system derived from 4-benzoyl-3-methyl-1-phenyl-2-pyrazolin-5-one (Figure 12). The related magnesium and zinc benzyloxy-bridged complexes are all active for the controlled polymerization of lactide, with

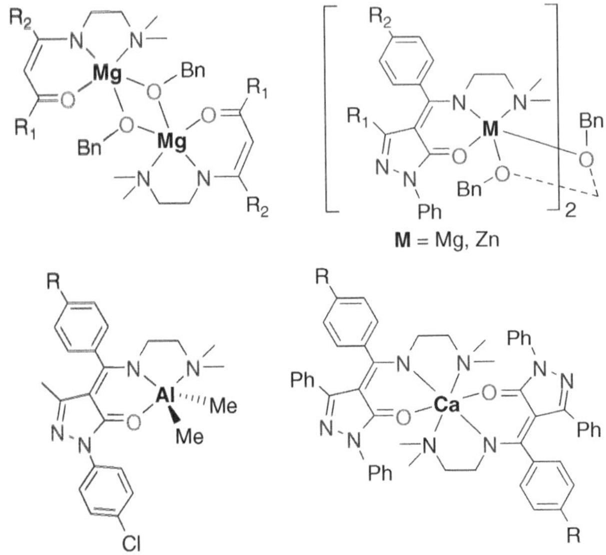

Figure 12 Metal complexes bearing tridentate ketiminate ligands

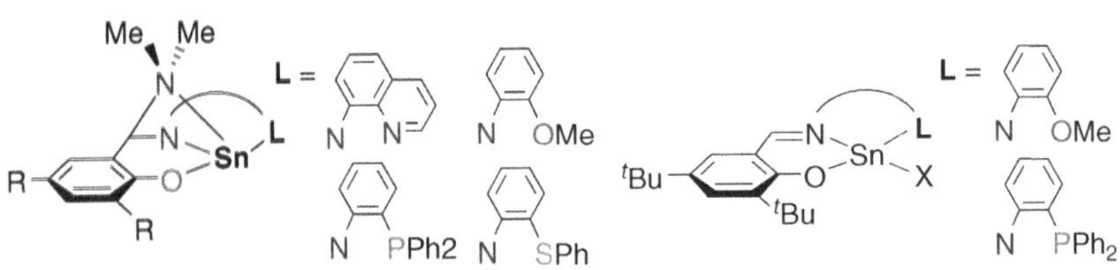

Figure 13 Tin(II) complexes ligated by iminophenolates and diamidophenolates

P_r values ranging from 0.6 to 0.87. No significant substituent effects were observed.[53b, 53d] Similar aluminum dimethyl and CaL_2 homoleptic complexes were active in the presence of benzyl alcohol and reach high conversions in 3–14 h, although they are considerably less active than the related Mg and Zn alkoxides.[53c, 53e]

10.1.3.5 Iminophenolates and related ligands

In 2004 Nimitsiriwat *et al.*[54] reported a series of Sn complexes bearing tridentate iminophenolate ligands formed either from reaction of $Sn(NMe_2)_2$ and iminophenol pro-ligands or from the respective iminophenolate tin chloride complexes and $LiNMe_2$ (Figure 13).[54] The systems show a range of reactivity based on ligand substitution, but the selectivity is unchanged by the different ligand motifs and mildly heterotactic PLA is obtained in all cases (P_r ~ 0.62).

Lin and co-workers have studied half-salen iminophenolate ligands with hemilabile side arms as supports for Zn and Mg complexes (Figure 14).[55] The Mg complex was

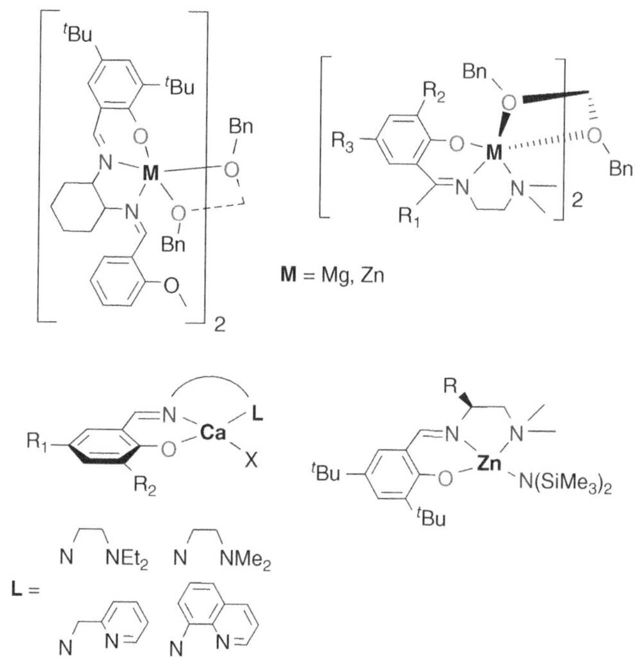

Figure 14 *Complexes with tridentate iminophenolate ligands*

more active than the Zn analog and shows a strong solvent dependence on selectivity; it is heteroselective (P_r=0.57) in THF at room temperature, but is isoselective (P_m=0.54–0.67) in toluene or dichloromethane at room temperature or below. The Zn analog is isoselective (P_m=0.75) in dichloromethane at room temperature.

A large family of analogous Mg and Zn half-salen complexes, without the hemilabile arm and bearing an ethylene diamine backbone, has also been reported by Lin and co-workers (Figure 14).[56] Darensbourg *et al.*[57] have reported closely related Ca, and Zn complexes bearing similarly functionalized iminophenolate ligands (Figure 14). The Zn and Mg complexes with ethylene diamine backbones show similar reactivity and substituent effects.[56] Substitution of the ligand backbone with various electron-donating or withdrawing groups has some effects on reactivity. Changing the imine substituent from H to Me/Ph in the Zn complexes increases the rates of polymerization, whereas in the related Mg complexes changing from Me to Ph substituents had a detrimental effect on the polymerization rate. Some Zn complexes with ethylene diamine or chiral diamine backbones showed moderate heterotacticity (P_r=0.59–0.83).[56a, 56b, 57c, 57d] The related Ca complexes were more active than the Zn or Mg complexes, although the polymerizations were carried out in the melt (110 °C) and are thus not comparable.[57a, 57b] The bulkier and more electron-donating ligand backbones led to slower polymerization rates. Bulky Ca compounds formed mildly heterotactic PLA in chloroform (P_r=0.66) at room temperature; less bulky systems produced only atactic PLA.

10.1.3.6 Tetradentate tripodal ligands

The facially coordinating, tetradentate, tripodal ligand design with three coordinating linkers attached to a central atom capable of binding to a metal center has been widely used to support metal-based catalysts in lactide polymerization, especially for group 4 metals.[26a] The donor atoms of the linkers and the central donor atom are usually nitrogen or oxygen.

Several reports in the literature describe the use of tripodal ligands that exclusively contain nitrogen donors (Figure 15). In 2009 Schwarz *et al.*[58] reported titanium and zirconium isopropoxide catalysts supported by sulfonamides for the ring-opening polymerization of *rac*-lactide and ε-caprolactone. The Zr catalyst was more active than the Ti analog for lactide polymerization. However, neither system was stereoselective and generated atactic PLA. Another example describes a five-coordinate aluminum alkoxide catalyst bearing the same ligand.[45] It has been shown spectroscopically that the amine moiety and the pyridyl group compete for the fifth coordination site. The catalyst was less active than the Zr analog but showed excellent molecular weight control for melt polymerization of lactide. An In analogue was reported by Blake *et al.*[46] This catalyst polymerized 300 equiv. of lactide in 2 h under melt conditions; however, no stereoselectivity was achieved.

Tripodal ligands containing both nitrogen and oxygen donors have been widely used to support group 4 metal catalysts in lactide polymerization, though other metals have also been investigated. In 2002 Kim *et al.*[59] reported the first use of titanium alkoxides in the ROP of lactide with a family of aliphatic and aromatic titanatranes (Figure 16). The authors rationalized the use of this ligand design by evoking possible *trans* effects by the nitrogen donor to labilize the *trans* axial alkoxide, which would promote high catalytic activity. While all the titanatranes polymerized lactide under melt conditions (130 °C) only the five-membered titanatranes polymerized

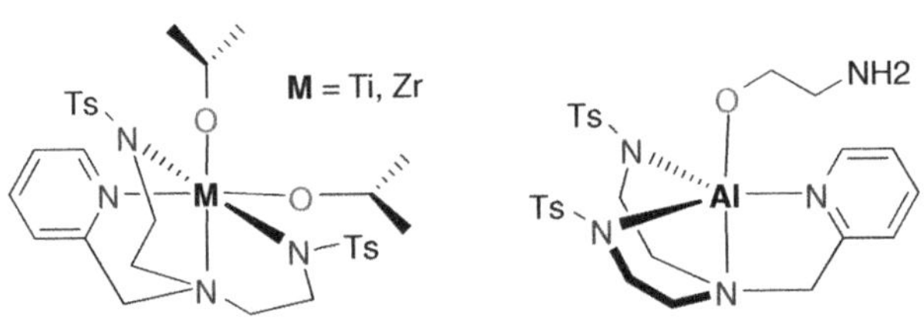

Figure 15 Metal catalysts bearing tripodal nitrogen donor ligands for lactide ROP

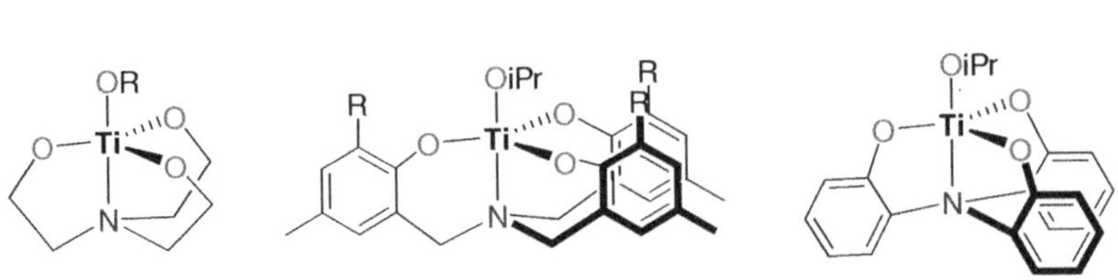

Figure 16 Examples of titanatranes for lactide ROP

lactide in solution (toluene, 70 °C). This indicated higher activity in the more strained compounds in this study. Overall this family of catalysts showed moderate activity requiring ~15 h under melt conditions to achieve >90% monomer conversion when polymerizing 300 equiv. of lactide.

The same amine tris(phenolate) ligand design was extended to Zr by Kol and co-workers who observed higher activity compared with the Ti analogs for lactide polymerization.[60] Chmura *et al.*[61] reported a family of tetravalent metal catalysts bearing amine tris(phenolate) ligands for the ROP of lactide (Figure 17). The authors reported that the Ti complex, which was active only under melt conditions, polymerized 300 equiv. of lactide in under an hour to generate atactic PLA. The Zr and Hf catalysts, while showing similar reactivity, gave heterotactic PLA with P_r values of 0.96 and 0.88, respectively. While both catalysts slowed down considerably in solution (toluene, 25 °C), requiring >48 h to polymerize 100 equiv. of lactide, the heterotacticity increased in both cases (P_r=0.97 and 0.98 for Hf and Zr, respectively). The Ge analog previously reported by the group was a slower initiator, requiring up to 24 h to polymerize 200 equiv. of lactide and achieved only moderate heteroselectivity (P_r~0.80) in comparison.[62] In another report of group 4 catalysts with amine bis(phenolate) ligands, this research group reported that while the Ti catalyst was not stereoselective, the Zr and Hf systems were able to generate PLA with an isotactic bias reaching up to P_m~0.75 under melt conditions.[63]

In an extension of this ligand architecture to group 3 and lanthanide catalysts, Mountford and co-workers reported several Y, La, Nd and Sm complexes with bridging borohydride and chloride ligands (Figure 17).[64] The Y, Sm and Nd complexes generated heterotactic PLA in THF with the Y and Sm catalysts imparting the highest stereoselectivity of P_r=0.87 and 0.72, respectively.

In 2004, Carpentier and co-workers reported several isostructural Y, La and Nd complexes with alkoxyamino bis(phenolate) ligands for the ROP of lactide.[65] These were highly active catalysts capable of polymerizing 200 equiv. of lactide in

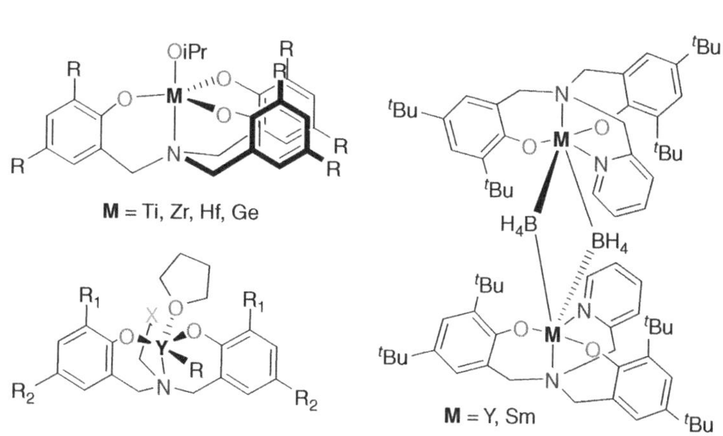

Figure 17 Examples of amine bis- and tris(phenolate)-ligand-based catalysts for lactide ROP

20 min. Several of these complexes generated heterotactic PLA with P_r values up to 0.90. The utility of these catalysts for immortal polymerization of lactide to grow multiple polymer chains by adding an exogenous alcohol has also been demonstrated.[66] In addition, they have been extended to the stereoselective ROP of β-butyrolactone.[67]

10.1.3.7 Macrocyclic ligands

While macrocycles have not been used widely as ligands in lactide polymerization, several examples of nitrogen- and oxygen-containing macrocycles used to support metal initiators have been reported (Figure 18). The use of metal porphyrin catalysts for cyclic ester polymerization, pioneered by Inoue and co-workers, represents one of the earliest examples of a macrocyclic ligand in metal-catalyzed lactide polymerization.[68] This is an extension of the work carried out by the same group where metalloporphyrins were used for the copolymerization of carbon dioxide and epoxides.[69] In 1987 an aluminum alkoxide complex supported by a porphyrin ligand was described by this group for lactide polymerization.[70] The authors employed rigorous conditions to facilitate polymerization of lactide (heated to 100 °C in a vacuum sealed tube with the reactants dissolved in CH_2Cl_2). The polymerizations were well controlled with good agreement between theoretical and experimental molecular weights and narrow $Đ_M$. There was broad monomer scope for aluminum porphyrin catalysts with other lactones.[71]

In 2011 Okuda and co-workers reported a series of metal catalysts supported by a tetradentate cyclen-derived ligand for lactide polymerization (Figure 18).[72] Notably, these catalysts were highly active at room temperature for the ROP of *meso*-LA and polymerized 100 equiv. of monomer in 30 min. However, the reactivity decreased when *rac*- or L-LA was used. The magnesium variant achieved modest isoselectivity with *rac*-LA (P_m = 0.64).[72]

An example of a macrocycle with oxygen donors to support a Ti catalyst for lactide polymerization has been reported by Frediani *et al.*[73] The authors described several titanium chloride complexes bearing calix[4]arene ligands, which act as catalysts for solvent-free lactide polymerization (Figure 18). These complexes acted as dual-site catalysts with two polymer chains growing from one metal center.

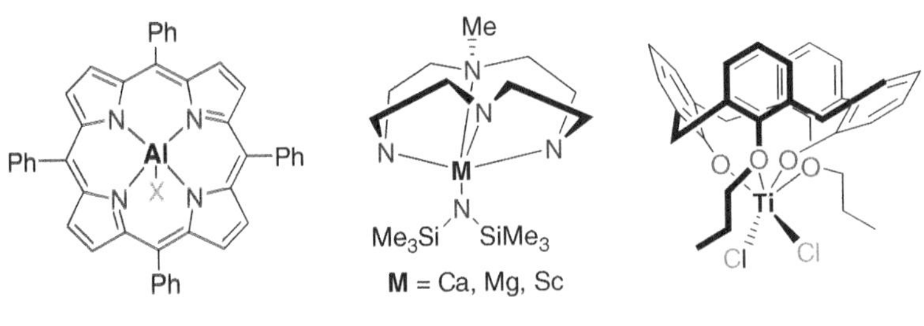

Figure 18 Examples of macrocyclic metal catalysts used for lactide polymerization

10.1.3.8 Salen-type ligands

Salen ligands are tetradentate, Schiff base bis(phenolate) compounds that are traditionally made via the condensation of a diamine and a salicylaldehyde (Figure 19).[74] They are widely used in asymmetric transformations, such as enantioselective epoxidations.[75] While both tripodal and macrocyclic ligands have been investigated for the stereoselective ROP of *rac*-lactide with some success, salen-type ligands have by far been much more successful in achieving isoselectivity with a variety of metals. Other derivatives of salen ligands, namely the reduced bis(aminophenolate) form (salan) and the asymmetrically reduced form (salalen), have also been used in lactide polymerization.[76] Since many variations of the classic salen ligand architecture incorporating different donor atoms have been reported in literature, the following discussion on salen-type ligands will be organized according to donor atoms.

The seminal work in using the classic salen ligand design in lactide polymerization was described by Spassky and co-workers, who had previously used a series of aluminum salen compounds in the polymerization of epoxides and β-butyrolactone (Figure 20).[77] They described an aluminum methoxide catalyst supported by an achiral salen ligand for the ROP of L- and *rac*-lactide. Although this system requires elevated temperatures (70–100 °C in toluene) and has poor control over the polymerization process with dispersities ranging from ~2 to 4, this catalyst is an important milestone in catalyst design for lactide polymerization. This report was a prelude to a landmark publication by this group, in which a highly stereoselective aluminum SalBinap (salen ligand with a Binap linker) catalyst for the ROP of lactide was described (Figure 20).[78] The enantiopure (*R*)-catalyst was highly competent at chiral resolution of *rac*-lactide, reaching ~50% conversion at 70 °C in toluene by almost exclusively polymerizing L-lactide. A study of the rates of polymerization of L- and D-lactide with the (*R*)

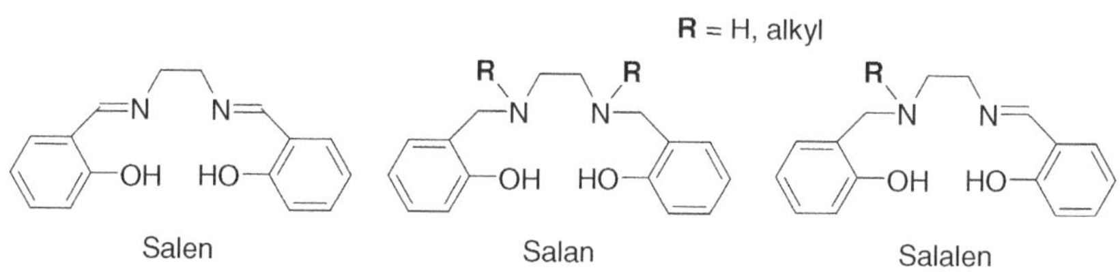

Figure 19 *Examples of salen, salan and salalen ligands*

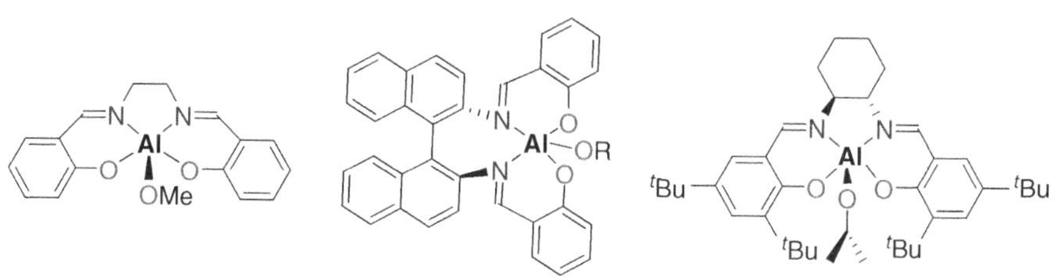

Figure 20 *Aluminum salen complexes*

enantiomer of the catalyst showed a k_D/k_L of ~20, which indicated a highly site-selective system. Subsequently, evidence of extensive transesterification during polymerization was reported using MALDI-TOF mass spectrometry.[79]

Subsequent to this work, a series of highly influential publications described studies of the polymerization of *rac*-lactide with aluminum SalBinap complexes in detail. Coates and co-workers improved the synthesis of the catalyst by substituting the methoxide ligand with an isopropoxide initiator to prevent the formation of unwanted aggregates.[12, 24a] This catalyst was stereoselective for the ROP of *rac*- and *meso*-LA.[12, 24a, 80] Though initially thought to form stereocomplex PLA, the polymerization of *rac*-lactide with the racemic aluminum SalBinap catalyst was shown to generate isotactic stereoblocks with a melting point of 179 °C.[80, 81] Duda and co-workers were able to use the same system to generate isotactic stereoblock PLA with a melting point approaching 210 °C through a two-step chiral ligand exchange mechanism.[82] Zhong *et al.*[83] reported an aluminum isopropoxide initiator bearing a Jacobsen salen ligand, which also generated highly isotactic PLA (Figure 20). The racemic catalyst generated isotactic stereoblock PLA with *rac*-lactide in toluene at 70 °C with a P_m ~ 0.93 at 85% monomer conversion. This high isoselectivity was also maintained in melt polymerization at 130 °C to generate PLA with a P_m ~ 0.88.

In a mechanistic study of this system Chisholm *et al.*[25, 84] reported significant solvent effects on isoselectivity. They highlighted the difficulty in assigning the mode of stereocontrol to exclusively enantiomorphic site-control or chain-end control, and argued that the chiral environment of the catalyst, the chirality of the chain-end, the helicity of the η^4-chelate, λ or δ, and the solvent all affect stereoselectivity to varying degrees. In a recent report, Pilone *et al.*[85] described an aluminum salalen catalyst for lactide polymerization and highlighted the contribution of chain-end control to systems with chiral ligands.

The mechanistic complexities of stereoselectivity is further evidenced by a recent report by Maudoux *et al.*[86] who describe a chiral aluminum salen catalyst that generates highly isotactic PLA from *rac*-lactide (P_m ~ 0.90). In this example, the kinetics indicated a dominant chain-end control mechanism, which contrasts to other chiral aluminum salen catalysts where enantiomorphic site control is thought to predominate.[80, 83b] All the previously mentioned chiral aluminum salen alkoxide systems require multiple days at elevated temperatures to polymerize ~200 equiv. of lactide. The low activity of chiral aluminum salen systems towards lactide polymerization is a major drawback of these systems.

Several Al catalysts supported by achiral salen ligands for isoselective lactide polymerization have been reported (Figure 21). After their initial publication,[77a] Spassky and co-workers reported a series of aluminum salen alkoxide catalysts that were used to generate crystalline PLA with T_m ~ 144–159 °C.[87] In one example, the authors also described decreased reactivity upon changing the ethylene bridge to a rigid phenyl moiety. Profound effects on reactivity and selectivity were observed by Nomura and co-workers when the ethylene linker was changed to a propylene functionality (Figure 21).[88] The ethylene-bridged catalysts polymerize 100 equiv. of lactide to 19% conversion (70 °C in toluene) in 3 days to generate isotactic PLA (P_m ~ 0.79, T_m ~ 163 °C).

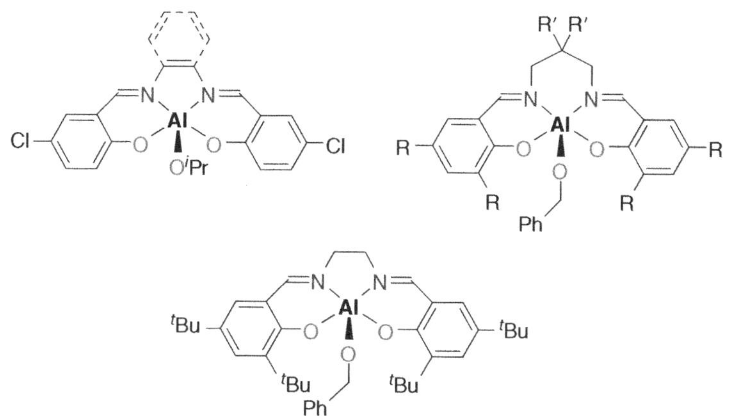

Figure 21 *Examples of achiral aluminum salen catalysts for enantioselective ROP of rac-lactide*

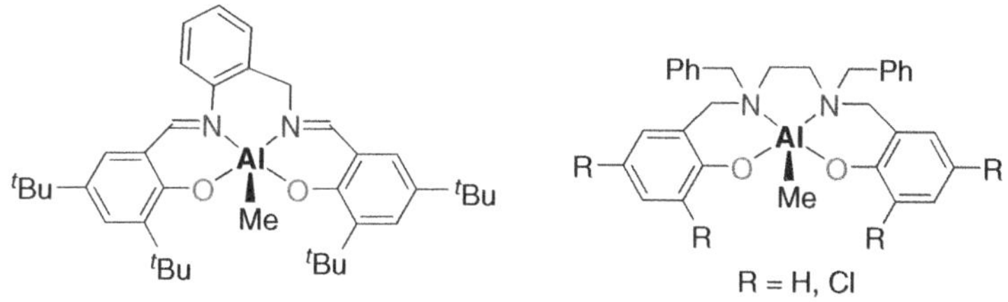

Figure 22 *Examples of aluminum salen and salan catalysts for lactide ROP reported by Hormnirun et al.*[91]

In contrast, the analog with a propylene linker not only achieves 95% monomer conversion under identical conditions in 14 h but also improves the isotacticity of the polymer ($P_m \sim 0.92$, $T_m \sim 192\,°C$). After a series of incremental modifications to the linker as well as the aromatic substituents, Nomura and co-workers discovered that geminal methyl groups on the propylene linker lead to a catalyst with very high isoselectivity ($P_m \sim 0.98$, $T_m \sim 210\,°C$), which is the highest isotacticity reported with an achiral catalyst.[89] Several highly isoselective aluminum salen isopropoxide catalysts, with the same 2,2-dimethylpropylene linker, but with different aromatic substituents, have been described by Tang *et al.*[90a, 90b] and more recently by Chen *et al.*[90c]

Hormnirun *et al.*[91] described a series of aluminum alkyl complexes supported by achiral salen ligands, which were used with benzyl alcohol for the ROP of lactide (Figure 22). The authors observed isoselectivity in many of their catalysts up to $P_m = 0.86$ and described the dependency of the degree of isoselectivity on the bridging ligand backbone and the aromatic substituents. A more profound change was observed when achiral salan (reduced-salen)-type ligands were used to prepare several aluminum catalysts (Figure 22).[92] When an unsubstituted salan ligand was used (R = H), PLA with $P_m \sim 0.79$ was produced. However, when chloride substituents were incorporated

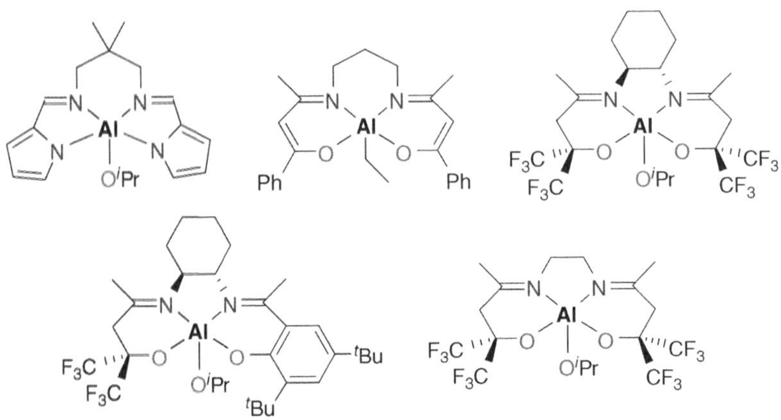

Figure 23 *Examples of Al catalysts supported by salen-like ligands for lactide ROP*

(R = Cl) the stereoselectivity switched to generate heterotactic PLA (P_r ~ 0.96). Du *et al.*[93] observed a similar effect with chiral aluminum salan complexes.

While the classic salen ligand design contains a bis(salicylidene) moiety, several recent examples describe the use of pyrrolic, enolic and alkoxide Schiff base frameworks to form tetradentate *salen-like* ligands, which were used to develop Al catalysts for lactide polymerization (Figure 23). Du *et al.*[94] reported an aluminum isopropoxide catalyst with a (*N,N,N,N*) salen-type ligand with pyrrole donors which generated isotactic PLA (P_m ~ 0.75) from *rac*-lactide. An aluminum salen catalyst with enolate donors, which showed isoselectivity (P_m ~ 0.80), was reported by Pang *et al*,.[95] Bouyahyi *et al.*[96] described several chiral and achiral aluminum salen-type catalysts comprising fluorinated alkoxide donors. These catalysts gave isotactically enriched PLA (P_m ~ 0.70–0.81) from *rac*-lactide in the melt. The less rigid achiral versions showed higher activity compared with the chiral analogs (30 min versus 72 h to achieve similar conversions under similar conditions). However, there was no change in selectivity between the two systems, which suggested a chain-end control mechanism for the catalysts. The mixed alkoxide-phenolate ligand based Al catalyst is also capable of isotactic enrichment of PLA in solution (P_m ~0.81 in toluene at 60 °C).[97]

Trivalent metals other than Al have also been used with salen/salen-type ligands in lactide ROP (Figure 24). Ovitt and Coates[12] reported a dinuclear yttrium SalBinap catalyst for the ROP of lactide. In stark contrast to the Al counterparts, no stereocontrol was achieved in the polymerization. A bismuth alkoxide complex bearing a Jacobsen salen ligand was reported by Balasanthiran *et al.*[98], which generates heterotactic PLA (P_r ~ 0.9) from *rac*-lactide. Recently, several indium catalysts bearing salen-type ligands have been reported by Carpentier and co-workers, but these, unlike their Al analogs, generated atactic polymer.[86, 99] A recent example by the same group described a bimetallic Li/Y catalyst bearing a fluorinated SalBinap ligand, which generated heterotactic PLA (P_r ~ 0.99).[100]

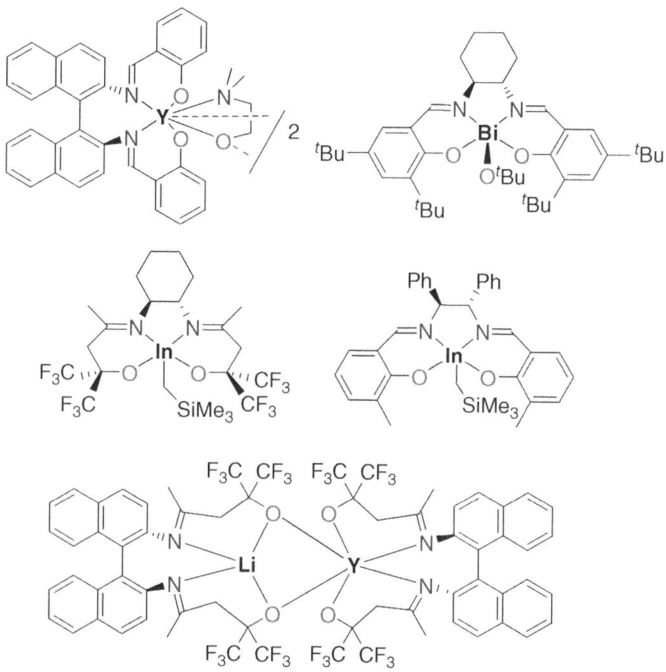

Figure 24 *Examples of trivalent metal catalysts with salen-type ligands for lactide ROP*

Group 4 metals have also been used widely in conjunction with salen-type ligands (Figure 25). In 2006 Gregson *et al.*[101] reported several chiral and achiral titanium salen alkoxide complexes for the ROP of lactide. All catalysts reported were modestly active and heteroselective (P_r ~ 0.51-0.57). Several achiral Ti and Zr salan catalysts were reported by Gendler *et al.*[60] for melt polymerization of lactide. While no stereoselectivity has been reported for either system, the Zr complexes were more active towards lactide ROP than the Ti analogs.

Saha *et al.*[102] reported zirconium and hafnium alkoxide complexes bearing a Jacobsen salen ligand (Figure 25). Notably, these complexes were bimetallic with each metal center coordinating to the ligand in a κ^2-coordination mode. While these systems were active for melt polymerization of lactide (200 equiv. in under 1 h) both generated atactic PLA. Several bimetallic and monometallic achiral Zr salen catalysts (Figure 25) reported by Tsai *et al.*[103] showed polymerization activity but again failed to achieve the high stereoselectivity of the Al analogs.

Broderick and Diaconescu[104] have reported a family of catalysts supported by ferrocene-based bis(phenolate) ligands (Figure 26). A Ce(IV) catalyst was highly active for the ROP of lactide with 300 equiv. of L-lactide being polymerized in under 20 min at room temperature. The Y analog showed redox-controlled polymerization behavior where a change in catalyst activity was observed depending on the oxidation state of ferrocene. Similar behavior was reported for a ferrocene-based titanium salen catalyst reported by Gregson *et al.*[105] (Figure 26). In 2012 Bakewell *et al.*[106]

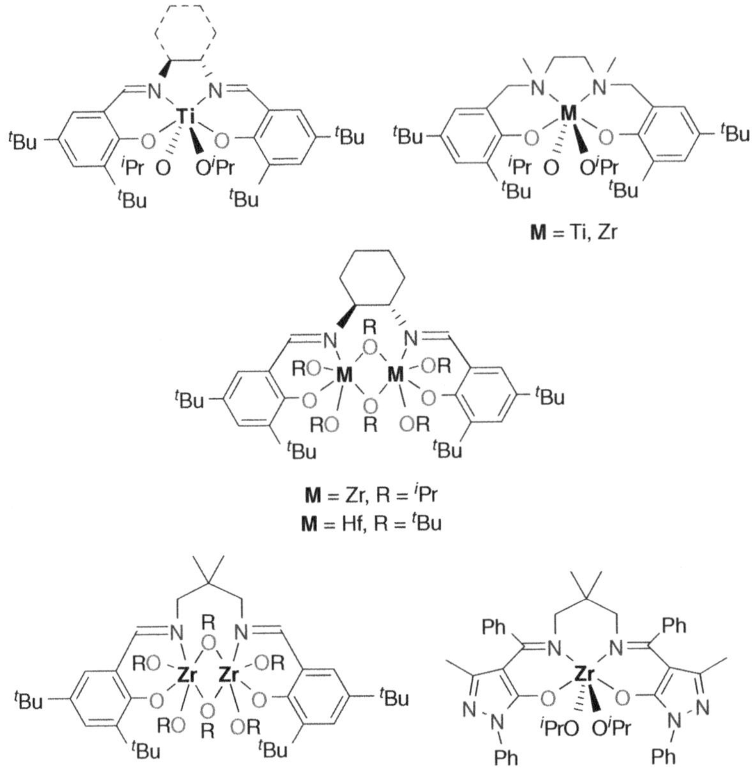

Figure 25 Examples of group 4 salen catalysts used in lactide polymerization

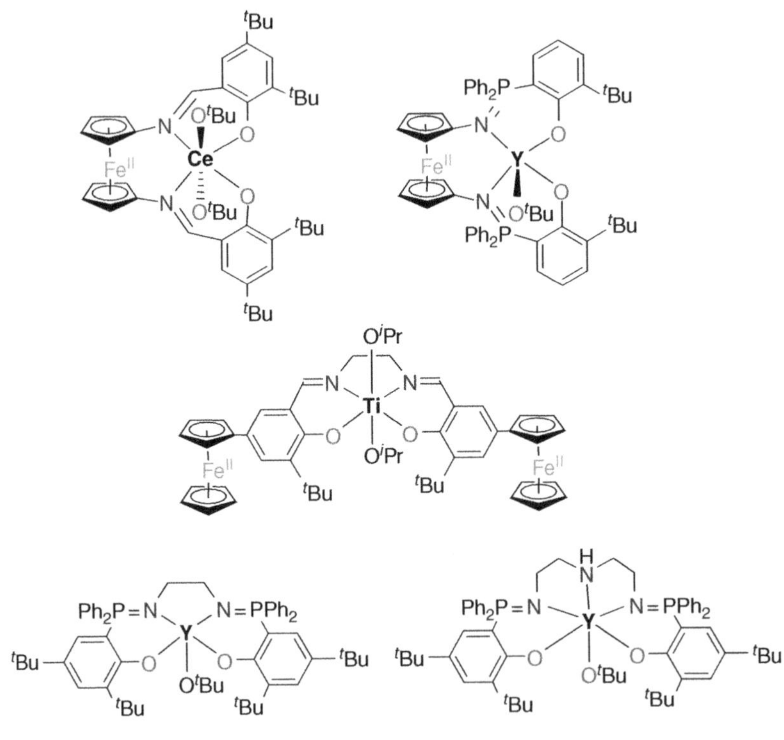

Figure 26 Variations of salen-type ligands used in metal-based lactide ROP catalysts

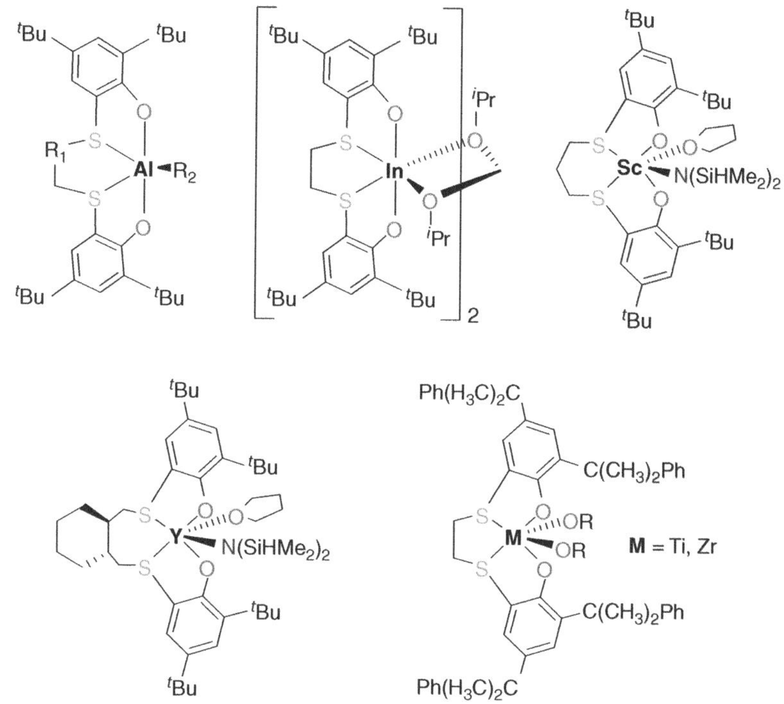

Figure 27 *Examples of (OSSO)-type-ligand-based catalysts for lactide polymerization*

reported the use of a Y catalyst bearing a P-containing Schiff base ligand, referred to as a *phosphasalen* ligand (Figure 26). Catalysts supported by such ligands containing an ethylene diamine linker showed high reactivity (polymerizing 1000 equiv. of lactide in 45 s at ambient temperature in THF) and high heteroselectivity ($P_r \sim 0.9$). The authors were able to maintain this reactivity and achieve isoselectivity ($P_m \sim 0.8$) by modifying the ligand backbone to prepare a pentadentate ligand system.[106] This ligand promoted isoselectivity with other metals such as Lu.[107]

Okuda and co-workers used a family of 1,ω-dithiaalkanediyl-bridged bis(phenolate) (OSSO)-type ligands with a range of different metals to develop catalysts for the ROP of lactide (Figure 27). While these ligands may not strictly fall into the category of salen-like ligands they coordinate to metals in a similar fashion. Several aluminum alkyl complexes bearing (OSSO)-type ligands were used as catalysts for lactide polymerization with an added alcohol.[108] While the complexes with an ethylene diamine backbone generated atactic PLA, the modification of the linker to a $-CH_2PhCH_2-$ moiety afforded modest heteroselectivity ($P_r \sim 0.65$) in the system. Dinuclear indium alkoxide analogs were active in the ROP of L-lactide.[109] In contrast, group 3 metal catalysts (such as Sc and Y) with these (OSSO)-type ligands were heteroselective. Scandium and yttrium amido analogs were highly heteroselective ($P_r = 0.95$ and 0.88, respectively).[110] When group 4 metals were used with achiral (OSSO)-type ligands, modest isotactic enrichment ($P_m \sim 0.6$) was achieved.[111]

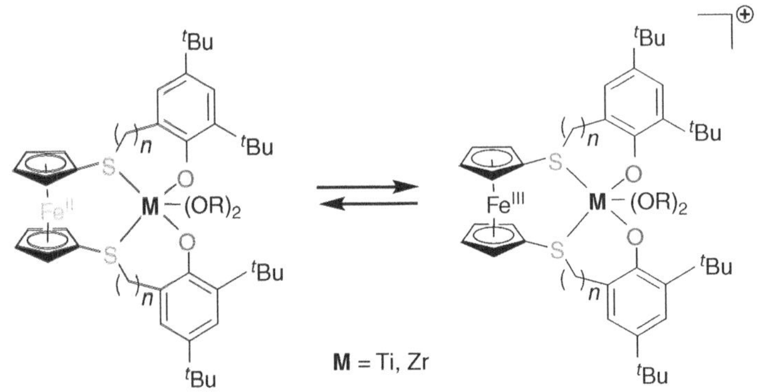

Figure 28 *Reduced and oxidized forms of thiolphan-ligand-based group 4 complexes*

In a recent report, Wang *et al.*[112] described two S-containing bis(phenolate) thiolphan ligands featuring ferrocene backbones (Figure 28). These ligands are used with Zr and Ti to form *redox switchable* catalysts, as discussed above, whose polymerization behavior changes based on the oxidation state of the ferrocene. The Zr complex was active for lactide polymerization in its reduced form but inactive in the oxidized form, which was in direct contrast to its polymerization of ε-caprolactone, where it showed the opposite behavior. While the Zr system failed to generate block copolymers of PLA and poly(caprolactone) in a one-pot synthesis via *redox switching*, the Ti analog was able to successfully carryout the block copolymerization.

10.2 Indium and zinc complexes bearing chiral diaminophenolate ligands

10.2.1 Zinc catalysts supported by chiral diaminophenolate ligands

In 2003 Williams *et al.*[113] reported a highly active Zn catalyst for the polymerization of lactide based on a tridentate diaminophenolate ligand (L) (Figure 29). This complex, synthesized from the protonolysis of a precursor zinc alkyl complex, was highly active for the controlled polymerization of *rac*-lactide at room temperature, reaching high conversions in ~5 min to produce atactic PLA. Mechanistic investigations supported a mononuclear propagating species, formed upon dissociation of the dimer in solution.

These promising results with Zn complexes bearing an achiral diaminophenolate ligand (L) encouraged us to develop a related chiral tridentate diaminophenolate proligand $H(N_{Me2}N_{Me}O_{tBu})$ (Figure 29) by adapting a synthetic methodology first reported by Mitchell and Finney.[114] The synthesis of Zn complexes using this ligand[115] was a first indication of the differences between the parent achiral ligand and the new chiral ligand set. The synthesis of the achiral zinc alkoxide analog was carried out via a simple alkane elimination reaction of the proligand HL and $Zn(Et)_2$ to yield (L)ZnEt (Figure 29). A subsequent alkane elimination reaction with ethanol formed the target ethoxy-bridged Zn catalyst.[113]

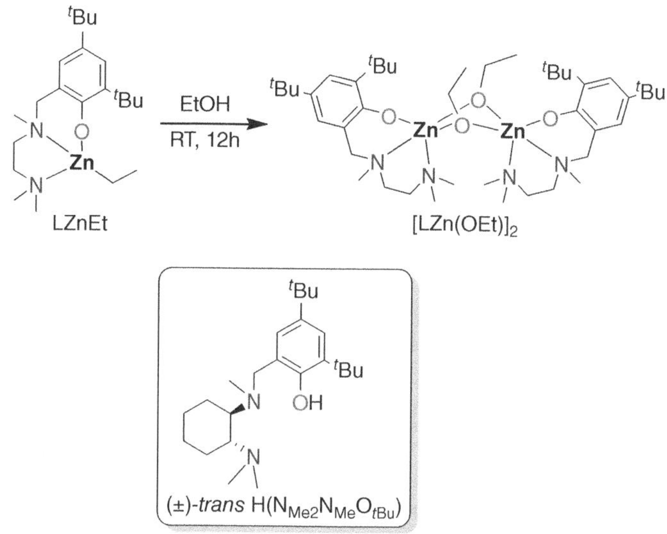

Figure 29 Synthesis of the highly active Zn catalyst reported by Williams et al.[113] and the first generation chiral diaminophenolate proligand (in box)

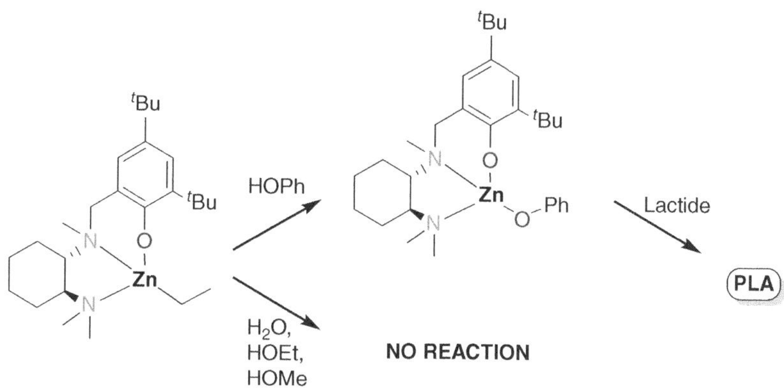

Figure 30 Synthesis and reactivity of Zn complexes bearing a chiral diaminophenolate ligand

When we utilized the chiral $H(N_{Me2}N_{Me}O_{tBu})$ proligand in an identical reaction, we discovered that the second protonolysis reaction does not proceed with ethanol or other aliphatic alcohols (Figure 30).[115] In fact, there was a direct correlation between the pK_a of the alcohol (in organic solvent) and its reactivity with $(N_{Me2}N_{Me}O_{tBu})$ZnEt. Thus, a reaction of $(N_{Me2}N_{Me}O_{tBu})$ZnEt with phenol formed the phenolate initiator $(N_{Me2}N_{Me}O_{tBu})$ZnOPh which was only mildly active for lactide polymerization. We attributed this lack of reactivity, in part, to the need for the terminal amine arm to dissociate in these compounds to open a coordination site for the incoming lactide.

New studies with analogous In complexes suggest that the central tertiary amine substituent in these complexes may also be important (see below).[116]

10.2.2 The first indium catalyst for lactide polymerization

In 2008, we reported the first indium catalyst for the ROP of lactide,[117] $[(N_{Me2}N_HO_{tBu})InCl]_2(\mu\text{-Cl})(\mu\text{-OEt})$ (**1**), supported by the chiral diaminophenolate ligand framework discussed above.[118] Subsequently, a family of compounds (Figure 31) was synthesized by expanding to different halides $[(N_{Me2}N_HO_R)InX]_2(\mu\text{-X})(\mu\text{-OEt})$ (X = Br, I) and forming bis-ethoxide-bridged dimers $[(N_{Me2}N_HO_R)InI(\mu\text{-OEt})]_2$.[119] These catalysts can be synthesized in two steps (Figure 31).[118, 119] A salt metathesis reaction of the diaminophenolate salts $K(N_{Me2}N_HO_R)$ with indium trihalides InX_3 forms $(N_{Me2}N_HO_R)InX_2$, which react with different equivalents of NaOEt in a second salt metathesis reaction to form the dimeric mono-ethoxide-bridged $[(N_{Me2}N_HO_R)InX]_2(\mu\text{-X})(\mu\text{-OEt})$ or bis-ethoxide-bridged $[(N_{Me2}N_HO_R)InX(\mu\text{-OEt})]_2$.

All the ethoxide complexes isolated in this series are dinuclear in the solid state.[118,119] Their solution structures were determined by pulsed gradient spin-echo (PGSE) NMR

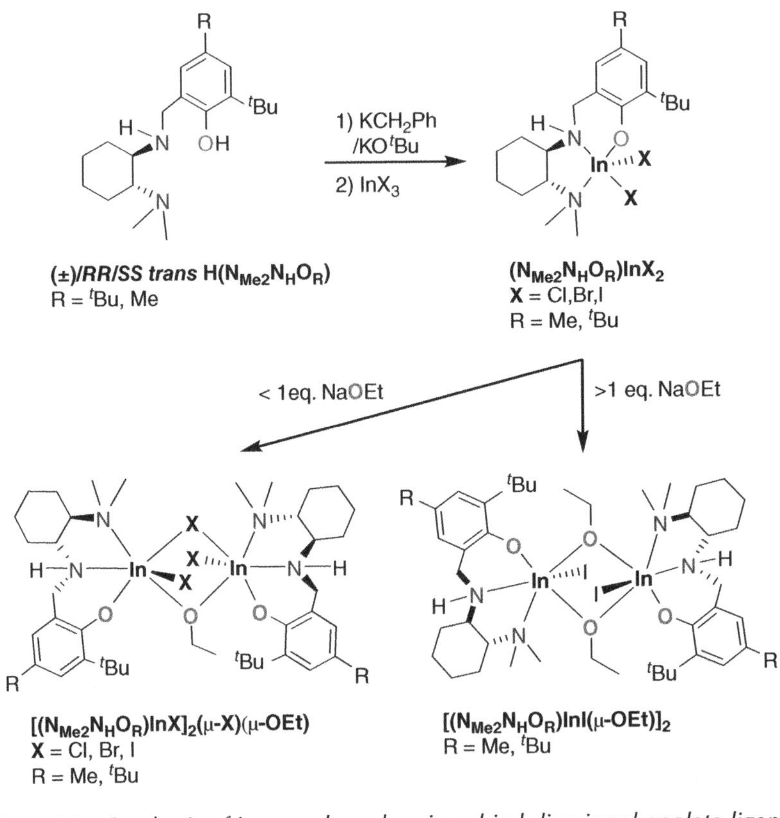

Figure 31 *Synthesis of In complexes bearing chiral diaminophenolate ligands*

spectroscopy and their calculated solution hydrodynamic radii, where available, are consistent with those obtained in the solid state.[119b] The dinuclear motif is the thermodynamically stable form of these ethoxide-bridged In compounds; although they are reactive and can undergo exchange with various species (H_2O, ROH), the resulting products are invariably dinuclear.[119]

Enantiopure (*RR* and *SS*) as well as racemic versions of the compounds were developed.[119b] When the racemic ligand is used in the synthesis of the compounds, the mono-ethoxide-bridged complexes are invariably a mixture of homochiral enantiomers (*RR/RR*)- and (*SS/SS*)-$[(N_{Me2}N_HO_R)InX]_2(\mu\text{-}X)(\mu\text{-}OEt)$, while the bis-ethoxide-bridged complexes are heterochiral (*RR/SS*)-$[(N_{Me2}N_HO_R)InI(\mu\text{-}OEt)]_2$. A mixture of homochiral bis-ethoxide-bridged complexes (*RR/RR*)- and (*SS/SS*)-$[(N_{Me2}N_HO_R)InI(\mu\text{-}OEt)]_2$ forms the heterochiral species in solution in a few hours at room temperature.

10.2.3 Polymerization of cyclic esters with first generation catalyst

The first reported complex in this series, $[(N_{Me2}N_HO_{tBu})InCl]_2(\mu\text{-}Cl)(\mu\text{-}OEt)$ (**1**), is highly active for the polymerization of lactide, showing first-order rates of polymerization, and displaying excellent control of PLA molecular weight and molecular weight distribution. Modestly isotactic PLA (P_m ~0.6) is obtained (Figure 32).[118] In particular, complex **1** is highly controlled for the living polymerization of lactide (LA) with two consecutive additions of 200 equiv. of LA resulting in similar rates of

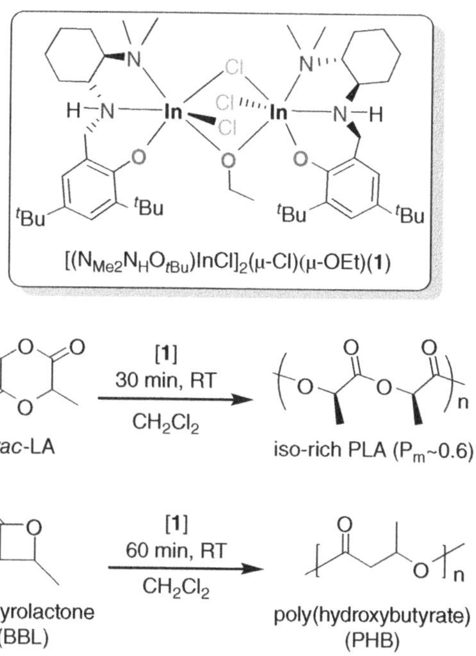

Figure 32 *Living polymerization of LA and BBL with* $[(N_{Me2}N_HO_{tBu})InCl]_2(\mu\text{-}Cl)(\mu\text{-}OEt)$ (**1**)

polymerization for both additions and a monomodal GPC trace for the resulting polymer with an M_n value corresponding to 400 LA units. Complex **1** is similarly active and controlled for the living polymerization of β-butyrolactone (BBL), yielding high molecular weight polymers with low molecular weight distributions (Figure 32).[120] The living behavior of catalyst **1** allows for the formation of block copolymers, such as PLA-PHB-PLA triblocks, by sequential addition of different monomer feedstocks.[19]

10.2.4 Ligand modifications

We sought to improve upon the modest isoselectivity (P_m ~0.6) of $[(N_{Me2}N_HO_{tBu})InCl]_2(\mu\text{-}Cl)(\mu\text{-}OEt)$ (**1**) in the polymerization of *rac*-LA by making a number of ligand modifications, including changes to the terminal amine substituents, central amine substituent and the phenolate substituents (Figure 33).[116, 121]

A number of important observations were made regarding the effects of these modifications on the activity and stereoselectivity of these complexes in the polymerization

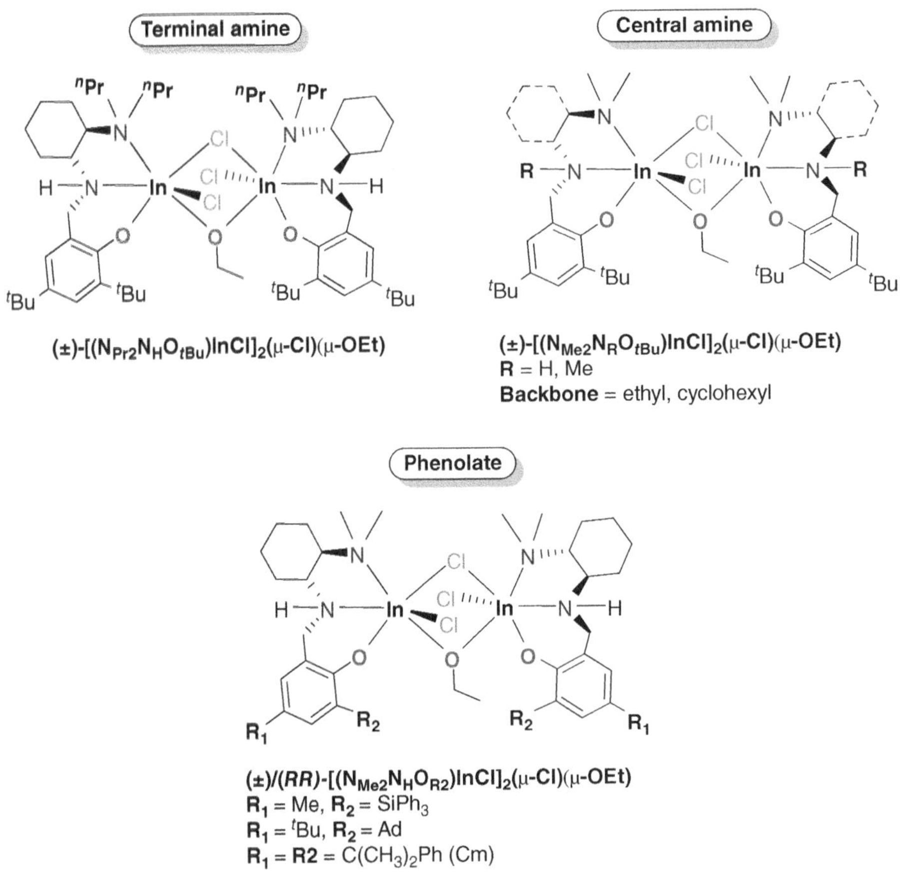

Figure 33 Ligand modifications made to dinuclear In complexes for rac-LA polymerization

of lactide: (1) increasing the steric bulk of the terminal amine substituents (Me to ^{n}Pr) led to dissociation of the catalyst during the polymerization of *rac*-LA and a subsequent decrease in the isoselectivity of the system ($P_m \sim 0.5$);[121a] (2) changing the central amine donor from a secondary (R=H) to a tertiary (R=Me) amine had a profound effect on the activity of the complexes (more than two orders of magnitude drop in activity for tertiary amine donors), which was independent of whether a chiral cyclohexyl or achiral ethyl backbone was utilized;[116] and (3) modifications of the phenolate substituents to include *ortho*-adamantyl or cumyl, $C(CH_3)_2Ph$, substituents had little effect on catalyst activity or stereoselectivity ($P_m \sim 0.6$), while increasing the steric bulk further to include *ortho*-$SiPh_3$ substituents again led to dissociation of the complex during polymerization and a corresponding decrease in the isoselectivity of the system ($P_m \sim 0.5$).[121b]

10.3 Dinuclear indium complexes bearing chiral salen-type ligands

10.3.1 Chiral indium salen complexes

As discussed above, Al catalysts supported by salen ligands are some of the most isoselective catalysts for the polymerization of racemic lactide.[24a,78,83a,88] However, slow polymerization rates and high water/air sensitivity preclude their use in an industrial setting.[12, 83b] We were interested in exploiting chiral indium salen complexes to achieve high activity, selectivity and control over the polymerization process. Although achiral indium salen complexes had been reported previously,[122] prior to our work there was only one example of a chiral indium salen complex, which was not used as a catalyst for polymerization.[123]

Deprotonation of the racemic or enantiopure salen proligand featuring a *trans* cyclohexyldiamine backbone (ONNO) and subsequent salt metathesis with indium trichloride and sodium ethoxide in a three-step process involving an indium chloride intermediate yields racemic or enantiopure indium ethoxide complex $[(ONNO)In(\mu\text{-}OEt)_2]_2$ (**2**) (Figure 34).[124]

Alternatively, a one-pot reaction can be used to synthesize chiral indium ethoxide complexes bearing a salen ligand with a Binap backbone (SalBinap) (Figure 35).[125] Similar to the corresponding Al analog,[12] a mixture of $[(\mu\text{-}\kappa^2\text{-}SalBinap)In(\mu\text{-}OEt)_2]_2$ (**3a**) and $[(\kappa^4\text{-}SalBinap)In(\mu\text{-}OEt)_2]_2$ (**3b**), with different ligand binding modes, is formed.

10.3.2 Polymerization studies

Complexes (±)- and (*R*,*R*)-**2** are highly active and controlled for the polymerization of *rac*-LA yielding a linear relationship between the observed PLA molecular weights and the added monomer, assuming both alkoxides initiate the polymerization, with low molecular weight distributions (Figure 36).[124] The rate of polymerization with these complexes is first order in lactide concentration with k_{obs} values comparable with the tridentate diaminophenolate indium complexes described in Section 2.[118, 119b] The rates of polymerization are much faster than any of the known chiral aluminum salen systems, which require elevated temperatures and days to reach full conversion.[12, 24a, 78, 80, 83]

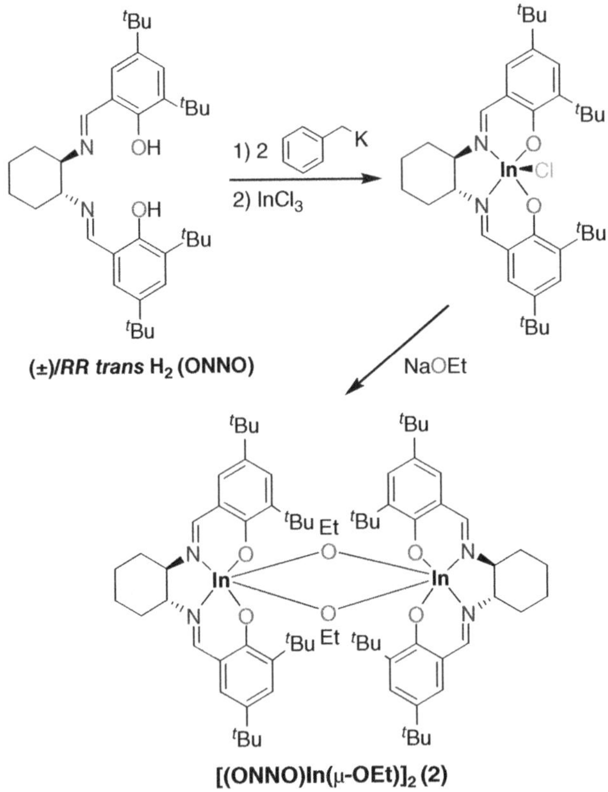

Figure 34 *Synthesis of chiral salen indium complexes with cyclohexyl backbones*

Polymerization of *rac*-LA with (±)- and (*R,R*)-**2** yields isotactic polymers (P_m ~ 0.75) with a stereogradient microstructure. Catalyst (*R,R*)-**2** shows a preference for the polymerization of L-LA over D-LA (k_L/k_D ~ 5), as well as a V-shaped P_m versus conversion plot for the polymerization of *rac*-LA, where the PLA tacticitiy reaches a minimum at 50% conversion. In contrast, catalyst (±)-**2** shows similar rate constants for *rac*-, L- and D-LA polymerization and shows no variation in P_m values as a function of conversion in the polymerization of *rac*-LA. These observations are indicative of enantiomorphic site control being the dominant mechanism of selectivity in this system and are in line with the analogous aluminum salen systems, which carry out the polymerization with greater selectivity but at lower rates.

In contrast, polymerization of *rac*-LA by the SalBinap complexes [(μ-κ^2-SalBinap)In(μ-OEt)$_2$]$_2$ (**3a**) and [(κ^4-SalBinap)In(μ-OEt)$_2$]$_2$ (**3b**) is slow, requiring elevated temperatures and 8 (**3b**) to 30 days (**3a**) to reach full conversion (Figure 37).[125] Complex **3a** is not well-controlled producing atactic PLA with higher than expected molecular weights consistent with poor initiation, which coupled with its low activity suggests that initiation and propagation are both very slow for this catalyst. This is similar to its aluminum analog [(*μ*-κ^2-Salbinap)Al(*μ*-OMe)]$_2$, which is inactive for lactide polymerization.[12]

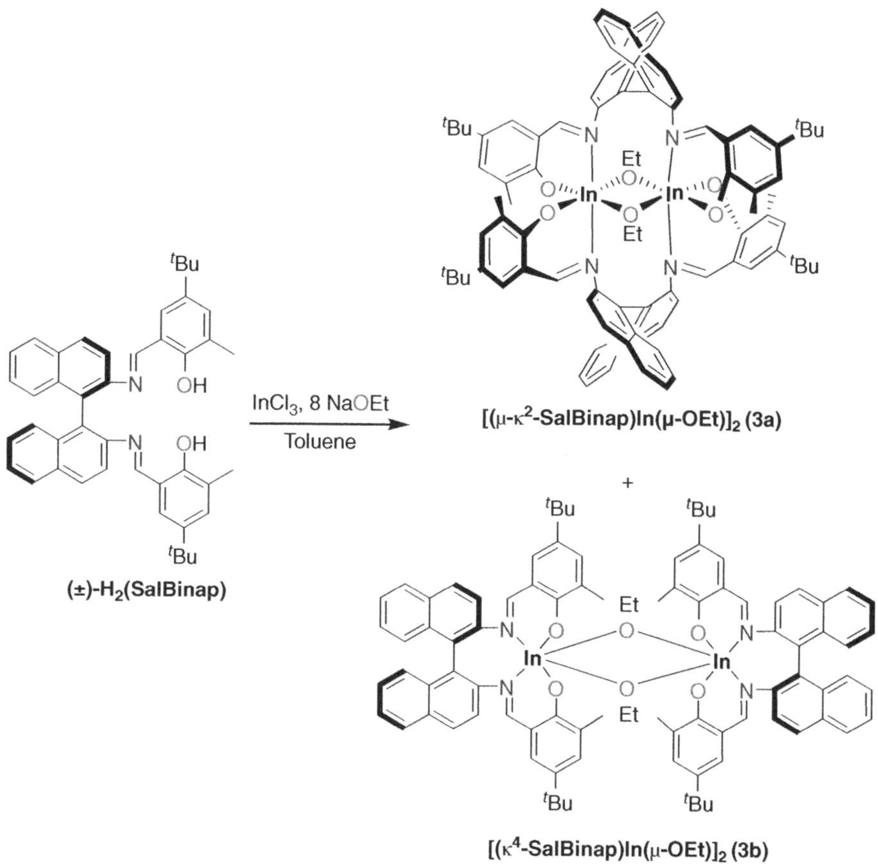

Figure 35 *Synthesis of chiral salen indium complexes with Binap backbones*

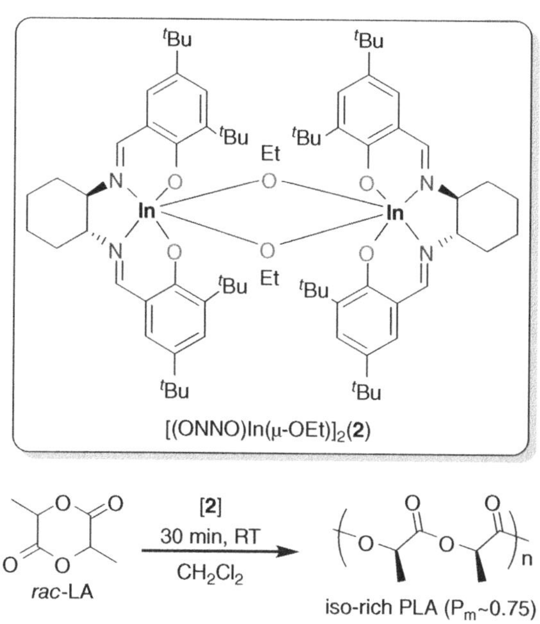

Figure 36 *Polymerization of rac-LA by complex $[(ONNO)In(\mu\text{-}OEt)_2]_2$* ***(2)***

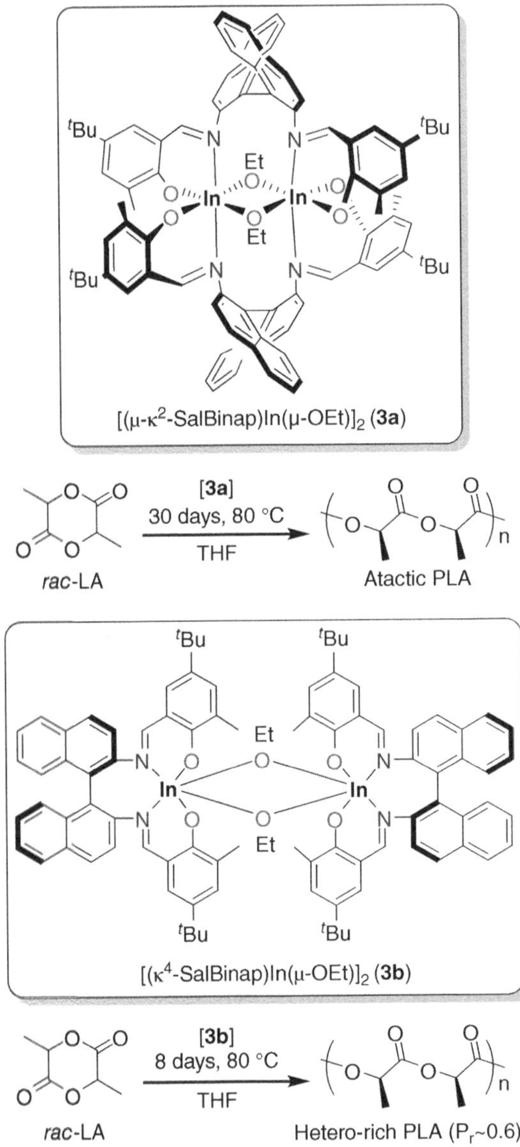

Figure 37 *Polymerization of rac-LA with [(**μ**-**κ**^{2}-SalBinap)In(**μ**-OEt)$_2$]$_2$ (**3a**) and [(**κ**^{4}-SalBinap)In(μ-OEt)$_2$]$_2$ (**3b**)*

Similar to its analog complex **2**, complex **3b** is well-controlled for lactide polymerization displaying good agreement between theoretical and experimental molecular weights, albeit with higher dispersity values. In contrast to complex **2**, complex **3b** shows a moderate heterotactic bias (P_r ~ 0.6), suggesting chain-end control, not enantiomorphic site control, is responsible for the moderate selectivity of the system

Table 1 *Tacticity data for the polymerization of rac-LA with indium and aluminum salen catalysts*

	Catalyst	P_m
1	$[(\mu\text{-}\kappa^2\text{-SalBinap})In(\mu\text{-OEt})_2]_2$ (**3a**)[a]	0.48
2	$[(\kappa^4\text{-SalBinap})In(\mu\text{-OEt})_2]_2$ (**3b**)[a]	0.40
3	$(\kappa^4\text{-SalBinap})Al(O^iPr)$[b]	>0.9
4	$(ONNO)Al(O^iPr)$[c]	0.93
5	$[(ONNO)In(\mu\text{-OEt})_2]_2$ (**2**)[d]	0.74

[a] Polymerizations were carried out in THF at 80 °C, [catalyst] ≈ 1 mM; P_m were determined by $^1H\{^1H\}$ NMR and Bernoullian statistics.
[b] Carried out in toluene at 70 °C, [catalyst] ≈ 1 mM, reported by Ovitt and Coates.[12]
[c] Carried out in toluene at 70 °C, [catalyst] ≈ 13 mM, reported by Zhong *et al.*[83a]
[d] Carried out in CH_2Cl_2 at 25 °C, [catalyst] ≈ 1 mM.[124]

(Table 1). This may be due to the increased steric bulk of the BINAP backbone relative to the cyclohexyl backbone in complex **2**, which may also be responsible for the low activity of the system. Interestingly, the same trend is not observed with the analogous Al complexes (κ^4-SalBinap)Al(O^iPr) and (ONNO)Al(O^iPr), which have polymerization rates within the same order of magnitude and produce highly isotactic PLA.[12, 83a]

10.4 Conclusions and future directions

The need for high performance biodegradable materials will grow in the future. New developments in this field must reconcile the need to identify new, inexpensive, and biorenewable sources of monomer and the possibility of converting them into biodegradable materials with comparable properties with the ubiquitous polyolefins. Metal-based catalysts for ROP of strained esters are an excellent tool in reaching this goal.

This chapter discussed different families of tridentate and tetradentate ligands for use in supporting various Lewis acidic catalysts for cyclic ester polymerization, and assessed their efficacy in controlling polymer macro- and microstructure. Although it is difficult to generalize, from the evidence obtained to date, the use of tridentate ligands tends to form heteroselective catalysts, whereas tetradentate ligands exert more control over catalyst selectivity and can generate both highly hetero- and isoselective catalysts. In particular, salen ligands and similar tetradentate ligands have shown some of the highest isoselectivity for lactide polymerization to date.

Despite all the progress in this field, it is clear that selectivity is unpredictable and highly metal dependent. For example, a comparison of Al and In complexes in our work shows that while the former have the potential for great selectivity, they are not suitable for commercialization due to slow rates of polymerization and low tolerance to monomer impurities. In contrast, In-based catalysts are more active than their Al counterparts and show great promise in the area of controlled and selective polymerization of cyclic esters, although they have not reached the selectivity of their Al counterparts.

The Lewis acidic nature of these catalysts often renders them prone to aggregation. New venues in ligand design in this field can focus on ligand architecture that controls aggregation, while promoting reactivity.

References

[1] J. A. Brydson, *Plastics Materials*, Butterworth-Heinemann, Oxford, **1999**.
[2] Nobelprize.org, *Vol. 2014*, Nobel Media AB (accessed April 11, **2016**).
[3] (a) W. Kaminsky, *Macromol. Chem. Phys.* **2008**, *209*, 459–466; (b) M. Gahleitner, L. Resconi, P. Doshev, *MRS Bull.* **2013**, *38*, 229–233.
[4] (a) G. Scott, *Polymers and the Environment*, The Royal Society of Chemistry, Letchworth, **1999**, pp. 19–37; (b) C. J. Moore, in *The New York Times* Aug 25, **2014**.
[5] (a) W. Amass, A. Amass, B. Tighe, *Polym. Int.* **1998**, *47*, 89–144; (b) R. A. Gross, B. Kalra, *Science* **2002**, *297*, 803–807.
[6] M. Jamshidian, E. A. Tehrany, M. Imran, M. Jacquot, S. Desobry, *Compr. Rev. Food Sci. Food Saf.* **2010**, *9*, 552–571.
[7] A. D. Jenkins, P. Kratochvíl, R. F. T. Stepto, U. W. Suter, *Pure Appl. Chem.* **1996**, *68*, 2287–2289.
[8] (a) N. E. Kamber, W. Jeong, R. M. Waymouth, R. C. Pratt, B. G. G. Lohmeijer, J. L. Hedrick, *Chem. Rev.* **2007**, *107*, 5813–5840; (b) R. F. T. Stepto, R. G. Gilbert, M. Hess, A. D. Jenkins, R. G. Jones, K. P., *Pure Appl. Chem.* **2009**, *81*, 351–353.
[9] S. Asano, T. Aida, S. Inoue, *J. Chem. Soc., Chem. Commun.* **1985**, 1148–1149.
[10] G. J. van Hummel, S. Harkema, F. E. Kohn, J. Feijen, *Acta Crystallogr. Sect. B: Struct. Sci.* **1982**, *38*, 1679–1681.
[11] V. J. Kleine, H.-H. Kleine, *Macromol. Chem. Phys.* **1959**, *30*, 23–38.
[12] T. M. Ovitt, G. W. Coates, *J. Am. Chem. Soc.* **2002**, *124*, 1316–1326.
[13] N. Othman, A. Acosta-Ramirez, P. Mehrkhodavandi, J. R. Dorgan, S. G. Hatzikiriakos, *J. Rheol.* **2011**, *55*, 987–1005.
[14] G. Perego, G. D. Cella, C. Bastioli, *J. Appl. Polym. Sci.* **1996**, *59*, 37–43.
[15] O. Dechy-Cabaret, B. Martin-Vaca, D. Bourissou, *Chem. Rev.* **2004**, *104*, 6147–6176.
[16] R. E. Drumright, P. R. Gruber, D. E. Henton, *Adv. Mater.* **2000**, *12*, 1841–1846.
[17] E. H. David, G. Patrick, L. Jim, R. Jed, in *Natural Fibers, Biopolymers, and Biocomposites*, CRC Press, Boca raton, FL, **2005**.
[18] A. C. Silvino, P. S. Corrêa, M. L. Dias, *J. Appl. Polym. Sci.* **2014**, *131*, 40771.
[19] D. C. Aluthge, C. Xu, N. Othman, N. Noroozi, S. G. Hatzikiriakos, P. Mehrkhodavandi, *Macromolecules* **2013**, *46*, 3965–3974.
[20] Y. Ikada, K. Jamshidi, H. Tsuji, S. H. Hyon, *Macromolecules* **1987**, *20*, 904–906.
[21] K. Fukushima, M. Hirata, Y. Kimura, *Macromolecules* **2007**, *40*, 3049–3055.
[22] W. Dittrich, R. C. Schulz, *Angew. Makromol. Chem.* **1971**, *15*, 109.
[23] A. Duda, S. Penczek, *Macromolecules* **1990**, *23*, 1636–1639.
[24] (a) T. M. Ovitt, G. W. Coates, *J. Am. Chem. Soc.* **1999**, *121*, 4072–4073; (b) B. M. Chamberlain, M. Cheng, D. R. Moore, T. M. Ovitt, E. B. Lobkovsky, G. W. Coates, *J. Am. Chem. Soc.* **2001**, *123*, 3229–3238; (c) P. J. Dijkstra, H. Z. Du, J. Feijen, *Polym. Chem.* **2011**, *2*, 520–527.
[25] M. H. Chisholm, N. J. Patmore, Z. P. Zhou, *Chem. Commun.* **2005**, 127–129.
[26] (a) M. J. Stanford, A. P. Dove, *Chem. Soc. Rev.* **2010**, *39*, 486–494; (b) S. Dutta, W.-C. Hung, B.-H. Huang, C.-C. Lin, in *Synthetic Biodegradable Polymers, Vol. 245*

(eds B. Rieger, A. Künkel, G. W. Coates, R. Reichardt, E. Dinjus, T. A. Zevaco), Springer, Berlin, **2012**, pp. 219–283.

[27] J. Dorgan, H. Lehermeier, M. Mang, *J. Polym. Environ.* **2000**, *8*, 1–9.

[28] C. M. Thomas, *Chem. Soc. Rev.* **2010**, *39*, 165–173.

[29] E. F. Connor, G. W. Nyce, M. Myers, A. Möck, J. L. Hedrick, *J. Am. Chem. Soc.* **2002**, *124*, 914–915.

[30] M. Myers, E. F. Connor, T. Glauser, A. Möck, G. Nyce, J. L. Hedrick, *J. Polym. Sci., Part A: Polym. Chem.* **2002**, *40*, 844–851.

[31] (a) M. Cheng, A. B. Attygalle, E. B. Lobkovsky, G. W. Coates, *J. Am. Chem. Soc.* **1999**, *121*, 11583–11584; (b) F. Drouin, P. O. Oguadinma, T. J. J. Whitehorne, R. E. Prud'homme, F. Schaper, *Organometallics* **2010**, *29*, 2139–2147.

[32] M. H. Chisholm, N. W. Eilerts, *Chem. Commun.* **1996**, 853–854.

[33] B. Chisholm, N. W. Eilerts, J. C. Huffman, S. S. Iyer, M. Pacold, K. Phomphrai, *J. Am. Chem. Soc.* **2000**, *122*, 11845–11854.

[34] (a) M. H. Chisholm, J. Gallucci, K. Phomphrai, *Chem. Commun.* **2003**, 48–49; (b) M. H. Chisholm, J. C. Gallucci, K. Phomphrai, *Inorg. Chem.* **2004**, *43*, 6717–6725.

[35] M. H. Chisholm, *Inorg. Chim. Acta* **2009**, *362*, 4284–4290.

[36] L. Clark, M. G. Cushion, H. E. Dyer, A. D. Schwarz, R. Duchateau, P. Mountford, *Chem. Commun.* **2010**, *46*, 273–275.

[37] M. G. Cushion, P. Mountford, *Chem. Commun.* **2011**, *47*, 2276–2278.

[38] (a) L. F. Sanchez-Barba, A. Garces, M. Fajardo, C. Alonso-Moreno, J. Fernandez-Baeza, A. Otero, A. Antinolo, J. Tejeda, A. Lara-Sanchez, I. Lopez-Solera, *Organometallics* **2007**, *26*, 6403–6411; (b) A. Garces, L. F. Sanchez-Barba, C. Alonso-Moreno, M. Fajardo, J. Fernandez-Baeza, A. Otero, A. Lara-Sanchez, I. Lopez-Solera, A. M. Rodriguez, *Inorg. Chem.* **2010**, *49*, 2859–2871; (c) L. F. Sánchez-Barba, A. Garcés, J. Fernández-Baeza, A. Otero, C. Alonso-Moreno, A. Lara-Sánchez, A. M. Rodríguez, *Organometallics* **2011**, *30*, 2775–2789.

[39] (a) C. Alonso-Moreno, A. Garces, L. F. Sanchez-Barba, M. Fajardo, J. Fernandez-Baeza, A. Otero, A. Lara-Sanchez, A. Antinolo, L. M. Broomfield, I. Lopez-Solera, A. M. Rodriguez, *Organometallics* **2008**, *27*, 1310–1321; (b) A. Otero, J. Fernandez-Baeza, L. F. Sanchez-Barba, J. Tejeda, M. Honrado, A. Garces, A. Lara-Sanchez, A. M. Rodriguez, *Organometallics* **2012**, *31*, 4191–4202; (c) M. Honrado, A. Otero, J. Fernández-Baeza, L. F. Sánchez-Barba, A. Lara-Sánchez, J. Tejeda, M. P. Carrión, J. Martínez-Ferrer, A. Garcés, A. M. Rodríguez, *Organometallics* **2013**, *32*, 3437–3440; (d) M. Honrado, A. Otero, J. Fernandez-Baeza, L. F. Sanchez-Barba, A. Garces, A. Lara-Sanchez, A. M. Rodriguez, *Organometallics* **2014**, *33*, 1859–1866.

[40] (a) A. Otero, A. Lara-Sanchez, J. Fernandez-Baeza, C. Alonso-Moreno, J. A. Castro-Osma, I. Marquez-Segovia, L. F. Sanchez-Barba, A. M. Rodriguez, J. C. Garcia-Martinez, *Organometallics* **2011**, *30*, 1507–1522; (b) J. A. Castro-Osma, C. Alonso-Moreno, I. Marquez-Segovia, A. Otero, A. Lara-Sanchez, J. Fernandez-Baeza, A. M. Rodriguez, L. F. Sanchez-Barba, J. C. Garcia-Martinez, *Dalton Trans.* **2013**, *42*, 9325–9337.

[41] (a) A. Otero, J. Fernandez-Baeza, A. Lara-Sanchez, C. Alonso-Moreno, I. Marquez-Segovia, L. F. Sanchez-Barba, A. M. Rodriguez, *Angew. Chem. Int. Ed.* **2009**, *48*, 2176–2179; (b) A. Otero, A. Lara-Sanchez, J. Fernandez-Baeza, C. Alonso-Moreno, I. Marquez-Segovia, L. F. Sanchez-Barba, J. A. Castro-Osma, A. M. Rodriguez, *Dalton Trans.* **2011**, *40*, 4687–4696.

[42] A. Garces, L. F. Sanchez-Barba, J. Fernandez-Baeza, A. Otero, M. Honrado, A. Lara-Sanchez, A. M. Rodriguez, *Inorg. Chem.* **2013**, *52*, 12691–12701.

[43] N. Emig, H. Nguyen, H. Krautscheid, R. Reau, J. B. Cazaux, G. Bertrand, *Organometallics* **1998**, *17*, 3599–3608.
[44] A. Dumitrescu, B. Martin-Vaca, H. Gornitzka, J. B. Cazaux, D. Bourissou, G. Bertrand, *Eur. J. Inorg. Chem. 2002*, 1948–1951.
[45] A. D. Schwarz, Z. Y. Chu, P. Mountford, *Organometallics* **2010**, *29*, 1246–1260.
[46] M. P. Blake, A. D. Schwarz, P. Mountford, *Organometallics* **2011**, *30*, 1202–1214.
[47] A. D. Schwarz, K. R. Herbert, C. Paniagua, P. Mountford, *Organometallics* **2010**, *29*, 4171–4188.
[48] B. Lian, C. M. Thomas, O. L. Casagrande, C. W. Lehmann, T. Roisnel, J. F. Carpentier, *Inorg. Chem.* **2007**, *46*, 328–340.
[49] (a) F. Hild, P. Haquette, L. Brelot, S. Dagorne, *Dalton Trans.* **2010**, *39*, 533–540; (b) F. Hild, N. Neehaul, F. Bier, M. Wirsum, C. Gourlaouen, S. Dagorne, *Organometallics* **2013**, *32*, 587–598.
[50] Y. Takashima, Y. Nakayama, K. Watanabe, T. Itono, N. Ueyama, A. Nakamura, H. Yasuda, A. Harada, J. Okuda, *Macromolecules* **2002**, *35*, 7538–7544.
[51] (a) C. Romain, L. Brelot, S. Bellemin-Laponnaz, S. Dagorne, *Organometallics* **2010**, *29*, 1191–1198; (b) C. Romain, B. Heinrich, S. Bellemin-Laponnaz, S. Dagorne, *Chem. Commun.* **2012**, *48*, 2213–2215.
[52] (a) E. Grunova, E. Kirillov, T. Roisnel, J. F. Carpentier, *Dalton Trans.* **2010**, *39*, 6739–6752; (b) J. S. Klitzke, T. Roisnel, E. Kirillov, O. D. L. Casagrande, J.-F. Carpentier, *Organometallics* **2014**, *33*, 309–321; (c) J. S. Klitzke, T. Roisnel, E. Kirillov, O. D. L. Casagrande, J.-F. Carpentier, *Organometallics* **2014**, *33*, 5693–5707.
[53] (a) H. Y. Tang, H. Y. Chen, J. H. Huang, C. C. Lin, *Macromolecules* **2007**, *40*, 8855–8860; (b) Y. Huang, W. C. Hung, M. Y. Liao, T. E. Tsai, Y. L. Peng, C. C. Lin, *J. Polym. Sci., Part A: Polym. Chem.* **2009**, *47*, 2318–2329; (c) H. J. Chuang, Y. C. Su, B. T. Ko, C. C. Lin, *Inorg. Chem. Commun.* **2012**, *18*, 38–42; (d) H. J. Chuang, H. L. Chen, B. H. Huang, T. E. Tsai, P. L. Huang, T. T. Liao, C. C. Lin, *J. Polym. Sci., Part A: Polym. Chem.* **2013**, *51*, 1185–1196; (e) M. W. Hsiao, G. S. Wu, B. H. Huang, C. C. Lin, *Inorg. Chem. Commun.* **2013**, *36*, 90–95.
[54] (a) N. Nimitsiriwat, E. L. Marshall, V. C. Gibson, M. R. J. Elsegood, S. H. Dale, *J. Am. Chem. Soc.* **2004**, *126*, 13598–13599; (b) N. Nimitsiriwat, V. C. Gibson, E. L. Marshall, M. R. J. Elsegood, *Inorg. Chem.* **2008**, *47*, 5417–5424.
[55] J. C. Wu, B. H. Huang, M. L. Hsueh, S. L. Lai, C. C. Lin, *Polymer* **2005**, *46*, 9784–9792.
[56] (a) H. Y. Chen, H. Y. Tang, C. C. Lin, *Macromolecules* **2006**, *39*, 3745–3752; (b) W. C. Hung, Y. Huang, C. C. Lin, *J. Polym. Sci., Part A: Polym. Chem.* **2008**, *46*, 6466–6476; (c) W. C. Hung, C. C. Lin, *Inorg. Chem.* **2009**, *48*, 728–734.
[57] (a) D. J. Darensbourg, W. Choi, C. P. Richers, *Macromolecules* **2007**, *40*, 3521–3523; (b) D. J. Darensbourg, W. Choi, O. Karroonnirun, N. Bhuvanesh, *Macromolecules* **2008**, *41*, 3493–3502; (c) D. J. Darensbourg, O. Karroonnirun, *Inorg. Chem.* **2010**, *49*, 2360–2371; (d) D. J. Darensbourg, O. Karroonnirun, *Macromolecules* **2010**, *43*, 8880–8886.
[58] A. D. Schwarz, A. L. Thompson, P. Mountford, *Inorg. Chem.* **2009**, *48*, 10442–10454.
[59] Y. Kim, G. K. Jnaneshwara, J. G. Verkade, *Inorg. Chem.* **2003**, *42*, 1437–1447.
[60] S. Gendler, S. Segal, I. Goldberg, Z. Goldschmidt, M. Kol, *Inorg. Chem.* **2006**, *45*, 4783–4790.
[61] A. J. Chmura, M. G. Davidson, C. J. Frankis, M. D. Jones, M. D. Lunn, *Chem. Commun.* **2008**, 1293–1295.
[62] A. J. Chmura, C. J. Chuck, M. G. Davidson, M. D. Jones, M. D. Lunn, S. D. Bull, M. F. Mahon, *Angew. Chem. Int. Ed.* **2007**, *46*, 2280–2283.

[63] A. J. Chmura, M. G. Davidson, M. D. Jones, M. D. Lunn, M. F. Mahon, A. F. Johnson, P. Khunkamchoo, S. L. Roberts, S. S. F. Wong, *Macromolecules* **2006**, *39*, 7250–7257.

[64] (a) F. Bonnet, A. R. Cowley, P. Mountford, *Inorg. Chem.* **2005**, *44*, 9046–9055; (b) H. E. Dyer, S. Huijser, N. Susperregui, F. Bonnet, A. D. Schwarz, R. Duchateau, L. Maron, P. Mountford, *Organometallics* **2010**, *29*, 3602–3621.

[65] (a) C.-X. Cai, A. Amgoune, C. W. Lehmann, J.-F. Carpentier, *Chem. Commun.* **2004**, 330–331; (b) A. Amgoune, C. M. Thomas, T. Roisnel, J. F. Carpentier, *Chem. Eur. J.* **2006**, *12*, 169–179.

[66] A. Amgoune, C. M. Thomas, J. F. Carpentier, *Macromol. Rapid Commun.* **2007**, *28*, 693–697.

[67] A. Amgoune, C. M. Thomas, S. Ilinca, T. Roisnel, J. F. Carpentier, *Angew. Chem. Int. Ed.* **2006**, *45*, 2782–2784.

[68] T. Aida, S. Inoue, *Acc. Chem. Res.* **1996**, *29*, 39–48.

[69] T. Aida, S. Inoue, *J. Am. Chem. Soc.* **1983**, *105*, 1304–1309.

[70] L. Trofimoff, T. Aida, S. Inoue, *Chem. Lett.* **1987**, *16*, 991–994.

[71] (a) T. Yasuda, T. Aida, S. Inoue, *Macromol. Rapid Commun.* **1982**, *3*, 585–588; (b) K. Shimasaki, T. Aida, S. Inoue, *Macromolecules* **1987**, *20*, 3076–3080; (c) M. Endo, T. Aida, S. Inoue, *Macromolecules* **1987**, *20*, 2982–2988.

[72] (a) J.-C. Buffet, J. P. Davin, T. P. Spaniol, J. Okuda, *New J. Chem.* **2011**, *35*, 2253–2257; (b) J.-C. Buffet, J. Okuda, *Dalton Trans.* **2011**, *40*, 7748–7754.

[73] M. Frediani, D. Sémeril, A. Mariotti, L. Rosi, P. Frediani, L. Rosi, D. Matt, L. Toupet, *Macromol. Rapid Commun.* **2008**, *29*, 1554–1560.

[74] P. G. Cozzi, *Chem. Soc. Rev.* **2004**, *33*, 410–421.

[75] W. Zhang, J. L. Loebach, S. R. Wilson, E. N. Jacobsen, *J. Am. Chem. Soc.* **1990**, *112*, 2801–2803.

[76] K. Matsumoto, B. Saito, T. Katsuki, *Chem. Commun.* **2007**, 3619–3627.

[77] (a) V. Vincens, A. Le Borgne, N. Spassky, *Macromol. Symp.* **1991**, *47*, 285–291; (b) A. Le Borgne, V. Vincens, M. Jouglard, N. Spassky, *Macromol. Symp.* **1993**, *73*, 37–46.

[78] N. Spassky, M. Wisniewski, C. Pluta, A. LeBorgne, *Macromol. Chem. Phys.* **1996**, *197*, 2627–2637.

[79] G. Montaudo, M. S. Montaudo, C. Puglisi, F. Samperi, N. Spassky, A. LeBorgne, M. Wisniewski, *Macromolecules* **1996**, *29*, 6461–6465.

[80] T. M. Ovitt, G. W. Coates, *J. Polym. Sci., Part A: Polym. Chem.* **2000**, *38*, 4686–4692.

[81] C. P. Radano, G. L. Baker, M. R. Smith, *J. Am. Chem. Soc.* **2000**, *122*, 1552–1553.

[82] K. Majerska, A. Duda, *J. Am. Chem. Soc.* **2004**, *126*, 1026–1027.

[83] (a) Z. Y. Zhong, P. J. Dijkstra, J. Feijen, *Angew. Chem. Int. Ed.* **2002**, *41*, 4510–4513; (b) Z. Y. Zhong, P. J. Dijkstra, J. Feijen, *J. Am. Chem. Soc.* **2003**, *125*, 11291–11298.

[84] M. H. Chisholm, J. C. Gallucci, K. T. Quisenberry, Z. Zhou, *Inorg. Chem.* **2008**, *47*, 2613–2624.

[85] A. Pilone, K. Press, I. Goldberg, M. Kol, M. Mazzeo, M. Lamberti, *J. Am. Chem. Soc.* **2014**, *136*, 2940–2943.

[86] N. Maudoux, T. Roisnel, V. Dorcet, J.-F. Carpentier, Y. Sarazin, *Chem. Eur. J.* **2014**, *20*, 6131–6147.

[87] D. Jhurry, A. Bhaw-Luximon, N. Spassky, *Macromol. Symp.* **2001**, *175*, 67–79.

[88] N. Nomura, R. Ishii, M. Akakura, K. Aoi, *J. Am. Chem. Soc.* **2002**, *124*, 5938–5939.

[89] N. Nomura, R. Ishii, Y. Yamamoto, T. Kondo, *Chem. Eur. J.* **2007**, *13*, 4433–4451.

[90] (a) Z. Tang, X. Chen, X. Pang, Y. Yang, X. Zhang, X. Jing, *Biomacromolecules* **2004**, *5*, 965–970; (b) Z. Tang, X. Chen, Y. Yang, X. Pang, J. Sun, X. Zhang, X. Jing, *J. Polym. Sci., Part A: Polym. Chem.* **2004**, *42*, 5974–5982; (c) H.-L. Chen, S. Dutta, P.-Y. Huang, C.-C. Lin, *Organometallics* **2012**, *31*, 2016–2025.

[91] P. Hormnirun, E. L. Marshall, V. C. Gibson, R. I. Pugh, A. J. P. White, *Proc. Natl. Acad. Sci. USA* **2006**, *103*, 15343–15348.

[92] P. Hormnirun, E. L. Marshall, V. C. Gibson, A. J. P. White, D. J. Williams, *J. Am. Chem. Soc.* **2004**, *126*, 2688–2689.

[93] H. Du, A. H. Velders, P. J. Dijkstra, J. Sun, Z. Zhong, X. Chen, J. Feijen, *Chem. Eur. J.* **2009**, *15*, 9836–9845.

[94] H. Du, A. H. Velders, P. J. Dijkstra, Z. Zhong, X. Chen, J. Feijen, *Macromolecules* **2009**, *42*, 1058–1066.

[95] X. Pang, X. Chen, H. Du, X. Wang, X. Jing, *J. Organomet. Chem.* **2007**, *692*, 5605–5613.

[96] M. Bouyahyi, E. Grunova, N. Marquet, E. Kirillov, C. M. Thomas, T. Roisnel, J. F. Carpentier, *Organometallics* **2008**, *27*, 5815–5825.

[97] M. Bouyahyi, T. Roisnel, J. F. Carpentier, *Organometallics* **2010**, *29*, 491–500.

[98] V. Balasanthiran, M. H. Chisholm, C. B. Durr, J. C. Gallucci, *Dalton Trans.* **2013**, *42*, 11234–11241.

[99] M. Normand, E. Kirillov, T. Roisnel, J. F. Carpentier, *Organometallics* **2012**, *31*, 1448–1457.

[100] N. Maudoux, T. Roisnel, J.-F. Carpentier, Y. Sarazin, *Organometallics* **2014**, *33*, 5740–5748.

[101] C. K. A. Gregson, I. J. Blackmore, V. C. Gibson, N. J. Long, E. L. Marshall , A. J. P. White, *Dalton Trans.* **2006**, 3134–3140.

[102] T. K. Saha, V. Ramkumar, D. Chakraborty, *Inorg. Chem.* **2011**, *50*, 2720–2722.

[103] C.-Y. Tsai, H.-C. Du, J.-C. Chang, B.-H. Huang, B.-T. Ko, C.-C. Lin, *RSC Advances* **2014**, *4*, 14527–14537.

[104] E. M. Broderick, P. L. Diaconescu, *Inorg. Chem.* **2009**, *48*, 4701–4706.

[105] C. K. A. Gregson, V. C. Gibson, N. J. Long, E. L. Marshall, P. J. Oxford, A. J. P. White, *J. Am. Chem. Soc.* **2006**, *128*, 7410–7411.

[106] C. Bakewell, T.-P.-A. Cao, N. Long, X. F. Le Goff, A. Auffrant, C. K. Williams, *J. Am. Chem. Soc.* **2012**, *134*, 20577–20580.

[107] C. Bakewell, A. J. P. White, N. J. Long, C. K. Williams, *Angew. Chem. Int. Ed.* **2014**, *53*, 9226–9230.

[108] H. Y. Ma, G. Melillo, L. Oliva, T. P. Spaniol, U. Englert, J. Okuda, *Dalton Trans.* **2005**, 721–727.

[109] I. Peckermann, A. Kapelski, T. P. Spaniol, J. Okuda, *Inorg. Chem.* **2009**, *48*, 5526–5534.

[110] (a) H. Y. Ma, T. P. Spaniol, J. Okuda, *Angew. Chem. Int. Ed.* **2006**, *45*, 7818–7821; (b) H. Ma, T. P. Spaniol, J. Okuda, *Inorg. Chem.* **2008**, *47*, 3328–3339.

[111] J.-C. Buffet, J. Okuda, *Chem. Commun.* **2011**, *47*, 4796–4798.

[112] X. Wang, A. Thevenon, J. L. Brosmer, I. Yu, S. I. Khan, P. Mehrkhodavandi, P. L. Diaconescu, *J. Am. Chem. Soc.* **2014**, *136*, 11264–11267.

[113] C. K. Williams, L. E. Breyfogle, S. K. Choi, W. Nam, V. G. Young, M. A. Hillmyer, W. B. Tolman, *J. Am. Chem. Soc.* **2003**, *125*, 11350–11359.

[114] J. M. Mitchell, N. S. Finney, *Tetrahedron Lett.* **2000**, *41*, 8431–8434.

[115] G. Labourdette, D. J. Lee, B. O. Patrick, M. B. Ezhova, P. Mehrkhodavandi, *Organometallics* **2009**, *28*, 1309–1319.

[116] K. M. Osten, D. C. Aluthge, B. O. Patrick, P. Mehrkhodavandi, *Inorg. Chem.* **2014**, *53*, 9897–9906.

[117] I. P. Hsieh, C. H. Huang, H. M. Lee, P. C. Kuo, J. H. Huang, H. I. Lee, J. T. Cheng, G. H. Lee, *Inorg. Chim. Acta* **2006**, *359*, 497–504.

[118] A. F. Douglas, B. O. Patrick, P. Mehrkhodavandi, *Angew. Chem. Int. Ed.* **2008**, *47*, 2290–2293.

[119] (a) A. Acosta-Ramirez, A. F. Douglas, I. S. Yu, B. O. Patrick, P. L. Diaconescu, P. Mehrkhodavandi, *Inorg. Chem.* **2010**, *49*, 5444–5452; (b) I. Yu, A. Acosta-Ramirez, P. Mehrkhodavandi, *J. Am. Chem. Soc.* **2012**, *134*, 12758–12773.
[120] C. Xu, I. Yu, P. Mehrkhodavandi, *Chem. Commun.* **2012**, *48*, 6806–6808.
[121] (a) K. M. Osten, I. Yu, I. R. Duffy, P. O. Lagaditis, J. C. C. Yu, C. J. Wallis, P. Mehrkhodavandi, *Dalton Trans.* **2012**, *41*, 8123–8134; (b) K. M. Osten, D. C. Aluthge, P. Mehrkhodavandi, *Dalton Trans.* **2015**, *44*, 6126–6139.
[122] D. A. Atwood, M. J. Harvey, *Chem. Rev.* **2001**, *101*, 37–52.
[123] F. X. Gao, C. J. Zhu, F. Yuan, Y. H. Zhu, Y. Pan, *Chin. Chem. Lett.* **2003**, *14*, 138–140.
[124] D. C. Aluthge, B. O. Patrick, P. Mehrkhodavandi, *Chem. Commun.* **2013**, *49*, 4295–4297.
[125] D. C. Aluthge, E. X. Yan, J. M. Ahn, P. Mehrkhodavandi, *Inorg. Chem.* **2014**, *53*, 6828–6836.

11

Modern Applications of Trispyrazolylborate Ligands in Coinage Metal Catalysis

Ana Caballero, M. Mar Díaz-Requejo, Manuel R. Fructos, Juan Urbano, and Pedro J. Pérez

Laboratorio de Catálisis Homogénea, Unidad Asociada al CSIC, Centro de Investigación en Química Sostenible (CIQSO) and Departamento de Química "Prof. José Carlos Vílchez Martín", Universidad de Huelva, Campus de El Carmen, 21007, Huelva, Spain

11.1 Introduction

The successful development of organometallic chemistry and modern coordination chemistry in the last half century is undoubtedly related to the design of a plethora of ligands that has provided the metal center with the electronic and/or steric properties envisioned by the researchers.[1] The most popular ligands in that time span have been phosphines (phosphanes, PR_3)[2] either mono- or polydentate, cyclopentadienyls (Cp^x),[1] *N*-heterocyclic carbenes[3] (NHCs), and trispyrazolylborates (Tp^x, Scheme 1).[4] However, from the perspective of the use of transition metal complexes as catalysts, a large difference in their uses is noted: the number of catalytic systems based on Tp^x-containing complexes is much lower than those involving the other three classes of ligands. This is probably the result of a more sterically congested metal center and stronger metal–ligand bonds in the case of the Tp^xM core, at variance with more

Ligand Design in Metal Chemistry: Reactivity and Catalysis, First Edition. Edited by Mark Stradiotto and Rylan J. Lundgren.

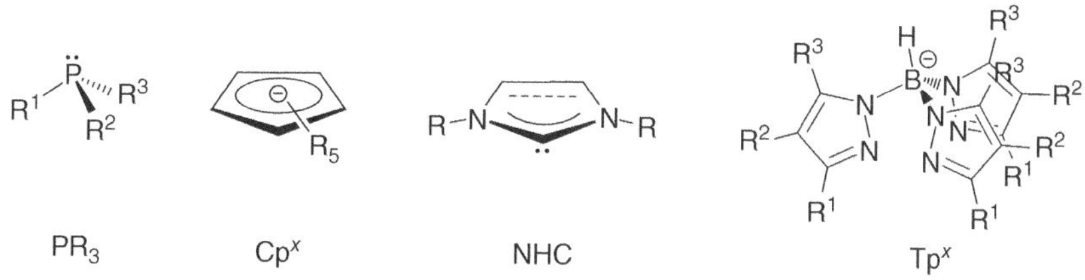

Scheme 1 *Types of ligands frequently employed in transition metal complexes*

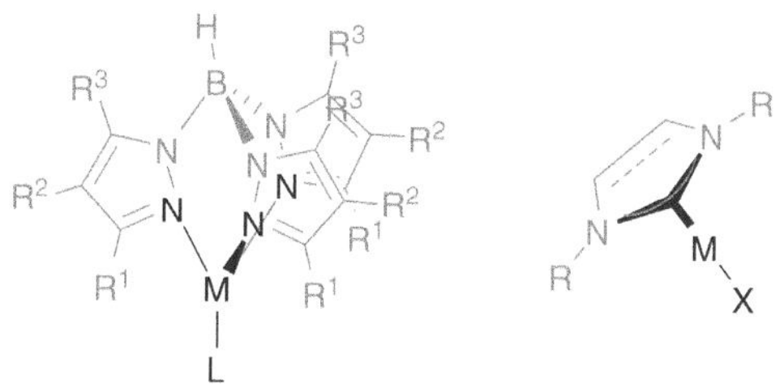

Scheme 2 *Geometries for group 11 Tp^{x}M and (NHC)M catalysts*

flexible M—P or M—Cp^x units that allow dissociation–association processes in the former or several hapticity equilibria in the latter.

Interestingly most of the catalytic systems based on Tp^{x}M units contain a coinage metal such as copper or silver.[5] This is also observed for NHC-based catalytic systems, with the addition of gold to those metals.[6] This similarity can be explained in terms of a common feature of both Tp^{x}M and (NHC)M moieties with those metals: they leave just one coordination site for the catalytic reaction to occur. A second coordination site may also accessible in some cases, for example with Cu(II)-based systems where five-coordinate geometries are available. In the case of (NHC)M systems (M = Cu, Ag, Au), the linear complexes can accept an incoming ligand (reactant) through the transient formation of three-coordinate intermediates (Scheme 2).

In this chapter the current state of the art of the use of copper and silver complexes containing trispyrazolylborate ligands in homogeneous catalysis is presented. As mentioned above, no related examples with the gold analogs have been described, in spite of the synthesis of several Tp^{x}AuL complexes.[7] After a brief introduction about the ligands, the use of Tp^{x}ML (M = Cu, Ag) complexes as catalysts will be presented according to the type of reaction: carbene transfer reactions from diazocompounds (addition or insertion); nitrene transfer reactions from hypervalent iodine compounds; oxo transfer reactions; and atom transfer radical reactions.

11.2 Trispyrazolylborate ligands: main features

Although this class of ligands is well known by the scientific community, a brief account of the main features is provided in this section for the sake of completeness. These ligands were designed and prepared by Trofimenko in the mid 1960s,[8] and since then more than 200 ligands have been reported.[4] The common route for their synthesis is the direct reaction of KBH_4 and the corresponding pyrazole in excess, although several variations have been described (Scheme 3).[4] The variety of groups that can be attached to the pyrazolyl rings provides a versatility that currently rivals those of Cp^x or NHC ligands, only phosphines being by far more developed in terms of substituents than Tp^x. The availability of an array of R groups in the ligand skeleton allows the control of the steric and electronic properties of the resulting metal complex, with the corresponding influence in the catalytic reaction outcome.

As shown in Scheme 3, the R^1 group, located in the vicinity of the metal center, influences markedly the catalytic pocket. The steric hindrance provided by the ligand has been defined in terms of the cone angle and the wedge angle, usually derived from X-ray data from $TlTp^x$ complexes. However, the nature of the metal center in each particular complex affects such angles, particularly the cone angle, and therefore those values must be employed with reservation, only providing a trend within the series of ligands.

The measurement of the electronic nature of Tp ligands can be readily performed using the well-known probe of metal carbonyls and infrared spectroscopy. Series of $Tp^xM(CO)$ (M=Cu, Ag) complexes have been prepared providing a collection of $\nu(CO)$ values from which the relative electron-donating capabilities of the pyrazolylborate

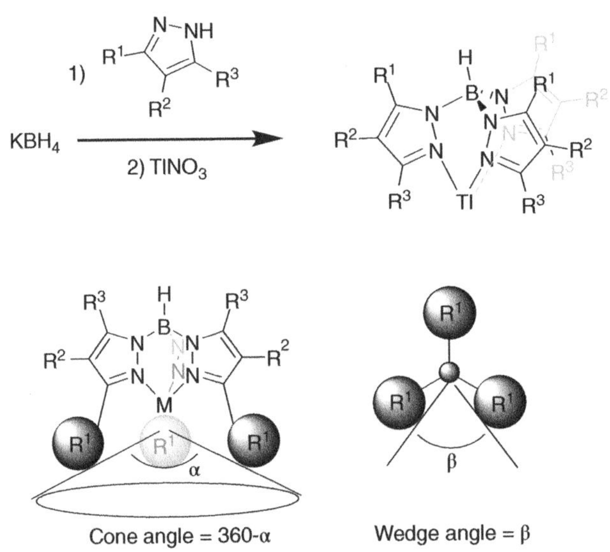

Scheme 3 *The syntheses and steric hindrance of Tp^x ligands*

ligands can be inferred.[9,10] The general trends can be described as in the following. First, with the same Tp^x ligand, the silver complex displays a higher ν(CO) value than the copper analog, that in some cases exceeds that of free CO. These are the so-called non-classic metal carbonyl complexes, where π-backbonding is very low or null. Values up to 2166 cm^{-1} have been found for some Tp^xAg(CO) compounds.[11] Within the same metal, the electronic nature of the R^1–R^3 substituents is transferred to the donor capabilities of the three N-donors. Thus, alkyl substitution in the pyrazolyl ring induces high electron density donation, as inferred from low ν(CO) values (ca. 2050–2060 cm^{-1} in the copper complexes).[12] The use of electron-withdrawing groups as R^1–R^3 affects the electronic nature at the metal center in such a way that when fluorinated[9] or brominated[13] substituents are employed, high ν(CO) values are found.

11.3 Catalytic systems based on Tp^xML complexes (M = Cu, Ag)

Herein we describe an account of the use of copper and silver and Tp-based catalytic systems from the perspective of the reaction class, in an attempt to organize this chemistry from a rational point of view. A large number of those systems correspond to the catalytic transfer of an E ligand bonded to the metal center by a multiple bond. This M=E moiety is generated in situ with the aid of the appropriate reagent. Carbene, nitrene or oxo ligands stand for E, their generation being induced by the direct reaction of the Tp^xML precursor and a diazo compound, a hypervalent I(III) reagent or an oxidant such as H_2O_2, respectively. Scheme 4 gives a very general picture of these transformations. The transient Tp^xM=E intermediate, electrophilic in nature, reacts with a nucleophile with the subsequent transfer of the X group and the net functionalization of that nucleophile. An array of substrates such as olefins, alkynes, arenes, alkanes (including the gaseous series C_1–C_4), polyolefins, both saturated and unsaturated, or haloalkanes have been functionalized in this manner.

A second type of reaction that involves the formal addition of a carbon–halogen bond to a double carbon–carbon, both inter- and intramolecularly, will also be discussed. These are the atom transfer radical reactions, and also include the polymerization of some olefins such as styrene or acrylates.

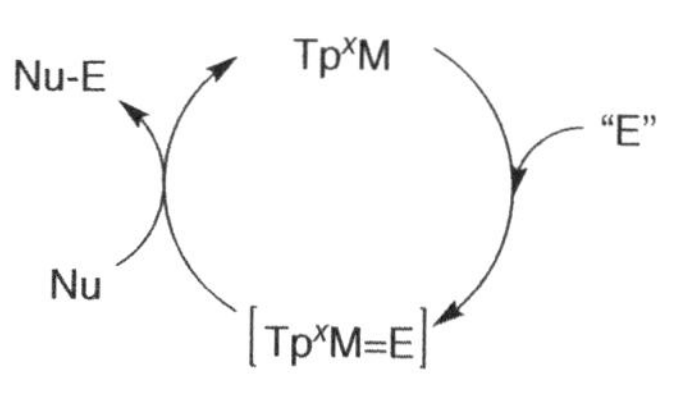

E	Precursor
CR^1R^2	$N_2CR^1R^2$
NR	LG=NR (LG = ArI, N_2)
O	H_2O_2, ROOR′

Scheme 4 *General functionalization of organic substrates by "E" transfer with Tp^xML catalysts*

11.3.1 Carbene addition reactions

There are examples of all metals from groups 8 to 11 to catalyze the transfer of a carbene group from a diazo compound to organic substrates.[5b] One of the most studied transformation is the olefin cyclopropanation reaction,[14] for which the use of Tp^xML catalysts has provided valuable improvement. Thus, the diastereoselectivity of this reaction, that usually leads to mixtures of both *cis* and *trans* isomers, was directed toward the *cis*-cyclopropane with the complex $Tp^{Ms}Cu(thf)$ (hydrotris [3-mesitylpyrazolyl]borate) as the catalyst, affording a 98:2 *cis:trans* mixture with styrene (Scheme 5) and ethyl diazoacetate (EDA) as the carbene source.[15] Other olefins were also cyclopropanated with the preferential formation of the *cis* isomer. The catalysts can be prepared in situ by mixing a Cu(I) source and the MTp^x salt.[16] Also, the $Tp^{Br3}Cu(NCMe)$ complex has been employed as catalyst in a fluorous phase for the styrene cyclopropanation reaction.[17]

The catalytic capabilities of the Tp^xML complexes for this transformation have also been applied to the functionalization of macromolecules such as polyolefins. Thus polybutadienes (Scheme 6a)[18] or styrene-butadiene rubbers (Scheme 6b) [19] have been modified upon addition of carbene units from EDA that were incorporated into the unsaturated C=C bonds of the polymeric chain, providing interesting features to the isolated materials: the incorporation of polar groups provided distinct properties regarding their potential use as adhesives, but maintaining the structure of the parent polymer.

Silver complexes have also been described for the cyclopropanation reaction. When using benzene, the use of $Tp^{(CF3)2}Ag(thf)$ (where $Tp^{(CF3)2}$=hydrotris(3, 5-bis(trisfluoromethyl)pyrazolyl)borate; for the rules of nomenclature of Tp^x ligands see ref[4a]) provided products derived from the addition of the carbene moiety to the arene ring (Scheme 7a), followed by ring expansion into a cycloheptatriene, in the

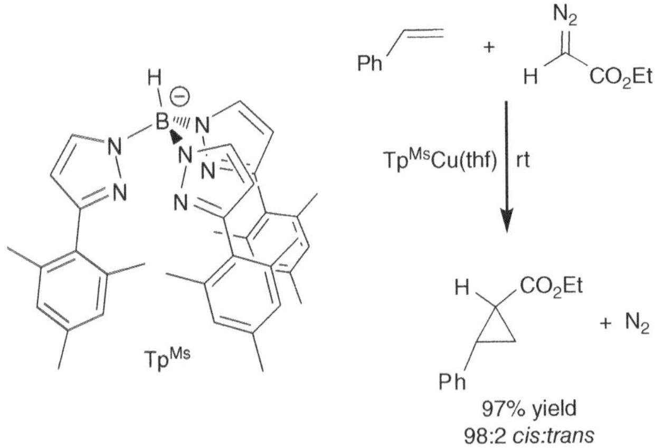

Scheme 5 *Preferential cis-cyclopropanation of olefins using $Tp^{Ms}Cu(thf)$ as catalyst.*

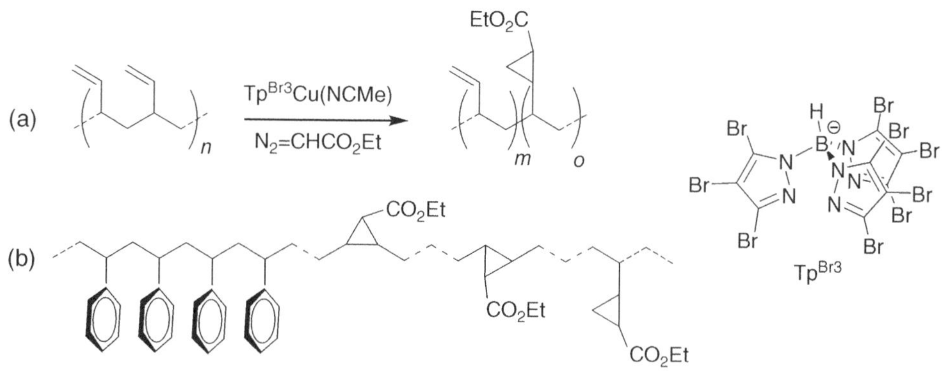

Scheme 6 *Modification of macromolecules by carbene addition*

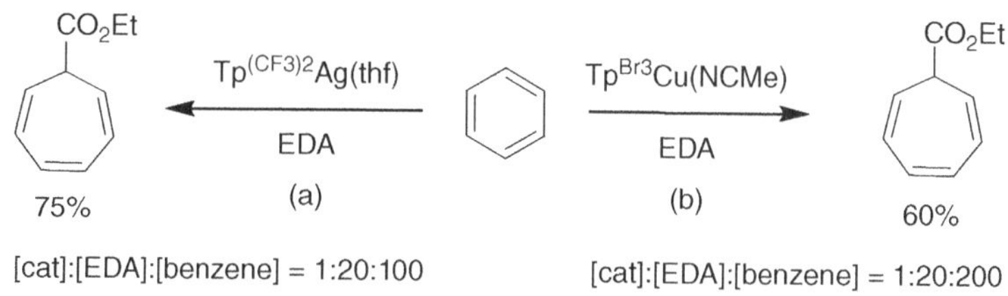

Scheme 7 *Addition of carbene units from ethyl diazoacetate (EDA) to benzene catalyzed with copper and silver Tp^x-containing complexes*

so-called Buchner reaction.[20] This reaction had been previously described with the $Tp^{Br3}Cu(NCMe)$ complex (where Tp^{Br3}=hydrotris(3,4,5-tribromo)pyrazolylborate) with benzene or alkylbenzene substrates (Scheme 7b).[21]

The asymmetric version of the olefin cyclopropanation reaction has also been described with a chiral *trans*-Tp^{pm}Cu complex (Scheme 8), that led to enantiomeric excess (ee) values in the 80–85% range for both *cis* and *trans* isomers in the reaction of styrene and EDA.[22]

In an example differing from regular olefins, furans have also been functionalized with this methodology. Several Tp^xCuL complexes were found to promote the cyclopropanation of one of the double bonds of these substrates that were obtained along with their ring-opening products.[23] The reaction could be driven toward the latter upon treatment with elemental iodine. Also, the synthesis of ostopanic acid was described as a practical example (Scheme 9a).

The related addition reaction to alkynes produces cyclopropenes, and it has also been developed with copper-based Tp^xCuL catalysts. Both terminal and internal alkynes were converted into these three-membered rings in good yields with activities at least comparable with those of the well-known $Rh_2(OOCR)_4$ catalysts (Scheme 9b).[24]

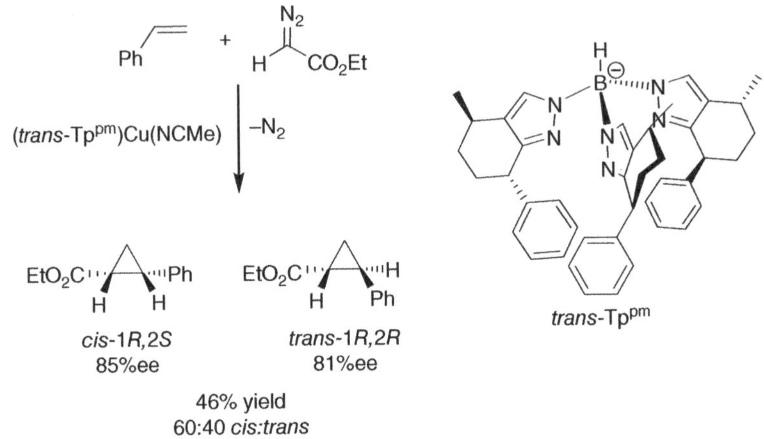

Scheme 8 *The asymmetric cyclopropanation of styrene with the chiral trans-Tp^{pm} ligand*

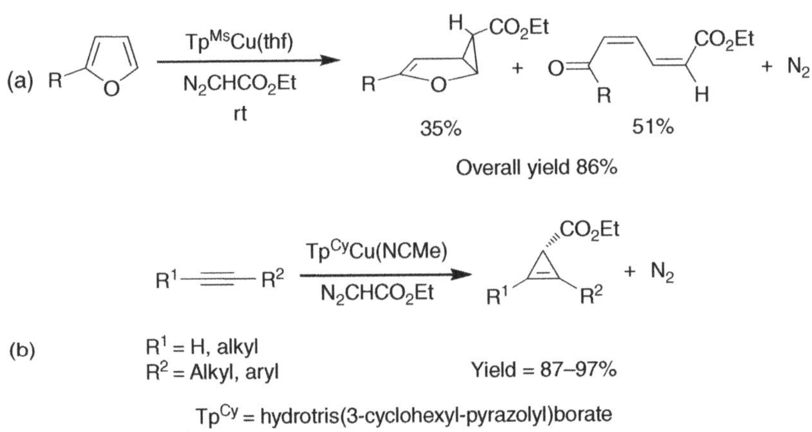

Scheme 9 *Furan functionalization (a) and alkyne cyclopropenation (b) by carbene addition*

11.3.2 Carbene insertion reactions

In addition to the unsaturated substrates, saturated bonds are also suitable for functionalization by formal carbene insertion, following the general process shown in Scheme 4. This strategy has been applied not only to highly polar N–H, O–H and C–Cl bonds, but also to the less nucleophilic C–H and Si–H bonds which have been modified with a series of Tp^xML catalysts (Scheme 10).

Amines[25] and alcohols[26] were readily functionalized with copper-based catalysts containing Tp^x ligands upon reaction of several diazo compounds under very mild conditions. In the case of the former, this strategy led to the formation of amino acids and peptides, the catalyst activity competing with that of the previous rhodium-based

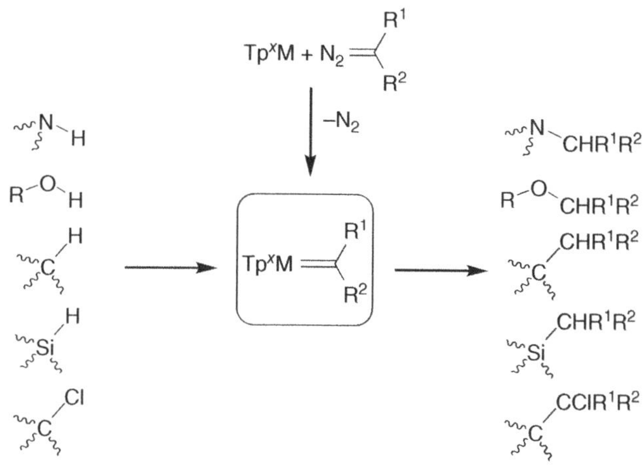

Scheme 10 *Functionalization of saturated bonds by carbene insertion with Tp^xML catalysts*

catalysts. In the case of alcohols, a series of ethers were obtained due to the insertion of the carbene group into the O—H bond, even with unsaturated alcohols. It is worth noting that for these two reactions no special requirements for the Tp^x ligand in the catalyst were observed: the reaction proceeded quite well with a variety of catalysts, a slight improvement being detected with those bearing less donating pyrazolyl groups.

In contrast with the examples above, less nucleophilic C—H and Si—H bonds required electron-poor metal centers to enable the targeted carbene insertion, with the best catalysts being those containing Tp^x ligands bearing electron-withdrawing groups, mainly halogenated. Thus, linear and branched as well as cyclic alkanes were functionalized in a general manner with EDA as the carbene source at room temperature with Tp^xM cores. $Tp^{Ms}CuL$[27] and $Tp^{Br3}CuL$[13] complexes provided the first series of results, although only secondary and tertiary sites of linear alkanes were modified with the latter. Primary sites were later consecutively modified with the $Tp^{Br3}Ag(NCMe)$[28] and $Tp^{(CF3)2}Ag(thf)$[29] catalysts. Scheme 11 provides a comparison of the activity and selectivity of these catalysts referred to hexane as substrate. The difference in reactivity was explained with the aid of theoretical calculations that showed a much lower barrier for the silver catalyst.[30] The mechanism of the EDA dimerization with these complexes has also been studied from a computational point of view.[31] The C—H functionalization reaction occurs with the formation of a metallocarbene intermediate that reacts with the alkane in a single, irreversible step where the regioselectivity is decided (Scheme 11). In all cases, and as the main drawback of this methodology, mixtures of products were obtained when more than one type of C—H bond was present in the substrate. The design of regioselective catalysts remains a goal in this chemistry.

The strategy of carbene insertion into C—H bonds as a functionalization tool was also employed with polyolefins such as polyethylene or polypropylene that preferentially underwent incorporation of the carbene moiety into the tertiary sites (Scheme 12).[32]

(a)

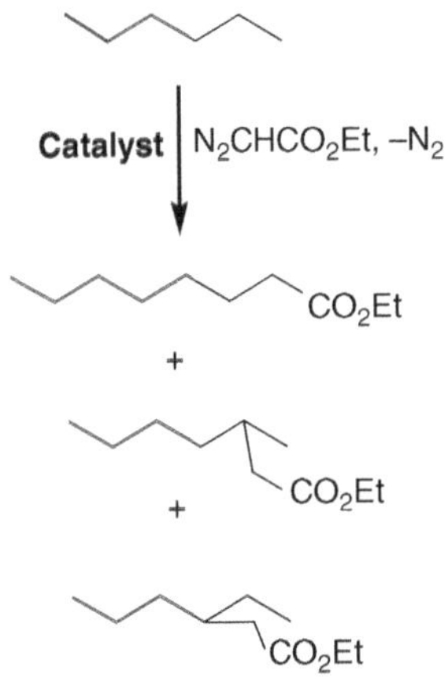

catalyst

$Tp^{Br3}Cu$	$Tp^{Br3}Ag$	$Tp^{(CF3)2}Ag$
n.d.	1	1
3	3.7	1.7
1	1.4	0.9

(b)

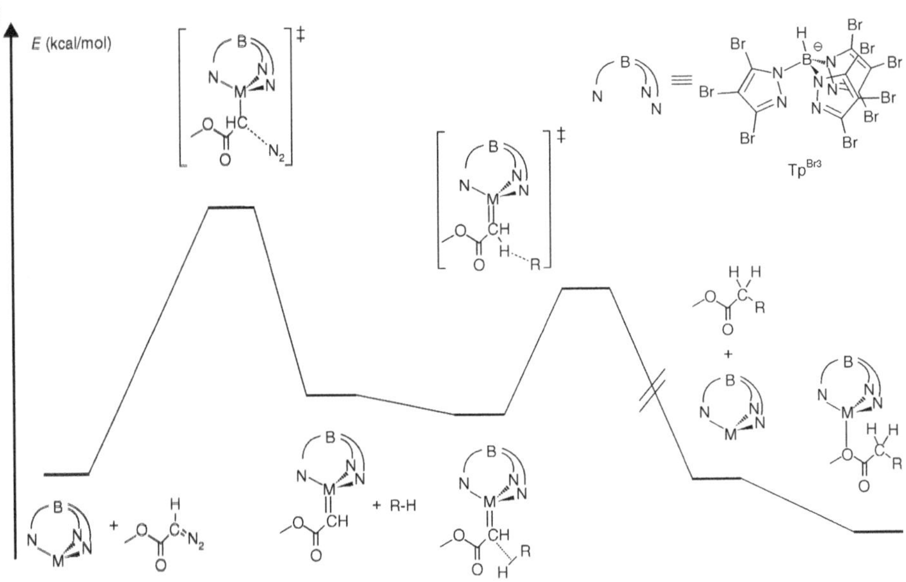

Scheme 11 *(a) The functionalization of hexane with EDA catalyzed by Tp^xM complexes. (b) Calculated reaction pathway for this transformation*

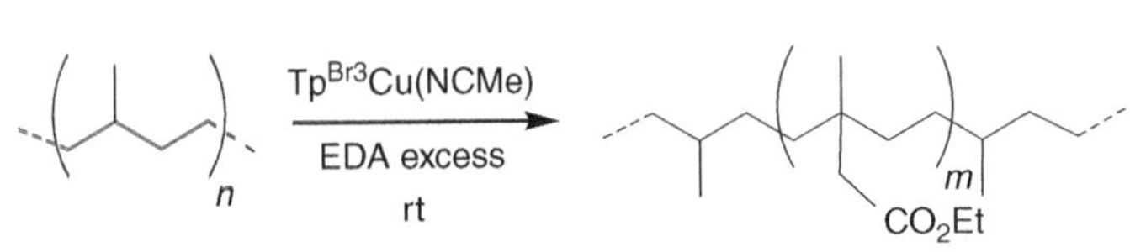

Scheme 12 *Polyolefin functionalization by carbene insertion catalyzed by $Tp^{Br3}Cu(NCMe)$*

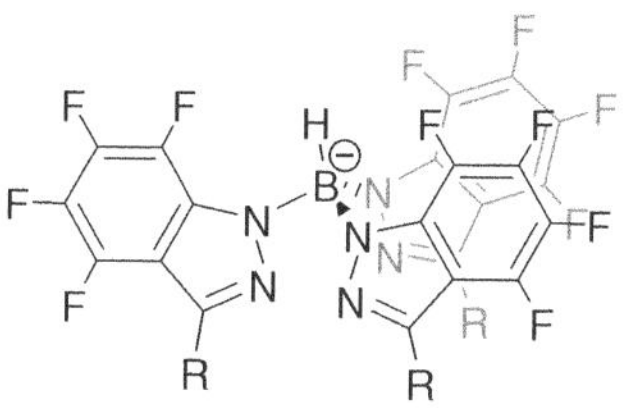

R	Tp^F	n(CO) in Tp^F Ag(CO)
CF_3	Tp^{F21}	2167 cm^{-1}
CF_2CF_3	Tp^{F27}	2166 cm^{-1}
$CF_2CF_2CF_3$	Tp^{F33}	2166 cm^{-1}
$CF_2CF_2CF_2CF_3$	Tp^{F39}	2166 cm^{-1}
$CF_2CF_2CF_2CF_2CF_2CF_3$	Tp^{F51}	2166 cm^{-1}

Scheme 13 *The second generation of poorly donating pyrazolylborates*

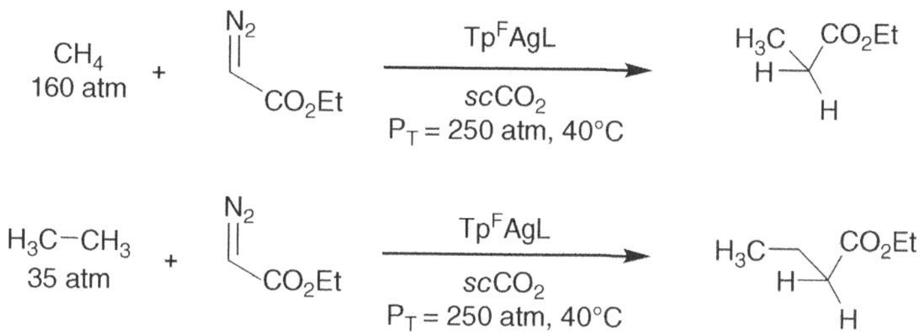

Scheme 14 *Functionalization of methane and ethane using silver-based catalysts and supercritical carbon dioxide as the solvent*

The second generation of catalysts for this transformation was based on fully fluorinated trisindazolylborates, leading to highly electrophilic metal centers. A series of five new ligands were prepared[33] (Scheme 13) and their silver complexes synthesized and tested as catalysts for the alkane functionalization reaction.[11] In addition to excellent activities, they were employed in fluorous biphasic conditions, allowing a ready separation of products from the catalyst, and the subsequent recycling of the latter.

These fluorinated catalysts were also employed in the functionalization of the first members of the series of alkanes, from methane to butane.[34,35] To that end, the reaction was carried out in supercritical carbon dioxide, mixed with the alkane, in a fluid mixture that also dissolved the catalyst and EDA (Scheme 14). In this manner, a significant amount of methane (and the other alkanes) was available to react with the in situ generated metallocarbene intermediate, with no other C—H bond in the reaction mixture. Methane has been converted into ethyl propionate with turnover numbers up to 750 at 40 °C, and a total pressure of 250 atm.

The intramolecular version of this methodology, that is the insertion of a carbene moiety into a C—H bond of the molecule also bearing the diazo group, has been achieved with Tp^xCuL catalysts, affording lactams and lactones (Scheme 15a, b).[36] The functionalization of Si—H bonds using this strategy has been induced with the $Tp^{Br3}M$ (M=Cu, Ag) complexes with good yields, the silver case being the first example of such a reaction (Scheme 15c).[37]

The highly electrophilic silver–carbene intermediates can react with carbon–halogen bonds. This reaction was first described with the $Tp^{(CF3)2}Ag$ catalyst,[38] and later with

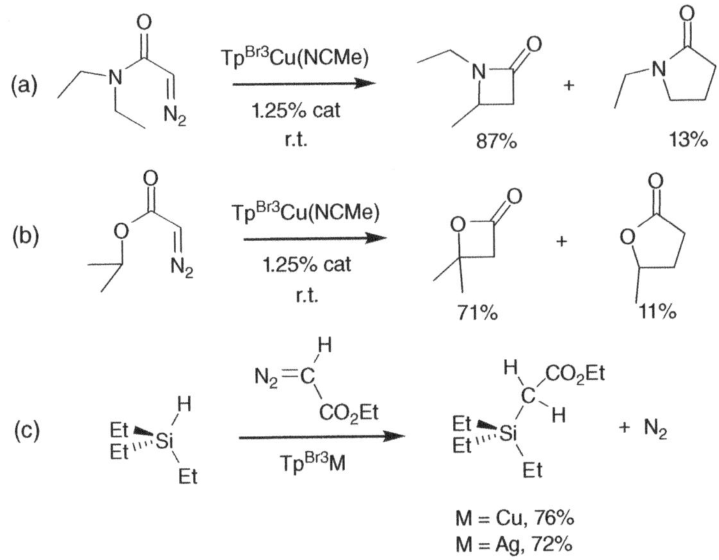

Scheme 15 *The intramolecular version of the C–H functionalization reaction by carbene insertion (a, b) and the intermolecular modification of silanes and EDA (c)*

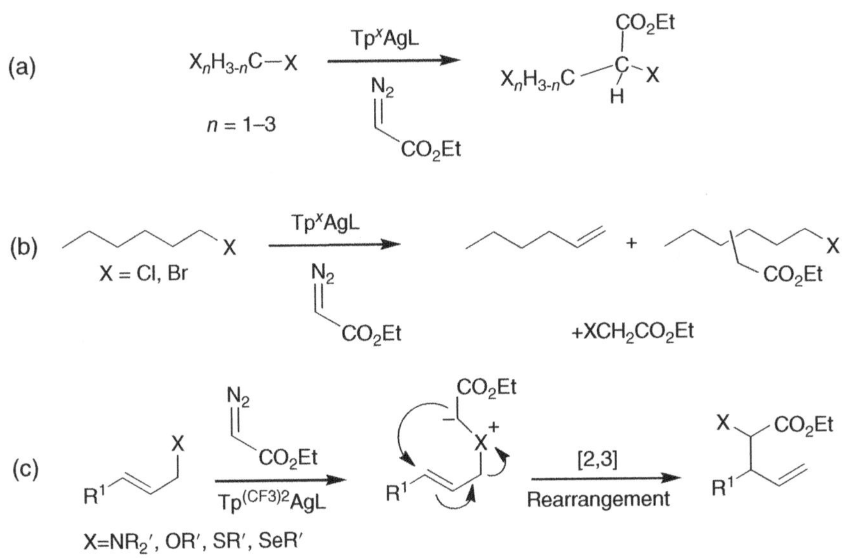

Scheme 16 *Haloalkanes as substrates in the reaction with ethyl diazoacetate*

the Tp^{Br3}-containing analog (See Scheme 6).[39] In this manner, the carbon–halogen bond was modified by the formal insertion of the $CHCO_2Et$ unit from EDA (Scheme 16a). Interestingly, mechanistic studies showed that the reaction proceeded through the transfer of the carbene group to the halogen atom and the formation of an ylide that later rearranged with the intermediacy of the silver center. The reaction was found to be general for several polyhalomethanes as well as for haloalkanes. For those bearing an alkyl chain, the dehydrohalogenation reaction was observed, thus producing an olefin and haloacetate (Scheme 16b). In the case of unsaturated substrates,[40] a 2,3-sigmatropic rearrangement of ylides has also been described (Scheme 16c).

11.3.3 Nitrene addition reactions

A related transformation to the previous carbene transfer reaction involves a nitrene ligand bonded to the metal center, in a metallonitrene intermediate in situ generated upon the appropriate selection of the catalyst and the nitrene precursor.[41] As shown in Scheme 17, some transition metal complexes react with such a precursor to generate an unsaturated intermediate, generally electrophilic in nature, which might react with olefins or C—H bonds affording aziridines or amines in a catalytic manner.[42] The most employed nitrene sources are hypervalent I(III) compounds such as PhI=NTs, chloramine-T or organic azides.

Tp^xCuL complexes catalyze both reactions shown in Scheme 17. The aziridination reaction with such catalysts was discovered using $Tp^{Me2}Cu(C_2H_4)$ and PhI=NTs as the nitrene source (Scheme 18).[24a] The influence of the hapticity of the Tp^x ligand and the oxidation state of the copper center were later studied demonstrating that tricoordination of the ligand and +1 as the copper oxidation state were the best choices.[43] The use of the fluorinated version of the above catalyst, that is $Tp^{(CF3)2}Cu(C_2H_4)$ also proved effective.[44] Moreover, the already mentioned $Tp^{Br3}Cu(NCMe)$ complex induced the aziridination reaction not only with the frequently employed olefins (styrene, 1-hexene, cyclooctene) but also with acrylates and using a stoichiometric mixture of olefin and PhI=NTs.[45]

The potential of this family of catalysts has been extended to more elaborated substrates. Tp^xML (M = Cu, Ag) complexes have been employed as catalysts in the aziridination of dien-1-ol substrates.[46] As shown in Scheme 19, with $Tp^{*,Br}Ag$ as the catalyst, the aziridination took place in the double bond vicinal to the OH group in a regioselective manner, with high conversion, and with complete stereospecificity. Further ring opening provided useful products, that with the appropriate side chain led to the synthesis of sphingosine.

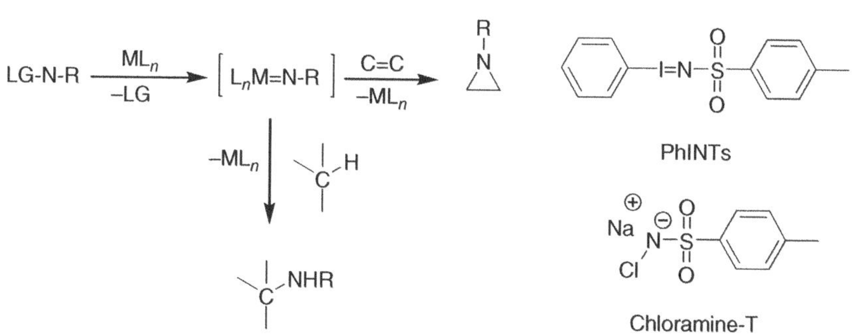

Scheme 17 The catalytic nitrene transfer reaction to olefins and C—H bonds

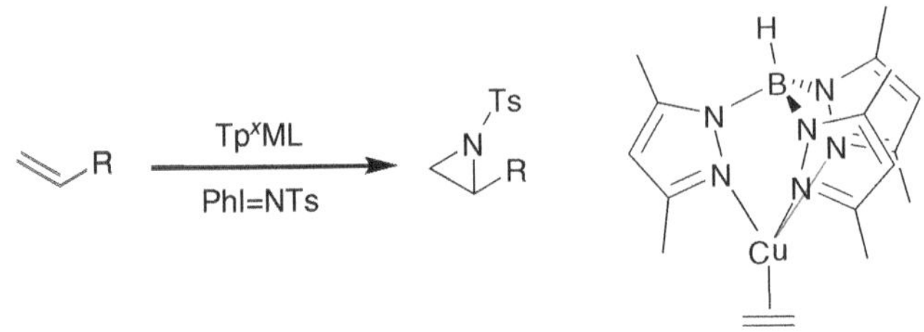

Scheme 18 The first example of an olefin aziridination reaction with a Tp^{x}CuL catalyst

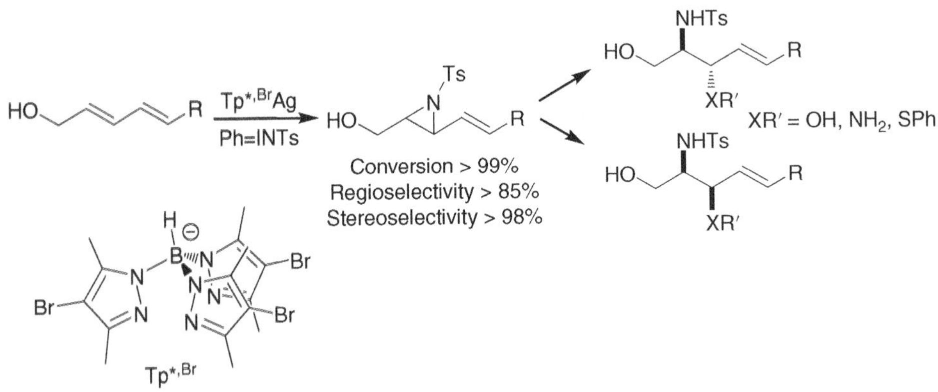

Scheme 19 The selective aziridination of dien-1-ols as a synthetic tool

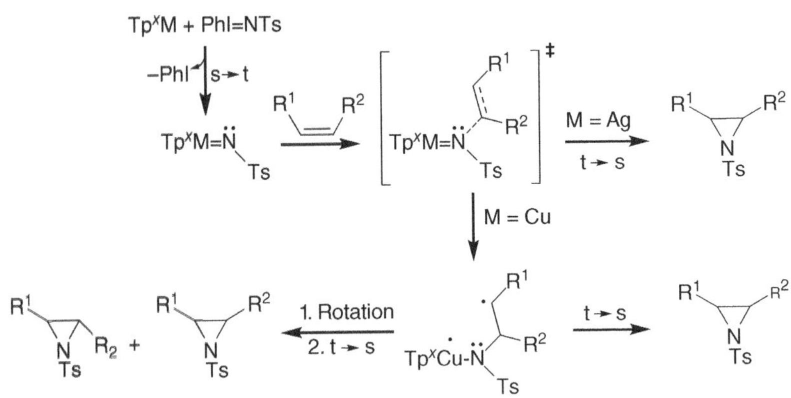

Scheme 20 The general mechanism of the Tp^{x}M-catalyzed aziridination reaction

It is worth mentioning that mechanistic studies on the olefin aziridination reaction with Tp^{x}ML complexes have shown that this transformation occurs throughout a complex pathway (Scheme 20) that involves metallonitrene intermediates in the triplet state as well as both singlet and triplet reaction pathways, which intercross several times along the

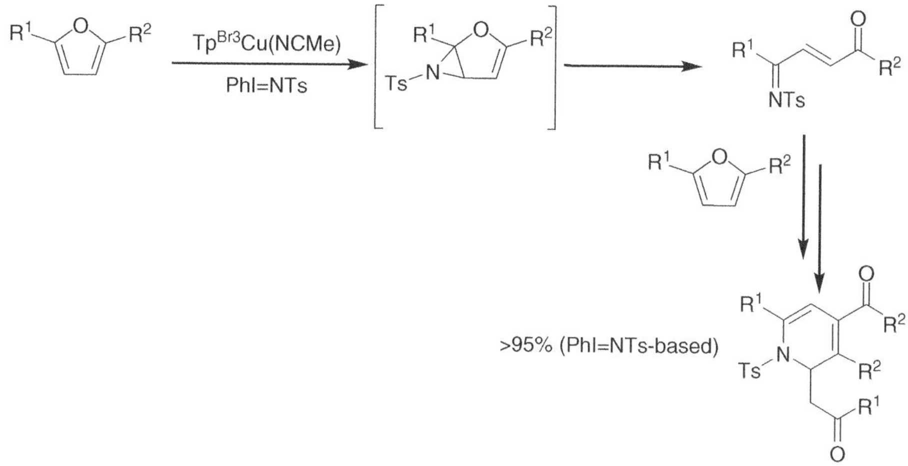

Scheme 21 *The synthetic route to 1,2-dihydropyridines through furan aziridination catalyzed by* $Tp^{Br3}Cu(NCMe)$ *as a representative example*

reaction coordinate. This proposal explained the somewhat contradictory experimental data obtained with different, traditionally employed mechanistic probes.[47]

The use of furans as substrates led to an unknown transformation while pursuing the synthesis of corresponding aziridines, using several Tp^xML catalysts, with both copper and silver derivatives. The best results in terms of yields were obtained with the $Tp^{Br3}Cu(NCMe)$ complex. The low stability of such strained molecules triggered the spontaneous ring opening and subsequent reaction with a second molecule of furan through an aza-Diels–Alder reaction with inverse electronic demand. The reaction provided 1,2-dihydropyridines[48] as the final products, in a complex sequence of reactions involving four consecutive catalytic cycles (Scheme 21), with at least three of them being promoted by the metal complex.

11.3.4 Nitrene insertion reactions

As shown in Scheme 17, C—H bonds are also prone to be activated by formal nitrene transfer from a metal center in a catalytic manner. Tp^xML complexes have also induced this transformation, with both sp^2 and sp^3 C—H bonds. The first results were obtained employing $Tp^{Br3}Cu(NCMe)$ as the catalyst for the functionalization of the C—H bonds of the alkyl substituents of arene substrates (Scheme 22). In addition to the benzylic sites, that can be considered as activated by the arene ring, the C—H bonds at the β-carbon in substrates such as ethylbenzene or cumene were also functionalized to a certain extent.[49] The use of the silver-based $Tp^{*,Br}Ag$ catalyst afforded[50] the functionalization of unactivated alkanes such as hexane or 2,3-dimethylbutane among others. These systems lack selectivity, a mixture of products derived from the insertion of the nitrene group into all available sites being obtained. It is worth noting that nitrene sources such as PhI=NTs, chloramine-T

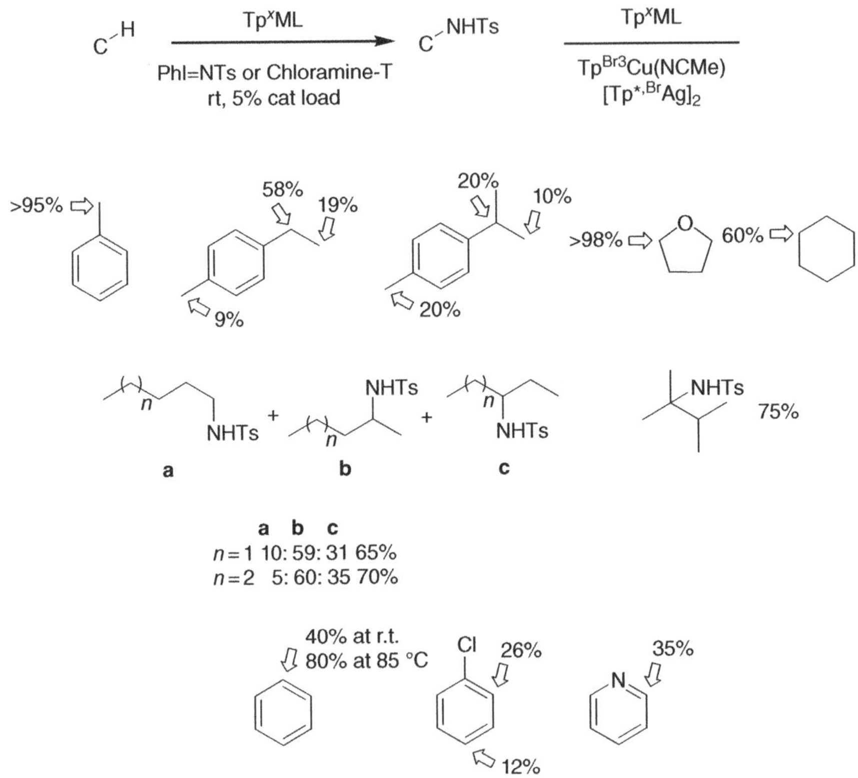

Scheme 22 *The functionalization of alkylic or aromatic C—H bonds by nitrene insertion catalyzed by TpˣML*

or even a mixture of $TsNH_2$ and $PhI(OAc)_2$[51] have been employed with these TpˣML catalysts. Benzene or other arenes with substituents without C—H bonds have also been functionalized leading to products resulting from the formal insertion of the nitrene unit into the sp^2 C—H bond (Scheme 22).[48]

11.3.5 Oxo transfer reactions

A third reaction related to the general process shown in Scheme 4 consists of the formal transfer of an oxo unit from sources such as a peroxodisulfate or hydrogen peroxide.[52] Compared with the previous carbene or nitrene counterparts, the examples with TpˣM-based catalyst remain scarce, and limited to copper as the metal center. The first example involved olefins as substrates, and potassium peroxodisulfate (Oxone®) as the oxidant (Scheme 23a).[53] Interestingly, the system could be rendered heterogeneous upon fixation of the catalyst onto silica gel, therefore providing a ready catalyst separation and recycling. This is a unique example of olefin epoxidation with a TpˣM core as catalyst with M being a coinage metal. Epoxides have been later obtained with $Tp^{Br3}Cu(NCMe)$ as the

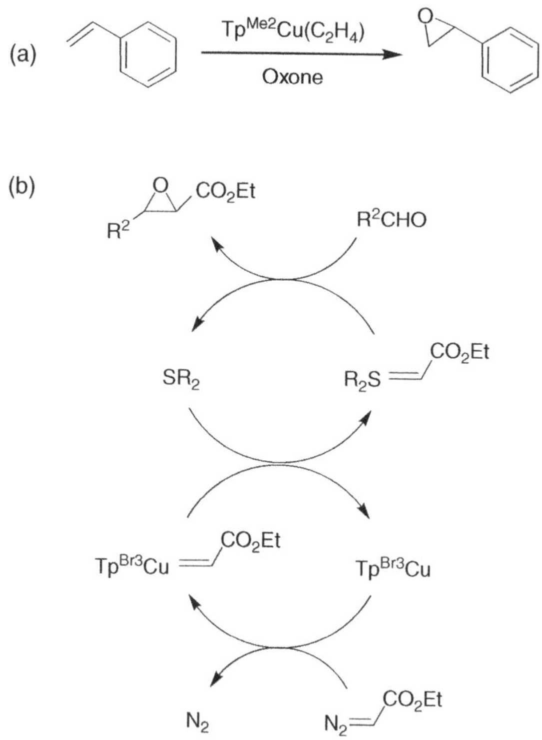

Scheme 23 *(a) The Tp^xCu-catalyzed styrene epoxidation with oxone. (b) The catalytic conversion of aldehydes and ethyl diazoacetate into epoxides*

catalyst in the reaction of benzaldehyde, phenyl diazomethane and Me_2S, although this reaction occurs through the catalytic transfer of a carbene group to the sulfide that further passes such a unit to the benzaldehyde (Scheme 23b).[54]

The direct oxidation of benzene into phenol constitutes one of the challenges in chemistry to substitute the cumene process at the industrial level. Such oxidation has also been achieved with several Tp^xCu complexes as catalysts, leading to moderate yields and high selectivity toward phenol, in a transformation using hydrogen peroxide as the oxidant and at moderate temperatures.[55] The same catalytic system has been employed for the selective oxidation of anthracenes into anthraquinones (Scheme 24).

The controlled oxidation of alkanes into alcohols also attracts attention from an industrial point of view. Copper-based catalysts containing Tp^x ligands have been employed as catalysts for this reaction that led to a very interesting as well as unprecedented transformation with copper. Thus, when cyclohexane was reacted with H_2O_2 in the presence of these catalysts, cyclohexane was partially converted into cyclohexanol and cyclohexanone, as expected. However, a certain amount of cyclohexane underwent dehydrogenation affording cyclohexene, in the first example of a copper-mediated alkane dehydrogenation process. Part of the cyclohexene was epoxidized in the reaction

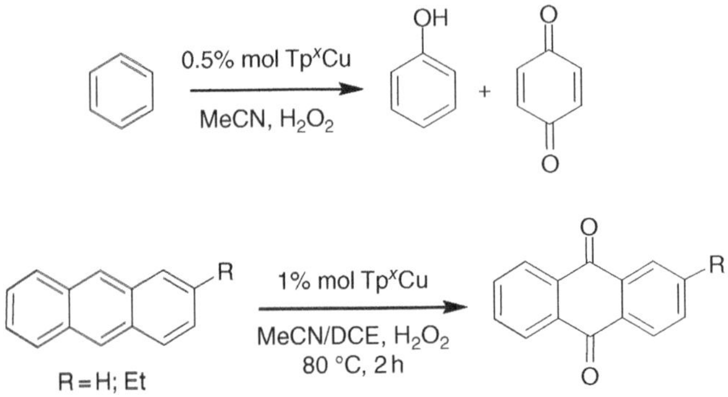

Scheme 24 *The direct oxidation of aromatic C—H bonds with hydrogen peroxide with Tp^{x}Cu catalysts*

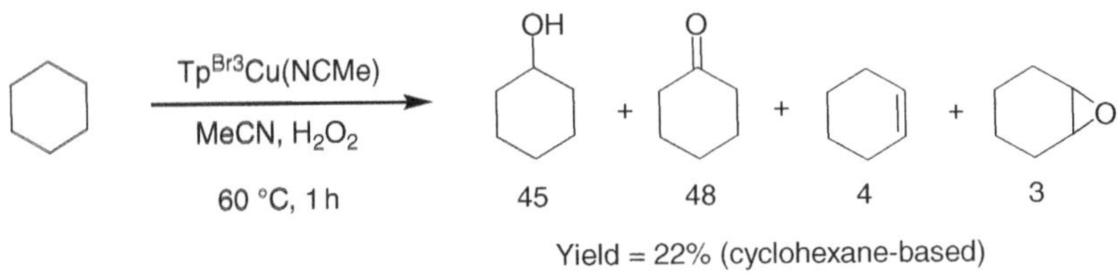

Scheme 25 *The alkane dehydrogenation reaction catalyzed by Tp^{x}Cu*

mixture (Scheme 25). A similar reaction, although at a lower yield, was observed with hexane.[56] It is worth noting that the silver analogs proved to be non-active in all these oxidation reactions.

11.3.6 Atom transfer radical reactions

This class of reactions corresponds to the current development of the initial Kharasch reaction, in which HBr is added to olefins in an anti-Markovnikov manner (Scheme 26a), with a peroxide as an initiator.[57] In the 1990s, this reaction was revived when transition metal complexes were employed as initiators or as a radical-controlling agent.[58] Scheme 26b contains a general cycle for the atom transfer radical addition (ATRA) reaction in which a C—X bond of a polyhalogenated molecule adds to a double bond, the metal center being responsible for halide abstraction, radical generation and final radical trapping. Related transformations are atom transfer radical polymerization (ATRP) and atom transfer radical cyclization (ATRC).

Complexes of type Tp^{x}Cu have been employed to promote this class of reactions. In contrast with the previous transformations where electrophilic centers were preferred, these ATRA and related reactions require electron-rich metal centers. Therefore, the best catalysts were those bearing donating alkyl groups in the pyrazolyl rings. Also, to avoid decomposition upon dinuclear interactions, bulky substituents at the 3-position of the pyrazolyl ring are also desirable. Catalyst design with such

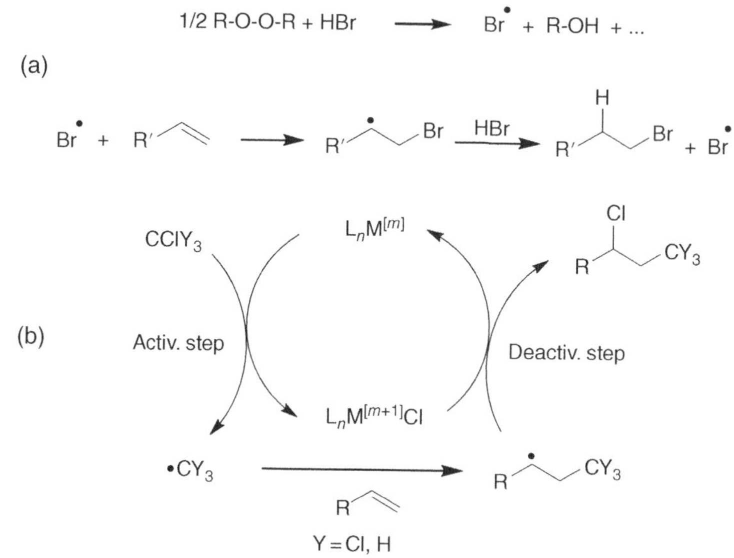

Scheme 26 (a) The Kharasch reaction. (b) The commonly proposed mechanism for atom transfer radical addition reactions

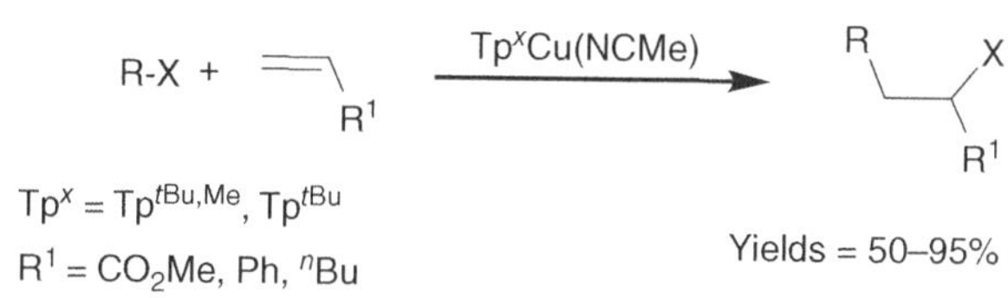

Scheme 27 The Tp^xCu-catalyzed atom transfer radical addition reaction

features are readily available when working with Tp^x ligands, due to the accessibility to different R groups in the pyrazolyl rings. This characteristic makes this family of catalyst easier for tuning. Also, the tridentate nature confers a certain stability, at variance with some of the diamine-based catalysts reported for these transformations. Thus, very good yields for the ATRA reaction of styrene or methyl methacrylate with CCl_4 were obtained with $Tp^{tBu,Me}Cu(NCMe)$[59] and $Tp^{tBu}Cu(NCMe)$ as the catalyst precursors (Scheme 27). The latter could be employed without an added reducing agent,[60] at variance with the majority of catalytic systems for this reaction. Also, a complete mechanistic study, including theoretical calculations, provided the overall picture for this transformation with these catalysts.[61]

The use of a Tp^xCu complex for the ATRP reaction was described with the simple $Tp^{Me2}Cu$ complex,[62] using styrene or *n*-butyl acrylate as the monomer. However, high values of the polydispersity were found ($M_w/M_n = 3.82–4.63$). Later, the bulkier $Tp^{iPr2}Cu(NCMe)$ catalyst provided much better results with acrylates, with polydispersity index values as low as 1.09 and lacking the need for additives (Scheme 28).[63]

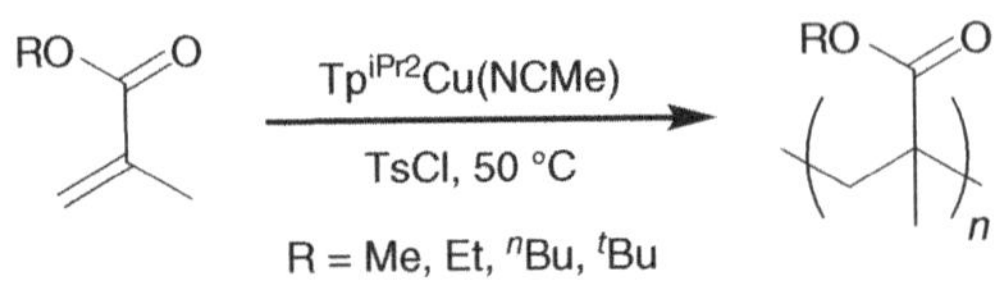

Scheme 28 *Atom transfer radical polymerization catalyzed by Tp^{x}Cu*

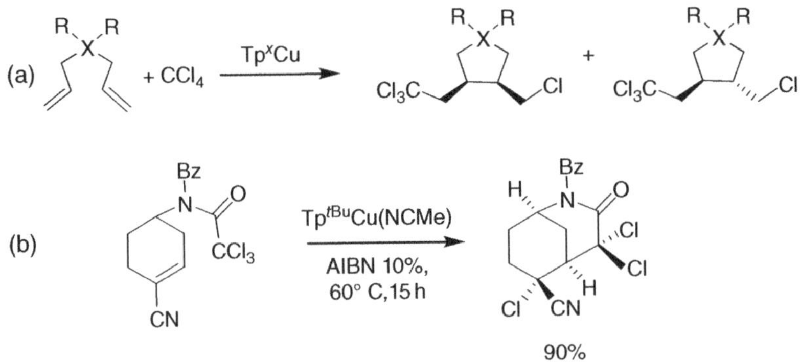

Scheme 29 *Atom transfer radical reactions involving the formation of cyclic molecules catalyzed by Tp^{x}Cu complexes*

A third type of reaction catalyzed by these compounds is the ATRC reaction in which the addition of the carbon–halogen bond to the C=C also triggers the formation of a ring (Scheme 29a).[64] The synthesis of more complex structures such as 2-azabiciclo[3.3.1] nonanes has also been achieved with these catalysts upon using reactants having both the double bond and the carbon–halogen bond (Scheme 29b).[65]

11.4 Conclusions

Copper and silver complexes bearing trispyrazolylborate ligands have shown catalytic activity towards several organic transformations involving the functionalization of unsaturated carbon–carbon bonds or several saturated E–H bonds, particularly of carbon–hydrogen bonds. The tunability of this class of ligands from both steric and electronic perspectives allows the control of those catalytic capabilities. On the basis of the work already described, the potential for their use on other, yet unreported transformations seems feasible.

Acknowledgments

The authors wish to thank the Plan Nacional de I+D of Spain for continuous support over the years (2000–2015) and current funding from MINECO (CTQ2011-28942-C02-01). The Junta de Andalucía (Grants P10-FQM-06292 and P12-FQM-01765) and the Universidad de Huelva are also acknowledged for continuous support.

References

[1] J. F. Hartwig, *Organotransition Metal Chemistry: From Bonding to Catalysis*, University Science Books, Sausalito, CA, **2009**.

[2] P. C. J. Cramer, P. W. N. M. van Leewen (eds), *Phosphorus(III) Ligands in Homogeneous Catalysis: Design and Synthesis*, John Wiley & Sons, Ltd, Chichester, **2012**.

[3] S. Díez-González (ed.), *N-Heterocyclic Carbenes: From Laboratory Curiosities to Efficient Synthetic Tools*, RSC Publishing, Cambridge, **2011**.

[4] (a) S. Trofimenko, *Scorpionates: The Coordination of Poly(pyrazolyl)borate Ligands*, Imperial College Press, London, **1999**; (b) C. Pettinari, *Scorpionates II: Chelating Borate Ligands*, Imperial College Press, London, **2008**.

[5] (a) M. M. Díaz-Requejo, P. J. Pérez, *Chem. Rev.* **2008**, *108*, 3379; (b) M. M. Díaz-Requejo, T. R. Belderraín, M. C. Nicasio, P. J. Pérez, *Dalton Trans.* **2006**, 5559.

[6] P. J. Pérez, M. M. Díaz-Requejo, in *N-Heterocyclic Carbenes in Synthesis* (ed. S. P. Nolan), Wiley-VCH, Weinheim, **2006**, Chapter 11.

[7] (a) H. V. R. Dias, W. Jin, *Inorg. Chem.* **1996**, *35*, 3687; (b) G. G. Lobbia, J. V. Hanna, M. Pellei, C. Pettinari, C. Santini, B. W. Skelton, A. H. White, *Dalton Trans.* **2004**, 951; (c) H. V. R. Dias, J. Wu, *Angew. Chem., Int. Ed.*, **2007**, *46*, 7814.

[8] S. Trofimenko, *J. Am. Chem. Soc.* **1966**, *88*, 1842.

[9] H. V. R. Dias, C. J. Lovely, *Chem. Rev.* **2008**, *108*, 3223.

[10] E. Despagnet-Ayoub, K. Jacob, L. Vendier, M. Etienne, E. Álvarez, A. Caballero, M. M. Díaz-Requejo, P. J. Pérez, *Organometallics* **2008**, *27*, 4779.

[11] M. A. Fuentes, B. K. Muñoz, K. Jacob, L. Vendier, A. Caballero, M. Etienne, P. J. Pérez, *Chem. Eur. J.* **2013**, *19*, 1327.

[12] M. A. Mairena, J. Urbano, J. Carbajo, J. J. Maraver, E. Álvarez, M. M. Díaz-Requejo, P. J. Pérez, *Inorg. Chem.* **2007**, *46*, 7428.

[13] A. Caballero, M. M. Díaz-Requejo, T. R. Belderrain, M. C. Nicasio, S. Trofimenko, P. J. Pérez, *J. Am. Chem. Soc.* **2003**, *125*, 1446.

[14] M. P. Doyle, M. A. McKervey, T. Ye, *Modern Catalytic Methods for Organic Synthesis with Diazo Compounds*, John Wiley & Sons, Inc., New York, **1998**.

[15] (a) M. M. Díaz-Requejo, T. R. Belderrain, S. Trofimenko, P. J. Pérez, *J. Am. Chem. Soc.* **2001**, *123*, 3167; (b) M. M. Díaz-Requejo, A. Caballero, T. R. Belderrain, M. C. Nicasio, S. Trofimenko, P. J. Pérez, *J. Am. Chem. Soc.* **2002**, *124*, 978.

[16] S. T. Handy, A. Ivanow, *Inorg. Chim. Acta* **2009**, *362*, 4468.

[17] J. Urbano, R. Izarra, J. L. Gómez-Ariza, S. Trofimenko, M. M. Díaz-Requejo, P. J. Pérez, *Chem. Commun.*, **2006**, 1000.

[18] J. Urbano, B. Korthals, M. M. Díaz-Requejo, S. Mecking, P. J. Pérez, *J. Polym. Chem., Part A* **2010**, *48*, 4439.

[19] A. Beltrán, B. P. Gómez-Emeterio, C. Marco, G. Ellis, M. D. Parellada, M. M. Díaz-Requejo, S. Corona-Galván, P. J. Pérez, *Macromolecules* **2012**, *45*, 9267.

[20] C. J. Lovely, R. G. Browning, V. Badarinaray, H. V. R. Dias, *Tetrahedron Lett.* **2005**, *46*, 2453.

[21] M. E. Morilla, M. M. Díaz-Requejo, T. R. Belderrain, M. C. Nicasio, S. Trofimenko, P. J. Pérez, *Organometallics* **2004**, *23*, 293.

[22] M. C. Keyes, B. M. Chamberlain, S. A. Caltagirone, J. A. Halfen, W. B. Tolman, *Organometallics* **1998**, *17*, 1984.

[23] A. Caballero, M. M. Díaz-Requejo, S. Trofimenko, T. R. Belderrain, P. J. Pérez, *J. Org. Chem.* **2005**, *70*, 6101.

[24] (a) P. J. Pérez, M. S. Brookhart, J. L. Templeton, *Organometallics* **1993**, *12*, 261; (b) M. M. Díaz-Requejo, M. A. Mairena, T. R. Belderrain, M. C. Nicasio, S. Trofimenko, P. J. Pérez, *Chem. Commun.* **2001**, 1804.

[25] M. E. Morilla, M. M. Díaz-Requejo, T. R. Belderrain, M. C. Nicasio, S. Trofimenko, P. J. Pérez, *Chem. Commun.* **2002**, 2998.

[26] M. E. Morilla, M. J. Molina, M. M. Díaz-Requejo, T. R. Belderrain, M. C. Nicasio, S. Trofimenko, P. J. Pérez, *Organometallics* **2003**, *22*, 2914.

[27] M. M. Díaz-Requejo, T. R. Belderrain, M.C. Nicasio, S. Trofimenko, P. J. Pérez, *J. Am. Chem. Soc.* **2002**, *124*, 896.

[28] J. Urbano, T. R. Belderrain, M. C. Nicasio, S. Trofimenko, M. M. Díaz-Requejo, P. J. Pérez, *Organometallics* **2005**, *24*, 1528.

[29] (a) H. V. R. Dias, R. G. Browning, S. A. Richey, C. J. Lovely, *Organometallics* **2004**, *23*, 1200; (b) H. V. R. Dias, R. G. Browning, S. A. Richey, C. J. Lovely, *Organometallics* **2005**, *24*, 5784.

[30] A. A. C. Braga, F. Maseras, J. Urbano, A. Caballero, M. M. Díaz-Requejo, P. J. Pérez, *Organometallics* **2006**, *25*, 5292.

[31] A. A. C. Braga, A. Caballero, J. Urbano, M. M. Díaz-Requejo, P. J. Pérez, F. Maseras, *ChemCatChem.* **2011**, *3*, 1646.

[32] M. M. Díaz-Requejo, P. Wehrmann, M. D. Leatherman, S. Trofimenko, S. Mecking, M. Brookhart, P. J. Pérez, *Macromolecules* **2005**, *38*, 4966.

[33] B. K. Muñoz, W.-S. Ojo, K. Jacob, N. Romero, L. Vendier, E. Despagnet-Ayoub, M. Etienne, *New J. Chem.* **2014**, *38*, 2451.

[34] A. Caballero, E. Despagnet-Ayoub, M. M. Díaz-Requejo, A. Díaz-Rodríguez, M. E. González-Núñez, R. Mello, B. K. Muñoz, W. Solo Ojo, G. Asensio, M. Etienne, P. J. Pérez, *Science* **2011**, *332*, 835.

[35] M. A. Fuentes, A. Olmos, B. K. Muñoz, K. Jacob, M. E. González-Núñez, R. Mello, G. Asensio, A. Caballero, M. Etienne, P. J. Pérez, *Chem. Eur. J.* **2014**, *20*, 11013.

[36] C. Martín, T. R. Belderrain, P. J. Pérez, *Org. Biomol. Chem.* **2009**, *7*, 4777.

[37] M. J. Iglesias, M. C. Nicasio, A. Caballero, P. J. Pérez, *Dalton Trans.*, **2013**, *42*, 1191.

[38] H. V. R. Dias, R. Greg Browning, S. A. Polach, H. V. K. Diyabalanage, C. J. Lovely, *J. Am. Chem. Soc.* **2003**, *125*, 9270.

[39] J. Urbano, A. A. C. Braga, F. Maseras, E. Álvarez, M. M. Díaz-Requejo, P. J. Pérez, *Organometallics* **2009**, *28*, 5968.

[40] P. Krishnamoorthy, R. G. Browning, S. Singh, R. Sivappa, C. J. Lovely, H. V. R. Dias, *Chem. Commun.* **2007**, 731.

[41] (a) J. W. W. Chang, T. M. U. Ton, P. W. H. Chan, *Chem. Rec.* **2011**, *11*, 331; (b) J. A. Halfen, *Curr. Org. Chem.* **2005**, *9*, 657; (c) P. Müller, C. Fruit, *Chem. Rev.* **2003**, *103*, 2905.

[42] (a) D.N. Zalatan, J. Du Bois, *Top. Curr. Chem.* **2010**, *292*, 347; (b) H. M. L. Davies, J. R. Manning, *Nature* **2008**, *451*, 417.

[43] (a) S. T. Handy, M. Czopp, *Org. Lett.* **2001**, *3*, 1423; (b) S. T. Handy, A. Ivanow, M. Czopp, *Tetrahedron Lett.* **2006**, *47*, 1821.

[44] H. V. R. Dias, H.-L. Lu, H.-J. Kim, S. A. Polach, T. K. H. H. Goh, R. G. Browning, C. J. Lovely, *Organometallics* **2002**, *21*, 1466.

[45] M. A. Mairena, M. M. Díaz-Requejo, T. R. Belderrain, M. C. Nicasio, S. Trofimenko, P. J. Pérez, *Organometallics* **2004**, *23*, 253.

[46] (a) J. Llaveria, A. Beltrán, M. M. Díaz-Requejo, M. I. Matheu, S. Castillón, P. J. Pérez, *Angew. Chem. Int. Ed.* **2010**, *49*, 7092; (b) J. Llaveria, A. Beltrán, W. M. C. Sameera,

A. Locati, M. M. Díaz-Requejo, M. I. Matheu, S. Castillón, F. Maseras, P. J. Pérez, *J. Am. Chem. Soc.* **2014**, *136*, 5342.
[47] L. Maestre, W. M. C. Sameera, M. M. Díaz-Requejo, F. Maseras, P. J. Pérez, *J. Am. Chem. Soc.* **2013**, *135*, 1338.
[48] (a) M. R. Fructos, E. Álvarez, M. M. Díaz-Requejo, P. J. Pérez, *J. Am. Chem. Soc.* **2010**, *132*, 4600; (b) L. Maestre, M. R. Fructos, M. M. Díaz-Requejo, P. J. Pérez, *Organometallics*, **2012**, *31*, 7839.
[49] (a) M. M. Díaz-Requejo, T. R. Belderrain, M. C. Nicasio, S. Trofimenko, P. J. Pérez, *J. Am. Chem. Soc.* **2003**, *125*, 12078; (b) M. R. Fructos, S. Trofimenko, M. M. Díaz-Requejo, P. J. Pérez, *J. Am. Chem. Soc.* **2006**, *128*, 17784.
[50] B. P. Gomez-Emeterio, J. Urbano, M. M. Díaz-Requejo, P. J. Pérez, *Organometallics*. **2008**, *27*, 4129.
[51] A. Beltran, C. Lescot, M. M. Díaz-Requejo, P. J. Pérez, P. Dauban, *Tetrahedron*. **2013**, *69*, 4488.
[52] A. Company, J. Lloret, L. Gómez, M. Costas, in *Alkane C-H Activation by Single-Site Metal Catalysis* (ed. P. J. Pérez), Springer, Dordrecht, **2012,** Chapter 5.
[53] M. M. Díaz-Requejo, T. R. Belderrain, P. J. Pérez, *Chem. Commun.* **2000**, 1853.
[54] A. Pereira, C. Martín, C. Maya, T. R. Belderráin, P. J. Pérez, *Adv. Synth. Catal.* **2013**, *355*, 2942.
[55] A. Conde, M. M. Díaz-Requejo, P. J. Pérez, *Chem. Commun.*, **2011**, *47*, 8154.
[56] A. Conde, L. Vilella, D. Balcells, M. M. Díaz-Requejo, A. Lledós, P. J. Pérez, *J. Am. Chem. Soc.* **2013**, *135*, 3887.
[57] M. S. Kharasch, H. Engelmann, F. R. Mayo, *J. Org. Chem.* **1937**, *2*, 288.
[58] J. M. Muñoz-Molina, T. R. Belderrain, P. J. Pérez, *Eur. J. Inorg. Chem.* **2011**, 3155.
[59] J. M. Muñoz-Molina, A. Caballero, M. M. Díaz-Requejo, S. Trofimenko, T. R. Belderrain, P. J. Pérez, *Inorg. Chem.* **2007**, *46*, 7725.
[60] J. M. Muñoz-Molina, T. R. Belderrain, P. J. Pérez, *Inorg. Chem.* **2010**, *49*, 642.
[61] J. M. Muñoz-Molina, W. M. C. Sameera, E. Álvarez, F. Maseras, T. R. Belderrain, P. J. Pérez, *Inorg. Chem.* **2011**, *50*, 2458.
[62] J. Gromada, K. Matyjaszewski, *Macromolecules* **2002**, *35*, 6167.
[63] J. M. Muñoz-Molina, T. R. Belderrain, P. J. Pérez, *Macromolecules* **2010**, *43*, 3221.
[64] J. M. Muñoz-Molina, T. R. Belderrain, P. J. Pérez, *Adv. Synth. Catal.* **2008**, *350*, 2365.
[65] F. Diaba, A. Martínez-Laporta, J. Bonjoch, A. Pereira, J. M. Muñoz-Molina, P. J. Pérez, T. R. Belderrain, *Chem. Commun.* **2012**, *48*, 8799.

12

Ligand Design in Modern Lanthanide Chemistry

David P. Mills and Stephen T. Liddle

School of Chemistry, The University of Manchester, Oxford Road, Manchester M13 9PL, UK

12.1 Introduction and scope of the review

The solution chemistry of the 4f series of elements, the lanthanide metals (Ce–Lu, Ln), is dictated by electrostatic interactions between Ln cations and coordinated ligands [1]. As Ln cations typically exhibit the +3 oxidation state and are electropositive, suitable ligands often exhibit highly electronegative donor atoms and are commonly anionic. Ln(III) cations are extremely oxophilic, hence Ln complexes lacking O-donor ligands are very oxygen and moisture sensitive and must be handled under strict anaerobic conditions. In electron donor–acceptor systems dominated by electrostatics the number of coordinated ligands is maximized to that allowed by steric constraints, with no directionality imposed by coordinated ligands.

The dominance of ionic bonding and the +3 oxidation state in 4f complexes is attributed to the "core-like" nature of the valence 4f orbitals [1]. From the parent Ln electronic configuration, $[Xe]6s^2 5d^1 4f^n$ ($n = 1$–14), the 6s and 5d electrons are easily ionized to give $[Xe]4f^n$ Ln^{3+} configurations. For several of the Ln series, notably Yb, Sm, and Eu, the molecular chemistry of the +2 oxidation state is rich and well

Ligand Design in Metal Chemistry: Reactivity and Catalysis, First Edition. Edited by Mark Stradiotto and Rylan J. Lundgren.

developed but this is not the case for most Lns. As 4f orbitals do not extend far enough from the nucleus, electrons in these orbitals are often chemically inaccessible and do not appreciably engage in covalent bonding. Ce is a notable exception as Ce(III) has a [Xe]$4f^1$ configuration and since it is at the start of the Ln series where the f orbitals are at their highest energy its valence electron is more easily removed to leave a stable closed shell electron configuration. As a result, Ce has a rich and well developed +4 oxidation state chemistry compared with the other Ln metals.

As Ln(III) cations are relatively large [Ce(III) coordination number, CN 8, 1.143Å; Lu(III) C N 8, 0.977Å] [2], high CNs are commonplace, typically 8–9 for small, neutral σ-donor ligands and up to 11–12 for multidentate ligands [1]. Ionic radii of Ln(III) cations decrease steadily across the Ln series as the valence 4f orbitals are filled, hence heavier Lns often have lower CNs than lighter Lns. Ligands in contemporary Ln chemistry are usually anionic and sterically demanding to impose low CNs and prevent unwanted decomposition and oligomerization pathways. Ln complexes of these ligands are often highly reactive and exhibit unusual bonding modes, which can be exploited in their further chemistry.

As the +3 oxidation state and ionic bonding dominates the solution chemistry of the group 3 elements, Sc, Y, and La, they are chemically similar to the Ln elements [1]. Further to this, Y(III) has a similar ionic radius to Ho(III) (C N 8; 1.019 and 1.015Å, respectively) and La(III) (CN 8, 1.16Å) is of a comparable size to Ce(III) (CN 8, 1.143Å) [2]. As a consequence, discussions of the chemistry of the group 3 metals are often amalgamated with Ln chemistry, as they will be in this chapter. Together with the Ln series, these are commonly referred to as the "rare earths" (REs) for historical reasons arising from the collective discovery of these elements. Sc(III), due to its high charge density (CN 8, 0.87Å) [2], has chemistry that is distinctive from most of the RE series but it will be discussed herein for completeness. Despite slight differences in their precise definitions, the terms RE and Ln will be used interchangeably throughout this chapter for brevity.

It follows from basic concepts that ligand design in Ln chemistry, to impart specific oxidation states, coordination numbers and geometries for future exploitation, has to be extremely well considered. Important attributes of Ln chemistry to consider when designing ligands are summarized as:

- Kinetics – As crystal field splitting energies are small, and the bonding is predominantly ionic, ligands in Ln complexes are typically labile and can undergo ligand exchange readily, hence hard electronegative anionic donor atoms are often utilized to increase electrostatic attraction and reduce lability.
- Thermodynamics – Ligands with multiple coordination sites are abundant as the gain in entropy from chelate and/or macrocyclic effects increase the thermodynamic stability of the system.
- Limited multiple bonding and absence of back-bonding.
- Large ionic radii and irregular coordination geometries that are dictated by sterics.

Sterically demanding and/or multidentate ligands are often employed to saturate the Ln coordination sphere, imparting control and thwarting common decomposition pathways, including:

- Oxophilicity – Ln complexes that do not contain O-donors are often highly air and moisture sensitive as a result of their latent oxophilicity, hence Schlenk line and glove box techniques are required for manipulation.
- Ligand scrambling – Heteroleptic complexes may convert to homoleptic complexes by Schlenk-type equilibria.
- Solvation – Vacant coordination sites may be occupied by donor organic solvents, increasing complex solubility.
- Aggregation – Dimerization and oligomerization can occur readily.
- Occlusion – Salt metathesis methodologies are often used to synthesize target Ln complexes, but in the presence of hard alkali metal cations ate complexes often form. Ate complexes are aggregates composed of the target Ln complex with one or more equivalents of the alkali metal salt by-product included in the primary Ln coordination sphere rather than eliminated. Occlusion occurs most readily for the hardest alkali metal cation Li^+, where electrostatic interactions between anions and cations are strongest, holding the ate complex together and preventing salt elimination.
- β-hydride elimination – Ln alkyls must not contain β-hydrogens as the formation of Ln hydrides and elimination of alkenes is usually facile.

The employment of judiciously selected ligands in non-aqueous Ln chemistry has transformed this discipline from one dominated by a small handful of ligands, bonding modes, and redox chemistry into a vibrant and dynamic field. The increasing speed of this progress is well illustrated by the rapid expansion of low oxidation state Ln chemistry. The +2 oxidation state was limited to Sm, Eu, and Yb for most of the last century [1] but in less than 20 years this has been achieved for all the remaining RE elements, save radioactive Pm [3]. Such remarkable and rapid advances are an underlying theme of this research area.

This chapter will focus on selected ligand systems that have imparted unusual properties and novel functionalities in Ln complexes. It is designed to be concise and representative of a burgeoning area rather than comprehensive and structurally characterized complexes are given preference over poorly defined systems. The preparation of Ln complexes without Ln—O bonds is a significant synthetic challenge in itself due to the oxophilic nature of Ln cations [1]. For this reason, most ligands included herein lack O-donors and the review is essentially divided into C-, N- and P-donor ligands, with a fourth section on multiple bonds separated for ease of reference. Review articles referenced throughout this chapter are included to stimulate further reading on specific areas to gain a more comprehensive knowledge of the literature if required.

The aqueous chemistry of the lanthanides is well developed and as such is covered comprehensively in general f-element textbooks [1]; therefore, a detailed discussion of this field is beyond the scope of this chapter. However, it should be noted that considerations of ligand design described herein are applicable and transferable to aerobic solution chemistry for the stabilization of unusual oxidation states and bonding regimes, leading to myriad applications in synthesis and as functional materials. These uses include, but are not limited to: reactivity towards unsaturated substrates; olefin polymerization; MRI contrast agents; NMR shift reagents; single molecule magnets; lasers; phosphors; luminescence; and imaging agents. Some of these applications will be discussed in more detail as appropriate as they are encountered throughout this chapter.

12.2 C-donor ligands

12.2.1 Silylalkyls

Ln alkyl complexes are typically prepared by salt metathesis methodologies from lanthanide halides and alkali metal transfer agents but are often prone to oligomerization and decomposition [4]. These reactions are often carried out in solvents of intermediate polarity such as THF, as a compromise between the low solubilities of Ln halides in non-polar solvents and the instability of group 1 and Ln alkyls in highly polar solvents. Bulky alkyl groups are employed to circumvent ate complex formation, whereby additional equivalents of group 1 alkyls are occluded within the coordination sphere of the Ln alkyl complex, blocking coordination sites and hindering future reactivity. Many of the decomposition pathways of Ln alkyls involve deprotonation and ring-opening of THF, hence Ln alkyls that have experienced most widespread use as both starting materials and ancillary ligands have some resistance to these unwanted side-reactions. The most frequently used Ln alkyl starting materials are $[Ln(CH_2Ph)_3(THF)_3]$ (Ln = Ce **1a**, Pr **1b**, Nd **1c**, Sm **1d**, Gd **1e**, Dy **1f**, Ho **1g**, Er **1h**, Lu **1i**, Sc **1j**, Y **1k**, La **1l**) [5] and $[Ln(CH_2SiMe_3)_3(THF)_2]$ (Ln = Sm **2a**, Tb **2b**, Er **2c**, Tm **2d**, Yb **2e**, Lu **2f**, Sc **2g**, Y **2h**) [6] and related derivatives. The latter series of complexes may only be isolated for relatively small REs as they are all prone to thermal decomposition by the α-elimination of $SiMe_4$ and the larger congeners are more vulnerable to this pathway as they are coordinatively unsaturated [6].

Alkyls with increased steric bulk have been developed to exclude THF from the coordination sphere of the Ln, generating homoleptic alkyl complexes [4]. The most frequently used alkyls of this class in Ln chemistry are $\{CH(SiMe_3)_2\}^-$ and $\{C(SiMe_3)_3\}^-$ (trisyl), with complexes of the general formula $[Ln\{CH(SiMe_3)_2\}_3]$ (Ln = Ce **3a**, Pr **3b**, Nd **3c**, Sm **3d**, Er **3e**, Lu **3f**, Sc **3g**, Y **3h**, La **3i**) [7] most often prepared by a circuitous route from Ln aryloxides for larger Ln(III) cations to prevent ate complex formation (Scheme 1). The first members of this series to be isolated, **3g** and **3h**, represented at the time the first neutral, donor solvent-free, homoleptic RE alkyl complexes and were instrumental in the development of organometallic Ln chemistry [7c]. XRD data revealed that **3a**, **3d**, **3h** and **3i** are structurally analogous, all exhibiting pyramidal geometries in the solid state with three additional short Ln···Cγ distances [7a,c]. Density functional theory studies on models of **3d** and **3i** concluded that these short distances were a result of agostic Ln···Si–Cβ interactions rather than agostic

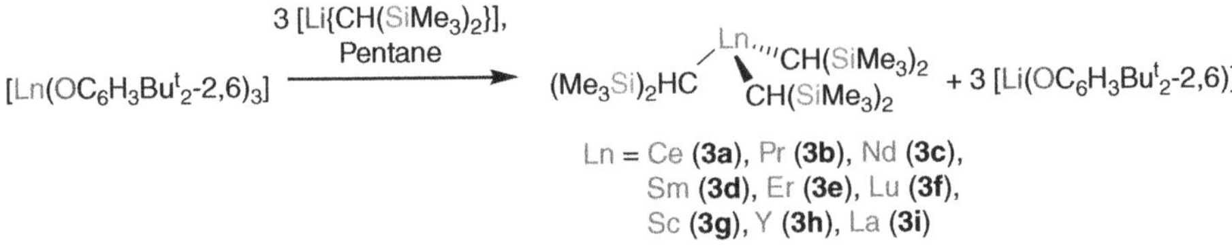

Scheme 1 *Synthesis of the homoleptic Ln(III) alkyls **3a–i** [7]*

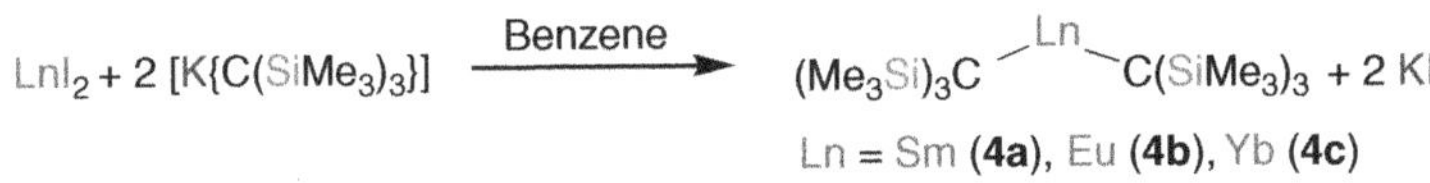

Scheme 2 *Synthesis of the homoleptic Ln(II) alkyls **4a–c** [9]*

Ln•••H–Cγ contacts [8]. These agostic interactions are considered to be the major influence on **3g** and **3h** adopting pyramidal rather than planar geometries, with minimizing steric repulsion between ligands a minor factor.

Perhaps the most remarkable series of homoleptic Ln alkyl complexes prepared to date are $[Ln\{C(SiMe_3)_3\}_2]$ (Ln = Sm **4a** [9a], Eu **4b** [9b], Yb **4c** [9c]), prepared directly from LnI_2 and $[K\{C(SiMe_3)_3\}]$ in benzene (Scheme 2). The choice of solvent is important in this synthesis as ethereal solvents such as diethyl ether are cleaved to give decomposition products such as dimeric $[Yb\{C(SiMe_3)_3\}(OEt)(Et_2O)]_2$ [9c]. At the time of publication, **4c** represented the first structurally characterized example of a 2-coordinate f-element complex, and this CN is still rare [9c]. Compounds **4a–c** all exhibit bent C—Ln—C geometries [**4a** ca. 143.43°; **4b** 136.0(2)°; **4c** 137.0(4)°], with two or three agostic interactions between the Ln and methyl group C—H bonds observed in the solid state depending on metal size, although ^{1}H NMR spectroscopy of **4c** indicated equivalent proton environments in solution even at −95 °C [9c]. The Ln—C distances in **4a–c** are relatively long to minimize interactions between some of the trisyl methyl groups.

The preference for bent rather than linear geometries in f-element complexes has been scrutinized by computational studies to determine whether this inclination stems from steric [10] or electronic [11] factors. All of these studies concluded that the energy difference between these geometries is small. The C—Ln—C angles are similar for **4b** and **4c**, despite discrepancies in mean Ln—C distances [**4b** 2.609(7) Å; **4c** 2.496(9) Å] that stem from differences in Ln(II) cation sizes. From this data it would appear that the deviation from linearity for **4a–c** is electronic in origin and this has been attributed to some involvement of Ln 5d orbitals in the bonding scheme [9b].

The synthetic utility of **4a–c** and closely related derivatives has been explored. Compound **4c** reacts with methyl iodide in diethyl ether to afford the dimeric "Ln Grignard" reagent $[Yb\{C(SiMe_3)_3\}(\mu\text{-}I)(OEt_2)]_2$ **5** by σ-bond metathesis [9c]. Compound **5** is remarkably stable to Schlenk equilibria in diethyl ether solutions and may alternatively be prepared by the treatment of **4c** with 1,2-diiodoethane or from the oxidative addition of trisyl iodide to Yb metal in diethyl ether [9c]. Compounds **4a–c** have proved effective in the polymerization of methyl methacrylate (MMA) and acrylonitrile, in common with related Ln(II) systems [9a,12]. Compound **4c** was found to give exceptional results in the production of poly(methyl methacrylate), yielding a high molecular weight (M_n, number average molecular weight = $5.1 \times 10^5\,g\,mol^{-1}$) polymer with low polydispersity index (PDI, $M_w/M_n = 1.1$; M_w = weight average molecular weight) in excellent yield with 97 % isotacticity [12].

12.2.2 Terphenyls

In contrast to the salt metathesis methodologies that dominate much of Ln organometallic synthesis, the most common routes to Ln(II) aryl complexes have been redox transmetalation reactions using organomercury reagents [13]. Although Ln(II) aryls had been known since the 1970s and the further chemistry of these complexes was being exploited, solid state characterization of this class of complex remained elusive for some time. The first structurally characterized Ln(II) terphenyl complexes were prepared by utilizing $\{C_6H_3Ph_2\text{-}2,6\}^-$ less than 15 years ago, where $[Yb(C_6H_3Ph_2\text{-}2,6)(I)(THF)_3]$ (**6**) was prepared by the oxidative addition of the terphenyl iodide $IC_6H_3Ph_2\text{-}2,6$ to Yb metal in THF [14] (Scheme 3). Donor solvents are required to prevent ligand scrambling and in aromatic solvents the Schlenk equilibrium shifts to favor the rearrangement of **6** to $[Yb(C_6H_3Ph_2\text{-}2,6)_2(THF)_2]$ (**7a**) and $[YbI_2(THF)_n]$. Heteroleptic terphenyl complexes of larger Ln such as Eu could not be isolated even in THF, with $[Eu(C_6H_3Ph_2\text{-}2,6)_2(THF)_2]$ (**7b**) and $[EuI_2(THF)_5]$ being the only tractable products due to the larger ionic radius of Eu(II) [Eu(II) CN 8, 1.25 Å vs. Yb(II) CN 8, 1.14 Å] [2] making the heteroleptic intermediate more susceptible to ligand scrambling (Scheme 3) [14, 15].

The first structurally characterized heteroleptic Ln(III) terphenyl complexes, $[Yb(C_6H_3Mes_2\text{-}2,6)(Cl)_2(N\text{-}MeIm)_2(py)]$ (**8**; Mes = $C_6H_2Me_3$-2,4,6, *N*-MeIm = *N*-methylimidazole) and $[Yb(C_6H_3Mes_2\text{-}2,6)\{N(SiMe_3)_2\}(\mu\text{-}Cl)_2\{Li(THF)_2\}]$ (**9**) were prepared, prior to the isolation of **6**, **7a**, and **7b**, by salt metathesis methodologies from $YbCl_3$ and group 1 transfer agents (Scheme 4) [16]. The sterically demanding supporting ligand in **9**, $\{N(SiMe_3)_2\}^-$, together with the terphenyl ligand, reduces the CN of Yb to 4, although LiCl occlusion has not been prevented. Terphenyl chemistry has been expanded to include *N*-MeIm- and LiCl-free heteroleptic terphenyl complexes of Sc(III), Y(III), and Tm(III) but is currently limited in scope to the smaller Ln [17].

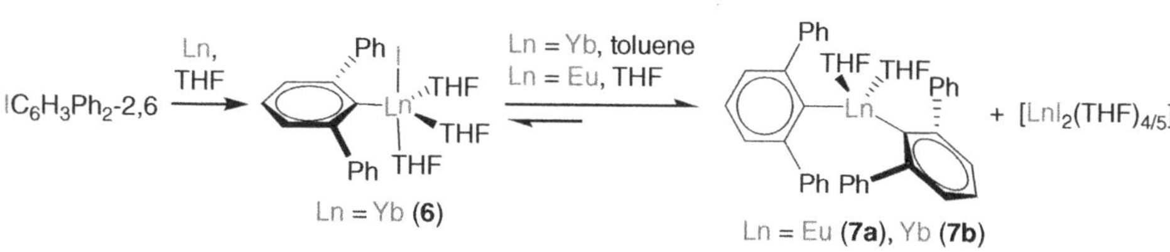

*Scheme 3 Synthesis of the Ln(II) terphenyls **6**, **7a**, and **7b**. Reproduced from [14] with permission of American Chemical Society and [15] with permission of John Wiley & Sons, Ltd*

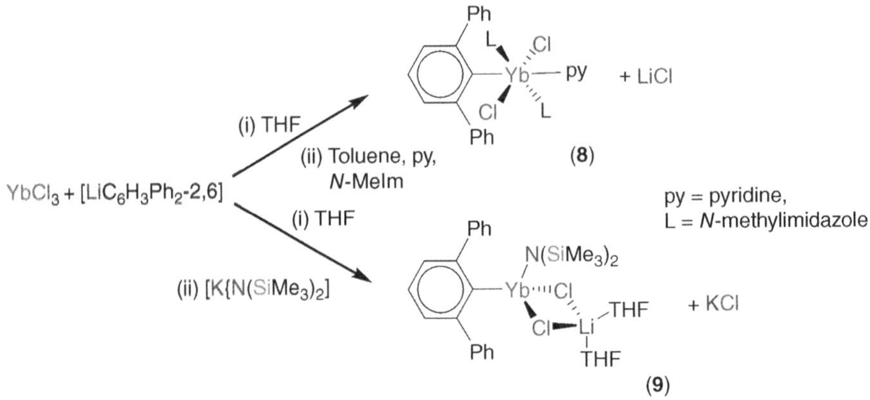

Scheme 4 *Synthesis of the Yb(III) terphenyls **8** and **9**. Reproduced from [16] with permission of American Chemical Society*

12.2.3 Substituted cyclopentadienyls

Ln cyclopentadienyl ($C_5H_5^-$, Cp) complexes have been known for over half a century and as such these ligands dominate non-aqueous Ln chemistry [1, 18]. In non-polar solvents, substituted Cp ligands are often employed, typically with alkyl or silyl groups, to impart higher solubility in non-polar solvents and prevent oligomerization. Notable examples include those substituted by multiple methyl groups (e.g., $\{C_5Me_5\}^-$, Cp*; $\{C_5Me_4H\}^-$), larger alkyl/aryl groups (e.g., $\{C_5H_3Bu^t_2\text{-}1,3\}^-$, Cp^{tt}; $\{C_5H_2Bu^t_3\text{-}1,2,4\}^-$, Cp^{ttt}) and trimethylsilyl groups (e.g., $\{C_5H_4SiMe_3\}^-$, Cp′; $\{C_5H_3(SiMe_3)_2\text{-}1,3\}^-$, Cp″; $\{C_5H_2(SiMe_3)_3\text{-}1,2,4\}^-$, Cp‴).

Substituted Cps have proved to be the most effective ligand systems in stabilising molecular Ln(II) complexes to date [19]. The chemistry of $[Sm(Cp^*)_2]$, **10**, and its THF-solvate, $[Sm(Cp^*)_2(THF)_2]$, are perhaps the most developed of any organometallic Ln(II) complex [19]. Compound **10** was originally prepared by the reaction of Sm vapor with 2 equiv. of Cp*H in hexane, followed by recrystallization from THF and sublimation to remove coordinated THF (Scheme 5) [20]. Compound **10** and its Lewis base adducts are all highly reactive reducing agents towards a wide variety of substrates but the unsaturated coordination sphere of **10** gives it an enhanced reactivity profile [19]. Compound **10** exhibits a bent structure in the solid state ($Cp_{centroid}$–Sm–$Cp_{centroid}$ = ca. 140.1°), postulated to derive from favourable dipole–dipole interactions that grow in contribution on deviation from linearity [10]. Noteworthy reactions of **10** and its adducts include the activation of N_2 to yield $[\{Sm(Cp^*)_2\}_2(\mu\text{-}\eta^2\text{:}\eta^2\text{-}N_2)]$, **11**, which exhibits side-on coordination of a bridging di-reduced N_2^{2-} unit [21], and the reaction with CO to afford ketene-carboxylate aggregates such as $[\{Sm(Cp^*)_2\}_4(\mu^3\text{-}O_2CC{=}C{=}O\text{-}\kappa^4\text{-}C,O,O',O'')_2(THF)_2]$, **12** by reductive homologation (Scheme 6) [22]. At the time of its publication, **11** represented the first structurally characterized example of an f-element dinitrogen complex.

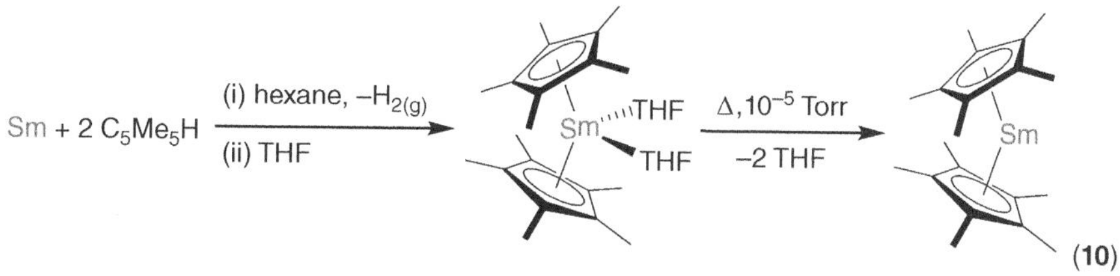

Scheme 5 *Synthesis of the Sm(II) metallocene* ***10****. Reproduced from [20] with permission of American Chemical Society*

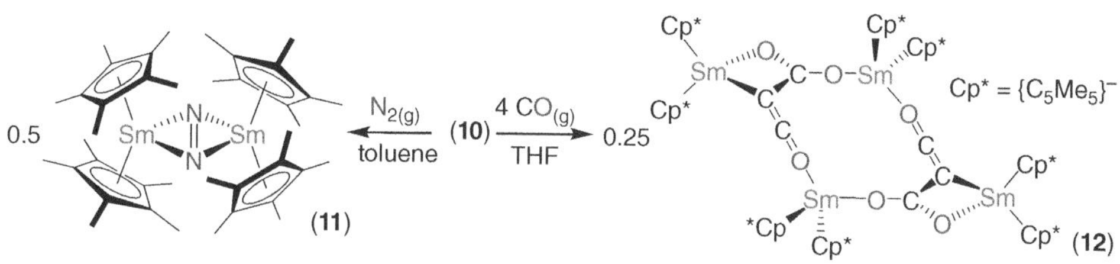

Scheme 6 *Reaction of* ***10*** *with N_2 and CO to afford* ***11*** *and* ***12****. Reproduced from [21, 22] with permission of American Chemical Society*

Molecular Sm(II), Eu(II) and Yb(II) chemistry is well developed as divalent halide starting materials are readily available [1, 19, 23]. The recent disclosure of solid state structures of Tm(II) [24], Dy(II) [25] and Nd(II) [26] halides has opened up their divalent chemistry but other molecular Ln(II) complexes may only be accessed by reduction of Ln(III) precursors and are typically highly reactive and unstable [1, 19, 23]. Strong reducing agents such as alkali metals are required to promote these reductions due to large negative Ln(III) → Ln(II) standard reduction potentials for most Lns [27]. The first structurally characterized La(II) and Ce(II) complexes, [K(L)][La(Cp″)$_3$] (L = 18-crown-6 and Et$_2$O or 2.2.2-cryptand, **13**) and [K(18-crown-6)(Et$_2$O)$_2$][Ce(Cp″)$_3$]·[Ce(Cp″)$_3$] **14** were prepared by reduction of [Ln(Cp″)$_3$] (Ln = La, Ce) with K mirrors in coordinating solvents (Scheme 7) [28].

This chemistry was extended over the next few years and [K(L)][Ln(Cp′)$_3$] (L = 18-crown-6 or 2.2.2-cryptand; Pr **15a**, Gd **15b**, Tb **15c**, Ho **15d**, Er **15e**, Lu **15f**, Y **15g**) were prepared by a modified procedure using low temperature potassium graphite columns to complete the series of structurally characterized Ln(II) complexes, with the exception of radioactive Pm (Scheme 7) [29]. Compounds **15a–g** are unstable above –30 °C and are prepared and stored under argon as they will react rapidly with dinitrogen in THF to afford [{Ln(Cp′)$_2$(THF)}$_2$(μ-η^2:η^2-N$_2$)] [19, 30]. Such complexes of the general formula [{Ln(A)$_2$(THF)$_x$}$_2$(μ-η^2:η^2-N$_2$)] (A = supporting anionic ligand, e.g., Cp*, C$_5$Me$_4$H, Cp′, Cp^{ttt}; x = 0–2) are typically synthesized by alkali metal reduction of [Ln(A)$_3$] precursors under dinitrogen atmospheres and have been in some cases been shown to form via Ln(II) intermediates [19, 30].

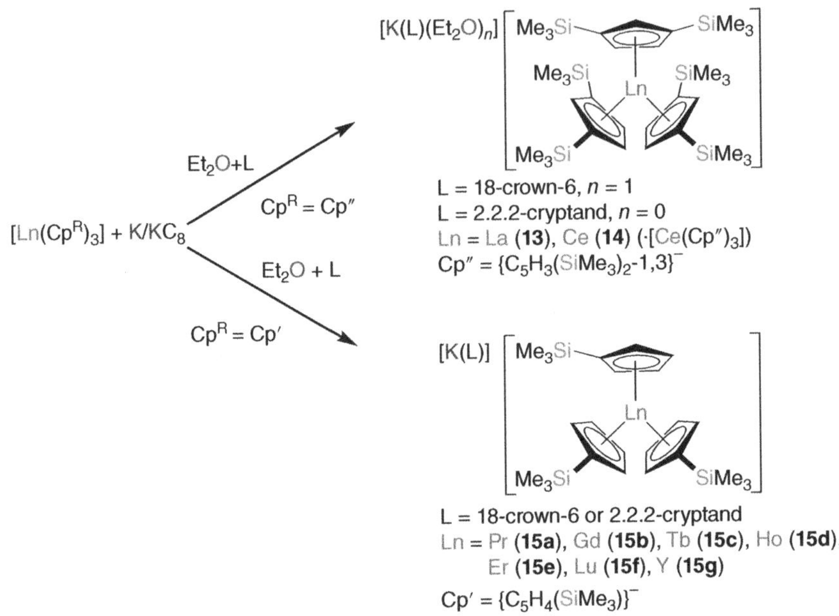

Scheme 7 *Synthesis of the Ln(II) complexes **13, 14,** and **15a–g** [28, 29]*

12.2.4 Constrained geometry cyclopentadienyls

Heterogeneous Ziegler–Natta catalytic olefin polymerization mediated by group 4 metal compounds has been of widespread industrial importance since the 1950s, despite the mechanism of polymer chain growth not being fully understood for decades [31]. Structural characterization of group 4 Ziegler–Natta catalysts was a long-term scientific target, to give information on this mechanism for scientific curiosity and to facilitate the design of more efficient and selective catalysts. Despite this interest, precise solid state structures of active olefin polymerization catalysts remained elusive despite developments in homogeneous catalysis. These advances included the preparation of *ansa*-metallocene early d-transition metal pre-catalysts, where the two Cp ligands are bridged by a linker, often a silyl group (e.g., $\{C_5R_4\text{-}SiMe_2\text{-}C_5R_4\}^{2-}$) [32]. In efforts to prepare RE analogs of *ansa*-metallocenes with less crowded coordination spheres, bridged amido-Cp ligands such as $\{C_5R_4\text{-}SiMe_2\text{-}NR'\}^{2-}$ were developed [31c]. These investigations furnished the first structurally characterized half-sandwich single component constrained geometry catalyst (CGC), $[Sc(\eta^5{:}\eta^1\text{-}C_5Me_4SiMe_2NBu^t)(PMe_3)(\mu\text{-}H)]_2$ **16**, prepared by hydrogenolysis of the precursor $[Sc(\eta^5{:}\eta^1\text{-}C_5Me_4SiMe_2NBu^t)\{CH(SiMe_3)_2\}]$ in the presence of trimethylphosphine (Scheme 8) [33].

It was found that **16** exhibited regiospecific oligomerization of propene, 1-butene, and 1-pentene to yield linear atactic polymers as "head-to-tail" coupling of monomer units is favored by this catalyst [33]. The polymer chain grows by a β-hydride elimination mechanism, which competes with phosphine coordination, leading to a relatively

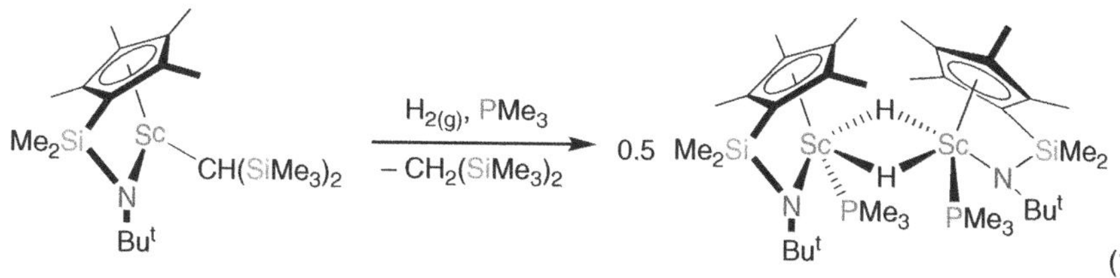

(16)

Scheme 8 *Synthesis of the Sc(III) olefin polymerization catalyst* ***16****. Reproduced from [33] with permission of American Chemical Society*

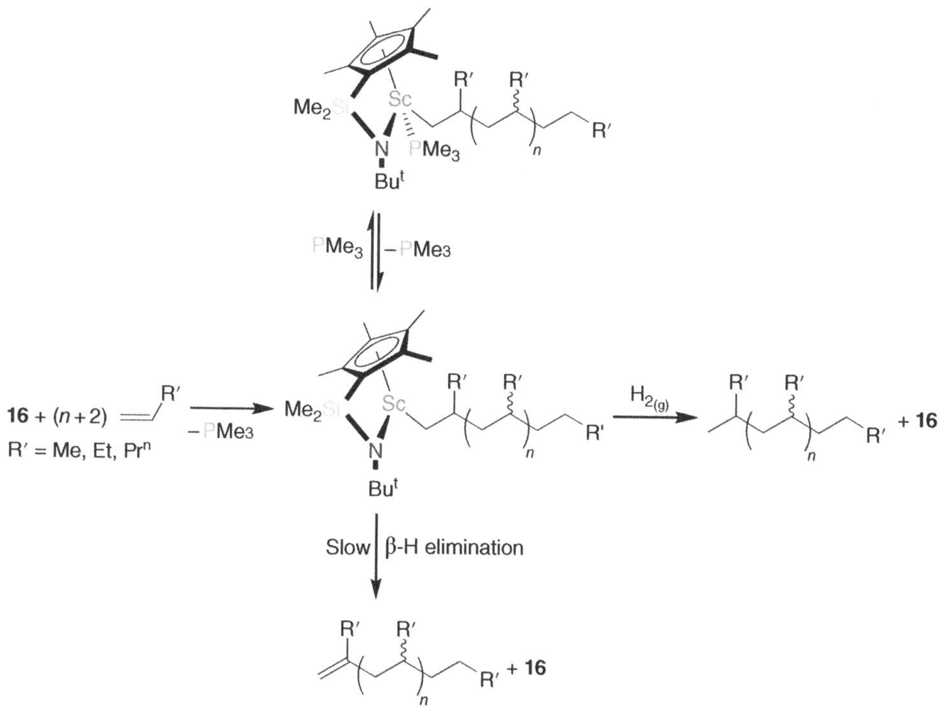

Scheme 9 *Mechanism of olefin oligomerization by* ***16****. Reproduced from [33] with permission of American Chemical Society*

slow chain growth (Scheme 9). Although propene oligomerization was found to be sluggish due to its low solubility in toluene at 25 °C, 1-butene was found to give polymers made up of around 71 monomer units (M_n = 4000 g mol^{-1}) with a PDI of 1.7 at 0.1 mol% of **16**.

Since the publication of the solid state structure of **16**, research in ligand design in RE and group 4 half-sandwich CGCs and well-defined *ansa*-metallocene Ziegler–Natta catalysts has been prolific due to their industrial importance [31, 34]. It has been shown that the metal coordination sphere may be altered by tuning the amido-Cp ligand framework, facilitating the preparation of specific polymers with useful properties. These advances have been reviewed thoroughly elsewhere so for the sake of brevity only the salient advantages of CGCs such as **16** over related catalysts are given here: (1) more accessible coordination spheres, that derive from relatively small $C_{centroid}$—M—N bite angles (typically 25–30° smaller than $C_{centroid}$—M—$C_{centroid}$ bite angles in analogous metallocene systems); (2) metal centers that are more Lewis acidic than those in *ansa*-metallocenes as amido-ligands are not as electron rich as Cps, which increases the activity of the catalyst and suppresses undesired chain transfer reactions; and (3) relatively high thermal stabilities, allowing polymerizations to be performed at higher temperature to increase rates [31].

12.2.5 Benzene complexes

The field of Ln-arene complexes has quickly blossomed since the first structurally characterized example, [Sm(η^6-C_6Me_6)$(AlCl_4)_3$] **17** was reported less than 30 years ago [35]. As such this area has already been reviewed thoroughly [36]. Compound **17** was obtained as a minor product by refluxing toluene mixtures of $SmCl_3$, $AlCl_3$, Al foil and hexamethylbenzene, following analogous procedures to those employed in the preparation of related U(III) mono-arene complexes (Scheme 10) [37]. This methodology was extended by several groups to prepare a wide range of complexes of the general formula [Ln(η^6-arene)$(AlX_4)_3$] (arene = C_6H_6, C_6H_5Me, $C_6H_4Me_2$-1,3 and $C_6H_2Me_4$-1,2,4,5; Ln = Pr, Nd, Sm, Gd, Eu, Yb, Y, La; X = Cl, Br, I) [38].

Although the formation of a Sm(II) complex by partial reduction of Sm(III) was postulated to explain the low yield of **17**, this could not be isolated [35]. In contrast, the tetrameric Eu(II) complex, [Eu(η^6-C_6Me_6)$(AlCl_4)$(μ-$AlCl_4$)$]_4$, **18**, could be prepared under analogous conditions (Scheme 10) [39]. Analysis of the solid state structures of **17** and **18** determined that the arene ligands are flat and carry no charge [**17** Sm(III)–C_{arene} 2.89(5) Å mean; **18** Eu(II)–C_{arene} 2.99(3) Å mean]. In conjunction with spectroscopic measurements it is evident that these systems are Lewis-base adducts of [Ln$(AlCl_4)_n$], with π-electron density donated into vacant Ln valence orbitals from the aromatic rings

Scheme 10 *Synthesis of the Ln-arene complexes* ***17*** *and* ***18*** *[35, 39]. Reproduced from [35] with permission of American Chemical Society*

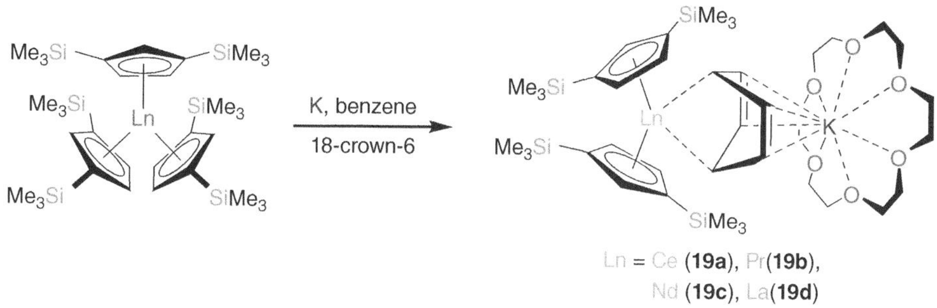

Scheme 11 *Synthesis of the Ln benzene dianion complexes* ***19a–d****. Reproduced from [40] with permission of American Chemical Society*

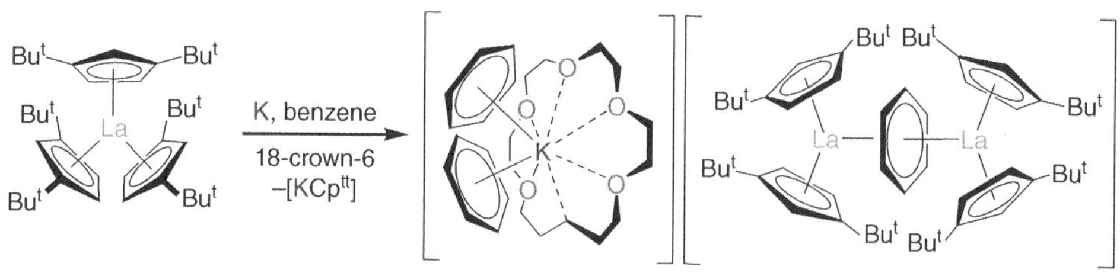

Scheme 12 *Synthesis of the formal La(II) complex* ***20****. Reproduced from [41] with permission of American Chemical Society*

[35, 36, 37, 38, 39]. The strength of the essentially electrostatic Ln–arene interactions in these systems is enhanced by the electron-withdrawing tetrahaloaluminate anions and electron-rich arenes, increasing the stability of **17** and **18** and their analogs.

Reduction of [Ln(Cp″)$_2$Cl] or [Ln(Cp″)$_3$] by excess potassium in benzene and 18-crown-6 afforded the first structurally characterized examples of Ln complexes coordinated by a benzene dianion, [K(18-crown-6)][Ln(Cp″)$_2$(μ,η^2:η^4-C$_6$H$_6$)] (Ln = Ce **19a**, Pr **19b**, Nd **19c**, La **19d**) (Scheme 11) [40]. The formation of **19d** was shown by EPR spectroscopy to proceed via a transient La(II) intermediate "[La(Cp″)$_2$]", which is relatively stabilized by its bulky Cp ligands [40b]. Compounds **19a–d** are intensely colored due to considerable ligand–metal charge transfer and their solid state structures reveal puckered benzenide dianions in boat conformations, with two short [e.g., **19a** 1.350(8) and 1.352(9) Å] and four long [e.g., **19a** 1.442(8)–1.464(7) Å] C–C distances. The {C$_6$H$_6$}$^{2-}$ unit is therefore best described as a cyclohexa-1,4-dienide, which is confirmed by hydrolysis of **19a–d** to liberate cyclohexa-1,4-diene [40].

As solid state characterization of Ln(II) complexes could not be obtained from **19a–d** reduction mixtures, attention switched to related Cp^{tt} systems. Reduction of [La(Cp^{tt})$_3$] with excess potassium in benzene and 18-crown-6 gave an intermediate complex, postulated as being analogous to **19a–d**. The reaction mixture was left for a further week, depositing crystals of the formal La(II) complex [K(18-crown-6)(η^2-C$_6$H$_6$)$_2$][{La(Cp^{tt})$_2$}$_2$(μ,η^6:η^6-C$_6$H$_6$)] **20** (Scheme 12) [41]. Compound **20** exhibits a planar bridging $\{C_6H_6\}^{n-}$ unit with six approximately equal C—C distances [1.42(1)–1.45(1) Å], and was assigned as a monoanion coordinated to La(II) cations based on a comparison of literature Ln(II)/Ln(III)···$C_{centroid}$ distances with those observed for **20** [La—C_{arene} 2.75(1)–2.79(1) Å].

12.2.6 Zerovalent arenes

A series of formally zerovalent Ln bis(arene) complexes, [Ln($C_6H_3Bu^t_3$-1,3,5)$_2$] (Ln = Pr **21a**, Nd **21b**, Sm **21c**, Gd **21d**, Tb **21e**, Dy **21f**, Ho **21g**, Er **21h**, Lu **21i**, Sc **21j**, Y **21k**, La **21l**) were prepared by the reaction of Ln vapors, generated using electron beams, at 77 K with 1,3,5-tri-*tert*-butylbenzene (super-mesitylene, Mes*H) (Scheme 13) [42]. The bulky Mes*H ligands are essential for the stabilization of **21a–l** and less sterically demanding arenes are not fit for this purpose with the exception of Sc complexes, as Sc(III) has a much smaller ionic radius than the other REs (see Section 12.1) [2]. Even with the support of Mes*H ligands, **21a**, **21c**, and **21l** were more unstable than others in the series and the Ce, Eu, Tm and Yb analogs could not be isolated at 77 K [42b]. In the case of **21j** C—H activation readily occurs and a formal Sc(II) complex, "[Sc($C_6H_3Bu^t_3$-1,3,5){η^6, η^1-C_6H_3(CMe_2CH_2)-1,Bu^t_2-3,5)(H)]", was postulated based on EPR spectroscopic data [42d].

The various stabilities of these complexes were later rationalized by analysis of differences in the Ln $4f^n6s^2 \rightarrow 4f^{n-1}5d^16s^2$ promotion energies, Ln—C_{arene} bond enthalpies and Ln ionic radii [43]. Compounds **21a–l** were characterized by various spectroscopic and magnetometric methods but solid state structures were only obtained for **21d** and **21g**, revealing eclipsed arene rings and staggered *tert*-butyl groups, and approximate D_{3d} symmetry. Metrical parameters confirmed that the two coordinated η^6-arenes are neutral donors, with planar, fully delocalized arene rings [e.g., **21d** C_{arene}—C_{arene} 1.400(6)–1.426(5) Å; Gd—C_{arene} 2.630(4) Å mean] [42a, 42b].

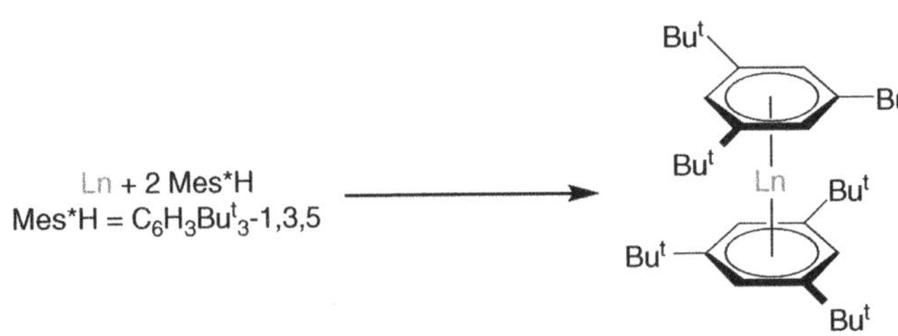

Ln = Pr **(21a)**, Nd **(21b)**, Sm **(21c)**, Gd **(21d)**, Tb **(21e)**, Dy **(21f)**, Ho **(21g)**, Er **(21h)**, Lu **(21i)**, Sc **(21j)**, Y **(21k)**, La **(21l)**

Scheme 13 *Synthesis of the Ln bis(arene) complexes **21a–l** [42]*

12.2.7 Tethered *N*-heterocyclic carbenes

Immediately following the disclosure of the first structurally characterized isolated *N*-heterocyclic carbene (NHC) [44], a raft of solid state structures of Ln-NHC complexes were reported [45]. Progress in this field withered briefly, due to the relatively weak binding of these neutral σ-donors to Ln cations, before the introduction to Ln chemistry of NHC ligands with pendant anionic groups [46]. The first structurally authenticated examples of tethered NHC Ln complexes, [Ln(T^1-NHC-κ^2-*C,N*)(N″)$_2$] (Ln = Sm **22b**, Y **22d**; T^1-NHC = {Bu^{t}NCH$_2$CH$_2$[1-C(Bu^{t}NCHCHN)]}$^-$; N″ = {N(SiMe$_3$)$_2$}$^-$) were synthesized by protonolysis of [Ln(N″)$_3$] by the LiBr-occluded pro-ligand T^1-NHC-H•LiBr (Scheme 14) [47]. Single crystal XRD studies showed relatively short Ln—C$_{carbene}$ distances in **22b** and **22d** [2.588(2) and 2.501(5) Å, respectively] compared with literature examples and the strength of these interactions was confirmed by the ^{13}C NMR spectrum of **22d**. This revealed a C$_{carbene}$ resonance at 186 ppm, with the $^1J_{YC}$ coupling constant (54 Hz) larger than any other C σ-donor ligand reported at that time in Y chemistry. The field of tethered NHC Ln complexes has blossomed and the Nd (**22a**) [48] and Eu (**22c**) [49] homologs and many other related Ln complexes have since been prepared by analogous procedures (Scheme 14) [46].

Oxidation of the Ce(III) tethered alkoxide-NHC complex [Ce(T^2-NHC-κ^2-*C,O*)$_3$] **23** (T^2-NHC = {OCMe$_2$CH$_2$[1-C(Pr^{i}NCHCHN)]}$^-$) with benzoquinone and 1 equiv. of [K(T^2-NHC)] yielded the first Ce(IV) carbene complex, [Ce(T^2-NHC-κ^2-*C,O*)$_2$(T^2-NHC-κ-*O*)$_2$] **24**, in excellent yield (Scheme 15) [50]. The solid state structure of **24** revealed two bound and two free carbenes [Ce—C$_{carbene}$ 2.673(7) Å mean], which were

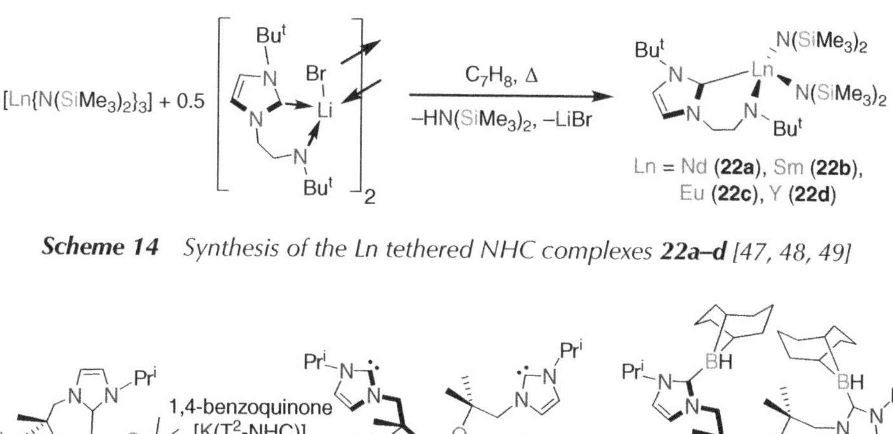

Scheme 14 Synthesis of the Ln tethered NHC complexes **22a–d** [47, 48, 49]

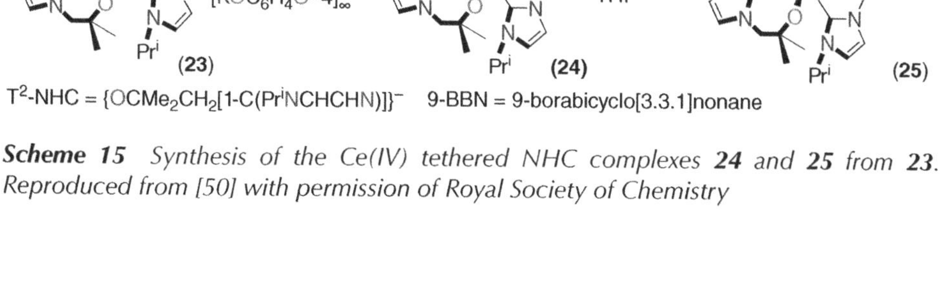

Scheme 15 Synthesis of the Ce(IV) tethered NHC complexes **24** and **25** from **23**. Reproduced from [50] with permission of Royal Society of Chemistry

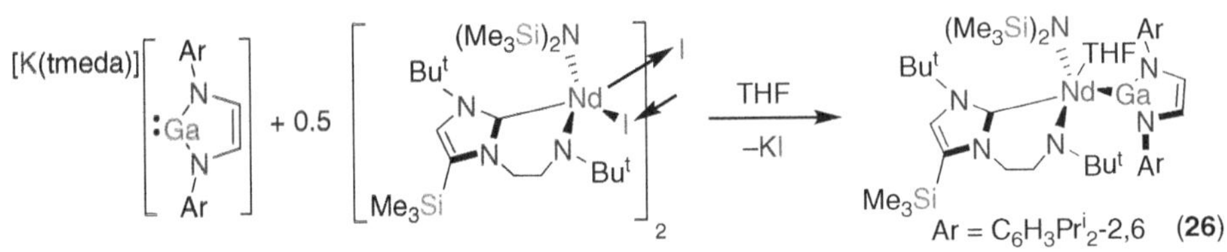

***Scheme 16** Synthesis of the neodymium-gallyl complex **26**. Reproduced from [52] with permission of American Chemical Society*

found to interchange rapidly in solution as one $C_{carbene}$ resonance was observed at 212 ppm in the ^{13}C NMR spectrum. The reaction of **24** with 2 equiv. of 9-BBN (9-borabicyclo[3.3.1] nonane) yielded [Ce(T^2-NHC-κ^2-*C*,*O*)$_2$(T^2-NHC-9-BBN-κ-*O*)$_2$] **25** (Scheme 15). Two of the carbenes in **25** bind strongly to 9-BBN [B–$C_{carbene}$ 1.639(3) Å mean], nullifying the dynamic fluxional behavior of free and bound carbenes in the solution phase that were observed for **24**, although $C_{carbene}$—B resonances were not observed in the ^{13}C NMR spectrum of **25** [50]. The highly oxidizing +4 oxidation state of Ce in **24** and **25** is stabilized by modification of the tethered NHC ligands to incorporate pendant alkoxide groups instead of amido moieties. The degree of hardness of the 4f Ce(IV) cation in **25** was emphasized by comparison with an analogous 5f U(IV) complex, which exhibited an additional relatively soft U—$C_{carbene}$ interaction in the solid state despite the similarities in ionic radii of Ce(IV) (CN 8, 0.97 Å) and U(IV) (CN 8, 1.0 Å) [51].

The further synthetic utility of Ln tethered-NHC complexes was demonstrated by the first structurally characterized bond between gallium and an f-element. [Nd{Ga[N(Dipp)CH)]$_2$}(T^3-NHC-κ^2-*C*,*N*)(N″)(THF)] (Dipp = $C_6H_3Pr^i_2$-2,6; T^3-NHC = {Bu^{t}NCH$_2$CH$_2$[1-C{Bu^{t}NC(SiMe$_3$)CHN}]}$^-$; **26**) was prepared by a salt metathesis reaction between [Nd(T^3-NHC-κ^2-*C*,*N*)(N″)(μ-I)]$_2$ and the Ga(I) complex [K(TMEDA)] [Ga{N(Dipp)CH}$_2$] (TMEDA = *N*,*N*,*N*′,*N*′-tetramethylethylenediamine) in THF (Scheme 16) [52]. The sterically demanding ligand frameworks in **26** are essential for kinetic stabilization of both the +1 oxidation state of gallium and the polarized Nd—Ga bond. The necessity of bulky supporting ligands in the stabilization of Ln—metal bonds is a common theme in the literature [53]. To date there are few structurally characterized examples of Ln—metal bonds for comparison, but the Nd—Ga bond distance in **26** [3.2199(3) Å] is longer than the sum of covalent radii for the two elements (Nd—Ga 2.89 Å) [54] and a computational study of a model complex showed that the bond is essentially comprised of donation of a formal 4s4p^2 hybridized Ga(I) lone pair into a vacant Nd 5d orbital, with a calculated Wiberg bond order of 0.827 [52].

12.3 N-donor ligands

12.3.1 Hexamethyldisilazide

The bulky silylamide hexamethyldisilazide (hmds, N″) has been the most frequently employed amide ligand in Ln chemistry to date [55]. These ligands were introduced to Ln chemistry in 1971, with a series of homoleptic complexes [Ln(N″)$_3$] (Ln = Ce **27a**,

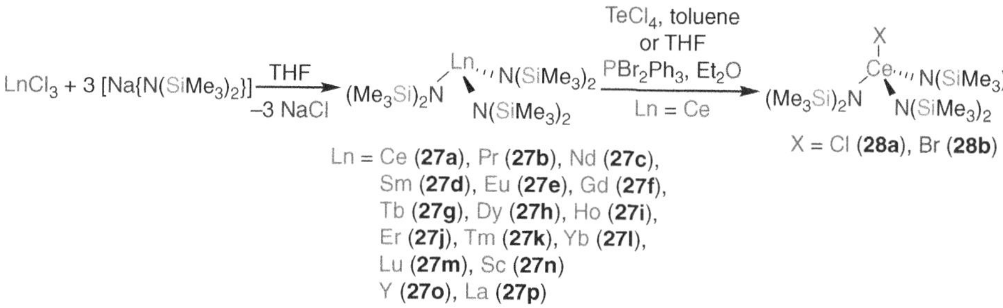

Scheme 17 *Synthesis of Ln silylamides* ***27a–p, 28a*** *and* ***28b*** *[56, 57, 58, 59, 60, 61, 62, 65]*

Pr **27b**, Nd **27c**, Sm **27d**, Eu **27e**, Gd **27f**, Ho **27i**, Yb **27l**, Lu **27m**, Sc **27n**, Y **27o**, La **27p**) prepared from $LnCl_3$ and 3 equiv. of LiN″ (Scheme 17) [56]. To avoid LiCl occlusion and ate complex formation [57], these procedures have since been modified to typically use sub-stoichiometric quantities of $LnCl_3$ and NaN″ or KN″ instead of LiN″, such as in the synthesis of $[Ln(N'')_3]$ (Ln = Tb **27g** [58], Dy **27h** [59], Tm **27k** [60]). Other strategies have included the use of $Ln(OTf)_3$ (OTf = triflate), such as in the synthesis of $[Er(N'')_3]$ (**27j**) [61] or protonolysis of alkyl precursors with HN″ [7c]. Structural characterization of most of **27a–p** followed subsequently [59, 62] and in all cases they were found to be three coordinate and pyramidal, with additional agostic interactions, even in the gas phase for **27a, 27b**, and **27n** [63]. As with **3a–i** the deviation from planarity of **27a–p** is attributed mainly to agostic Ln···Si–C_β interactions, which predominate over steric effects [8, 62b, 64]. These complexes and closely related variants have been used extensively as starting materials and are important precursors to a wide range of complexes, such as in the synthesis of **22a–d** (see above) [47, 48, 49] and the rare Ce(IV) amides $[Ce(N'')_3(X)]$ (X = Cl **28a**; Br **28b**), which are prepared by oxidation of **27a** with 0.25 equiv. of $TeCl_4$ or equimolar PBr_2Ph_3, respectively (Scheme 17) [65].

Ln silylamide chemistry was expanded by the discovery that reduction mixtures of **27a–p** and KC_8 in the presence of N_2 in THF gave the corresponding bridged dinitrogen Ln(III) complexes, $[\{Ln(N'')_2(THF)\}_2(\mu\text{-}\eta^2{:}\eta^2\text{-}N_2)]$ (Ln = Nd **29a**, Gd **29b**, Tb **29c**, Dy **29d**, Ho **29e**, Er **29f**, Tm **29g**, Lu **29h**, Y **29i**) (Scheme 18), which have N=N bond distances that are consistent with a N_2^{2-} dianion [N–N distance range 1.172(6)–1.305(6) Å] [66]. In the case of the largest Ln, dinitrogen complexes could not be isolated and $[M(THF)_n(Et_2O)_m][Ln(N'')_4]$ (Ln = Ce, M = Na, n = 5, m = 1; Ln = Pr, La, M = K, n = 6, m = 0) complexes formed except for in the case of samarium, where the previously reported Sm(II) ate complex $[Sm(N'')_3(K)]$ was generated [67]. In most cases, these reactions proceed through short-lived Ln(II) intermediates, though when stable Ln(II) precursors are available, an alternative synthetic route direct from LnI_2 is possible (Scheme 18) [68]. For some of the Ln dinitrogen complexes with relatively small Ln ionic radii and high magnetic anisotropy, further reduction by KC_8 was performed in the presence of 18-crown-6 to yield $[K(18\text{-crown-6})(THF)_n][\{Ln(N'')_2(THF)\}_2(\mu\text{-}\eta^2{:}\eta^2\text{-}N_2)]$ (Ln = Gd **30a**, Dy **30c**, n = 0;

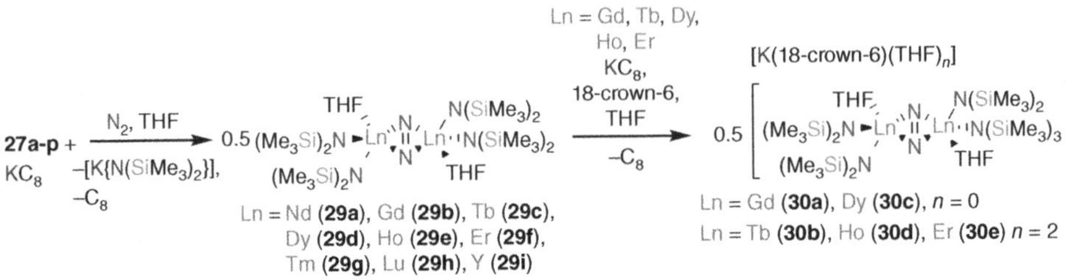

Scheme 18 *Synthesis of Ln silylamide N_2 complexes **29a–i** and **30a–e** [66, 69]*

Ln = Tb **30b**, Ho **30d**, Er **30e**, $n=2$), which contain N_2^{3-} radical trianionic bridges [N—N distance range 1.3940(3)–1.4085(3) Å] and exhibit interesting magnetic properties (Scheme 18) [69].

It has been proposed that Ln single molecule magnets (SMMs) have numerous potential magnetic applications as they can exhibit high magnetic anisotropies and therefore increased barriers to molecular spin inversion [70]. Fast quantum tunneling relaxation pathways, which result from predominantly electrostatic bonding and consequently weak magnetic exchange coupling, tend to dominate in these systems. This leads to decreased blocking temperatures, defined as the temperature at which magnetization is lost over a set time period (typically arbitrarily set at 100 s). These blocking temperatures are often at cryogenic temperatures, precluding many of the magnetic applications of Ln SMMs [70]. The enhanced magnetic properties of **30a–e** derive from the diffuse radical spin orbitals in the bridging N_2^{3-} unit mediating strong magnetic exchange coupling between Ln centers [69]. Comparison of magnetic data for **30a–e** with their parent N_2^{2-} complexes demonstrated the extent of this increased coupling directly. The strong coupling promoted by the N_2^{3-} unit increases the energy barrier towards fast zero-field relaxation pathways, favoring a thermally induced relaxation mechanism at low temperatures [69]. The strongest magnetic exchange coupling was observed for **30a**, which has a calculated intramolecular coupling constant, J, of $-27\,cm^{-1}$, the strongest coupling observed for any gadolinium complex at the time of publication [69a]. Compound **30a** is not an SMM as it contains two isotropic Gd(III) f^7 ions but **30b–e** all contain anisotropic ions and are all SMMs with relatively high blocking temperatures, the highest being the 100 s blocking temperature of 13.9 K for **30b** [69].

Recently, bulkier analogs of N″ have been introduced to Ln(II) chemistry, with THF-free $[Sm\{N(SiPr^i_3)_2\}_2]$ (**31**) prepared from $[SmI_2(THF)_5]$ and 2 equiv. of $[K\{N(SiPr^i_3)_2\}]$ (Scheme 19) [71]. Compound **31** is the first near-linear bis(amide) f-block complex, with an N—Sm—N angle of 175.52(18)° in the solid state, and four electrostatic interactions are observed between methine groups and the Sm(II) center. The authors performed calculations on a theoretical Dy(III) cationic analog of **31**, $[Dy\{N(SiPr^i_3)_2\}_2]^+$, and found that the model Dy(III) complex would be an SMM with a blocking temperature higher than 77 K, providing a synthetically feasible target for a liquid nitrogen-stable Ln SMM.

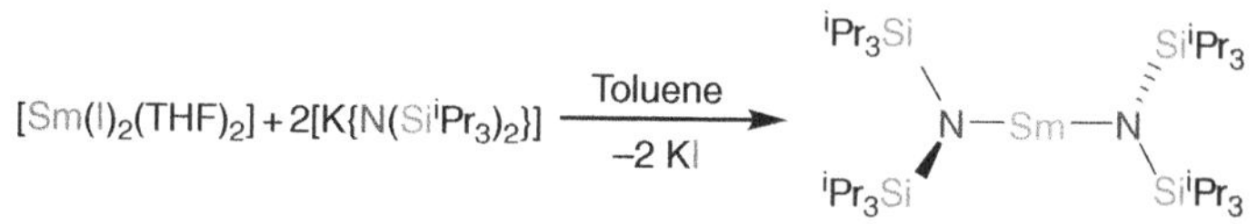

Scheme 19 *Synthesis of the Sm(II) silylamide complex* ***31****. Reproduced from [71]. http://pubs.rsc.org/en/content/articlehtml/2014/cc/c4cc08312a CC BY 3.0 public domain http://creativecommons.org/licenses/by/3.0/*

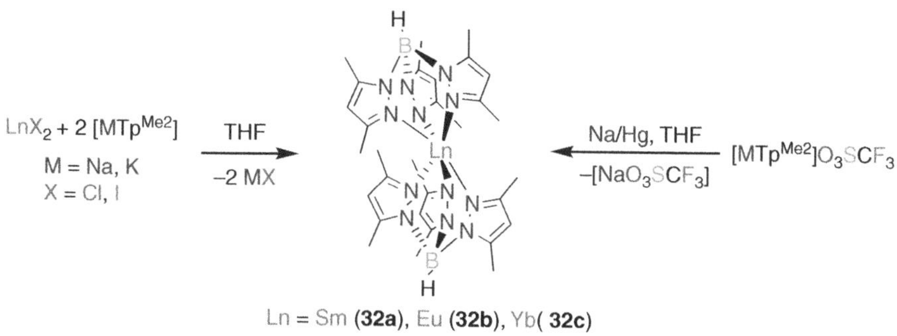

Scheme 20 *Synthesis of the homoleptic Ln(II) complexes* ***32a–c*** *[73, 74]*

12.3.2 Substituted trispyrazolylborates

A review of stabilizing N-donor ligands in Ln chemistry would be incomplete without a section on the widespread trispyrazolylborate (Tp) ligand class, but these are only discussed briefly here as they have been reviewed thoroughly elsewhere [72]. Ln complexes containing unsubstituted Tp ligands have been known for nearly 50 years and are often even stable in water, but these complexes are typically insoluble in non-polar solvents, reducing their further synthetic utility. Tuning the substituents at the 3- and 5- positions of the pyrazolyl rings has allowed the synthesis of a wide range of substituted Tp Ln complexes with increased solubilities and steric restraints, promoting low CNs [72]. One of the simplest modifications, tris(3,5-dimethylpyrazolyl) borate, Tp^{Me2}, has been widely utilized and is effective at stabilizing Ln(II) complexes [72]. Perhaps the most widely utilized Ln(II) Tp^{Me2} complexes are $[Ln(Tp^{Me2})_2]$ (Ln = Sm **32a**, Eu **32b**, Yb **32c**), which are easily synthesized by salt metathesis strategies from LnX_2 (X = Cl, I) and MTp^{Me2} (M = Na, K) (Scheme 20) [73]. Compounds **32b** and **32c** may alternatively be prepared by a sodium amalgam reduction of Ln(III) triflate precursors $[Ln(Tp^{Me2})_2]O_3SCF_3$ (Scheme 20) [74].

Compounds **32a–c** are insoluble in non-polar solvents and in the solid state the Ln(II) centers of **32a** and **32c** are bound by six N-donors and exhibit trigonal antiprismatic geometries and S_6 symmetry elements [73, 74]. Reactivity studies of **32a** have been extensive and include the reduction of unsaturated organic substrates and of d-transition metal carbonyl complexes [72]. The facile oxidation of **32a** is highlighted by its reaction with

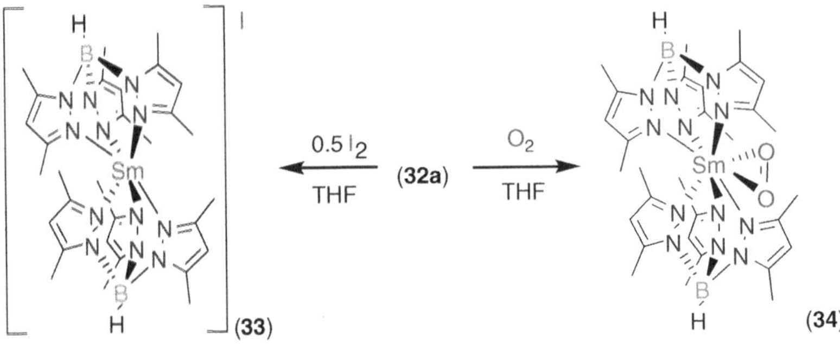

Scheme 21 *Reactivity of **32a** with I_2 and O_2 to form **33** and **34** [75, 76]. Reproduced from [76] with permission of American Chemical Society*

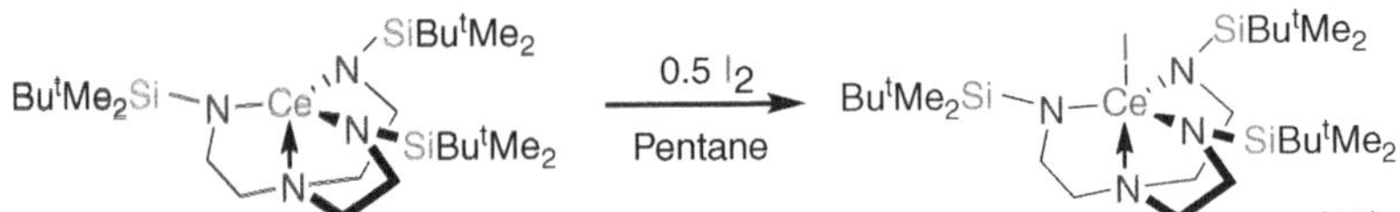

Scheme 22 *Synthesis of the heteroleptic Ce(IV) amide complex **35**. Reproduced from [77] with permission of American Chemical Society*

relatively mild oxidants such as 0.5 equiv. of I_2 yielding the salt $[Sm(Tp^{Me2})_2][I]$ (**33**), with a non-coordinating iodide anion (Scheme 21) [75]. The reaction of **32a** with O_2 yielded the first structurally characterized Ln superoxo complex $[Sm(Tp^{Me2})_2](\eta^2\text{-}O_2)]$ (**34**), in which the O_2^- anion was found to coordinate to Sm(III) in a side-on fashion and the O—O distance [1.319(5) Å] and various UV-Vis, Raman and IR spectroscopic ^{18}O labeling studies confirmed mono-reduction of dioxygen (Scheme 21) [76].

12.3.3 Silyl-substituted triamidoamine, $[N(CH_2CH_2NSiMe_2Bu^t)_3]^{3-}$

The oxidation of the Ce(III) complex [Ce(NN′)] (NN′ = $\{N(CH_2CH_2NSiMe_2Bu^t)_3\}^{3-}$) with 0.5 equiv. of molecular iodine yielded [Ce(NN′)(I)] (**35**), which at the time of publication was the first structurally characterized Ce(IV) amide complex (Scheme 22) [77]. This reaction is in itself exceptional, given that iodine is not an overly strong oxidant (standard reduction potential in acidic solution, $E^\circ\ I_2 \rightarrow I^- = 0.535\,V$) and Ce(IV) is itself a strong oxidant (E° Ce(IV) → Ce(III) = 1.76V) [27]. These potentials were overcome by the large Ce—I bond enthalpy and it was proposed that the rigid trigonal pyramidal arrangement of the NN′ scaffold in [Ce(NN′)] renders the Ce(III) center more oxidizable as minimal ligand reorganization is required upon oxidation to **35** [77]. Apart from the sterically demanding silyl groups providing kinetic protection of the Ce(IV) oxidation state in **35**, the NN′ ligand has stabilized Ce(IV) electronically as it has donated considerable electron density to Ce, preventing its reduction by I^-. Compound **35** is also stabilized by the strong Ce—I bond enthalpy, which would not be

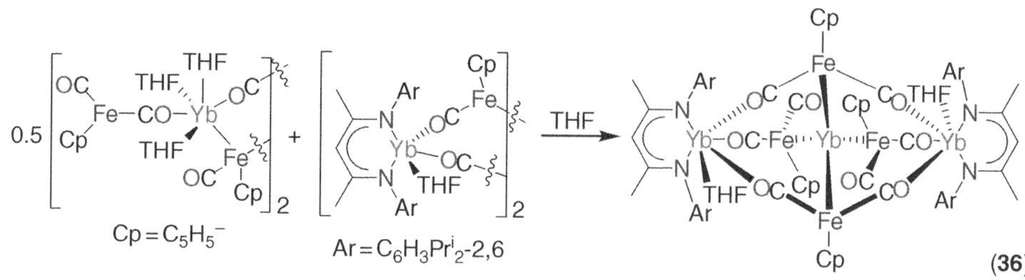

(36)

Scheme 23 *Synthesis of the Yb(II) complex* ***36***. *Reproduced from [79] with permission of Royal Society of Chemistry*

compensated by weak I→Ce(III) dative bonds formed upon reduction. In the solid state the unique Ce(IV)—I distance of **35** is 3.1284(6) Å and the complex was found to be diamagnetic, confirming its closed shell $[Xe]4f^0$ conformation.

12.3.4 NacNac, {N(Dipp)C(Me)CHC(Me)N(Dipp)}$^-$

The NacNac ligand, {N(Dipp)C(Me)CHC(Me)N(Dipp)}$^-$, has been extensively employed as an ancillary ligand in f-element chemistry and this area has been reviewed elsewhere [78, 79]. Recently, the NacNac ligand has been used to stabilize an approximately tetrahedral Yb(II) cluster complex $[\{Yb(NacNac)(THF)\}_2\{\mu\text{-}Yb(Fp)_4\}]$ (**36**; Fp = $[Fe(Cp)(CO)_2]^-$), that unusually exhibits four Fe—Yb bonds to the same Yb(II) center. Compound **36** was initially isolated as a minor product from the decomposition of $[Yb(NacNac)(I)(THF)(\mu\text{-}Fp)]_2$ in a non-coordinating solvent, however it was found that **36** could be rationally synthesized by the 2:1 reaction of $[Yb(NacNac)(I)(THF)(\mu\text{-}Fp)]_2$ with $[Yb(Fp)_2(THF)_3]_2$ [80] (Scheme 23).

12.4 P-donor ligands

12.4.1 Phospholides

Given the preference of Ln cations for hard donor atoms, it would be anticipated that the Ln coordination chemistry of P-donor ligands is poorly developed in comparison with O-, N- and C- donor ligands. Although this is generally the case, the phospholide ligands, $\{PC_4R_4\}^-$ (R = variously H, Me, Bu^t, etc.), are noteworthy exceptions that have rich and well developed Ln coordination chemistry that has been comprehensively reviewed previously [81]. A comprehensive discussion is beyond the scope of this chapter but Ln phospholide coordination chemistry is included briefly for completeness. Phospholides are monophosphorus-substituted analogs of Cp ligands and interest in Ln phospholide chemistry derives from both their similarity to Cp ligands and ease of synthesis. Phospholides may bind either through the heteroatom lone pair (η^1) or π-system (η^5), with the preferred binding mode often dictated by substitution of bulky groups in the aromatic ring [81]. As phospholides are relatively soft, they have found specific utility

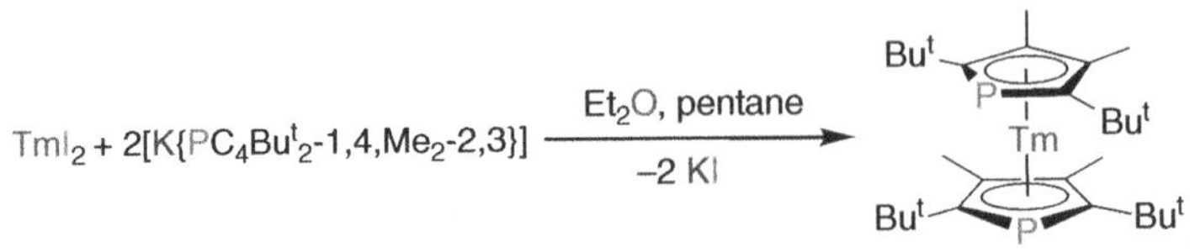

(37)

Scheme 24 *Synthesis of the Tm(II) homoleptic phospholide **37**. Reproduced from [82] with permission of American Chemical Society*

in the stabilization of Ln(II) complexes. With sufficiently bulky substituents, mononuclear Ln metallocene complexes may be isolated, for example [Tm{P[C(Bu^t)C(Me)]$_2$}$_2$] (**37**) [82]. Compound **37** was synthesized by a salt metathesis methodology from TmI_2 and [K{P[C(Bu^t)C(Me)]$_2$}] in diethyl ether followed by removal of coordinated solvent *in vacuo*, and at the time of publication **37** was the first structurally characterized homoleptic Tm(II) complex (Scheme 24). In the solid state the phospholide ligands in **37** coordinate in a staggered conformation in an approximate η^5- fashion with two additional agostic interactions between Tm(II) and methyl carbons of the *tert*-butyl groups. The phospholide ligands adopt a $\mu,\eta^1:\eta^5$ binding mode in the dimeric Sm(II) analog of **37** as the larger Sm(II) cations would have unsaturated coordination spheres in the absence of an additional dative interaction.

12.5 Multiple bonds

The preparation of terminal unsupported Ln=E (E=p-block element, e.g., C, N, O, P) multiple bonds is of considerable synthetic interest, given the wealth of applications of TM=E (TM=d-transition metal) complexes in synthetic transformations and catalysis [83]. However, highly polarized Ln=E bonding and orbital energy mismatch (e.g., Ln 5d, 4f and O 2s, 2p) often renders these species highly reactive and unstable with respect to oligomerization and various decomposition pathways due to the build-up of electron density on formal E^{2-} fragments [84]. As a result, the field of Ln=E multiple bonding is poorly developed to date, although this field has started to gather considerable momentum in the last ten years. A variety of sterically demanding and electronically stabilizing bespoke ligand frameworks have been employed to prevent decomposition of Ln=E groups, and some highlights in the literature are discussed below [84].

12.5.1 Ln=CR$_2$

Although the field of Ln carbene chemistry lags far behind that of related early TM systems, there are more examples of Ln complexes exhibiting formal Ln=C multiple bonds than any other Ln=E group [84,85]. In the absence of stabilizing R groups, Ln=CR$_2$ complexes (Ln alkylidenes/carbenes), tend to form cluster complexes, with methylidene CH_2^{2-} units bridging to multiple metal centers [85]. This bonding motif is analogous to early TM methylidene chemistry, such as that observed in Tebbe's reagent,

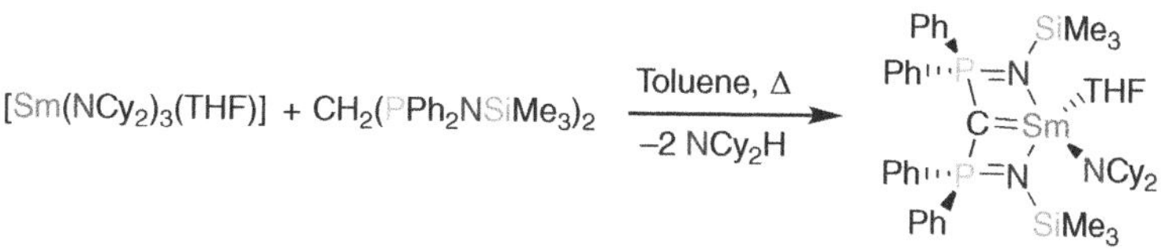

(38)

Scheme 25 *Synthesis of the Sm methanediide complex **38**. Reproduced from [87] with permission of American Chemical Society*

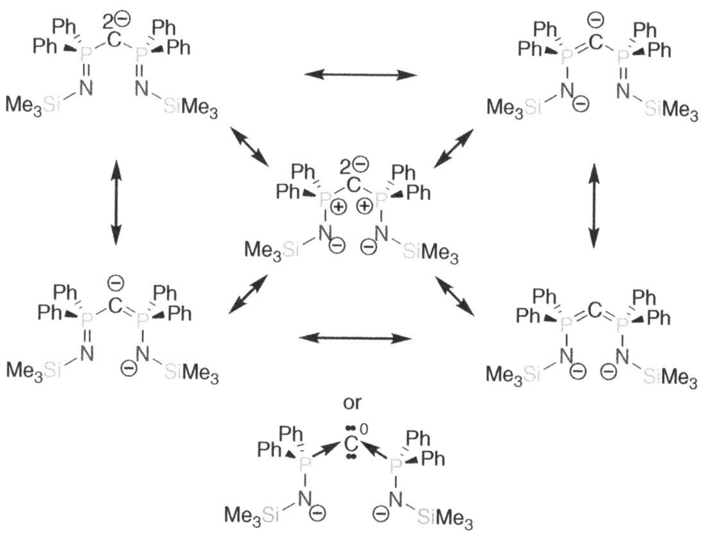

Scheme 26 *Resonance forms of the {BIPM^{TMS}}$^{2-}$ ligand [85, 88]*

$[Ti(Cp)_2(\mu\text{-}CH_2)(\mu\text{-}Cl)(AlMe_2)]$, which contains a "trapped" reactive Ti=CH_2 fragment [86]. The first structurally characterized supported Ln carbene complex, [Sm(BIPM^{TMS})(NCy_2)(THF)] (BIPM^{TMS} = {$CPPh_2NSiMe_3$}$^{2-}$, **38**) was prepared by a double deprotonation of $BIPMH_2$ by $[Sm(NCy_2)_3]$ at elevated temperatures (Scheme 25) [87].

In the solid state **38** exhibits a typical "open-book" conformation of the {BIPM^{TMS}}$^{2-}$ framework and a formal Sm=C double bonding interaction [Sm—C 2.467(4) Å], which was much shorter than the majority of previously reported Sm—C distances (mean 2.743 Å) [87]. A number of different resonance forms can be depicted for the {BIPM^{TMS}}$^{2-}$ ligand (Scheme 26), meaning that the ligand can be described alternatively as a carbene, carbodiphosphorane, or methanediide, but the dipolar geminal methanediide resonance form with two anionic amides dominates [88]. The {BIPM^{TMS}}$^{2-}$ ligand in **38** fixes the formal methanediide R_2C^{2-} dianion in close proximity to the Sm(III) cation by additionally bonding to samarium through its iminophosphorano N lone pairs. The carbanion stabilization energy of the PH_2 group has been calculated to be −89.1 kJ mol^{-1} hence by extension the P(V) substituents in the {BIPM^{TMS}}$^{2-}$ scaffold

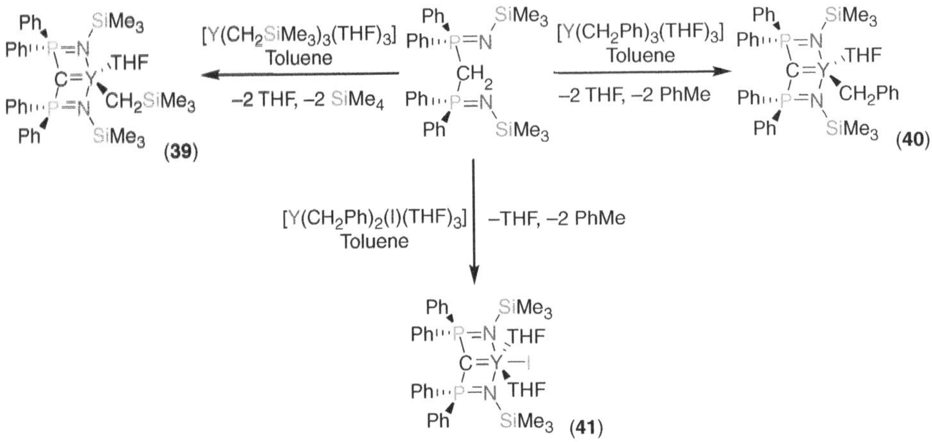

*Scheme 27 Synthesis of the Y methanediide complexes **39–41** [90]*

are vitally important for stabilization of the methanediide as they provide considerable electronic stabilization through negative hyperconjugation as well as kinetic stabilization from their steric bulk [89].

As no reactivity studies of **38** were reported, facile and general routes to [Ln(BIPM^{TMS})(R)] complexes were required and this was addressed by the syntheses of [Y(BIPM^{TMS})(CH_2SiMe_3)(THF)] (**39**), [Y(BIPM^{TMS})(CH_2Ph)(THF)] (**40**) and [Y(BIPM^{TMS})(I)$(THF)_2$] (**41**) from yttrium alkyl precursors at room temperature (Scheme 27) [90]. It was shown that **41** could alternatively be synthesized by deprotonation of [Y(BIPM^{TMS}H)$(I)_2$(THF)] with benzyl potassium [91]. These synthetic techniques have since been employed to synthesize a wide range of Ln methanediide complexes and this area has been reviewed thoroughly [85]. Computational studies on **39–41** revealed as expected localized electron density at the methanediide centers corresponding to approximately two lone pairs of electrons in frontier orbitals that do not interact appreciably with vacant Y orbitals (<5 % Y orbital contribution) [90]. The Y—$C_{methanediide}$ distances in **39–41** [range 2.356(3)–2.406(3) Å] are similar to the Y—$C_{methanide}$ distances [range 2.406(4)–2.408(3) Å] but computational studies revealed higher Y—$C_{methanediide}$ Wiberg bond orders (range 0.61–0.66) than Y—$C_{methanide}$ (range 0.35–0.51) [90]. These studies concluded that the Y—$C_{methanediide}$ linkages in **39–41** can be regarded as latent Y=C bonds in an analogous fashion to the latent Ti=C bond in Tebbe's reagent.

Detailed reactivity studies were carried out on **39–41**, showing that the polarized bonding in these systems engenders a reactivity profile that complements and contrasts related TM carbene systems [90, 92]. The highlight of these studies showed that **39** effects a site-specific, regioselective and sequential C—H activation and C—C and C—O bond formation with 3 equiv. of benzophenone, Ph_2CO, to form the coordinated substituted *iso*-benzofuran complex [Y(BIPM^{TMS}H){OC(CH_2SiMe_3)Ph_2}O-{(CPh_2)(OCPh)C_6H_4}] (**42**) (Scheme 28) [92a]. Detailed mechanistic studies showed that **42** is formed by initial 1,2-migratory insertion of Ph_2CO into the Y—$C_{methanide}$ bond to form [Y(BIPM^{TMS}){OC(CH_2SiMe_3)Ph_2})(THF)], which could be isolated,

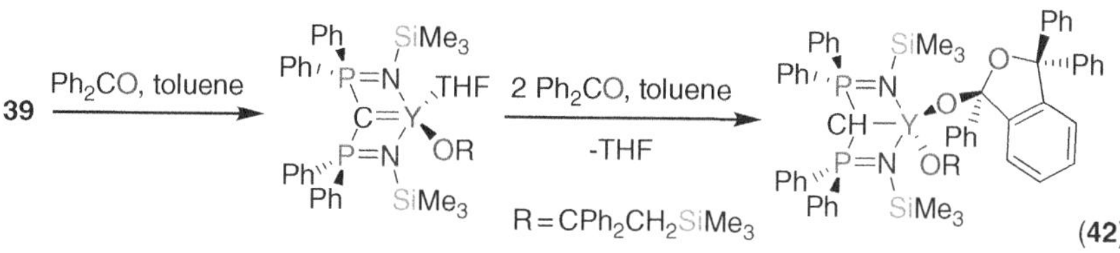

(42)

Scheme 28 *Reaction of **39** with Ph_2CO to form **42**. Reproduced from [92a] with permission of American Chemical Society*

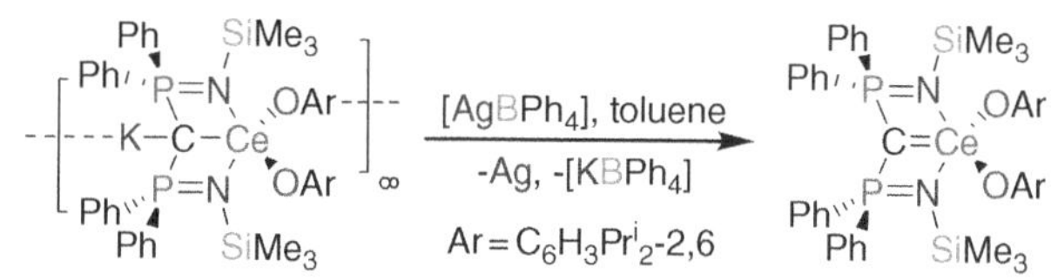

(43)

Scheme 29 *Synthesis of the Ce(IV) carbene complex **43**. Reproduced from [93] with permission of John Wiley & Sons, Ltd*

showing that the most reactive site on **39** is the metal-alkyl group. The second Ph_2CO molecule coordinates and an *ortho*-H on a phenyl ring is deprotonated by the methanediide, forming a methanide and a carbanion that attacks the ketyl carbon of the third molecule of Ph_2CO, which rearranges to form **42**. The reactivity of **39** is in contrast to related TM carbenes, which undergo Wittig-type reactivity with ketones to form alkenes and metal oxides [83], therefore the further investigation of $Ln{=}CR_2$ complexes is expected to generate more novel metal–carbene reactivity profiles in future.

The first structurally characterized $Ce(IV){=}CR_2$ complex $[Ce(BIPM^{TMS})(ODipp)_2]$, **43**, was recently prepared by the oxidation of a Ce(III) ate complex, $[Ce(BIPM^{TMS})(ODipp)_2(K)]_\infty$, with $AgBPh_4$ (Scheme 29) [93]. The Ce(IV) oxidation state in **43** is stabilized by both the bulky ligand frameworks and the aryloxide donor atoms and at the time of its publication, **43** represented the first structurally authenticated Ln(IV)=E multiple bond. The Ce(IV)=C bond in **43** was predicted to have a greater degree of covalency than the Ln(III)=C bonds in related complexes, rendering its description as a carbene more appropriate than a methanediide, and all characterization data concur with this hypothesis. A resonance at 324.6 ppm ($^1J_{PC}$ = 148.7 Hz) in the ^{13}C NMR spectrum of **43** was assigned to the carbene, which is similar to those observed in TM carbene complexes (range δ 200–400 ppm) [83] but is far downfield of those observed for ionic **39–41** (range δ 10–40 ppm) [90]. The Ce—$C_{carbene}$ distance of **43** [2.441(5) Å] is one of the shortest Ce—C bonds reported to date and computational studies yielded a Nalewajski–Mrozek bond index of 1.1, much higher than the Y—$C_{methanediide}$ bond orders calculated for **39–41** (see above) and significant Ce orbital contributions to both σ (13%) and π (12%) bonds. This covalent contribution is still modest and the bonding in **41** is predominantly electrostatic and best described as ionic-covalent [93].

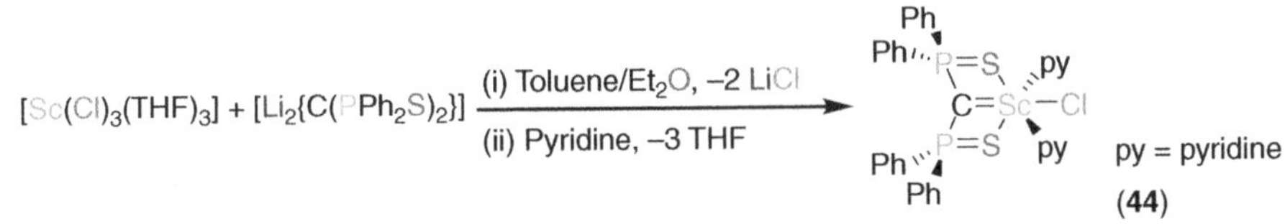

Scheme 30 *Synthesis of the Sc carbene complex* ***44****. Reproduced from [94] with permission of American Chemical Society*

In parallel with studies of Ln alkylidenes stabilized by the $\{BIPM^{TMS}\}^{2-}$ ligand, numerous investigations into the synthesis of Ln alkylidenes using the analogous $\{C(PPh_2S)_2\}^{2-}$ (SCS) dianion have been undertaken [85]. The SCS scaffold exhibits similar stabilization properties to $\{BIPM^{TMS}\}^{2-}$, though the thiophosphinoyl arms offer less kinetic protection. The SCS ligand was utilized to prepare the first structurally characterized Sc-carbene complex [Sc(SCS)(Cl)(py)] (py = pyridine, **44**), prepared directly by a salt metathesis strategy from $[Sc(Cl)_3(THF)_3]$ and Li_2SCS using pyridine to displace THF (Scheme 30) [94]. The $Sc—C_{carbene}$ distance of **44** [2.2072(1) Å] is comparable with Sc—C single bond lengths in the literature but a comparatively high Wiberg bond order of 0.60 was calculated as well as considerable σ and π overlap of the formal carbene lone pairs with the Sc 3d orbital. Direct experimental evidence for a significant covalent contribution to the bonding in **44** was evidenced by its reaction with benzophenone to afford the expected metallo-Wittig products $(Ph)_2C{=}C(PPh_2S)_2$ and "ScOCl". This reaction was found to be sluggish, taking several days at elevated temperatures to reach completion and an unusual Sc μ^3-oxo tetranuclear intermediate, $[\{Sc(\mu\text{-}SCS)(THF)(\mu\text{-}Cl)_2(\mu^3\text{-}O)\}\{Sc(THF)\}]_2$, could be isolated from reaction mixtures if sub-stoichiometric quantities of Ph_2CO were added [94].

12.5.2 Ln=NR

To date, no terminal unsupported Ln-imido complexes, Ln=NR, (Ln=Ce-Lu) have been isolated and characterized in the solid state, despite the comparative fertility of this area for related early TMs [83]. Structurally authenticated Ln-imido and -phosphinidene complexes in the literature, typically prepared by the deprotonation of Ln—NHR and Ln—PHR precursors, form dimers in the solid state with bridging $\{NR\}^{2-}$ and $\{PR\}^{2-}$ fragments [84]. It is noteworthy that the transient existence of unsupported Ln=NR and Ln=PR linkages had previously been observed, though could not be isolated [95]. In the case of the RE, the first terminal Sc-imido complex, [Sc(=NDipp)(T^1-NacNac-κ^3-*N,N′,N″*)(DMAP)] (T^1-NacNac = {N(Dipp)C(Me)CHC(Me)N($CH_2CH_2NMe_2$)}$^-$; DMAP = 4-dimethylaminopyridine, **45**), was recently prepared, which is stabilized by a bespoke tri-coordinate variant of the ubiquitous NacNac ligand [96].

Compound **45** was synthesized from the [Sc(T^1-NacNac)$(Cl)_2$] starting material by alkylation with 2 equiv. of MeLi followed by protonolysis with 1 equiv. of $DippNH_2$ to

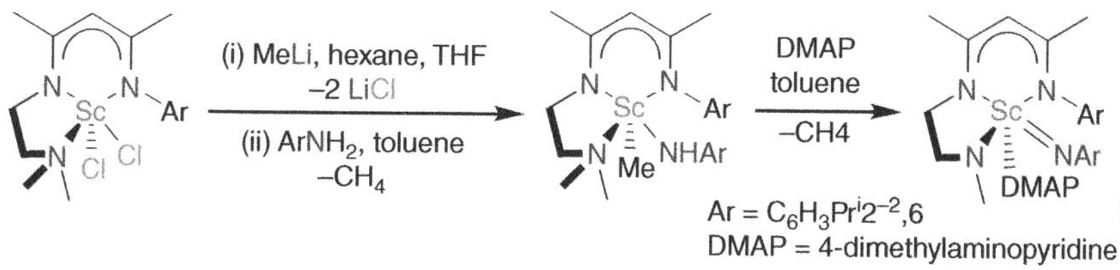

(45)

Scheme 31 *Synthesis of the Sc imido complex* ***45****. Reproduced from [96] with permission of Royal Society of Chemistry*

form the [Sc(NHDipp)(T^1-NacNac)(Me)] precursor, which yielded **45** by methane elimination upon addition of DMAP (Scheme 31) [96]. The Sc=N bond length of **45** [1.881(5) Å] was at the time the shortest Sc—N distance ever recorded and the almost linear Sc—N$_{imido}$—C angle [169.6(5)°] also suggested considerable covalent overlap as the two 2p-orbitals on N need to be approximately orthogonal to the Sc—N axis for efficient π-overlap with the Sc 3d orbitals. Computational analysis of the Sc=N bond of a model of **45** showed two π bonds compared with one in a model of [Sc(NHDipp)(T^1-NacNac)(Me)] and a large Wiberg bond order for **45** (1.32) that was approximately double the value of that calculated for the parent anilide (0.66). Since the isolation of **45**, Sc-imido chemistry has started to flourish, with Sc=N bonds being supported by a variety of ligand systems and their further chemistry being investigated, though a detailed analysis is beyond the scope of this chapter [84b].

12.5.3 Ln=O

The chemistry of unsupported terminal Ln=O bonds remains unexplored to date, as unlike Ln=CR_2 and Ln=NR chemistry there is an absence of bulky R substituents in these linkages to prevent dimerization to μ-oxo species [84]. The first supported Ln=O bond was recently reported, however, in [Ce(L_{OEt})$_2$(H_2O){μ-(O)···HN(H)C(O)Me}]$_2$ (L_{OEt}=[Co(Cp){P(O)(OEt)$_2$}$_3$]$^-$, **46**), in which the terminal oxo linkage is supported by hydrogen bonding to bridging molecules of acetamide [97]. The bulky oxygen-rich L_{OEt} tripodal ligands are ideally suited to stabilizing the trapped Ce=O group in **46** as they have proven utility in stabilizing M(IV) complexes in aqueous conditions [98]. Compound **46** was initially prepared in low yield (<5%) by the treatment of [Ce(L_{OEt})$_2$(Cl)$_2$] with Ag_2O in acetonitrile, but it was realized that the acetamide derived from oxidative hydrolysis of acetonitrile. As such, yields of **46** greatly improved upon addition of water and acetamide to reaction mixtures (Scheme 32) [97]. The Ce–O distance of **46** [1.857(3) Å] is the shortest reported to date and computational analysis verifies covalent contributions from Ce to the σ (8.6% Ce 5d, 16.0% Ce 4f) and π (5.4% Ce 5d, 7.6% Ce 4f) components of the predominantly electrostatic and localized bonding in the Ce=O linkage in **46**. Preliminary reactivity studies of **46** with the unsaturated substrates CO, CO_2 and Bu^{t}NCO led in each case to the formation of Ce(IV) carbonate complexes, highlighting the oxidizing reactivity profile of the supported Ce=O bond [97].

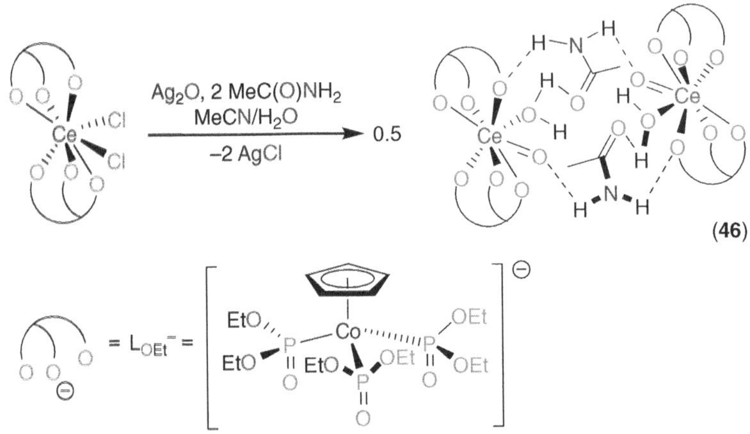

Scheme 32 *Synthesis of the Ce(IV) oxo complex* ***46****. Reproduced from [97] with permission of John Wiley & Sons, Ltd*

12.6 Conclusions

Given the wide variety of ligands in modern non-aqueous Ln chemistry, it would be an enormous task to discuss the merits of each system in detail. As such, this Chapter has focused on selected ligands that have engendered remarkable properties to Ln complexes to illustrate what can be achieved through considered ligand design and selection. The ligands discussed within this chapter have variously stabilized:

- unusual Ln oxidation states;
- low CNs;
- remarkable coordination geometries;
- novel bonding regimes.

The synthesis of new lanthanide–element single and multiple bonds has opened up new classes of complexes and huge vistas to explore in future. Reactivity studies on these complexes have shown interesting and unusual profiles that complement and contrast with related s- and d-block systems. These advances are of fundamental interest and potential industrial applications for these complexes in catalysis and materials are being realized, driving future investigations.

Within the last 50 years, major advances in non-aqueous Ln chemistry have been achieved with increasing regularity. This can be directly correlated to the growing number of research groups exploring the frontiers of this vibrant and exciting research field. Despite this rapid progress, our understanding of non-aqueous Ln chemistry still lags behind that of the d-transition metals. Most of the ligands that have found most success to date in Ln chemistry have first been of widespread utility in s- and d-block chemistry. The ligands that have proved most transferable are those that are anionic and sterically demanding and have multiple coordination sites and at present, examples of bespoke ligand systems for the Lns are few. As Ln—E bond energies are well established and are crucial in predicting the stability of complexes, the design and selection of suitable stabilizing ligands should utilize these data [99]. To conclude, further fundamental studies on complexes of existing and modified ligands are essential to deepen our understanding so that rational ligand design in Ln chemistry can be performed more readily in future.

Notes

Since this Chapter was submitted there have been many notable advances in molecular Ln chemistry. Several highlights are listed here to provide some context for the reader, in the approximate order that ligands have been discussed in this Chapter:

i) The synthesis of homologues of compound **15** for the entire Ln series (save Pm) have allowed the preference for $4f^{n+1}$ or $4f^n5d^1$ configurations for various Ln(II) systems to be rationalized, and their magnetic anisotropy has been investigated [100].
ii) Dinuclear La(II) and Ce(II) complexes supported by a bridging direduced benzene unit have extended the range of ligands that stabilize La(II) and Ce(II) centers [101].
iii) The synthesis and physical properties of near-linear Eu(II), Tm(II) and Yb(II) homologues of compound **31** have been reported, with the Tm(II) complex exhibiting remarkable stability towards dinitrogen [102].
iv) Ln $BIPM^{TMS}$ chemistry has been extended to include a Dy(III) SMM with a large energy barrier and a Ce(IV)=C complex with comparable covalency to a U(IV)=C homologue [103].
v) The first terminal unsupported Y=NR and Lu=NR bonds have been stabilized by a bulky Tp ligand [104] and a Ce(IV)=NR bond has been supported by a tris(hydroxylaminato) ligand and a capping potassium cation [105].
vi) The reactivity of compound **46** with a range of substrates has now been reported [106].

References

[1] (a) S. A. Cotton, *Lanthanide and Actinide Chemistry*, John Wiley & Sons, Ltd, Chichester, **2006**, pp. 1–144; (b) H. C. Aspinall, *Chemistry of the f-Block Elements*, Gordon and Breach, Amsterdam, **2001**, pp. 61–156; (c) S. T. Liddle, *Lanthanides: Organometallic Chemistry, Encyclopedia of Inorganic and Bioorganic Chemistry*, John Wiley & Sons, Ltd, Chichester, **2012**; (d) D. A. Atwood (ed.), *The Rare Earth Elements: Fundamentals and Applications*, John Wiley & Sons, Ltd, Chichester, **2012**.

[2] R. D. Shannon, *Acta Crystallogr., Sect. A: Cryst. Phys., Diffr., Theor. Gen. Crystallogr.* **1976**, *A32*, 751–767.

[3] (a) G. Meyer, *Angew. Chem. Int. Ed.* **2014**, *53*, 3550–3551; (b) F. T. Edelmann, *Coord. Chem. Rev.* **2014**, *261*, 73–155; and references cited therein.

[4] (a) M. Zimmermann, R. Anwander, *Chem. Rev.* **2010**, *110*, 6194–6259; (b) S. A. Cotton, *Coord. Chem. Rev.* **1997**, *160*, 93–127.

[5] (a) A. J. Wooles, D. P. Mills, W. Lewis, A. J. Blake, S. T. Liddle, *Dalton Trans.* **2009**, 500–510; (b) C. Döring, R. Kempe, *Z. Kristallogr., NCS* **2008**, *223*, 397; (c) W. Huang, B. M. Upton, S. I. Khan, P. L. Diaconescu, *Organometallics* **2013**, *32*, 1379–1386; (d) W. Huang, B. M. Upton, S. I. Khan, P. L. Diaconescu, *Organometallics* **2013**, *32*, 2275; (e) N. Meyer, P. W. Roesky, S. Bambirra, A. Meetsma, B. Hessen, K. Saliu, J. Takats, *Organometallics* **2008**, *27*, 1501–1505; (f) S. Ge, A. Meetsma, B. Hessen, *Organometallics* **2009**, *28*, 719–726; (g) S. Harder, C. Ruspic, N. N. Bhriain, F. Berkermann, M. Schurmann, *Z. Naturforsch.* **2008**, *63b*, 267–274; (h) D. P. Mills, O. J. Cooper, J. McMaster, W. Lewis, S. T. Liddle, *Dalton Trans.* **2009**, 4547–4555; (i) S. Bambirra, A. Meetsma, B. Hessen, *Organometallics* **2006**, *25*, 3454–3462.

[6] (a) H. Schumann, D. M. M. Freckmann, S. Dechert, *Z. Anorg. Allg. Chem.* **2002**, *628*, 2422–2426; (b) J. L. Atwood, W. E. Hunter, R. D. Rogers, J. Holton, J. McMeeking, R. Pearce, M. F. Lappert, *J. Chem. Soc., Chem. Commun.* **1978**, 140–142; (c) H. Schumann, J. Müller, *J. Organomet. Chem.* **1978**, *146*, C5–C7; (d) M. F. Lappert, R. Pearce, *J. Chem. Soc., Chem. Commun.* **1973**, 126.

[7] (a) A. G. Avent, C. F. Caro, P. B. Hitchcock, M. F. Lappert, Z. Li, X.-H. Wei, *Dalton Trans.* **2004**, 1567–1577; (b) C. Guttenberger, H. –D. Amberger, *J. Organomet. Chem.* **1997**, *545-546*, 601–606; (c) P. B. Hitchcock, M. F. Lappert, R. G. Smith, R. A. Bartlett, P. P. Power, *J. Chem. Soc., Chem. Commun.* **1988**, 1007–1009; (d) H. Reddmann, C. Guttenberger, H.–D. Amberger, *J. Organomet. Chem.* **2000**, *602*, 65–71; (e) C. J. Schaverien, A. G. Orpen, *Inorg. Chem.* **1991**, *30*, 4968–4978; (f) G. K. Barker, M. F. Lappert, *J. Organomet. Chem.* **1974**, *76*, C45–C46.

[8] (a) D. L. Clark, J. C. Gordon, P. J. Hay, R. L. Martin, R. Poli, *Organometallics* **2002**, *21*, 5000–5006; (b) L. Perrin, L. Maron, O. Eisenstein, M. F. Lappert, *New J. Chem.* **2003**, *27*, 121–127.

[9] (a) G. Qi, Y. Nitto, A. Saiki, T. Tomohiro, Y. Nakayama, H. Yasuda, *Tetrahedron* **2003**, *59*, 10409–10418; (b) C. Eaborn, P. B. Hitchcock, K. Izod, Z.-R. Lu, J. D. Smith, *Organometallics* **1996**, *15*, 4783–4790; (c) C. Eaborn, P. B. Hitchcock, K. Izod, J. D. Smith, *J. Am. Chem. Soc.* **1994**, *116*, 12071–12072.

[10] T. K. Hollis, J. K. Burdett, B. Bosnich, *Organometallics* **1993**, *12*, 3385–3386.

[11] (a) R. L. De Kock, M. A. Peterson, L. K. Timmer, E. J. Baerends, P. Vernooijs, *Polyhedron* **1990**, *9*, 1919–1934; (b) M. Kaupp, P. v. R. Schleyer, M. Dolg, H. Stoll, *J. Am. Chem. Soc.* **1992**, *114*, 8202–8208.

[12] (a) H. Yasuda, *J. Polym. Sci., Part A: Polym. Chem.* **2001**, *39*, 1955–1959; (b) H. Yasuda, *J. Organomet. Chem.* **2002**, *647*, 128–138.

[13] (a) G. B. Deacon, A. J. Koplick, *J. Organomet. Chem.* **1978**, *146*, C43–C45; (b) G. B. Deacon, A. J. Koplick, W. D. Raverty, D. G. Vince, *J. Organomet. Chem.* **1979**, *182*, 121–141.

[14] G. Heckmann, M. Niemeyer, *J. Am. Chem. Soc.* **2000**, *122*, 4227–4228.

[15] M. Niemeyer, *Acta Crystallogr., Sect. E: Struct. Rep. Online* **2001**, *E57*, m578–m580.

[16] G. W. Rabe, C. S. Strissel, L. M. Liable-Sands, T. E. Concolino, A. L. Rheingold, *Inorg. Chem.* **1999**, *38*, 3446–3447.

[17] (a) G. W. Rabe, C. D. Bérubé, G. P. A. Yap, *Inorg. Chem.* **2001**, *40*, 2682–2685; (b) G. W. Rabe, C. D. Bérubé, G. P. A. Yap, K.-C. Lam, T. E. Concolino, A. L. Rheingold, *Inorg. Chem.* **2002**, *41*, 1446–1453.

[18] (a) M. Ephritikhine, *Organometallics* **2013**, *32*, 2464–2488; (b) H. Schumann, J. A. Meese-Marktscheffel, L. Esser, *Chem. Rev.* **1995**, *95*, 865–986; (c) F. T. Edelmann, D. M. M. Freckmann, H. Schumann, *Chem. Rev.* **2002**, *102*, 1851–1896.

[19] (a) Z. Hou, Y. Wakatsuki, *J. Organomet. Chem.* **2002**, *647*, 61–70; (b) W. J. Evans, *J. Alloys Compd.* **2009**, *488*, 493–510; (c) W. J. Evans, *J. Organomet. Chem.* **2002**, *652*, 61–68; (d) W. J. Evans, *J. Alloys Compd.* **1993**, *192*, 205–210.
[20] (a) W. J. Evans, I. Bloom, W. E. Hunter, J. L. Atwood, *J. Am. Chem. Soc.* **1981**, *103*, 6507–6508; (b) W. J. Evans, L. A. Hughes, T. P. Hanusa, *J. Am. Chem. Soc.* **1984**, *106*, 4270–4272.
[21] W. J. Evans, T. A. Ulibarri, J. W. Ziller, *J. Am. Chem. Soc.* **1988**, *110*, 6877–6879.
[22] W. J. Evans, J. W. Grate, L. A. Hughes, H. Zhang, J. L. Atwood, *J. Am. Chem. Soc.* **1985**, *107*, 3728–3730.
[23] (a) F. Nief, *Dalton Trans.* **2010**, 6589–6598; (b) M. N. Bochkarev, *Coord. Chem. Rev.* **2004**, *248*, 835–851; (c) M. N. Bochkarev, *Mater. Sci. Forum* **1999**, *315–317*, 144–153.
[24] M. N. Bochkarev, I. L. Fedushkin, A. A. Fagin, T. V. Petrovskaya, J. W. Ziller, R. N. R. Broomhall-Dillard, W. J. Evans, *Angew. Chem. Int. Ed. Engl.* **1997**, *36*, 133–135.
[25] W. J. Evans, N. T. Allen, J. W. Ziller, *J. Am. Chem. Soc.* **2000**, *122*, 11749–11750.
[26] M. N. Bochkarev, I. L. Fedushkin, S. Dechert, A. A. Fagin, H. Schumann, *Angew. Chem. Int. Ed.* **2001**, *40*, 3176–3178.
[27] L. J. Nugent, R. D. Baybarz, J. L. Burnett, J. L. Ryan, *J. Phys. Chem.* **1973**, *77*, 1528–1539.
[28] P. B. Hitchcock, M. F. Lappert, L. Maron, A. V. Protchenko, *Angew. Chem. Int. Ed.* **2008**, *47*, 1488–1491.
[29] (a) M. R. MacDonald, J. W. Ziller, W. J. Evans, *J. Am. Chem. Soc.* **2011**, *133*, 15914–15917; (b) M. R. MacDonald, J. E. Bates, M. E. Feiser, J. W. Ziller, W. J. Evans, *J. Am. Chem. Soc.* **2012**, *134*, 8420–8423; (c) M. R. MacDonald, J. E. Bates, J. W. Ziller, F. Furche, W. J. Evans, *J. Am. Chem. Soc.* **2013**, *135*, 9857–9868.
[30] M. E. Fieser, J. E. Bates, J. W. Ziller, F. Furche, W. J. Evans, *J. Am. Chem. Soc.* **2013**, *135*, 3804–3807; and references cited therein.
[31] (a) J. Okuda, *Dalton Trans.* **2003**, 2367–2378; (b) Y. Nakayama, H. Yasuda, *J. Organomet. Chem.* **2004**, *689*, 4489–4498; (c) H. Braunschweig, F. M. Breitling, *Coord. Chem. Rev.* **2006**, *250*, 2691–2720.
[32] G. Jeske, L. E. Schock, P. N. Swepston, H. Schumann, T. J. Marks, *J. Am. Chem. Soc.* **1985**, *107*, 8103–8110.
[33] P. J. Shapiro, E. Bunel, W. P. Schaefer, J. E. Bercaw, *Organometallics* **1990**, *9*, 867–869.
[34] H. G. Ali, A. Köppl, *Chem. Rev.* **2000**, *100*, 1205–1221.
[35] F. A. Cotton, W. Schwotzer, *J. Am. Chem. Soc.* **1986**, *108*, 4657–4658.
[36] M. N. Bochkarev, *Chem. Rev.* **2002**, *102*, 2089–2117.
[37] (a) M. Cesari, U. Pedretti, A. Zazetta, G. Lugli, N. Marconi, *Inorg. Chim. Acta* **1971**, *5*, 439–444; (b) F. A. Cotton, W. Schwotzer, C. Q. Simpson, *Angew. Chem. Int. Ed. Engl.* **1986**, *25*, 637–639.
[38] (a) P. Biagini, G. Lugli, R. Millini, *Gazz. Chim. Ital.* **1994**, *124*, 217–225; (b) P. Biagini, G. Lugli, L. Abis, R. Millini, *New. J. Chem.* **1995**, *19*, 713–722; (c) H. Liang, Q. Shen, S. Jin, Y. Lin, *Zhongguo Xitu Xuebao* **1994**, *12*, 193–196; (d) H. Liang, Q. Shen, J. Guan, Y. Lin, *J. Organomet. Chem.* **1994**, *474*, 113–116; (e) H. Liang, J. Guan, Y. Lin, Q. Shen, *Youji Huaxue* **1994**, *14*, 380–382; (f) B. Fan, Y. Lin, Q. Shen, *Yingyong Huaxue* **1990**, *7*, 23–27; (g) B. Fan, Q. Shen, Y. Lin, *J. Organomet. Chem.* **1989**, *377*, 51–58.
[39] H. Liang, Q. Shen, S. Jin, Y. Lin, *J. Chem. Soc., Chem. Commun.* **1992**, 480–481.
[40] (a) M. C. Cassani, Y. K. Gun'ko, P. B. Hitchcock, M. F. Lappert, *Chem. Commun.* **1996**, 1987–1988; (b) M. C. Cassani, Y. K. Gun'ko, P. B. Hitchcock, M. F. Lappert, F. Laschi, *Organometallics* **1999**, *18*, 5539–5547.

[41] M. C. Cassani, D. J. Duncalf, M. F. Lappert, *J. Am. Chem. Soc.* **1998**, *120*, 12958–12959.
[42] (a) D. M. Anderson, F. G. N. Cloke, P. A. Cox, N. Edelstein, J. C. Green, T. Pang, A. A. Sameh, G. Shalimoff, *J. Chem. Soc., Chem. Commun.* **1989**, 53–55; (b) P. L. Arnold, M. A. Petrukhina, V. E. Bochenkov, T. I. Shabatina, V. V. Zagorskii, G. B. Sergeev, F. G. N. Cloke, *J. Organomet. Chem.* **2003**, *688*, 49–55; (c) J. G. Brennan, F. G. N. Cloke, A. A. Sameh, A. Zalkin, *J. Chem. Soc., Chem. Commun.* **1987**, 1668–1669; (d) F. G. N. Cloke, K. Khan, R. N. Perutz, *J. Chem. Soc., Chem. Commun.* **1991**, 1372–1373; (e) F. G. N. Cloke, K. A. E. Courtney, A. A. Sameh, A. C. Swain, *Polyhedron* **1989**, 13–14, 1641–1648.
[43] (a) W. A. King, T. J. Marks, D. M. Anderson, D. J. Duncalf, F. G. N. Cloke, *J. Am. Chem. Soc.* **1992**, *114*, 9221–9223; (b) W. A. King, S. Di Bella, G. Lanza, K. Khan, D. J. Duncalf, F. G. N. Cloke, I. L. Fragala, T. J. Marks, *J. Am. Chem. Soc.* **1996**, *118*, 627–635.
[44] A. J. Arduengo III, R. L. Harlow, M. J. Kline, *J. Am. Chem. Soc.* **1991**, *113*, 361–363.
[45] (a) A. J. Arduengo III, M. Tamm, S. J. McLain, J. C. Calabrese, F. Davidson, W. J. Marshall, *J. Am. Chem. Soc.* **1994**, *116*, 7927–7928; (b) W. A. Herrmann, F. C. Munck, G. R. J. Artus, O. Runte, R. Anwander, *Organometallics* **1997**, *16*, 682–688; (c) H. Schumann, M. Glanz, J. Winterfield, H. Hemling, N. Kuhn, T. Kratz, *Angew. Chem. Int. Ed. Engl.* **1994**, *33*, 1733–1734; (d) H. Schumann, M. Glanz, J. Winterfield, H. Hemling, N. Kuhn, T. Kratz, *Chem. Ber.* **1994**, *127*, 2369–2372.
[46] (a) P. L. Arnold, S. T. Liddle, *Chem. Commun.* **2006**, 3959–3971; (b) P. L. Arnold, I. J. Casely, *Chem. Rev.* **2009**, *109*, 3599–3611.
[47] P. L. Arnold, S. A. Mungur, A. J. Blake, C. Wilson, *Angew. Chem. Int. Ed.* **2003**, *42*, 5981–5984.
[48] P. L. Arnold, S. T. Liddle, *Chem. Commun.* **2005**, 5638–5640.
[49] P. L. Arnold, S. A. Mungur, A. J. Blake, C. Wilson, CCDC #605446, Cambridge Crystallographic Data Centre, Cambridge, UK, **2006**, http://www.ccdc.cam.ac.uk/(accessed April 13, 2016).
[50] I. J. Casely, S. T. Liddle, A. J. Blake, C. Wilson, P. L. Arnold, *Chem. Commun.* **2007**, 5037–5039.
[51] P. L. Arnold, A. J. Blake, C. Wilson, *Chem. Eur. J.* **2005**, *11*, 6095–6099.
[52] P. L. Arnold, S. T. Liddle, J. McMaster, C. Jones, D. P. Mills, *J. Am. Chem. Soc.* **2007**, *129*, 5360–5361.
[53] (a) S. T. Liddle, D. P. Mills, *Dalton Trans.* **2009**, 5592–5605; (b) D. Patel, S. T. Liddle, *Rev. Inorg. Chem.* **2012**, *32*, 1–22.
[54] P. Pyykkö, M. Atsumi, *Chem. Eur. J.* **2009**, *15*, 186–197.
[55] (a) M. F. Lappert, P. P. Power, A. R. Sanger, R. C. Srivastava, *Metal and Metalloid Amides*, Ellis Horwood-Wiley, Chichester, **1980**; (b) R. Anwander, *Top. Curr. Chem.* **1996**, *179*, 33–112; (c) M. F. Lappert, A. Protchenko, P. P. Power, A. Seeber, *Metal Amide Chemistry*, John Wiley & Sons, Ltd, Chichester, **2008**, pp. 79–120.
[56] (a) E. C. Alyea, D. C. Bradley, R. G. Copperthwaite, *J. Chem. Soc., Dalton Trans.* **1972**, 1580–1584; (b) D. C. Bradley, J. S. Ghotra, F. A. Hart, *J. Chem. Soc., Dalton Trans.* **1973**, 1021–1023.
[57] (a) F. T. Edelmann, A. Steiner, D. Stalke, J. W. Gilje, S. Kagner, H. Håkansson, *Polyhedron* **1994**, *13*, 539–546; (b) M. Westerhausen, M. Hartmann, A. Pfitzner, W. Schwarz, *Z. Anorg. Allg. Chem.* **1995**, *621*, 837–850.
[58] M. Allen, H. C. Aspinall, S. R. Moore, M. B. Hursthouse, A. I. Karvalov, *Polyhedron* **1992**, *11*, 409–413.
[59] W. A. Herrmann, R. Anwander, F. C. Munck, W. Scherer, V. Dufaud, N. W. Huber, G. R. Artus, *Z. Naturforsch.* **1994**, *49b*, 1789–1797.

[60] C. Hagen, H. –D. Amberger, *Z. Naturforsch.* **1993**, *48b*, 1365–1371.

[61] S. A. Schuetz, V. W. Day, R. D. Sommer, A. L. Rheingold, J. A. Belot, *Inorg. Chem.* **2001**, *40*, 5292–5295.

[62] (a) W. S. Rees Jr, O. Just, D. S. Van Derveer, *J. Mater. Chem.* **1999**, *9*, 249–252; (b) R. A. Andersen, D. H. Templeton, A. Zalkin, *Inorg. Chem.* **1978**, *17*, 2317–2319; (c) E. D. Brady, D. L. Clark, J. C. Gordon, P. J. Hay, D. W. Keogh, R. Poli, B. L. Scott, J. G. Watkin, *Inorg. Chem.* **2003**, *42*, 6682–6690; (d) J. Sundermeyer, A. Khvorost, K. Harms, *Acta Crystallogr., Sect. E: Struct. Rep. Online* **2004**, *E60*, m1117–m1119; (e) J. S. Ghotra, M. B. Hursthouse, A. J. Welch, *J. Chem. Soc., Chem. Commun.* **1973**, 669–670; (f) P. B. Hitchcock, A. G. Hulkes, M. F. Lappert, Z. Li, *Dalton Trans.* **2004**, 129–136; (g) M. Niemeyer, *Z. Anorg. Allg. Chem.* **2002**, *628*, 647–657; (h) S. Jank, H. Reddmann, C. Apostolidis, H. –D. Amberger, *Z. Anorg. Allg. Chem.* **2007**, *633*, 398–404.

[63] (a) T. Fjeldberg, R. A. Andersen, *J. Mol. Struct.* **1985**, *128*, 49–57; (b) T. Fjeldberg, R. A. Andersen, *J. Mol. Struct.* **1985**, *129*, 93–105.

[64] L. Maron, O. Eisenstein, *New J. Chem.* **2001**, *25*, 255–258.

[65] (a) O. Eisenstein, P. B. Hitchcock, A. G. Hulkes, M. F. Lappert, L. Maron, *Chem. Commun.* **2001**, 1560–1561; (b) P. B. Hitchcock, A. G. Hulkes, M. F. Lappert, *Inorg. Chem.* **2004**, *43*, 1031–1038.

[66] (a) W. J. Evans, D. S. Lee, J. W. Ziller, *J. Am. Chem. Soc.* **2004**, *126*, 454–455; (b) W. J. Evans, D. S. Lee, D. B. Rego, J. M. Perotti, S. A. Kozimor, E. K. Moore, J. W. Ziller, *J. Am. Chem. Soc.* **2004**, *126*, 14574–14582.

[67] W. J. Evans, M. A. Johnston, R. D. Clark, R. Anwander, J. W. Ziller, *Polyhedron* **2001**, *20*, 2483–2490.

[68] W. J. Evans, G. Zucchi, J. W. Ziller, *J. Am. Chem. Soc.* **2003**, *125*, 10–11.

[69] (a) J. D. Rinehart, M. Fang, W. J. Evans, J. R. Long, *Nat. Chem.* **2011**, *3*, 538–542; (b) J. D. Rinehart, M. Fang, W. J. Evans, J. R. Long, *J. Am. Chem. Soc.* **2011**, *133*, 14236–14239.

[70] J. D. Rinehart, J. R. Long, *Chem. Sci.* **2011**, *2*, 2078–2085.

[71] N. F. Chilton, C. A. P. Goodwin, D. P. Mills, R. E. P. Winpenny, *Chem. Commun.* **2015**, *51*, 101–103.

[72] N. Marques, A. Sella, J. Takats, *Chem. Rev.* **2002**, *102*, 2137–2159.

[73] (a) M. A. J. Moss, R. A. Kresinski, C. J. Jones, W. J. Evans, *Polyhedron* **1993**, *12*, 1953–1955; (b) J. Takats, X. W. Zhang, V. W. Day, T. A. Eberspacher, *Organometallics* **1993**, *12*, 4286–4288.

[74] G. H. Maunder, A. Sella, D. A. Tocher, *J. Chem. Soc., Chem. Commun.* **1994**, 885–886.

[75] A. C. Hillier, S. Y. Liu, A. Sella, M. R. J. Elsegood, *Angew. Chem. Int. Ed. Engl.* **1999**, *38*, 2745–2747.

[76] X. Zhang, G. R. Loppnow, R. McDonald, J. Takats, *J. Am. Chem. Soc.* **1995**, *117*, 7828–7829.

[77] C. Morton, N. W. Alcock, M. R. Lees, I. J. Munslow, C. J. Sanders, P. Scott, *J. Am. Chem. Soc.* **1999**, *121*, 11255–11256.

[78] L. Bourget-Merle, M. F. Lappert, J. R. Severn, *Chem. Rev.* **2002**, *102*, 3031–3065.

[79] M. P. Blake, N. Kaltsoyannis, P. Mountford, *Chem. Commun.* **2013**, *49*, 3315–3317.

[80] M. P. Blake, N. Kaltsoyannis, P. Mountford, *J. Am. Chem. Soc.* **2011**, *133*, 15358–15361.

[81] (a) F. Nief, *Coord. Chem. Rev.* **1998**, *178–180*, 13–81; (b) F. Nief, *Eur. J. Inorg. Chem.* **2001**, 891–904; (c) P. Le Floch, *Coord. Chem. Rev.* **2006**, *250*, 627–681.

[82] D. Turcitu, F. Nief, L. Ricard, *Chem. Eur. J.* **2003**, *9*, 4916–4923.

[83] (a) C. Elschenbroich, *Organometallics*, 3rd edn, Wiley-VCH, Weinheim, **2006**; (b) W. A. Nugent, J. M. Mayer, *Metal-Ligand Multiple Bonds*, Wiley-Interscience, New York, **1988**.

[84] (a) G. R. Giesbrecht, J. C. Gordon, *Dalton Trans.* **2004**, 2387–2393; (b) O. T. Summerscales, J. C. Gordon, *RSC Adv.* **2013**, *3*, 6682–6693.
[85] (a) S. T. Liddle, D. P. Mills, A. J. Wooles, *Organomet. Chem.* **2010**, *36*, 29–55; (b) S. T. Liddle, D. P. Mills, A. J. Wooles, *Chem. Soc. Rev.* **2011**, *40*, 2164–2176.
[86] (a) J. Scott, D. J. Mindiola, *Dalton Trans.* **2009**, 8463–8472; (b) F. N. Tebbe, G. W. Parshall, G. S. Reddy, *J. Am. Chem. Soc.* **1978**, *100*, 3611–3613.
[87] K. Aparna, M. Ferguson, R. G. Cavell, *J. Am. Chem. Soc.* **2000**, *122*, 726–727.
[88] A. Cornish, D. P. Mills, W. Lewis, A. J. Blake, S. T. Liddle, *C. R. Chim.* **2010**, *13*, 593–602.
[89] (a) P. V. R. Schleyer, T. Clark, A. J. Kos, G. W. Spitznagel, C. Rohde, D. Arad, K. N. Houk, N. G. Rondan, *J. Am. Chem. Soc.* **1984**, *106*, 6467–6478; (b) A. M. El-Nahas, P. V. R. Schleyer, *J. Comput. Chem.* **1994**, *15*, 596–626; (c) B. Romer, G. G. Gatev, M. Zhong, J. I. Brauman, *J. Am. Chem. Soc.* **1998**, *120*, 2919–2924.
[90] (a) S. T. Liddle, J. McMaster, J. C. Green, P. L. Arnold, *Chem. Commun.* **2008**, 1747–1749; (b) D. P. Mills, O. J. Cooper, J. McMaster, W. Lewis, S. T. Liddle, *Dalton Trans.* **2009**, 4547–4555; (c) D. P. Mills, A. J. Wooles, J. McMaster, W. Lewis, A. J. Blake, S. T. Liddle, *Organometallics* **2009**, *28*, 6771–6776.
[91] S. T. Liddle, D. P. Mills, B. M. Gardner, J. McMaster, C. Jones, W. D. Woodul, *Inorg. Chem.* **2009**, *48*, 3520–3522.
[92] (a) D. P. Mills, L. Soutar, W. Lewis, A. J. Blake, S. T. Liddle, *J. Am. Chem. Soc.* **2010**, *132*, 14379–14381; (b) D. P. Mills, W. Lewis, A. J. Blake, S. T. Liddle *Organometallics* **2013**, *32*, 1239–1250; (c) D. P. Mills, L. Soutar, A. J. Wooles, W. Lewis, A. J. Blake, S. T. Liddle *Organometallics* **2013**, *32*, 1251–1264.
[93] M. Gregson, E. Lu, J. McMaster, W. Lewis, A. J. Blake, S. T. Liddle, *Angew. Chem. Int. Ed.* **2013**, *52*, 13016–13019.
[94] M. Fustier, X. F. Le Goff, P. Le Floch, N. Mézailles, *J. Am. Chem. Soc.* **2010**, *132*, 13108–13110.
[95] (a) J. Scott, F. Basuli, A. R. Fout, J. C. Huffman, D. J. Mindiola, *Angew. Chem. Int. Ed.* **2008**, *47*, 8502–8505; (b) J. D. Masuda, K. C. Jantunen, O. V. Ozerov, K. J. T. Noonan, D. P. Gates, B. L. Scott, J. L. Kiplinger, *J. Am. Chem. Soc.* **2008**, *130*, 2408–2409; (c) B. F. Wicker, J. Scott, J. G. Andino, X. Gao, H. Park, M. Pink, D. J. Mindiola, *J. Am. Chem. Soc.*, **2010**, *132*, 3691–3693.
[96] (a) E. Lu, Y. Li, Y. Chen, *Chem. Commun.* **2010**, 4469–4471; (b) J. Chu, X. Han, C. E. Kefalidis, J. Zhou, L. Maron, X. Leng, Y. Chen, *J. Am. Chem. Soc.* **2014**, *136*, 10894–10897; (c) W. Rong, J. Cheng, Z. Mou, H. Xie, D. Cui, *Organometallics*, **2013**, *32*, 5523–5529.
[97] Y.-M. So, G.-C. Wang, Y. Li, H. H.-Y. Sung, I. D. Williams, Z. Lin, W.-H. Leung, *Angew. Chem. Int. Ed.* **2014**, *126*, 1652–1655.
[98] W.-H. Leung, Q.-F. Zhang, X.-Y. Yi, *Coord. Chem. Rev.* **2007**, *251*, 2266–2279.
[99] S. P. Nolan, D. Stern, D. Hedden, T. J. Marks, *ACS Symp. Ser.* **1990**, *428*, 159–174.
[100] (a) M. E. Feiser, M. R. MacDonald, B. T. Krull, J. E. Bates, J. W. Ziller, F. Furche, W. J. Evans, *J. Am. Chem. Soc.* **2015**, *137*, 369–382; (b) K. R. Meihaus, M. E. Feiser, J. F. Corbey, W. J. Evans, J. R. Long, *J. Am. Chem. Soc.* **2015**, *137*, 9855–9860.
[101] C. M. Kotyk, M. E. Feiser, C. T. Palumbo, J. W. Ziller, L. E. Darago, J. R. Long, F. Furche, W. J. Evans, *Chem. Sci.*, **2015**, *6*, 7267–7273.
[102] C. A. P. Goodwin, N. F. Chilton, G. F. Vettese, E. Moreno Pineda, I. F. Crowe, J. W. Ziller, R. E. P. Winpenny, W. J. Evans, D. P. Mills, *Inorg. Chem.* **2016**, DOI: 10.1021/acs.inorgchem.6b00808.

[103] (a) M. Gregson, N. F. Chilton, A.-M. Ariciu, F. Tuna, I. F. Crowe, W. Lewis, A. J. Blake, D. Collison, E. J. L. McInnes, R. E. P. Winpenny, S. T. Liddle, *Chem. Sci.* **2016**, *7*, 155–165; (b) M. Gregson, E. Lu, F. Tuna, E. J. L. McInnes, C. Hennig, A. C. Scheinost, J. McMaster, W. Lewis, A. J. Blake, A. Kerridge, S. T. Liddle, *Chem. Sci.* **2016**, *7*, 3286–3297.

[104] D. Schädle, M. Meermann-Zimmermann, C. Schädle, C. Maichle-Mössmer, R. Anwander, *Eur. J. Inorg. Chem.* **2015**, *8*, 1334–1339.

[105] L. A. Solola, A. V. Zabula, W. L. Dorfner, B. C. Manor, P. J. Carroll, E. J. Schelter, *J. Am. Chem. Soc.* **2016**, DOI: 10.1021/jacs.6b03293.

[106] (a) Y.-M. So, Y. Li, K.-C. Au-Yeung, G.-C. Wang, K.-L. Long, H. H.-Y. Sung, P. L. Arnold, I. D. Williams, Z. Lin, W.-H. Leung, *Inorg. Chem.* **2016**, DOI: 10.1021/acs.inorgchem.6b00480; (b) G.-C. Wang, Y.-M. So, K.-L. Wong, K.-C. Au-Yeung, H. H.-Y. Sung, I. D. Williams, W.-H. Leung, *Chem. -Eur. J.*, **2015**, *21*, 16126–16135.

13

Tight Bite Angle *N*,*O*-Chelates. Amidates, Ureates and Beyond

Scott A. Ryken, Philippa R. Payne, and Laurel L. Schafer

2036 Main Mall, Department of Chemistry, The University of British Columbia, Vancouver, British Columbia, Canada V6T 1Z1

13.1 Introduction

Ligand design and development for early transition metals has resulted in controlled and selective chemistry with these oxophilic, reactive metal centers. The most predominant ligand scaffold utilized with early transition metals is cyclopentadienyl (Cp), as well as its common derivatives (Cp*, indenyl, etc.). [1] Monoanionic Cp ligands can be up to six-electron donors while formally occupying up to three coordination sites, as is the case when this ligand is in the most common η^5-binding mode. These electron-rich ligands are ideal for the generation of organometallic complexes of these electropositive, high oxidation state metal centers. The electronic features and steric bulk of this ligand can be varied with the installation of substituents, although only monosubstituted variants of Cp are easily prepared. Unfortunately, further modification of this ligand framework can be synthetically challenging. [2] While Cp ligands have yielded a bountiful amount of productive chemistry, alternative ligand motifs are being explored, especially with an emphasis on monoanionic ligands that are easily prepared and are of modular design to allow for facile variation of steric and electronic properties. [3]

Ligand Design in Metal Chemistry: Reactivity and Catalysis, First Edition. Edited by Mark Stradiotto and Rylan J. Lundgren.

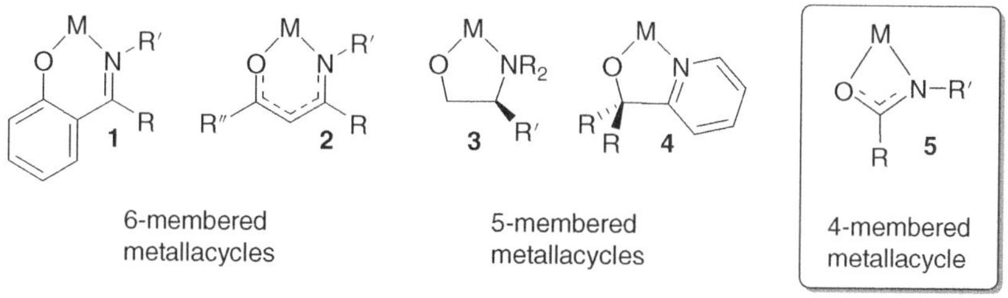

Figure 1 *Bidentate, monoanionic N,O-chelates are commonly used ligands for the generation of six-, five- and four-membered metallacycles*

Bidentate, monoanionic *N,O*-chelating ligands are ideal supporting scaffolds for catalytically active early transition metal complexes, as they can accommodate the changes in coordination number as catalytic transformations proceed. The hard, electron-rich nitrogen and oxygen atoms are very effective at stabilizing the high oxidation states of hard, Lewis acidic, oxophilic early transition metal centers. Such *N,O*-chelating ligands commonly generate six-, five- and four-membered metallacycles (Figure 1). One particularly notable ligand set that has been extensively investigated for olefin polymerization catalysis is the Salen-type phenoxy-imine chelate (**1**), which, upon coordination, generates six-membered metallacycles. [4] The related β-ketoiminate (**2**) motif has been extensively investigated for supporting a variety of electrophilic, early transition metal centers. [5] Five-membered metallacycles can be readily accessed from amino acid precursors (**3**), [6] donor-substituted alkoxides (**4**), [7] and amido ether ligands. [6, 8] Such supporting ligands have been explored for a variety of early transition metal catalyzed reactions including olefin polymerization and hydroamination. [3, 9] Both five- and six-membered metallacycles are favored cyclic structures with minimal ring strain. Thus, such ligands strongly prefer a bidentate binding mode and the sufficiently long tethers between the donor atoms ensure that they can coordinate to metals with minimal distortion of the anticipated bond angles in classical coordination geometries, such as octahedral metal centers. [4c]

Of particular interest here is a class of *N,O*-chelating ligands that generate four-membered metallacycles with tight bite angles upon chelation. These *N,O*-chelates are an underexplored class of ligands that have recently been used to advantage in several early transition metal catalyzed transformations described here: terminal alkene polymerization; [10] ring-opening polymerization (ROP); [11] hydroamination; [12] hydroaminoalkylation; [13] and hydroalkynylation (enyne formation). [14] This class of tight bite angle ligands is directly related to bidentate, monoanionic *N,N*-chelates that have been extensively investigated (Figure 2): amidinate (**6**), [15] aminopyridinate (**7**), [16] and guanidinate (**8**) ligands. [15] While the chemistry of such *N,N*-chelates has been under extensive investigation by a variety of research groups around the world for over two decades, it is surprising that *N,O*-variants have been largely overlooked until the early 2000s. [10a]

The *N,O*-chelating motif does offer some inherent challenges in that an asymmetric ligand set affords the opportunity to access multiple isomers (see below). [17]

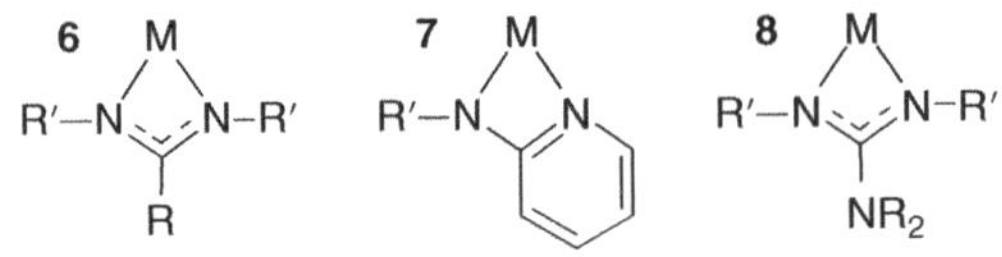

Figure 2 Related tight bite angle N,N-chelates that have been extensively explored in combination with early transition metals

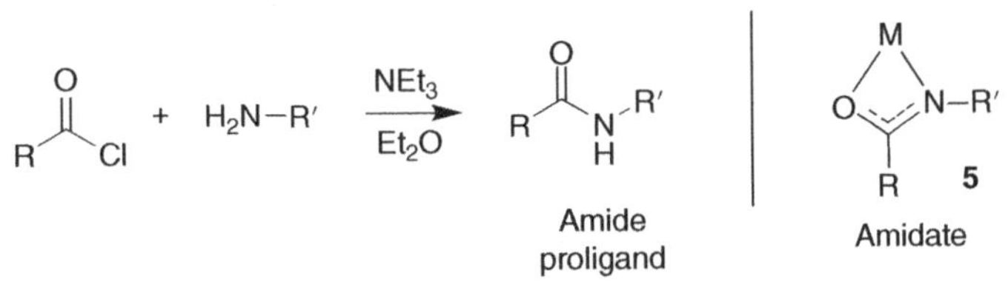

Scheme 1 Modular one-step proligand syntheses ensure that steric bulk and electronic properties can be easily varied

Furthermore, there are minimal opportunities for providing steric protection about the *O*-donor of the chelate, which is known to engage in bridging interactions (see below), thereby promoting the formation of ill-defined aggregate species. [18] As a core aspect of our work in this field we sought to explore the fundamental coordination chemistry of this simple *N,O*-chelating ligand set on early transition metals.

13.1.1 *N,O*-Proligands

One of the most attractive features of this ligand set is the fact that they are readily synthesized and have tunable steric and electronic parameters by varying the R and R′ substituents in the proligand. [19] For example, the amidate ligand (**5**) can be generated by deprotonation of the organic amide proligand, which in turn is easily synthesized from simple and commercially available acid chlorides and amines (Scheme 1). By varying the nature of the amine and acid chloride reagents, a broad range of amide proligands can be rapidly prepared.

Initial work focused on the exploration of just amidate complexes, however, a variety of related *N,O*-chelates can also be accessed (Figure 3). [20] More recent work has exploited the use of complementary monoanionic species including ureate, [21] pyridonate, [22] phosphoramidate, [23] and sulfonamidate ligands. [24] The various scaffolds contain multiple sites for ligand modification and optimization (e.g., R, R′, Figure 3) and therefore the steric and electronic properties of these ligands can probe a broad range of parameters. The straightforward modification of these modular ligand scaffolds is a particularly attractive quality for catalyst development work, in that such flexible approaches allow for rapid tuning and optimization of a particular metal–ligand combination to suit the transformation, substrate combination, or selectivity required.

Figure 3 *Related bidentate, monoanionic ligands with N,O-donors*

The monoanionic ligands in Figure 3 have similar characteristics but are sterically and electronically unique. The ureate ligand class, closely related to the amidates, contains an amino substituent on the backbone that is capable of engaging in resonance interactions. This alters the electronic properties of the system, improving electron-donating abilities of these ligands via π-donation of the backbone nitrogen lone pair. In the 2-pyridonate ligand system, the nitrogen is part of an aromatic ring and is, therefore, both sterically and electronically distinct from the acyclic amidate and ureate ligands. Furthermore, this cyclic ligand motif reduces one degree of conformational freedom as the ligand itself is conformationally rigid, unlike the parent amide proligand that can adopt cisoid- and transoid-geometries. Finally, the substituents R and R′ are slightly more removed from the N and O donor atoms of the chelate of pyridonate in comparison with the amidate. Thus the steric bulk imposed by the ligand is farther away from the metal center than in the parent amidate system, resulting in a more sterically accessible reactive site. The sulfonamidate and phosphoramidate ligands contain the tetrahedral phosphorus or sulfur atom in the backbone, rather than the planar carbon of the amidate, ureate, and pyridonate backbones. These phosphorus- and sulfur-containing ligands are attractive since they have increased electron-withdrawing properties compared with the amidate, ureate, and pyridonate ligands and thus they modify the electronic features of the resultant complexes from those previously mentioned. These ligands also have increased steric shielding ability due to the bulky substituents on the tetrahedral S or P atoms in comparison with the planar amidates.

Synthesis and characterization of this class of tight bite angle *N,O*-chelating ligands will be described with an emphasis on the amidate ligand set, which has been most extensively explored to date.

13.1.2 Preparing metal complexes

The modular synthesis of complexes is a desirable quality for ancillary ligands in catalytic systems. Previous to our investigations, however, tight bite angle *N,O*-chelating ligand classes were not extensively investigated as tunable auxiliary ligands; instead, many of the initial examples of amidate or ureate metal complexes were products of insertion reactions. Examples of this include the preparation of ureate complexes by insertion into metal–amido bonds [25] or amidate complexes from insertion into metal–alkyl bonds (Scheme 2). [26] However, this is a laborious approach which demands the synthesis of reactive metal amido or metal alkyl reagents as precursors, significantly limiting the modularity of the synthetic approach.

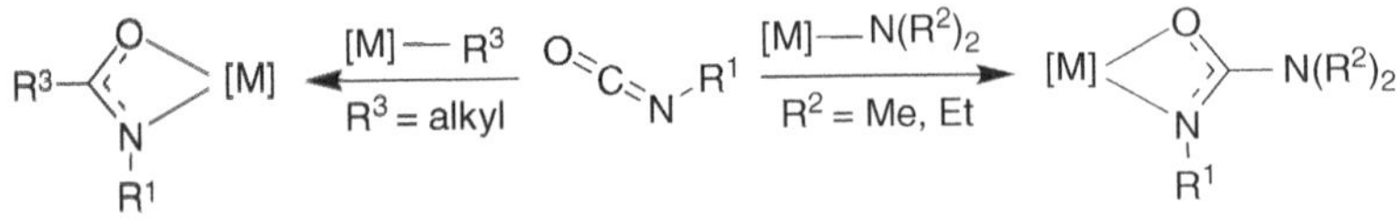

Scheme 2 *Isocyanate insertion into a metal–element bond to generate amidate (left) or ureate (right) complexes*

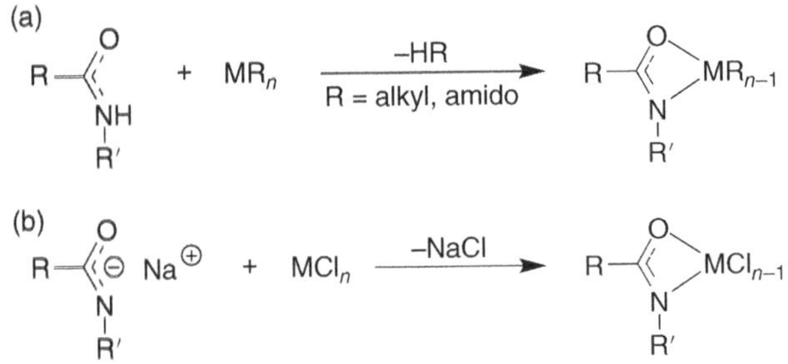

Figure 4 *Protonolysis (a) and salt metathesis (b) approaches can be used, with protonolysis being the most generally applicable*

Furthermore, the complexes formed are also susceptible to further isocyanate insertions leading to product mixtures. [27]

A more attractive and modular route to the synthesis of complexes bearing *N,O*-chelating ligands is through protonolysis or salt metathesis procedures (Figure 4). The neutral proligand for all of the aforementioned classes of ligands contains a proton of sufficient acidity for a protonolysis reaction with metal alkyl or metal amido starting materials. This approach can be used to advantage for the one-step synthesis of metal complexes (Figure 4a). The by-products of the protonolysis approach are volatile alkanes or amines that can be easily removed from the reaction mixture under reduced pressure, thereby simplifying the workup and purification procedures. This reaction can be carried out in a broad range of solvents including non-coordinating hexanes, [19] aprotic donor solvents such as tetrahydrofuran (THF), [17] and polar solvents such as dichloromethane. [28] Furthermore, many early transition metal amido complexes are cost-effective, commercially available precursors.

Another commonly used synthetic strategy involves deprotonation of the proligand with alkali metal bases to generate a ligand salt that can then react with metal halide precursors via salt metathesis (Figure 4b). [23] Unfortunately, this traditional approach for the synthesis of organometallic complexes is of limited usefulness for the preparation of this class of compounds with early transition metals. This synthetic limitation may rationalize why such *N,O*-chelated complexes have not been extensively explored in the past. Over time we have observed that this method for installing *N,O*-chelating ligands is very sensitive to the steric bulk of the ligand, counter-ion effects and solvent effects. As a result, at present, reactivity trends remain unpredictable. Notably, groups 3 and 4 metal salts often yield ill-defined aggregate species with alkali metal salts remaining in the coordination sphere. [29] Successful syntheses have employed sodium counter ions rather than lithium, and bulky bases, such as hexamethyldisilylamide (e.g., NaHMDS) over the more nucleophilic metal alkyl reagents. [21b] While groups 3 and 4 metal complexes have presented challenges when employing salt metathesis, group 5 complexes can be reliably prepared using this approach. [23] This is fortuitous as group 5 metal salts are commercially available but their metal alkyl and amido starting materials are less common and often must be prepared. [30]

A significant challenge with these tight bite angle *N,O*-chelates is the fact that the resultant metal complexes can contain the ligands bound in a variety of ways. The most common of these binding modes include monodentate, *O*- or *N*-bound, chelating *N,O*-bound, and a bridging motif involving multiple metal centers (Figure 5). [20] While such variable binding modes can be identified by X-ray crystallography and even NMR spectroscopy (by using the diagnostic ^{13}C signal of the chelate backbone), [31] it must be noted that equilibria can exist and it can be difficult to rigorously assign the coordination environment about the metal center in the solution phase. While such equilibria present challenges for the careful characterization of the resultant complexes, we postulate that such coordinative flexibility is advantageous in catalysis (see below).

Groups on both the carbonyl and the nitrogen of the amidate ligand contribute to the coordination geometry of the resultant metal complex, exemplified by ligands with lower degrees of steric bulk having a greater likelihood of dimer and oligomer formation. Consequently, to minimize this possibility, a sufficient degree of steric bulk must be maintained in the ligand design to encourage discrete single-site metal catalysts.

The electronic and steric asymmetry induced by the presence of the *N,O*-donor atoms can also result in ligand hemilability (Figure 6). [32] Hemilabile ligands display dynamic coordination behavior of the more weakly bound donor atom allowing for the generation and control of vacant sites at the metal center. The use of specifically κ^2-bound, four-membered metallacycles in this family of complexes results in a propensity for such hemilability, due to the significant ring strain of the chelated

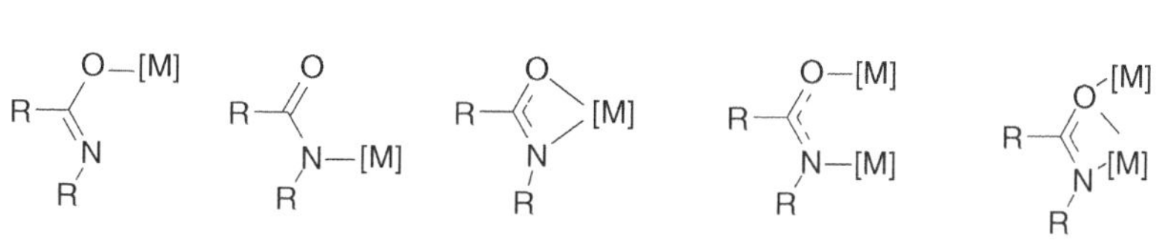

Figure 5 *Variable coordination modes of N,O-chelating ligands accessible*

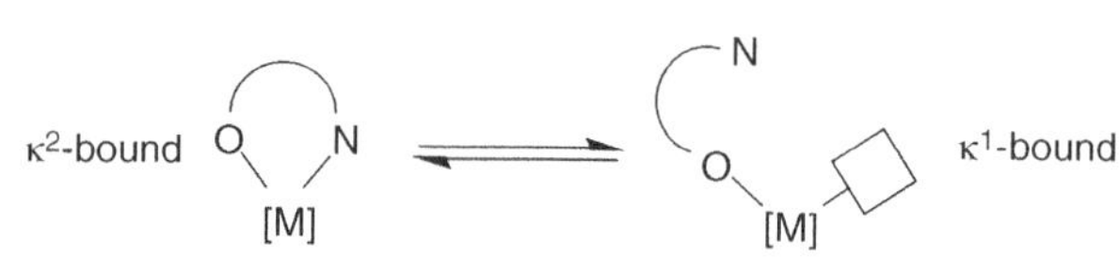

Figure 6 *Hemilability schematic for metal complexes supported by an asymmetric bidentate ligand, leading to protected coordination sites*

heterometallacycle. [17] Furthermore, the oxophilic metal centers of the early transition metal series and the strategic inclusion of steric bulk into the N-donor atom, suggests that these ligands are most likely to be labile while retaining contact to the metal center preferentially through oxygen (κ^1-bound). This fluxional behavior of the ligand is particularly advantageous for catalytic applications; the hemilabile ligands add stability to the metal catalyst while retaining the potential for revealing open coordination sites required for desirable reactivity. The ligand can also assist in the associative displacement of coordinated products to regenerate the active catalyst. Ligand hemilability can be difficult to observe, as it is ruled by subtle energy differences and is difficult to control; however, it provides yet another facet of ligand design that can be tuned and optimized. We postulate that ligand hemilability is a key feature in the observed reactivity of these systems and is a powerful ligand design approach to consider when generating new systems for catalysis. [33]

13.1.2.1 Amidate ligands

With these aspects of ligand synthesis and resultant coordination chemistry taken into consideration, a summary of reactivity trends within the early transition metal series is provided. [20] Although the salt metathesis synthetic route has been used with limited success, it has successfully generated mixed bis(amidate)dichloro complexes of group 4 metals (Scheme 3). However, this methodology can result in intractable mixtures of products, including bridged, dimeric, or ill-defined multi-metallic species. [11, 20, 21b] To date, only bis(ligated)dichloro complexes have been prepared (Figure 7) and unfortunately all attempts to use such bis(amidate)dichloro complexes as starting materials for the preparation of alkyl reagents via salt metathesis have not been successful.

Alternatively, protonolysis is a reliable route to prepare a variety of groups 3, 4, and 5 metal amidate complexes. In most cases, adjusting the ligand to metal stoichiometry allows for the selective generation of targeted early transition metal complexes. Group 3 mono-, bis-, and tris(amidate) (Figure 8) yttrium complexes have all been prepared and characterized. The amidate ligands are all bound in a κ^2-*N,O*-chelating motif with a neutrally bound THF solvent molecule in the coordination sphere. In the case of group 4 complexes tetrakis-, tris- and bis-ligated complexes can be prepared by controlling the amide proligand stoichiometry; [34] however, all attempted syntheses of mono(amidate) complexes of group 4 metals have not been successful and ligand

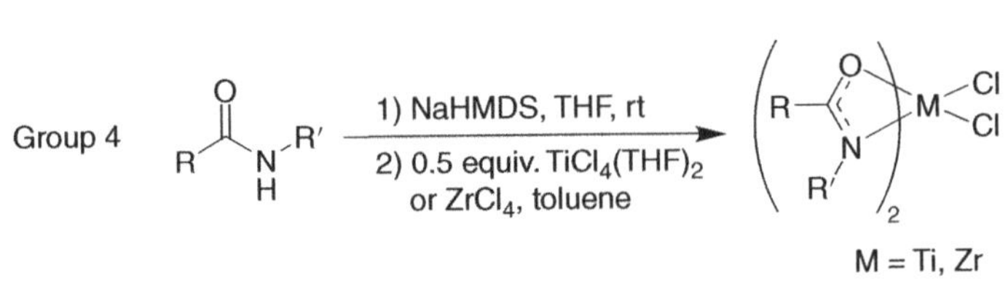

Scheme 3 *Bis(amidate)dichloro complexes can be prepared via salt metathesis*

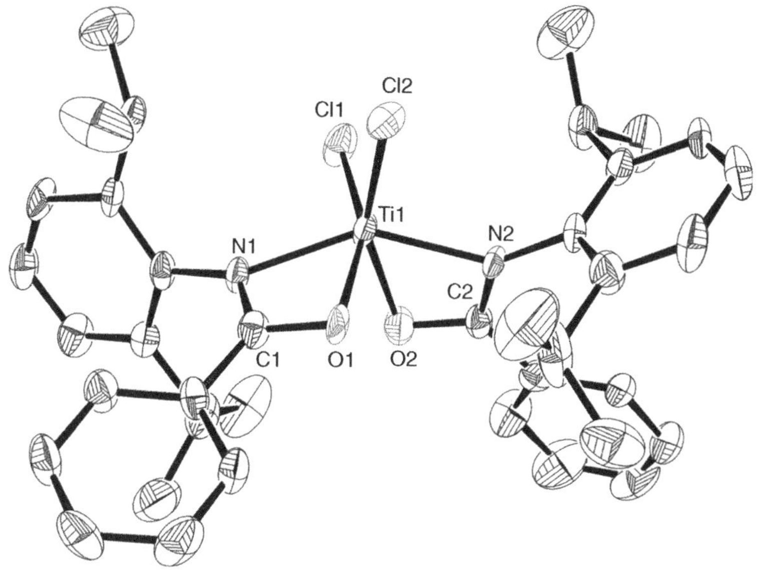

Figure 7 *ORTEP representation of one of two independent molecules (ellipsoids plotted at 50% probability, toluene molecule and hydrogen atoms omitted for clarity) with selected bond lengths (Å) and angles (°) averaged between the two molecules: Ti–N1, 2.074(5); Ti–O1, 2.038(4); Ti–N2, 2.062(4); Ti–O2, 2.036(5); Ti–Cl1, 2.246(2); Ti–Cl2, 2.235(2); C1–N1, 1.315(8); C1–O1, 1.306(7); C2–N2, 1.314(8); C2–O2, 1.308(7); N1–Ti–O1, 63.9(2); N2–Ti–O2, 63.9(2); Cl1–Ti–Cl2, 98.81(8); N1–C1–O1, 112.4(6); N2–C2–O2, 111.4(5). (See insert for color/color representation of this figure)*

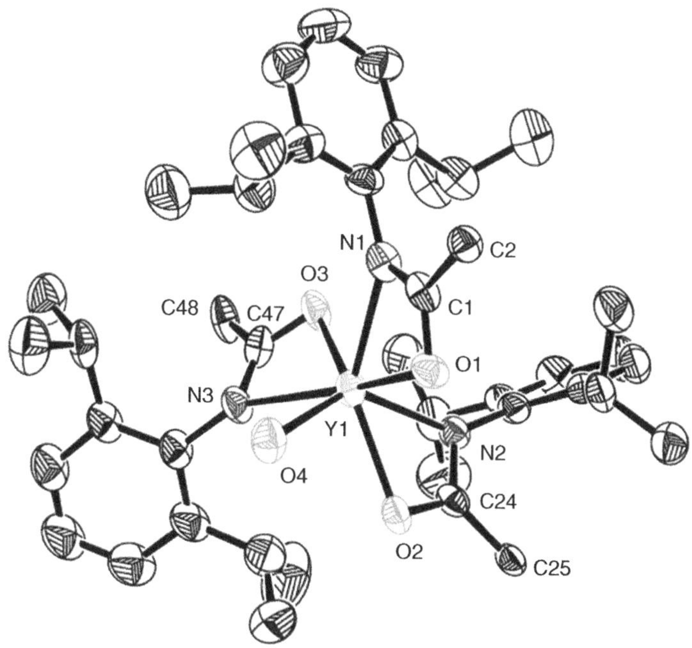

Figure 8 *ORTEP diagram of the solid-state molecular structure of one independent molecule of [(THF)Y(Nap[O,N](i-Pr)$_2$Ph)$_3$ with the probability ellipsoids drawn at the 50% level. Naphthyl groups (except for ipso-carbon), carbon atoms of the THF groups, and hydrogen atoms omitted for clarity. (See insert for color/color representation of this figure)*

redistribution is observed. Under such conditions of ligand redistribution, the bis(amidate)bis(amido) and the homoleptic metal amido complexes result preferentially, although a mixture of products is observed. [34]

Another complicating feature of these asymmetric ligands is the fact that multiple geometric isomers can result. For example, for the preferentially formed monomeric bis(amidate) group 4 complexes, there are five possible geometric isomers that can result, as shown in Figure 9. This large number of potential isomers complicates spectral assignments and could be another reason why tight bite angle *N,O*-chelates have been largely overlooked as auxiliary ligands. However, with the incorporation of large substituents, we typically obtain crystalline products that display characterization data consistent with the formation of a single C_2-symmetric product, generally with *cis*-labile ligands (L) and either *N,N*-trans or, as shown, *O,O*-trans geometries (Figure 10). [19]

Very sterically crowded group 4 metal complexes bearing three or four amidate ligands can be synthesized via protonolysis (Figure 11). [34] The tris(amidate)

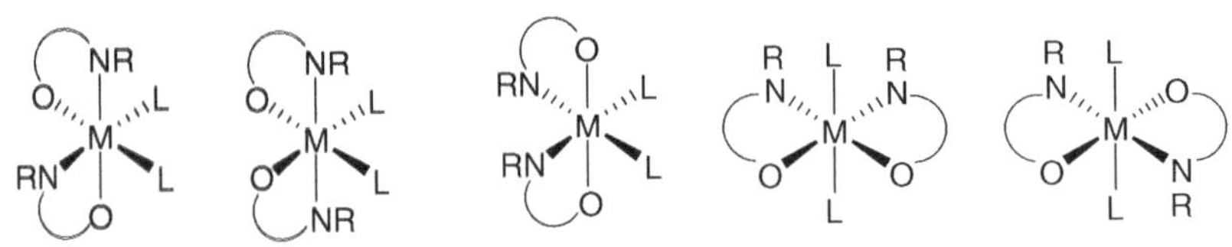

Figure 9 Possible isomers of bis(amidate) complexes. L = chloro, alkyl, amido

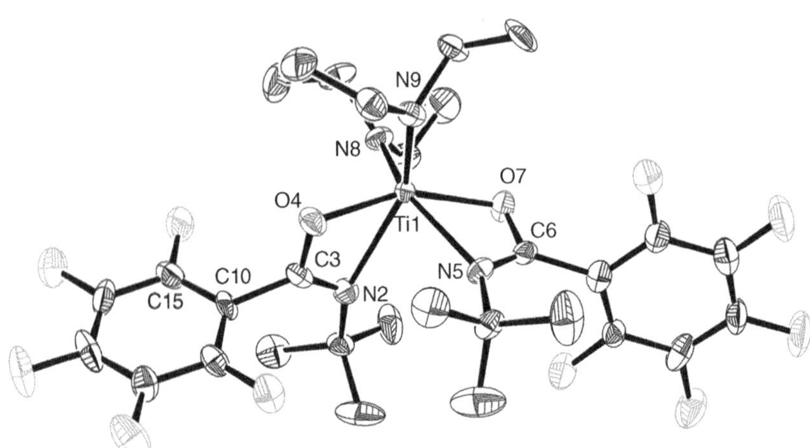

Figure 10 *ORTEP representation of the structure of bis(N-t-butylperfluorophenylamidate) titanium-bis(diethylamide) at 50% probability ellipsoids. Selected bond lengths (Å), and bond and torsion angles (°): Ti—N(2), 2.356(7); Ti—O(4), 2.044(6); Ti—N(9), 1.887(7); N(2)—C(3), 1.272(11); O(4)—C(3), 1.307(10); O(4)—Ti-N(2), 59.6(2); N(2)—C(3)—O(4), 117.4(7); O(4)—Ti—O(7), 155.6(3); N(2)—Ti—N(9), 147.9(3); N(2)—C(3)—C(10)—C(15), 94.2(12). (See insert for color/color representation of this figure)*

complexes can display different coordination modes depending upon the size of the metal center. The complex containing the smaller titanium metal, which is less able to accommodate expanded coordination numbers and high steric crowding, contains two κ^2- and one κ^1-*O*-bound amidate ligands. The larger zirconium and hafnium metal centers are capable of accommodating three and even four κ^2-bound amidate ligands, resulting in complexes with expanded coordination numbers.

Tantalum bis- and mono(amidate) complexes have been prepared using a protonolysis route from the commercially available $Ta(NMe_2)_5$ starting material (Figure 11). In this case the mono(amidate) complex bears a κ^2-bound ligand (**9**) while the bis(amidate) complex has both a κ^2-bound and a κ^1-*O*-bound amidate ligand. [35] Variable temperature NMR spectroscopy has shown that these ligands are fluxional and change binding mode on the NMR timescale. [35]

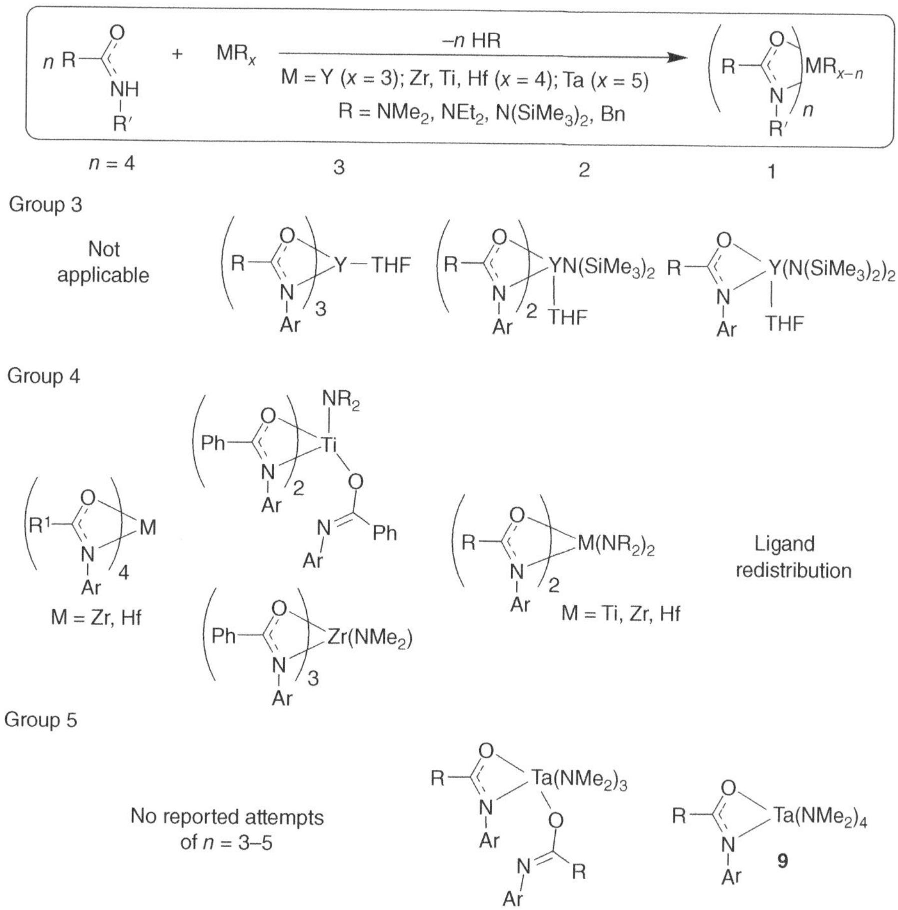

Figure 11 *Observed coordination modes for early transition metal complexes containing amidate ligand(s) synthesized via protonolysis with organic amide proligands*

13.1.2.2 Ureate ligands

To date, work with ureate ligands has been limited to group 4 metals. The salt metathesis and protonolysis routes with urea proligands are analogous to those utilized for preparing amidate complexes. The synthesis, structure, and reactivity of titanium and zirconium bis(ureate) complexes bearing alkyl, [36] chloro, [21b] and amido ligands have all been reported (Scheme 4). [21a] The ureate ligands examined include both tethered and untethered bis(ureate) complexes. Studies show that the tethered ligand is often much more successful in generating well behaved coordination complexes by eliminating the fluxional behavior and coordination isomerism observed with the untethered bis(ureate) complexes. [21b] The tethered motif also increases steric accessibility to the metal center (**10**), which is particularly attractive for catalytic applications, by enforcing an equatorial disposition of the ligand. The chloro complexes formed are electron-deficient complexes and often retain a neutral donor ligand; either the dimethylamine by-product when mixed amidochloride metal precursors are used or, alternatively, coordinating solvent. [37] Larger alkyl ligands, for example benzyl groups, allow for the synthesis of complexes without a coordinating neutral donor ligand by affording greater steric protection of the metal center. [37]

These complexes have been extensively characterized and the metrical parameters in the solid-state molecular structures provide firm evidence that the ureate ligands bind more tightly to the metal center than their amidate counterparts, resulting in shortened metal–ligand interactions (J. Pacheco *et al*, manuscript in preparation). The ureate ligands are observed to be consistently bound in the κ^2-chelating motif irrespective of the steric bulk of the ligand. The solid-state data for the tethered complexes reveal planar sp^2 geometry of the backbone nitrogen, consistent with lone-pair donation into

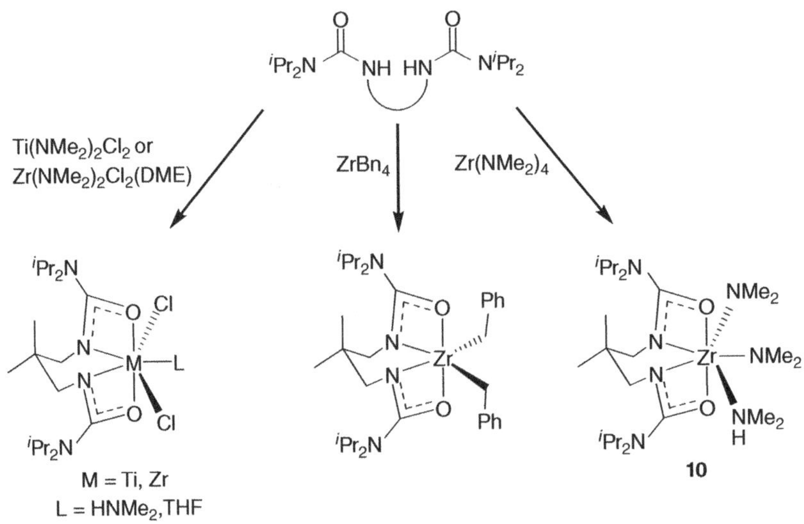

Scheme 4 *Synthesis of bis(ureate) chloro, alkyl, or amido complexes of group 4 metals. DME = 1,2-dimethoxyethane*

the π-system. This is not always consistent with solution-phase NMR spectroscopy that shows magnetic equivalence of the *iso*-propyl methyl groups in **10**, suggesting weak electron donation by the distal nitrogen, allowing for free rotation about the C(carbonyl)–N(^{i}Pr)$_2$ bond in solution. [21a]

13.1.2.3 Pyridonate ligands

Until recently, use of the aromatic 2-pyridonate ligand motif with early transition metals has been scarce, [38] though extensively investigated with late transition metals, [39] including in the assembly of classic paddlewheel dinuclear complexes that feature variable metal–metal bonding motifs. Preliminary examples of early transition metal systems include mixed pyridonate/cyclopentadienyl complexes that display both chelating and κ^1-*O*-binding. [39b] Using the protonolysis methodology, a variety of group 4 complexes have been synthesized, including titanium alkoxide complexes supported by 3-substituted and 6-substituted 2-pyridonates, [22b] as well as bis(pyridonate) bis(amido) zirconium and titanium complexes (Scheme 5). [40] These complexes all display a κ^2-*N,O*-chelating motif with long metal–nitrogen contacts resulting in a formal assignment of the bonding in these systems being best described as aryloxide neutral imine donors. [22b] This is also consistent with the multiple bond character of the C—N bond, due to the aromatic pyridonate ring. One of the key features of pyridonate ligands, and indeed any cyclic *N,O*-chelate, is the fact that *cis/trans* isomerization of the κ^1-*O*-bound species is not possible due to the fact that the carbonyl and N-substituents are tethered. This removes one degree of freedom from dynamic isomerization processes.

Group 5 pyridonate complexes are also readily prepared by salt metathesis and protonolysis routes to access both mono- and bis(pyridonate) tantalum complexes (Figure 12). [41] Such complexes can be difficult to rigorously characterize in the solid state as they are much more soluble than their group 4 pyridonate, or their group 5 amidate, counterparts.

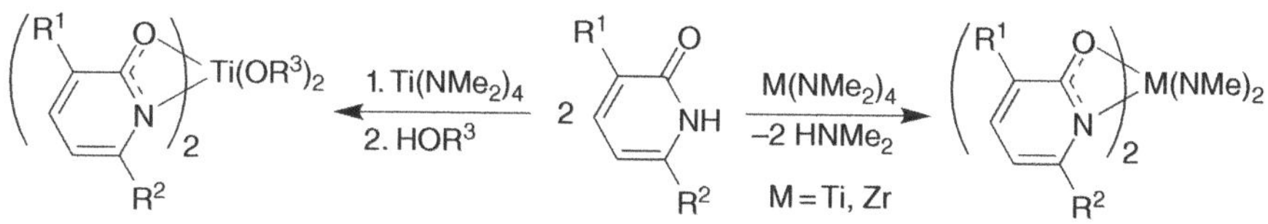

Scheme 5 *Synthesis of bis(pyridonate) alkoxide or amido complexes of group 4 metals*

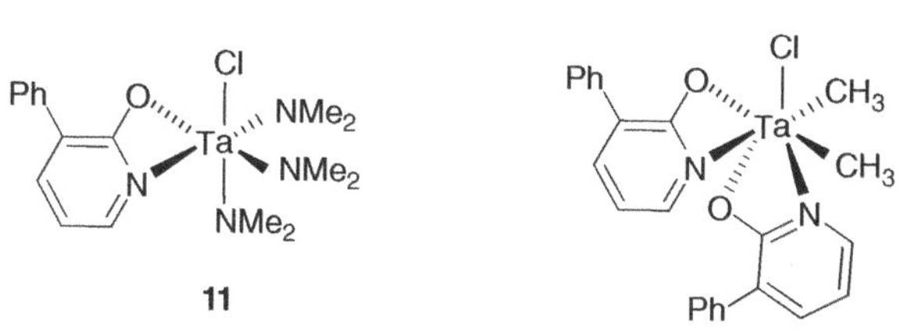

Figure 12 *Tantalum mono(pyridonate) (**11**, left) and bis(pyridonate) (right) complexes [22d, 41]*

13.1.2.4 Sulfonamidate ligands

Sulfonamidates, also termed sulfonamide and sulfonamido ligands, have a rich history as ancillary ligands in titanium complexes for application in synthesis. [24a] The majority of these ligand motifs are the bis(tosyl) ligands based on a chiral diamine backbone (Figure 13). These tethered ligands are chiral and, therefore, have the potential to afford enantioenriched products through catalytic asymmetric transformations. Indeed, these ligands have been extensively studied for the titanium-catalyzed asymmetric addition of dialkylzinc reagents to aldehydes. [42]

The initial studies used an in situ based procedure to generate the chiral catalyst and proposed an active bis(sulfonamidate) titanium species. Extensive solution phase and mechanistic investigations have since been performed, as well as extensive structural studies on the bonding of these bis(sulfonamidate) systems with titanium; [24a, 43] common coordination motifs observed during these studies include κ^3-bound complexes (Figure 13). The addition of the sulfur into the backbone does significantly alter the coordination geometries of these complexes compared with the amidate, ureate, and pyridonate complexes discussed in the above sections. The most evident difference is the preferential binding to nitrogen in these systems; typically the Ti—N bond distance is shorter than that of the Ti—O bond. [43b] Notably in these systems the *N,O*-chelate can be highly fluxional as the tetrahedral sulfur center is substituted by two oxygen atoms. This fluxionality contributes to an even more coordinatively flexible system than the previously presented *N,O*-chelates. Zhang *et al.* [44] have also described axially chiral bis(sulfonamidate) group 5 complexes.

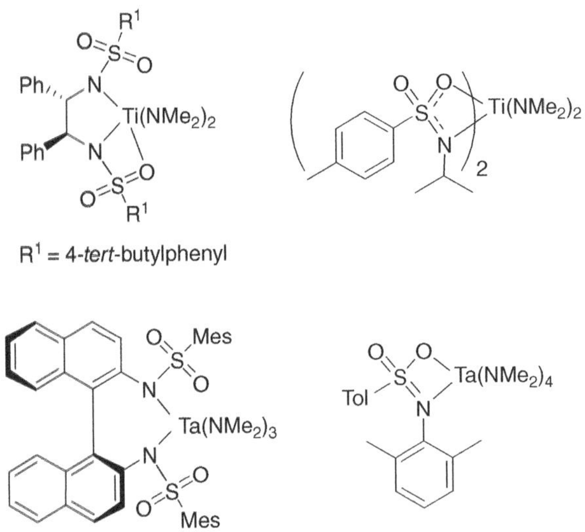

Figure 13 *Representative early transition metal systems containing sulfonamidate and sulfonamide ligands. Mes = mesityl*

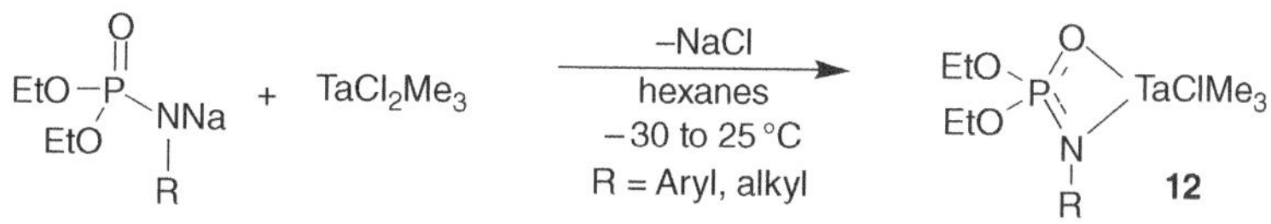

Scheme 6 *Salt metathesis can be used for the preparation of phosphoramidate-supported organometallic complexes*

13.1.2.5 *Phosphoramidates*

Phosphoramidate ligands afford another *N,O*-chelating motif that incorporates a tetrahedral heteroatom into the ligand backbone. In this case, however, there is only one phosphorus–oxygen multiple bond, thereby defining the nature of the *N,O*-chelate. These ligands can be prepared from readily available chlorophosphates and amines via nucleophilic substitution or phosphites and azides via a Staudinger reaction. [23] The resultant phosphoramidate proligands are acidic with pK_a values of approximately 20 in DMSO; [45] thus, the proligand can be used in combination with homoleptic amido complexes to access phosphoramidate-supported amido complexes by protonolysis. Alternatively, salt metathesis can be used effectively with these ligands. [23] Using this approach, mixed phosphoramidate-chloro-ligated species can be formed and, most interestingly, by using organometallic precursors as shown in Scheme 6, phosphoramidate-supported organometallic species have been isolated (**12**). [23] Such complexes can be prepared on multigram scale, although the methyl derivatives have proven to be both thermally and light sensitive.

The preceding overview provides a summary of a variety of *N,O*-chelating ligands that can be readily prepared from simple commercially available starting materials. The key features of modular synthesis, hemilability, and coordinative flexibility have been used to advantage in developing complexes that can be reliably prepared on multigram scale. The rigorous characterization of these systems in the solid state as well as in solution phase has provided valuable insight into reactivity trends. These tunable, electrophilic, early transition metal complexes have been utilized in the development of new reactivity profiles and desirable catalytic activity for the selective synthesis of C–C and C–N bonds. Such reactivity has been applied toward industrially relevant polymerizations as well as selective small molecule synthesis, specifically amine synthesis. Specifically, this chapter will present advances in catalytic ROPs of lactones, alkene polymerizations and hydrofunctionalization reactions including hydroamination and hydroaminoalkylation.

13.2 Applications in reactivity and catalysis

13.2.1 Polymerizations

The ability of early transition metals to initiate polymerizations has been known and utilized since the mid-20th century. [46] Such Ziegler–Natta catalyst systems with homoleptic and simple Ti and Al compounds have paved the way to the development of new catalysts, in large part consisting of metallocene complexes of Ti, Zr, and Hf, but also in recent years branching into complexes bound by a variety of ligands, including *N,O*-chelates. [4c, 47] In addition to these early transition metal promoters, alkyl aluminum compounds, such as MAO (methylaluminoxane), are necessary to form active catalytic species. [48]

These advances in reactivity have encouraged the plastic industry of the mid-1900s to further expand potential polyolefin products. [49] More recently, alternative biodegradable polymeric materials that can be degraded and destroyed in an environmentally benign manner have been targeted. [50] Just as early transition metals played a key role in alkene polymerization, they have been employed in the ROP of cyclic esters to prepare biodegradable polymers. [51] Notably, complexes bearing amidate, pyridonate, and sulfonamidate ligands have been applied in the synthesis of these environmentally friendly materials. [11, 22b, 24b, 52]

13.2.1.1 Alkene polymerization

The polymerization of α-olefins progressed beyond homoleptic chloro and metallocene complexes to *N,O*-chelates in the 1990s by employing the phenoxyimine class of ligand, forming a six-membered metallacycle (Figure 1). [53] Mono- and bis-ligated phenoxyimine complexes of group 4 metals have experienced success in this application, as outlined elsewhere. [4c] In addition to those wider bite angle *N,O*-chelates, there are recent examples of four-membered metallocycle *N,O*-chelated titanium complexes that have been utilized for this transformation. [10]

An early example of amidate ligands being utilized with early transition metals was reported in 2001 by Giesbrecht *et al.* [10a] A tethered bis(amidate) ligand was used to prepare a dimeric species (Figure 14), which was tested for ethylene polymerization. In the presence of 500 equiv. of MAO at 1 atm ethylene gas only modest catalytic activity was observed. [10a]

A more recent and in-depth polymerization study of a tight bite angle *N,O*-chelated early transition metal being utilized for alkene polymerization was presented by Sun's group [10b] in 2010. A set of bis(chloro)mono(cyclopentadienyl)mono(amidate) titanium complexes, in a half-sandwich configuration, were synthesized. Electronic

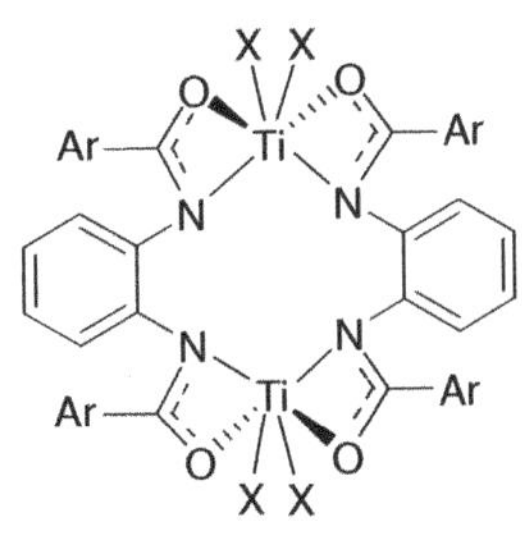

Figure 14 *Tethered bis(amidate) titanium complex for preliminary studies as initiators for ethylene polymerization. Ar = 4-tert-butylphenyl, X = NMe_2 or Cl [10a]*

effects were probed as the *para*-position of the phenyl group attached to the carbonyl carbon of the amidate was modified with electron-donating and electron-withdrawing groups. Additionally, both Cp and Cp* (C_5H_5 and C_5Me_5) were utilized in select cases (Scheme 7).

These complexes were used as catalysts for ethylene polymerization with MAO as a cocatalyst under optimized conditions of 1500:1 Al:Ti ratio and 10 atm of ethylene at 30 °C. [10b] The ligands had a clear effect on the catalytic activities: the Cp* analogs outperformed their Cp counterparts, while electron-donating substituents on the amidate phenyl led to higher catalyst activity. Thus, the compound with the more electron-donating *para*-methoxyphenylamidate and Cp* ligands was found to be the ideal candidate with the highest activity [2170 kg (polyethylene)/mol (Ti)/h]. [10b] Notably, the preferred complex highlighted above, with MAO, could also catalyze the copolymerization of ethylene and either 1-hexene or 1-octene with approximately 15 mol% incorporation of the larger terminal alkene. However, the related larger bite angle *N,O*-chelates, such as phenoxyimine ligands, commonly achieve activities two orders of magnitude higher than these titanium amidate catalysts. [4c]

In 2013, Harkness *et al.* [10c] reported a set of titanium ureate catalysts that were found to have activity for ethylene polymerization. Both a mono(cyclopentadienyl) mono(ureate)dichloro titanium complex and a bis(ureate)dichloro titanium complex of the same ureate ligand were synthesized. An interesting ureate ligand was employed, bearing the *N*-heterocyclic carbene IMes on the backbone ureate nitrogen, in an imine-type moiety (Scheme 8).

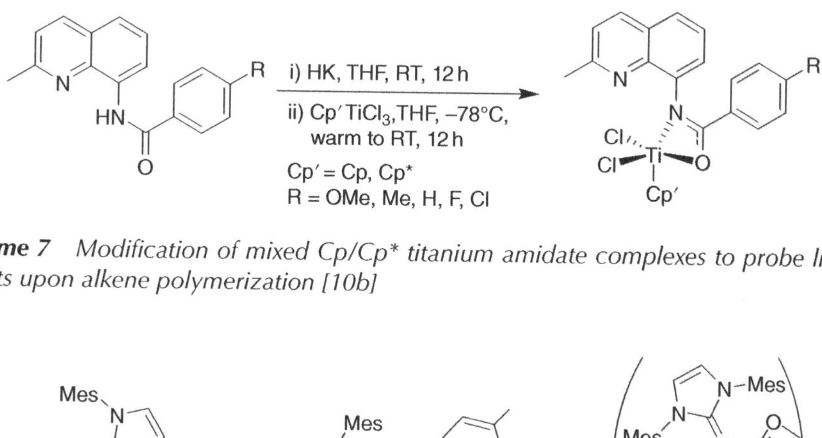

Scheme 7 *Modification of mixed Cp/Cp* titanium amidate complexes to probe ligand effects upon alkene polymerization [10b]*

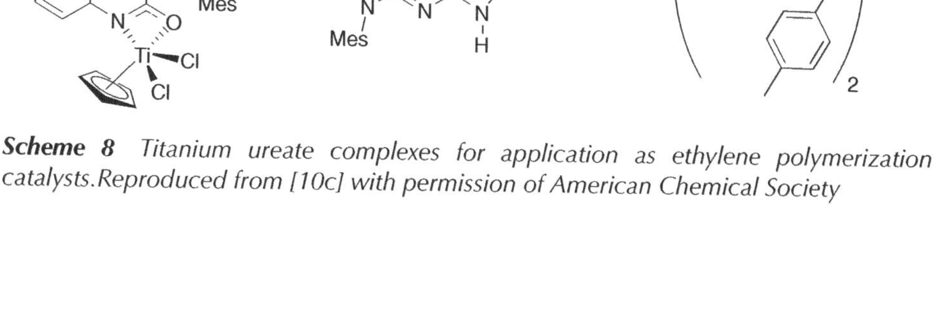

Scheme 8 *Titanium ureate complexes for application as ethylene polymerization catalysts. Reproduced from [10c] with permission of American Chemical Society*

The polymerization of ethylene was tested with 1000 equiv. of MAO, with only 1 atm of C_2H_4 and at room temperature. Under these conditions however, both the half-sandwich ureate complex and the bis(ureate) complex showed markedly lower activity than the half-sandwich titanium amidates previously described, with only 60 kg polymer/mol catalyst. [10c]

These examples demonstrate that tight bite angle *N,O*-chelates of titanium can be used for alkene polymerization. However, the area is largely unexplored and investigations have not extended to other early transition metals that have also been commonly used for this application such as zirconium and hafnium.

13.2.1.2 Ring-opening polymerization of cyclic esters

Alternatively, syntheses of specialty biodegradable polymers, prepared by ROP, have also been explored using tight bite angle *N,O*-chelating ligands on a variety of early transition metals. The prominent monomers used for ROP are lactide, a cyclic six-membered ring diester, and ε-caprolactone, a seven-membered ring ester (Figure 15). Lactide is also a biorenewable polymer monomer, derived from plant sources such as corn. [51] Lactide has two stereocenters and therefore, depending on the stereochemistry of the monomer, the resultant polylactide material can have distinct materials properties that can be varied through control of the microstructure of the polymer backbone. Additionally, copolymerization of lactide and ε-caprolactone can be used for varying polymeric properties. [22b]

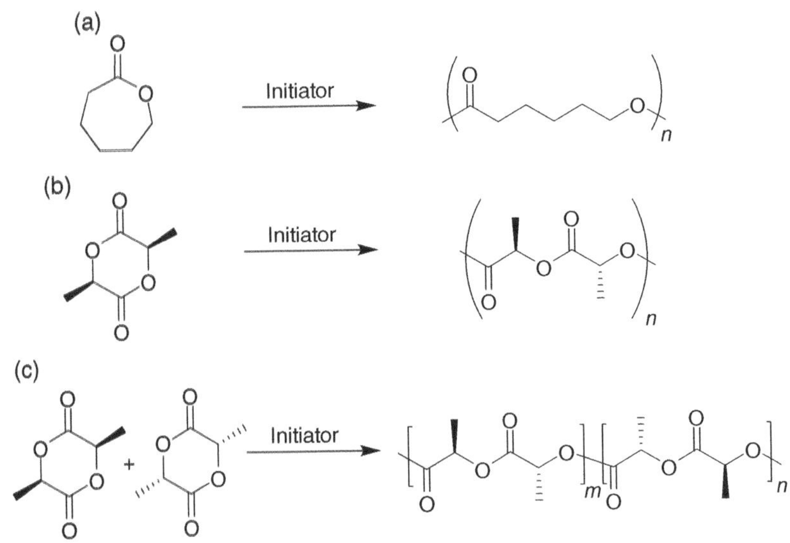

Figure 15 Ring-opening polymerization of cyclic esters to make biodegradable polymers: (a) ROP of ε-caprolactone; (b) ROP of one enantiomer of lactide to give isotactic polymer; and (c) ROP of rac-lactide to form polymers of variable tacticity

Early transition metals utilized for this transformation include groups 3 and 4 metals such as yttrium, titanium, and zirconium. A number of *N,O*-chelates are also already being used to access this type of reactivity, including some wider bite angle *N,O*-chelates, [51] but also amidate, pyridonate, and sulfonamidate ligand sets (see below).

Group 3 elements have emerged as being useful Lewis-acidic metal centers for the ROP of cyclic esters. While these elements are highly reactive for this transformation, one significant challenge remains – the extreme water sensitivity of yttrium and other group 3 metals. [52b]

The first report of a group 3 metal bearing a tight bite angle *N,O*-chelate for the ROP of a cyclic ester was by Schafer's group in 2008. [11] These yttrium amidate complexes were used for ROP of ε-caprolactone. The trivalent metal was used to make homoleptic yttrium complexes, however depending on the synthetic route, monomeric or dimeric species were obtained. When starting with a tris(amido) complex and using 3 equiv. of amide proligand in THF, a monomeric yttrium amidate is formed (Figure 11). The THF-coordinated monomeric complexes of three different amidate ligands were screened for ROP of ε-caprolactone, with a monomer:initiator ratio of 225:1, at 25 °C for 15 min. Moderate to good yields were obtained for these reactions (up to 91%) with average molecular weights of around $3–4 \times 10^5$ g/mol. Comparing the amidate ligands involved for the three complexes, it was found that electron-withdrawing groups on the *para*-position of a phenyl group off the carbonyl carbon of the amidate detracted from the effectiveness of the initiator.

In a follow up report from Schafer's group [52a] in 2012, largely focused on studying the viscoelastic behavior of the polymer, a modified tris(amidate) yttrium complex was used to initiate ROP of ε-caprolactone. This slightly bulkier complex with a diisopropylphenyl on the nitrogen instead of dimethylphenyl, gave lower molecular weight polymers of just under 1×10^5 g/mol, but of a narrower polydispersity index (PDI) of around 1.5 compared with the values of between 2 and 2.5 achieved in previous studies.

In 2011, Zi's group [52b] reported two different yttrium amidate systems based on the idea of incorporating a neutral pendant donor off the nitrogen substituent of the amidate (Figure 16). Polymerization studies with *rac*-lactide of these two complexes showed comparable reactivities, yielding high to quantitative yields over 0.5–2 h at 20–40 °C, with molecular weights of $6.5–7.0 \times 10^4$ g/mol and PDI values of 1.2–1.3. The tacticity of the polylactide could be determined by NMR spectroscopy. These two initiators could be used to make polymers with a homotactic bias.

In a follow up report from Zi's group [52c], a very related set of two more yttrium mono- and bis(amidate) complexes were reported for their ROP reactivity. Not surprisingly their reactivities as initiators led to very similar polylactide formation, however with slightly less control over the stereoregularity of the polymer.

In 2012, Sun's group [10b] reported the use of the same amidate ligands that they had used on titanium for terminal alkene polymerization, on yttrium for ROP of ε-caprolactone. [52d] Going through a salt metathesis route, a highly aggregating series of complexes were isolated, involving three yttrium atoms. Four differently substituted amidate ligands were separately installed and all reactivities were found to be comparable. Interestingly, addition of benzyl alcohol in a 1:1 ratio with the initiator led

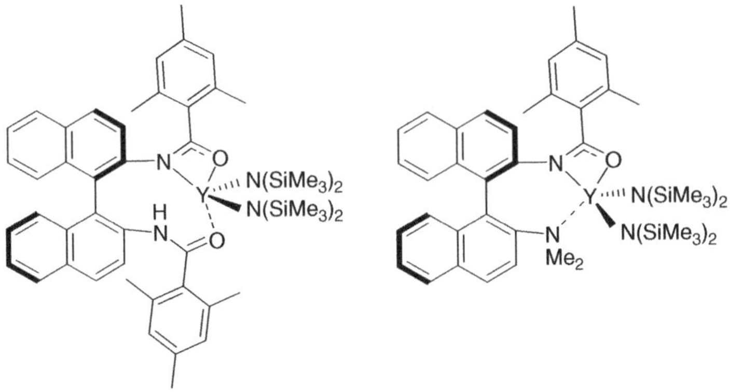

Figure 16 Yttrium amidate complexes for ring-opening polymerization give polylactide. Reproduced from [52b, 52c] with permission of Elsevier

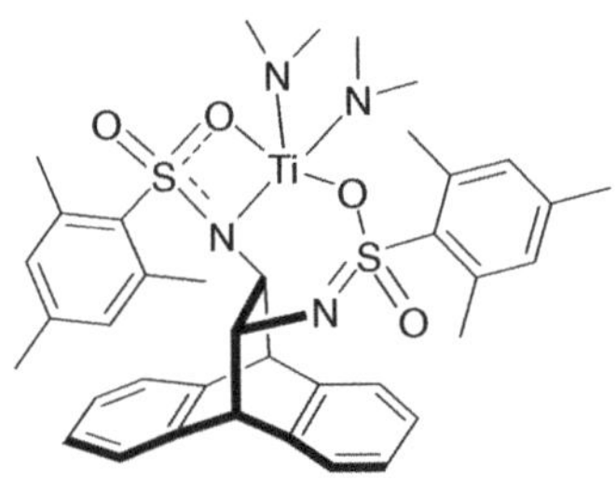

Figure 17 A titanium sulfonamidate complex for polylactide synthesis

to slightly decreased average molecular weights (9.3×10^4 g/mol down to 5.8×10^4 g/mol) but largely improved PDIs (from 2.7 to approximately 1.6). [52d]

A report in 2013 by Zhao's group [52e] described similar types of yttrium amidate aggregate species for the ROP of *rac*-lactide. While the molecular weights and distributions were comparable with previous yttium amidate initiated ROP of *rac*-lactide, the selectivity interestingly contrasted with other reports. These initiators showed a higher likelihood of heterotactic bias.

In addition to the more generally reactive group 3 elements, examples of group 4 metals with amidate, pyridonate, and sulfonamidate ligands have been reported for ROP of cyclic esters. Such group 4 metals, and in particular titanium, are attractive due to their low cost, low toxicity, and high earth abundance. Furthermore, such complexes are known to be more robust than rare earth element complexes and thus less sensitive to the purity of the monomeric feedstock that is used for ROP.

In 2010, Zi's group [24b] reported ethanoanthracene-based ligands bound through two sulfonamidate arms, with two different binding modes (Figure 17). Polymerization of *rac*-lactide formed polylactide in quantitative yield with 0.4 mol% catalyst loading at 70 °C for 24 h. Average molecular weights were 3.2×10^4 g/mol depending upon reaction solvent with consistent PDI values of 1.23. [24b] Additionally, a good homotactic bias was measured, making these titanium complexes selective for isotactic polylactide but with less than half the average molecular weights than those initiated by the yttrium complexes developed by the same group, as described above.

In two reports, Mountford's group [52f, 52g] reported a series of sulfonamidate titanium and zirconium complexes and their reactivities toward ROP of both ε-caprolactone and *rac*-lactide. In this series of complexes, the nitrogen atom of the sulfonamidate ligand is bound to the metal center; however, only zirconium displays structural evidence for *N,O*-chelation of the sulfonamidate in the solid state as seen in an example in Figure 18. These multipodal ligands were varied and the labile ligands were substituted to access a wide variety of complexes and identify their reactivities. General trends unrelated to the sulfonamidate ligand include noting that isopropoxide rather than dimethylamido labile ligands results in a more controlled polymerization of ε-caprolactone, and zirconium analogs bearing the same ligands are more reactive than their titanium counterparts. [52f] The zirconium amido complexes bearing these sulfonamidate ligands achieved average molecular weights of 5.2–7.8 kg/mol with PDI values of 1.18–1.19 after 1 h at 100 °C.

For the ROP of *rac*-lactide, the titanium complexes were poor initiators, not leading to high conversions, while zirconium complexes were more successful, managing high

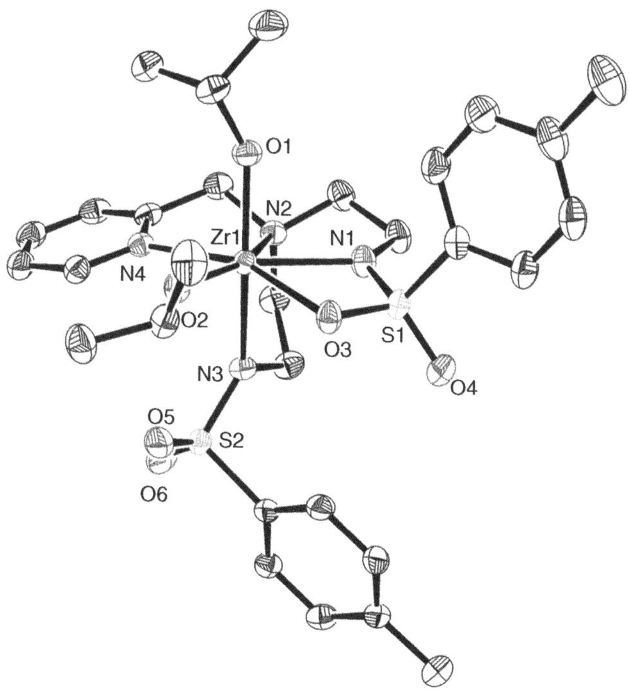

Figure 18 *Zirconium sulfonamidate complex for ring-opening polymerization of ε-caprolactone and rac-lactide. Reproduced from [52f] with permission of American Chemical Society. (See insert for color/color representation of this figure)*

conversions with molecular weights of 1.3–2.9 × 10^4 g/mol and PDI values from 1.38 down to as low as 1.08 with 1 mol% initiator. The researchers concluded the least bulky sulfonyl substituents led to the most active initiators for the ROP of *rac*-lactide. [52g] For all studies with *rac*-lactide, the polymer material was found to be predominantly atactic.

In 2013, Schafer's group [22b] reported titanium bis(amidate) and bis(pyridonate) complexes for the homopolymerization of *rac*-lactide and ε-caprolactone, and also the formation of a random copolymer of the two. These complexes form pseudo-octahedral six-coordinate species, which were characterized in the solid state. Complexes were synthesized by first installing 2 equiv. of the ligand on homoleptic $Ti(NMe_2)_4$ followed by protonolysis of dimethylamido ligands with 2 equiv. of alcohol (Figure 19).

All complexes were used for homopolymerization studies where moderate to good yields were obtained with a 1:300 [Ti]:monomer ratio, at 100–130 °C over 16–24 h. [22b] For the homopolymerization of *rac*-lactide, the atactic polymer was obtained with average molecular weights of 1.4–2.5 × 10^4 g/mol and PDI values of 1.16–1.22, and for the homopolymerization for ε-caprolactone, average molecular weights of 2.0–3.9 × 10^4 g/mol were obtained with PDI values of 1.28–1.48. Interestingly, for polylactide, titanium initiators bearing amidate ligands led to higher molecular weights while a trend existed within the 2-pyridonate ligands where more bulk in the 3-position and less bulk in the 6-position next to nitrogen led to higher molecular weights. Curiously, this trend was reversed for the polymerization of ε-caprolactone.

The efficiency that this set of initiators showed for both types of polymerization reactions, and the displayed influence of the bound *N,O*-chelating ligand, led to testing for copolymerization of the two monomers. Impressively, the titanium pyridonate complexes led to a random copolymer with a nearly 1:1 ratio of monomers. These random copolymers have average molecular weights of 1.8–2.2 × 10^4 g/mol with PDI values of 1.29–1.41.

While the differences in reactivities of the *N,O*-chelated complexes used for polymerization initiators is often influenced by the metal (e.g., yttrium is typically more active than group 4 metals), the reactivity trends for the *N,O*-chelates show that the ligand has a profound effect on the polymerization. These examples show that

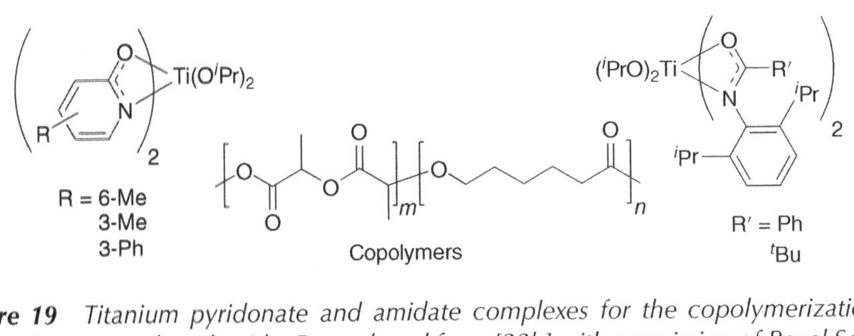

***Figure 19** Titanium pyridonate and amidate complexes for the copolymerization of ε-caprolactone and rac-lactide. Reproduced from [22b] with permission of Royal Society of Chemistry*

N,O-chelating ligands offer the support to fine tune initiators to not only give improved activities but, just as importantly, access to more challenging products through stereocontrol and copolymerizations.

13.2.2 Hydrofunctionalization

The most high-profile application of these complexes is as precatalysts for hydrofunctionalization reactions. These are atom-economic transformations that proceed via E-H activation and selective addition across carbon–carbon multiply bonded species. In these reactions all of the atoms in the reagents are retained in the product, eliminating the production of wasteful by-products. With the push toward more environmentally benign methodologies and the inherent economic advantages of minimized waste production, such technologies are attractive alternatives to catalytic transformations with expensive, toxic late transition metals. Furthermore, the challenge of realizing regioselective and stereoselective transformations when reacting across carbon–carbon multiple bonds presents opportunities for expanding reactivity profiles in hydrofunctionalization reactions. Such reactions with early transition metals have been characterized to proceed through the generation of M—E bonded species, or may function simply as Lewis acids, as in ROP as illustrated above. Here we summarize the application of such *N,O*-chelated complexes in hydroamination, [52b, 54] hydroaminoalkylation, [31, 55] and hydroalkynylation, [14] and provide mechanistic insights that illustrate the usefulness of *N,O*-chelating ligands for modulating and expanding reactivity and selectivity in these transformations.

13.2.2.1 Hydroamination

Early catalytic investigations with this family of complexes focused on the C—N bond-forming reaction, hydroamination, which proceeds via N—H addition across alkynes and alkenes (Figure 20). [9, 56]

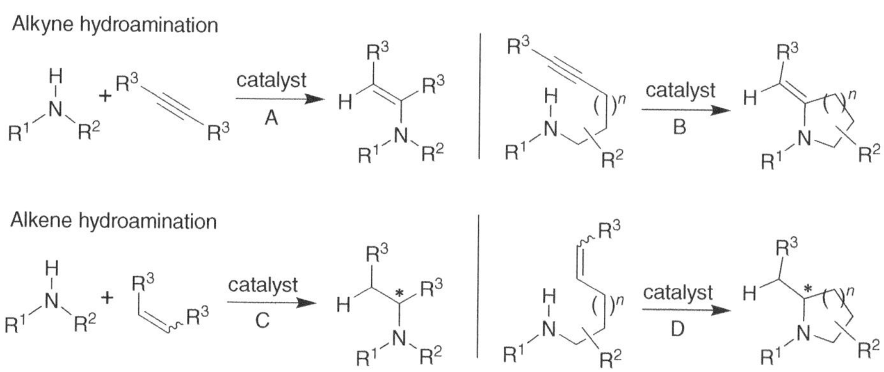

Figure 20 *Catalytic inter- and intramolecular alkyne and alkene hydroamination (A–D)*

Hydroamination is an attractive methodology for the synthesis of higher value amine products, with increased molecular complexity, from readily available starting materials. The amine compounds produced are relevant to a variety of applications including pharmaceutical drugs, agrochemicals, and natural product synthesis. Hydroamination is kinetically challenging, due to the difficulty of bringing together the electron-rich amine and alkene/alkyne reaction partners and the intermolecular variant of this transformation has a minimal thermodynamic driving force (A and C in Figure 20). Thus, initial investigations focused on intermolecular variants of the reaction and more recently have grown to include expanded substrate scope and realizing stereoselective transformations.

Transition metal catalysts from across the periodic table have been investigated for this transformation. [56b, 57] Early transition metal catalysts [58] are of particular interest due to their high reactivities, with reduced air and moisture sensitivity compared with the rare earth metal systems, and lower cost and toxicity compared with the late transition metal catalysts. The *N,O*-ligands generating tight four-membered metallacycles described above have been studied as precatalysts for hydroamination methodologies that display promising substrate scope and reactivity.

The bis(amidate)bis(amido) complexes of titanium are a broadly applicable class of precatalysts for hydroamination. The titanium complexes are the most active for the hydroamination of alkynes [19] and complex **13** has been identified as an efficient catalyst for the intramolecular [19] reaction and a regioselective system for the intermolecular hydroamination of alkynes, [59] giving the aldimine products selectively (Scheme 9). This catalyst can also be used with internal alkynes to give good regioselectivity for installation of the amine at the more sterically accessible position. [60] This titanium catalyst functions well with both aryl- and alkylamines, displays good tolerance to esters, silyl-protected alcohols, and aryl halide functional groups, and promotes excellent regioselectivity. The reactive aldimine products can be further elaborated using one-pot procedures to synthesize aldehydes, [59] substituted amines, [59] α-cyano amines, [61] α-amino acids, [61] piperazines and morpholines, [62] and tetrahydroisoquinoline and benzoquinolizine alkaloids. [59b] The catalyst is air and moisture sensitive, but can be prepared and stored in a glovebox or can be prepared in

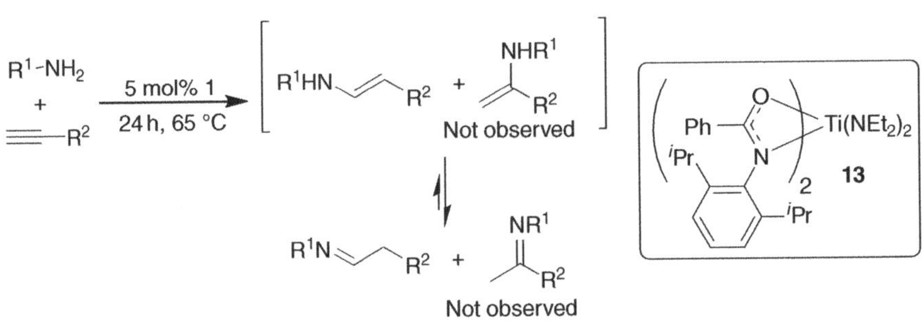

Scheme 9 *Regioselective hydroamination of terminal alkynes promoted by precatalyst **13***

situ using simple syringe techniques to give the desired products in excellent yield. [60] Complex **13** is now commercially available. [63]

The majority of group 4 catalyzed hydroamination reactivity has been proposed to occur via a [2+2] cycloaddition mechanism (Scheme 10) involving a catalytically active metal-imido species (**A**). [64] Mechanistic investigations for the bis(amidate) catalysts are consistent with this proposal, supported by the lack of reactivity observed with secondary amine substrates. [64e, 64f, 64g, 65] Furthermore, titanium terminal imido complexes supported by amidates have been stoichiometrically prepared (Scheme 11) and have shown identical catalytic reactivity to the precatalyst **13**. [59b] The regioselectivity observed with complex **13** is unique among terminal alkyne hydroamination catalysts and has been postulated as arising due to the hemilabile amidate ligand, which can become κ¹-bound during the catalytic cycle and accommodate the sterically demanding intermediate metallacyclic product (**B**). [60]

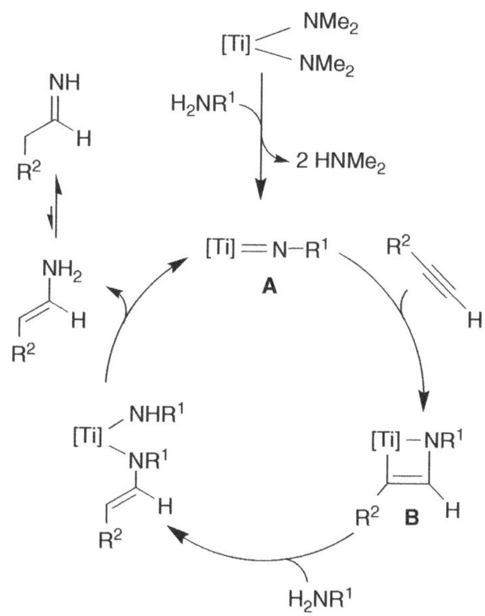

Scheme 10 *[2+2] Cycloaddition mechanism for alkyne hydroamination catalysis*

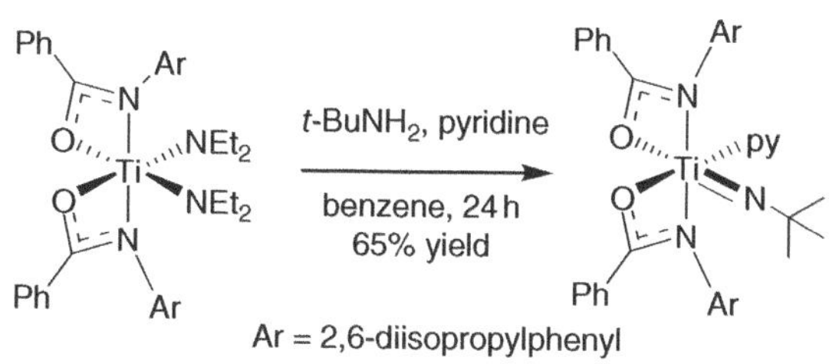

Scheme 11 *Preparation of catalytically competent terminal titanium imido complex*

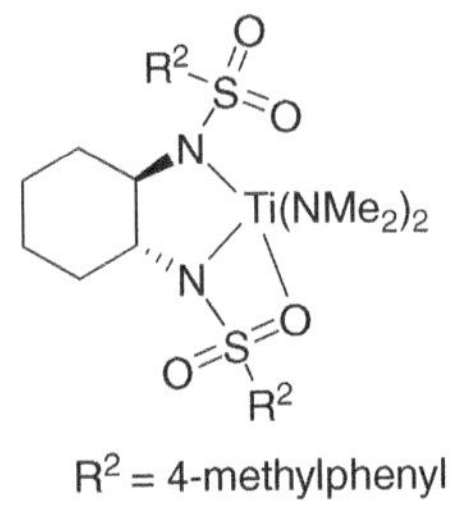

Figure 21 *Reported early transition metal complex for hydroamination containing the sulfonamidate ligand*

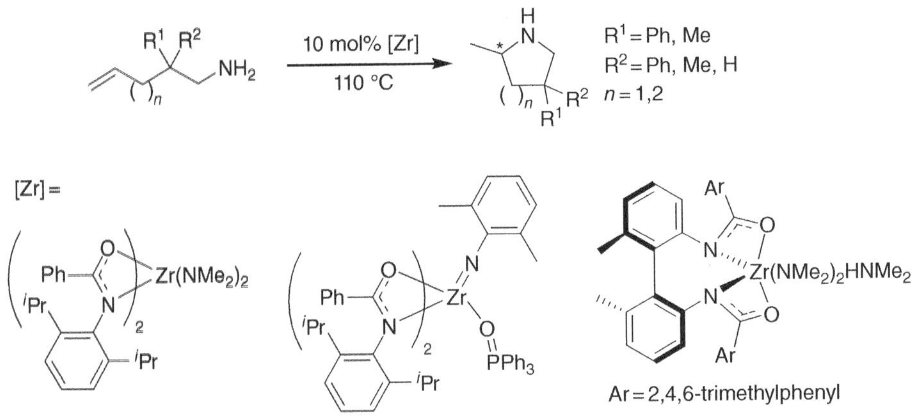

Scheme 12 *Intramolecular hydroamination of primary aminoalkenes with neutral bis(amidate) zirconium complexes. Reproduced from [64e] with permisssion of American Chemical Society and from [65a] with permission of John Wiley & Sons, Ltd*

Bergman and co-workers investigated the application of a sulfonamidate complex (Figure 21) for the hydroamination of alkynes and allenes. [66] These complexes are significantly more reactive and regioselective compared with $Ti(NMe_2)_4$ or Cp_2TiMe_2 precatalysts, [64b, 67] which can be attributed to the increased electron-withdrawing properties of the sulfonamidate ligand. The sulfonamidate titanium complexes are also proposed to proceed via the [2+2] cycloaddition mechanism shown in Scheme 10. [66b]

Group 4 bis(amidate)bis(amido) complexes have also been identified as precatalysts for the more challenging hydroamination of alkenes. The majority of investigations in this field focus on the intramolecular cyclization of aminoalkenes with zirconium-based catalysts. [64e] Neutral group 4 bis(amidate) zirconium amido or imido complexes are efficient precatalysts for the intramolecular cyclization of primary amines to form pyrrolidine and piperidine products (Scheme 12). The monomeric imido complex can be generated by reaction of the bis(amido) complex with 2,6-dimethylaniline and trapped with triphenylphosphine oxide. [64e] The bis(amido) and imido complexes show comparable half-lives for the cyclization reactions, which is consistent with both precatalysts sharing a common catalytically active species.

The asymmetric version of this transformation, producing enantioenriched α-chiral amines, is an attractive goal. Following a report by Bergman's group [68] of neutral bis(amido) zirconium precatalysts displaying enantiomeric excess values of up to 80%, Schafer's group [65, 69] described the use of neutral biaryl bis(amidate) zirconium complexes for this transformation that are proficient in the cyclization of aminoalkenes with enantiomeric excesses up to 93%. Related chiral tethered bis(amidate) complexes of Ti and Zr have been reported by others, [70] although no broadly useful enantioselective catalyst has been developed using this family of catalysts. The rigid nature of the chiral tether was determined to be critical to accessing high enantioselectivities, as chiral biphenyl backbones afforded improved enantioselectivities over more conformationally flexible chiral binapthyl tethers.

The reaction involves initial formation of the zirconium-imido species, followed by [2+2] cycloaddition with the C–C unsaturation (Scheme 13). This is consistent with the observation that bis(amidate) complexes do not mediate hydroamination with secondary amine containing substrates. The cyclic transition state of the intramolecular reaction determines the regioselectivity of the reaction followed by successive protonation of the intermediate metallacycle and release of product to regenerate the catalytically active imido species.

Interestingly, by switching from bis(amidate) to bis(ureate) bis(amido) complexes, a broader scope of reactivity can be realized in intramolecular alkene hydroamination. [28] Reactivity studies indicate that the tethered zirconium bis(ureate) precatalysts are more reactive for intramolecular alkene hydroamination than the titanium analogs,

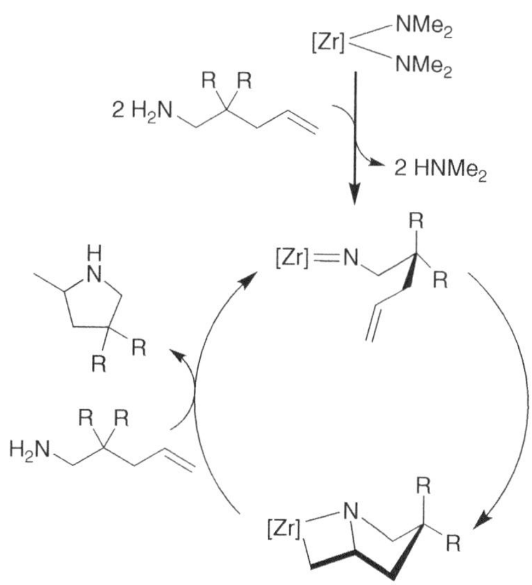

Scheme 13 *Simplified [2+2] cycloaddition mechanism for the bis(amidate) zirconium-catalyzed intramolecular cyclization of aminoalkenes*

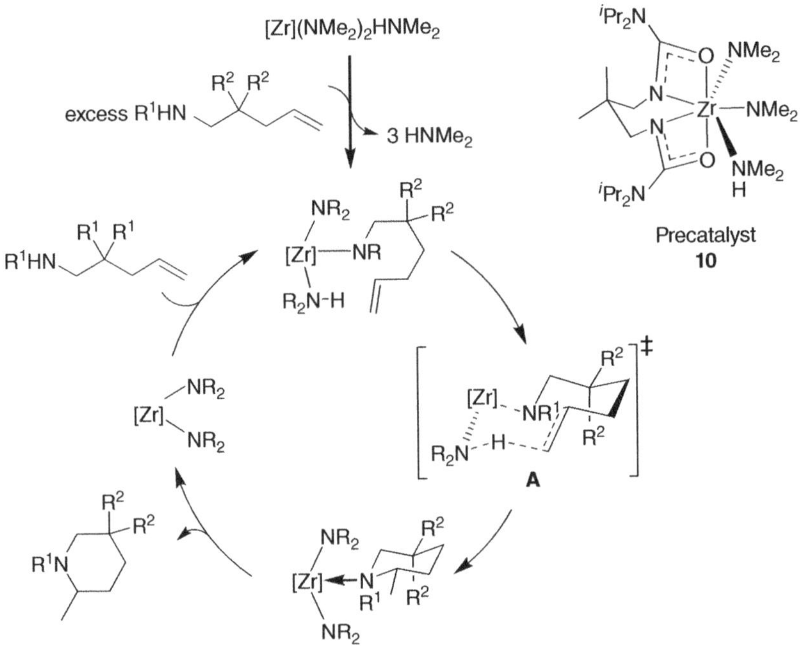

*Scheme 14 Proposed catalytic cycle for hydroamination using zirconium precatalyst **10** in which the key step is a proton-assisted C—N bond formation*

consistent with what has been observed with the amidate systems. [64e] The tethered systems show drastically improved activity compared with the untethered systems. [21a, 21b] Complex **10** (Scheme 14) was determined to be a broadly applicable hydroamination precatalyst. [21a] This system is applicable for the intramolecular hydroamination of alkynes as well as the intramolecular hydroamination of aminoalkenes. Heteroatoms are also tolerated and most notably this system functions well with a broad variety of primary and secondary amine substrates. The reactivity with secondary amines is particularly interesting, as this implies that the bis(ureate) ligand is active via a different mechanistic pathway than that followed by most group 4 hydroamination catalysts. Extensive mechanistic studies support the catalytic cycle shown (Scheme 14) which has been independently corroborated by computational studies. [37, 71] The key step of this mechanism is a proton-assisted C—N bond formation via transition state **A** (Scheme 14). This novel reactivity has been attributed to increased nucleophilicity of the equatorial amido ligand because of the electron-rich ureate ligand, as well as the sterically accessible metal center that can readily accommodate seven-coordinate species. This expanded coordination number is critical for coordinating a neutral amine that participates in proton transfer. [37, 71]

Bis(pyridonate) zirconium complex **14**, bearing sterically demanding pyridonate ligands, can be synthesized via protonolysis (Scheme 15), and is an active precatalyst for the intramolecular hydroamination of aminoalkenes to generate pyrrolidine and

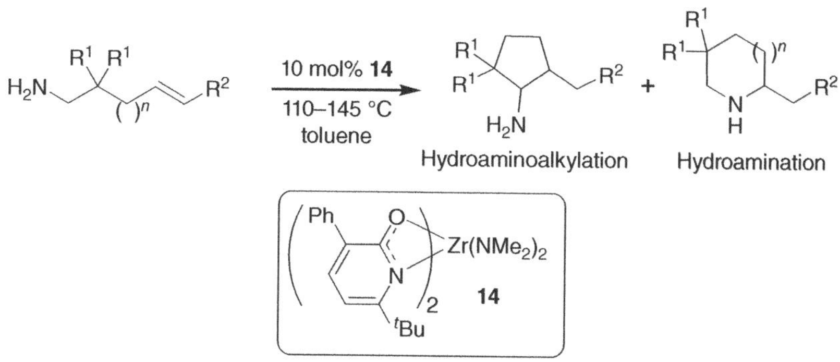

Scheme 15 *Bis(pyridonate) bis(dimethylamido) zirconium complex **14** that promotes both hydroamination and hydroaminoalkylation*

piperidine products. Substrates that are more challenging for hydroamination, such as those without *gem*-disubstituents, can undergo both hydroamination and an unexpected intramolecular hydroaminoalkylation reaction (see below) resulting in undesirable product mixtures. [72]

By exploiting the variability of the *N,O*-chelating ligands, a family of hydroamination catalysts have been developed. To date, these systems display impressive substrate scope, accommodating intra- and intermolecular alkyne hydroamination and intramolecular alkene hydroamination. Outstanding regioselectivities can be realized and impressive enantioselective transformations have been achieved. Intermolecular alkene hydroamination is an outstanding challenge in the field broadly, and has not been reported for *N,O*-chelated complexes. With the synthetic flexibility accessible with these complexes, and the fact that recent mechanistic insights have shown that both [2+2] cycloaddition and proton-assisted C—N bond formation are accessible to these complexes, new approaches for enhancing reactivity are under investigation.

13.2.2.2 Hydroaminoalkylation

Catalytic hydroaminoalkylation is an emerging 100% atom economic $C(sp^3)$–$C(sp^3)$ bond forming strategy, involving the addition of a $C(sp^3)$—H bond adjacent to nitrogen across a C=C bond (Scheme 16). [55b] While the early transition metal community refers to this transformation as hydroaminoalkylation, this transformation is the same as the late transition metal catalyzed C—H alkylation reaction which has attracted significant attention over the past decade. [16d, 73] One notable difference between the early transition metal variant and the late transition metal reaction is the fact that late transition metals require the installation of directing and/or protecting groups to direct the site of C—H activation (presumably via oxidative addition), while early transition metals require secondary amines to first form M—N bonded species which then undergo C—H activation α to nitrogen. The use of inexpensive early transition metal (Ti, Zr, Ta, Nb) catalysts of low toxicity hold promise in the development of hydroaminoalkylation catalysts.

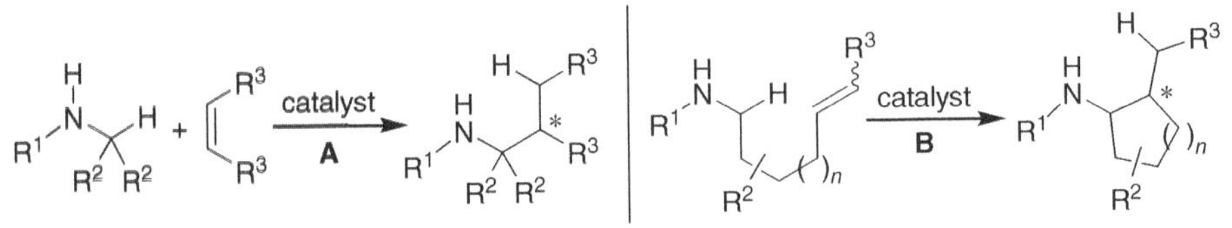

Scheme 16 *Inter- and intramolecular hydroaminoalkylation reactions for the atom-economic synthesis of selectively substituted amines*

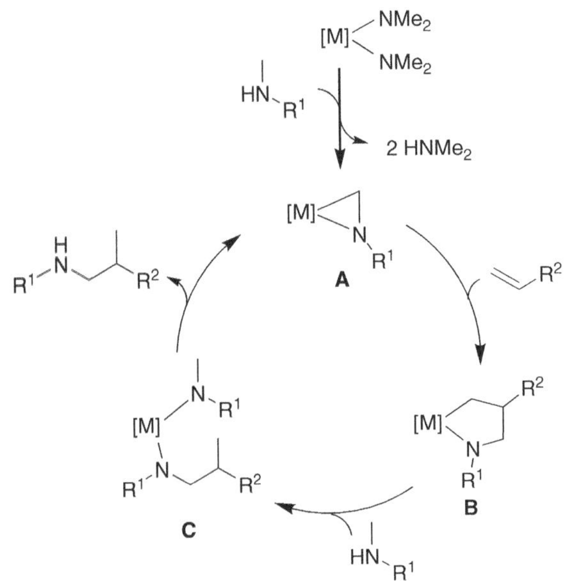

Scheme 17 *Proposed catalytic cycle for the hydroaminoalkylation of secondary amines with tantalum precatalysts*

Such transformations were first reported in the 1980s, [74] although it was not until 2007 that the synthetic potential of the reaction was further explored. [75]

Catalytic hydroaminoalkylation was first reported in 1980 by Clerici and Maspero [74b], when dimethylamine or diethylamine was observed to undergo α-alkylation at high temperatures (160–200 °C) in the presence of simple alkene substrates with catalytic amounts of homoleptic dialkylamido zirconium, niobium and tantalum complexes. Later, Nugent *et al.* [74a] used deuterium labeling studies to show that this reaction proceeds via a catalytically active metallaziridine complex (**A**, Scheme 17), which is formed upon hydrogen abstraction. Alkene insertion into the reactive M—C bond of the metallaziridine results in the formation of the five-membered metallacycle **B**, which can then undergo protonolysis to give a new bis(amido) poised for hydrogen abstraction and product elimination.

This mechanistic proposal is consistent with recent reported results for hydroaminoalkylation. [35, 76] Furthermore, metallaziridines are an interesting class of organometallic

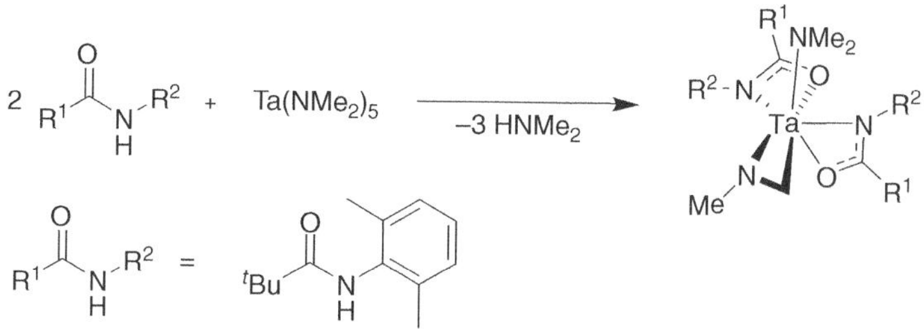

Scheme 18 *Synthesis of bis(amidate) tantalaziridines*

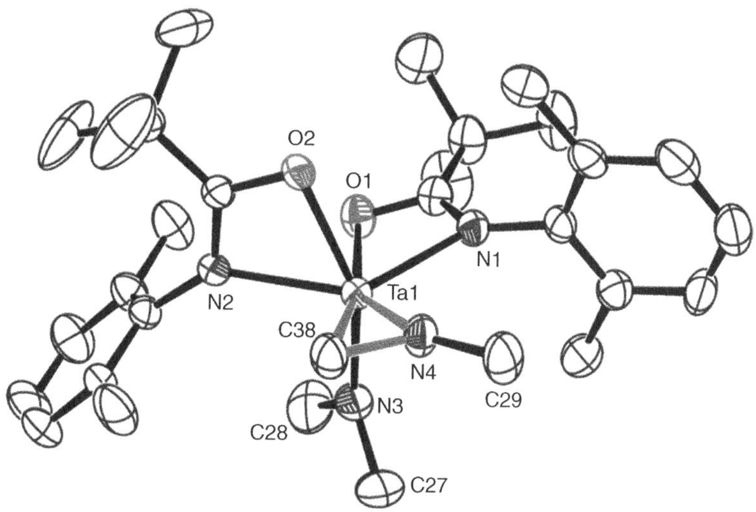

Figure 22 *ORTEP plot of the solid-state molecular structure of a bis(amidate) supported tantalaziridine. Reproduced from [31] with permission of John Wiley & Sons, Ltd. (See insert for color/color representation of this figure)*

compounds in their own right. [77] As shown in Scheme 18, the protonolysis reaction of sufficiently bulky amide proligands with pentakis(dimethylamido) tantalum results in the generation of 3 equiv. of dimethylamine and results in the room temperature formation of the tantalaziridine (Figure 22). Notably, this tantalaziridine is also a competent intermolecular hydroaminoalkylation catalyst, although reactivity is sluggish for this sterically bulky system. [31]

Considering the sluggish reactivity of these systems, and the promising reported reactivity of simple $Ta(NMe_2)_5$ [75] and $[TaCl_2(NMePh)_3]_2$ [78] for intermolecular hydroaminoalkylation, less sterically congested mono(amidate) complexes were targeted. The axially chiral biaryl-based bis(amidate) ligands support catalytic systems that are generally less reactive than their mono(amidate) counterparts but promote enantioselective hydroaminoalkylation with enantiomeric excesses reported of up to 93% (Scheme 19). Group 5 mono(amidate) complexes have been successfully

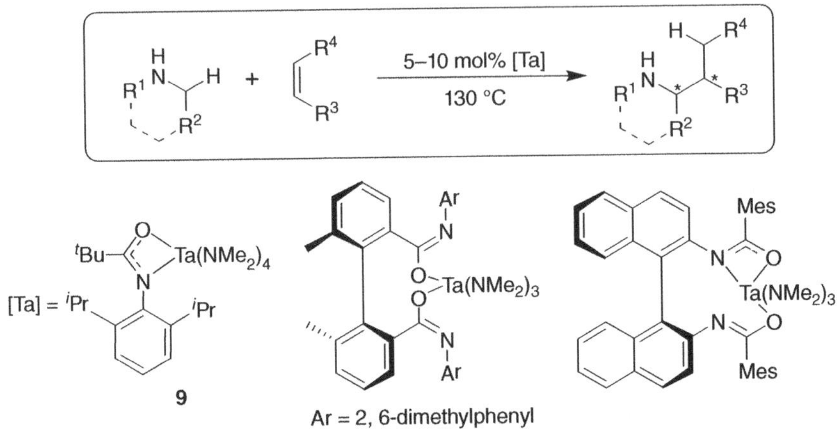

Scheme 19 Intramolecular hydroaminoalkylation of unactivated olefins with secondary amines catalyzed by mono- and bis(amidate) tantalum amido complexes

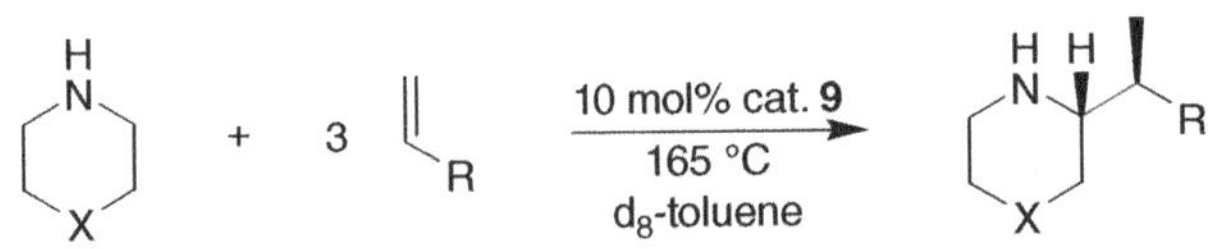

Scheme 20 The diastereoselective direct C–H alkylation of six-membered N-heterocycles

prepared and in particular, mono(amidate) tantalum complex **9** is a broadly applicable precatalyst for the intermolecular hydroaminoalkylation of terminal alkenes.

Complex **9** remains one of the most broadly applicable catalysts for this reaction, catalyzing hydroaminoalkylation with *N*-arylalkylamines, including controlled monoalkylation of dienes. This reactive catalyst **9** even displays tolerance of oxygen-containing substrates. Most importantly, this remains the only catalytic system capable of the direct C–H alkylation α to N of unprotected heterocyclic amine substrates. Such products are potentially important structural motifs for exploration in medicinal chemistry. In all cases this precatalyst shows regioselective hydroaminoalkylation to generate the branched product, and excellent diastereoselectivity when applicable (Scheme 20). [79]

An investigation of factors in *N,O*-chelate design and how it affects hydroaminoalkylation have resulted in some general trends that suggest that the hemilabile chelating motif is favorable for enhancing reactivity. [35] Furthermore, the nature of the *N,O*-chelate can be varied and can include amidate, ureate, pyridonate and sulfonamidate ligands. [80] The comparison of various ligands with different steric bulk and electronic features in the preferred test reaction of 1-octene with *N*-methylaniline showed that sufficient steric bulk must be incorporated into the reaction, presumably to promote dissociation of the product upon hydrogen abstraction of the bis(amido) intermediate **C** (Scheme 17). [31] Also, increasing the electron-withdrawing character of the

N,O-chelate resulted in improved reactivity. [31] Throughout these investigations, elevated temperatures were required to realize meaningful reactivity within 24 h. Thus, by taking the insights provided through the ligand structure/catalyst function investigations described above, we targeted an *N,O*-chelate that could realize catalytic turnover at reduced reaction temperatures. Such catalysts would be more amenable toward application in practice as temperatures in excess of boiling points of common solvents (e.g., toluene, 110 °C) are undesirable.

New catalyst development efforts have exploited the incorporation of a tetrahedral phosphorus heteroatom in the backbone of the *N,O*-chelate to generate a new modular class of tantalum precatalysts. [23] The phosphoramidate tantalum complex **15** (Figure 23) displays significantly improved reactivity for the hydroaminoalkylation of secondary amines eliminating the need for high reaction temperatures, catalyzing the hydroaminoalkylation of 1-octene with *para*-methoxyaniline in 86% yield after 20 h at room temperature. [23] This reactivity has been attributed to the increased electron-withdrawing nature of the phosphoramidate ligand, combined with the chloride ligand to give a more electropositive and reactive metal center.

Complex **15** illustrates the importance of incorporating an *N,O*-chelate of sufficient steric bulk on an electrophilic metal center to achieve overall improved reactivity. While these room temperature transformations are promising, this complex is not thermally robust and thus displays limited substrate scope. In particular, heating to promote reactivity with unstrained internal alkenes or challenging *N*-heterocyclic substrates results in no appreciable product formation and catalyst decomposition, as noted by ^{31}P NMR spectroscopy.

Doye's group [81] showed that a dinuclear titanium-sulfonamidate complex (Scheme 21), with a tetrahedral sulfur in the ligand backbone, can be used for intermolecular hydroaminoalkylation as well. This system gives mixtures of branched and linear products, although to date there has been no mechanistic rationale provided for the reduced regioselectivity of group 4 metal complexes in this transformation. There has been one report by Zi's group [44] that describes axially chiral bis(sulfonamidate) tantalum and niobium complexes for application as precatalysts for hydroamination and hydroaminoalkylation. Unfortunately, these complexes did not show any reactivity for either of these reactions.

Our attention turned toward the challenge of developing an alternative, robust *N,O*-chelated complex for improved substrate scope. A mixed 2-pyridonate-$Ta(NMe_2)_3Cl$

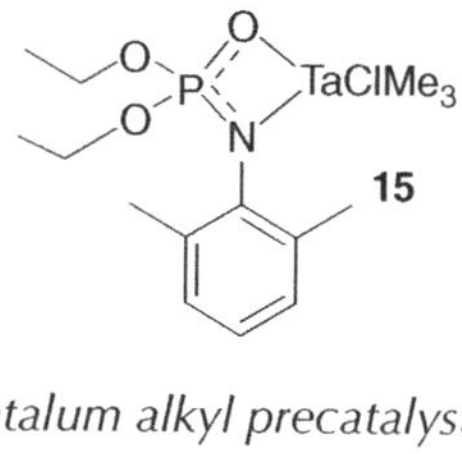

15

Figure 23 *Phosphoramidate tantalum alkyl precatalyst for room temperature hydroaminoalkylation*

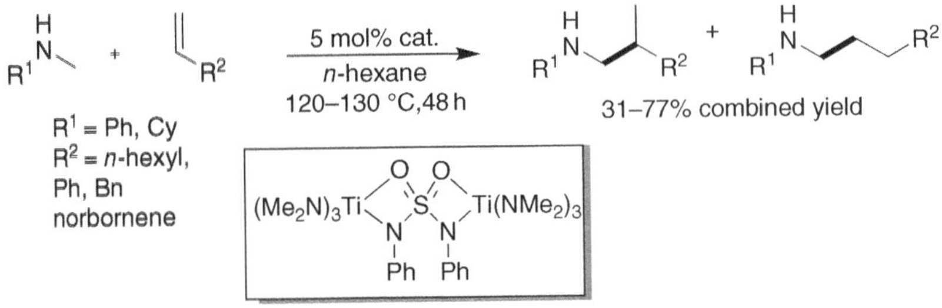

Scheme 21 *Hydroaminoalkylation with a dinuclear titanium-sulfonamide complex*

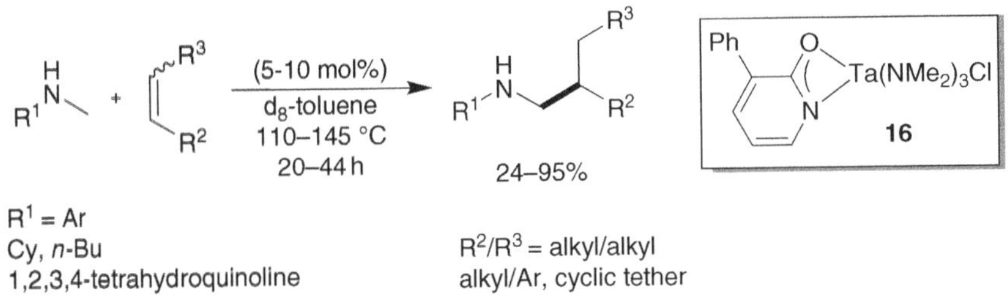

Scheme 22 *2-Pyridonate Ta complex for the first example of hydroaminoalkylation with unstrained internal alkenes to give branched products*

complex (**16**, Scheme 22) was prepared using a simple salt metathesis route with $Ta(NMe_2)_3Cl_2$ as a starting material. [22d] This mixed chloride pyridonate is sufficiently robust to tolerate elevated temperatures and most importantly, it can handle challenging substrates such as internal alkenes. [22d] To date, this *N,O*-chelated complex is the only example of hydroaminoalkylation (or C—H alkylation) using internal alkenes to give branched products. The increased accessibility to the metal center in this complex is postulated to result in a system that can handle the increased steric demands of cyclic alkenes and both *cis* and *trans* internal alkenes. [22d]

Catalytic challenges that are currently being targeted in intermolecular hydroaminoalkylation include the development of catalyst systems that require only modest reaction temperatures for the direct alkylation of *N*-heterocycles. Another area of interest is the preparation of new chiral complexes suitable for enantioselective transformations. The development of a broadly useful, yet regioselective and stereoselective catalyst will be of interest to the synthetic community.

An interesting observation in these investigations is the fact that intermolecular hydroaminoalkylation catalysts are not viable for application in intramolecular reactions. Furthermore, there are no catalysts that are effective for the intramolecular hydroaminoalkylation of secondary aminoalkenes and all high yielding examples of intramolecular hydroaminoalkylation are restricted to primary aminoalkenes. Meanwhile, intermolecular hydroaminoalkylation with primary amines has not yet been reported.

Early examples of intramolecular hydroaminoalkylation were reported as side-products accessed unexpectedly during hydroamination investigations using group 4

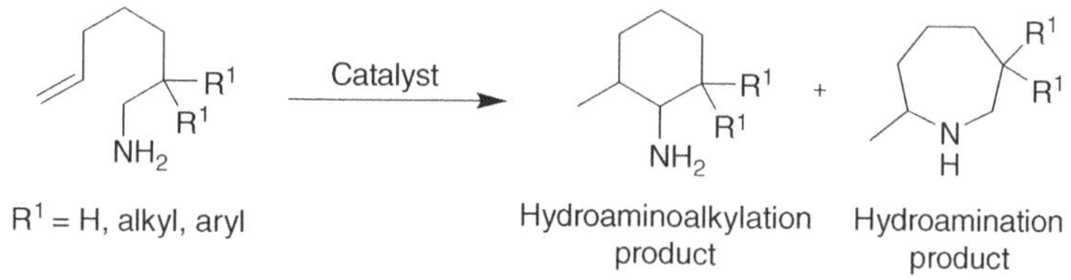

Scheme 23 *Cyclization of primary aminoalkenes can result in mixtures of hydroamino-alkylation and hydroamination products*

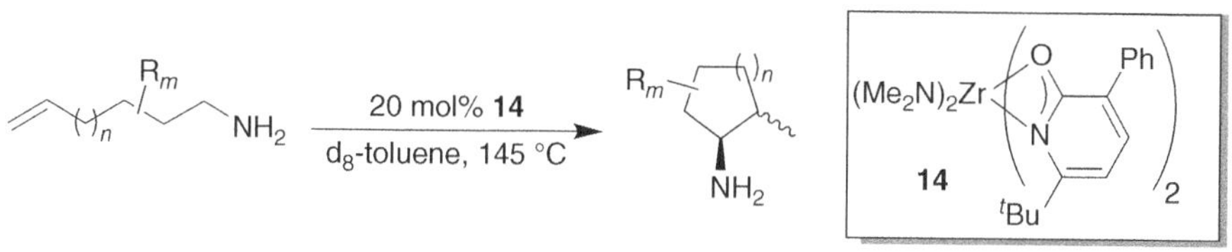

Scheme 24 *Bis(2-pyridonate) Zr-catalyzed hydroaminoalkylation*

metals. [82] In fact, when intermolecular hydroaminoalkylation catalysts are explored for intramolecular hydroaminoalkylation, hydroamination products result (Scheme 23). [44,83] Early efforts to explore the application of *N,O*-chelated early transition metal complexes using 2-pyridonate ligated zirconium metal complexes resulted in mixtures of hydroamination and hydroaminoalkylation products. [72] The ratio of hydroamination to hydroaminoalkylation products is substrate controlled, such that substrates with reduced activation barriers to cyclization undergo hydroamination to give a racemic mixture of substituted piperidines and azapanes, while other more challenging substrates undergo hydroaminoalkylation to give amine-substituted cyclopentanes and hexanes as a mixture of diastereomers (Scheme 24).

As mentioned above, only primary aminoalkenes are suitable substrates for hydroaminoalkylation and it is known that many group 4 complexes invoke catalytically active imido complexes for hydroamination. Such imido complexes are also known to access bridged bonding motifs, which are commonly observed in a broad range of early transition metal complexes, including some examples of rigorously characterized *N,O*-chelated early transition metal bridged imido complexes (Figure 24). [37] Furthermore, such bridged species are proposed to be catalytically inactive for hydroamination. However, an unexpected hydrogen abstraction from such a bridged imido resulted in the preparation of a bridged metallaziridine complex (Figure 25). [72]

The formation of such bridged metallaziridine species rationalizes the selectivity for primary amines and suggests that dimeric species may be key catalytic intermediates. This is further supported by experiments that illustrated that increasing catalyst concentration results in an increase in hydroaminoalkylation product versus hydroamination product. [72] Therefore, we proposed that catalyst-controlled chemoselectivity for hydroaminoalkylation (C—C bond formation) versus hydroamination (C—N bond formation), could be achieved by designing catalyst systems that promote the formation of bridged species.

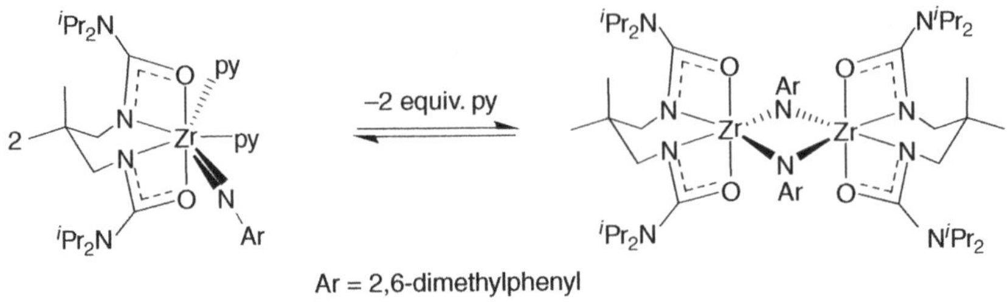

Figure 24 *Equilibria between terminal and bridged imido species can be observed in solution phase*

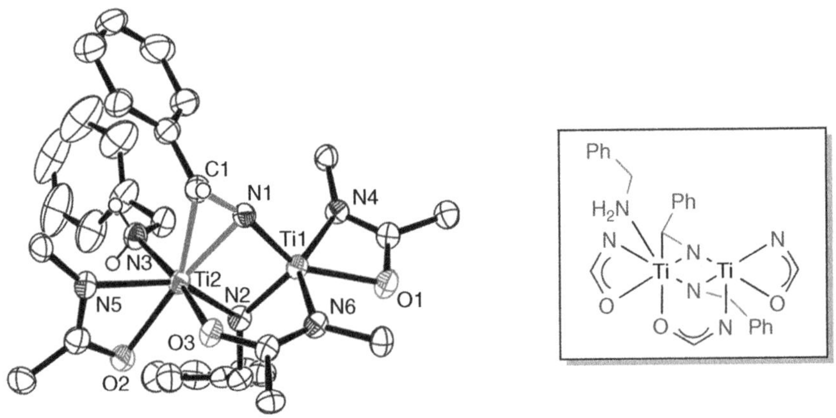

Figure 25 *A bridged metallaziridine with another bridging imido, μ-amidate and two terminal κ²-amidate ligands. (See insert for color/color representation of this figure)*

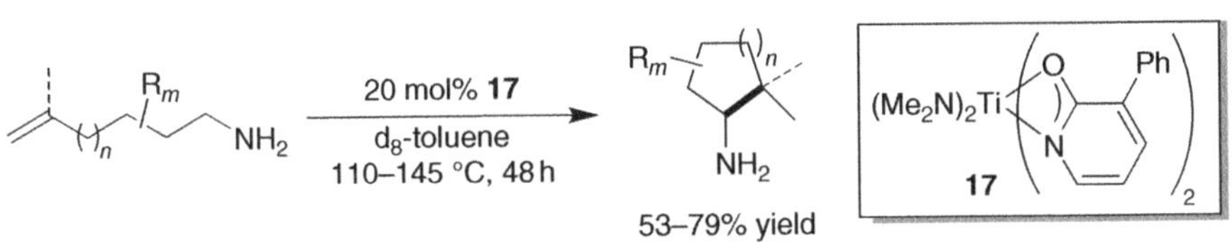

Scheme 25 *Intramolecular hydroaminoalkylation catalyst with excellent chemoselectivity*

To this end, a titanium bis(2-pyridonate) complex (**17**) has been developed. [22c] These catalytic systems promote catalyst-controlled selectivity for hydroaminoalkylation over hydroamination and for the first time amine-substituted cyclopentanes can be prepared preferentially over piperidine hydroamination products (Scheme 25).

The working hypothesis for this catalyst-controlled selectivity is presented in Scheme 26. Catalyst systems that favor the formation of terminal imido complexes are proposed to promote the formation of hydroamination products with substrates that can do either hydroamination or hydroaminoalkylation.

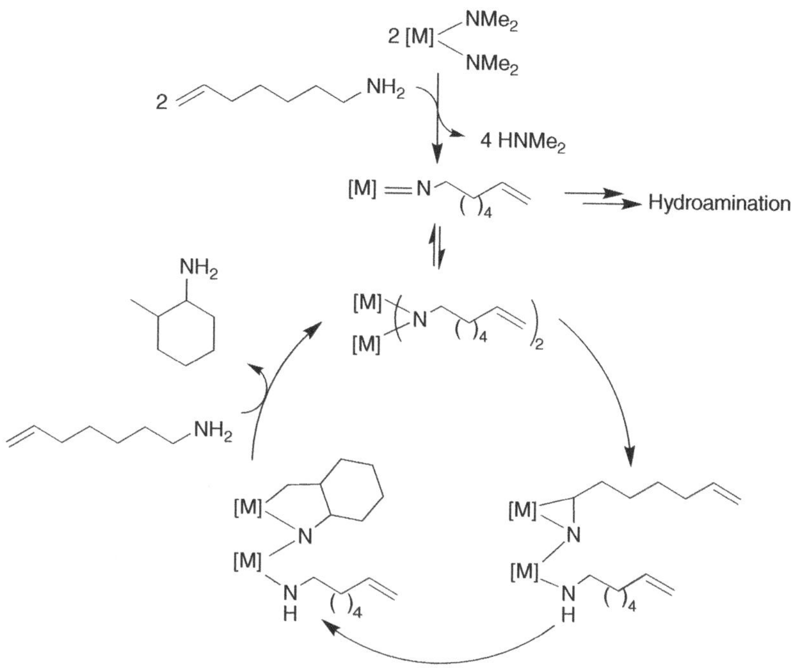

Scheme 26 *Simplified proposed mechanism for hydroaminoalkylation promoted by dimeric complexes via a bridging metallaziridine*

Intramolecular hydroaminoalkylation presents a unique challenge in that the two targeted reactions, hydroamination and hydroaminoalkylation, are proposed to be accessed from the same reactive intermediate. Mechanistic investigations provide insight in how to toggle between the two reactivity profiles. The flexibility of the *N,O*-chelating ligands, with their easily tunable features, provide opportunity for designing catalyst systems with specific structural features and in this case, promoting the formation of dimeric species in situ. This is another example of how these simple ligands that are easily prepared can be used to advantage in targeting selected transformations.

13.2.2.3 Hydroalkynylation – enyne formation

Terminal alkynes can undergo dimerization via hydroalkynylation to give enynes. [14] While many metal-based catalysts are known to mediate such reactivity, [84] control of regiochemistry and stereochemistry remains a challenge, as many reactions give mixtures of products (Scheme 27). Interestingly, the Zr complex **10** that was originally developed for hydroamination catalysis was noted to dimerize phenylacetylene to give selectively the *Z*-enyne product in high yield when aniline was used as an additive in the reaction (Scheme 28). [14] In the absence of amine additive there is no catalytic reaction.

The selective formation of specifically *Z*-enynes has been rarely reported, [84b, 85] and presents an interesting mechanistic question, as previous catalytic syntheses of

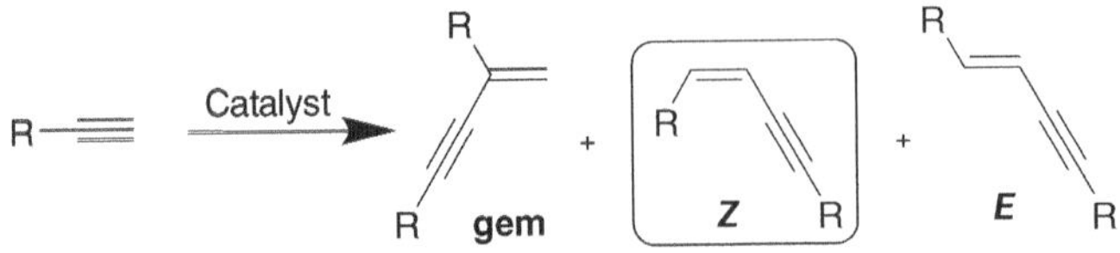

Scheme 27 *Catalytic synthesis of enynes from terminal alkynes can result in the formation of mixtures of products*

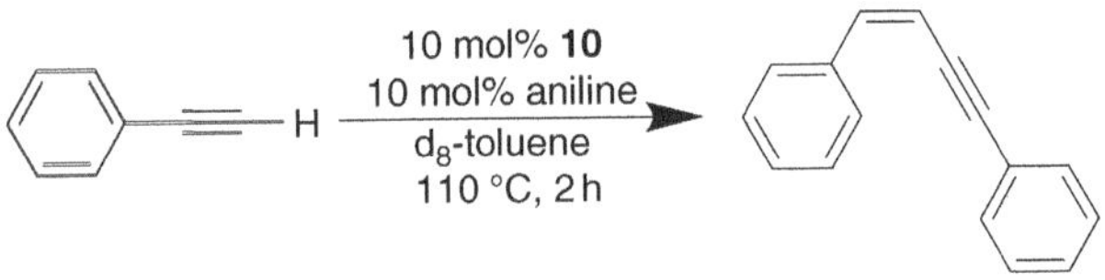

Scheme 28 *Selective Z-enyne formation using Zr ureate catalyst **10** with aniline as an additive*

enynes using group 4 metal catalysts have yielded mixtures of products with *gem*-enynes being preferentially formed. [84e, 84f] Investigations of this transformation are on-going and preliminary work suggests that dimeric bridging imido species may be key reactive intermediates here as well. [14] Similar dimeric species have been postulated as key reactive intermediates in the selective formation of Z-enynes with rare earth metal catalysts. [84b]

Z-Enynes are useful synthetic building blocks that can be assembled in an atom-economic fashion by hydrofunctionalization. Future work on this transformation aims to develop further mechanistic insight and use these observations to enhance catalyst development efforts. With a more detailed understanding of the nature of the Zr—C bonded intermediates involved in this transformation and the exact role of the aniline additive, alternative reactivity profiles will be targeted.

13.3 Conclusions

The *N,O*-chelating ligands such as amidates, ureates, phosphoramidates, sulfonamidates, and pyridonates represent modular ligands with easily varied steric and electronic properties. While the ligands themselves are derived from common organic small molecules, this family of early transition metal complexes has only been carefully investigated over the past decade. Fundamental insights into the limitations in preparing complexes, and the complicated equilibria that have been observed for these systems have been established in fundamental coordination chemistry investigations. Organometallic reactivity investigations have provided foundational insights toward reactivity trends with these coordinatively flexible complexes. The ease of synthesis of a variety of related ligands allows for facile tuning and optimization of catalytic reactivity. Applications of these systems range from terminal alkene and cyclic ester ROPs through to various hydrofunctionalization transformations. Notably, these complexes have been used to address challenges in the catalytic synthesis of selectively substituted amines from simple amine and alkene starting materials. Such 100% atom-economic transformations are attractive targets for developing new synthetic methodologies.

References

[1] R. H. Crabtree, *The Organometallic Chemistry of the Transition Metals*, John Wiley & Sons, Inc., Hoboken, NJ, **2005**, pp. 125–158.

[2] (a) I. R. Butler, *Specialist Periodical Reports, Volume 30: Organometallic Chemistry*, Royal Society of Chemistry, Cambridge, **2002**, pp. 385–434; (b) R. L. Halterman, *Chem. Rev.* **1992**, *92*, 965–994.

[3] G. J. P. Britovsek, V. C. Gibson, D. F. Wass, *Angew. Chem. Int. Ed.* **1999**, *38*, 428–447.

[4] (a) H. Makio, N. Kashiwa, T. Fujita, *Adv. Synth. Catal.* **2002**, *344*, 477–493; (b) H. Makio, T. Fujita, *Acc. Chem. Res.* **2009**, *42*, 1532–1544; (c) H. Makio, H. Terao, A. Iwashita, T. Fujita, *Chem. Rev.* **2011**, *111*, 2363–2449.

[5] (a) S. Doherty, R. J. Errington, N. Housley, J. Ridland, W. Clegg, M. R. J. Elsegood, *Organometallics* **1999**, *18*, 1018–1029; (b) G. Xie, Y. Li, J. Sun, C. Qian, *Inorg. Chem. Commun.* **2009**, *12*, 796–799.

[6] J. M. Hoover, J. R. Petersen, J. H. Pikul, A. R. Johnson, *Organometallics* **2004**, *23*, 4614–4620.

[7] T. Tsukahara, D. C. Swenson, R. F. Jordan, *Organometallics* **1997**, *16*, 3303–3313.

[8] (a) J. Unruangsri, H. Morgan, A. D. Schwarz, A. D. Schofield, P. Mountford, *Organometallics* **2013**, *32*, 3091–3107; (b) R. Baumann, W. M. Davis, R. R. Schrock, *J. Am. Chem. Soc.* **1997**, *119*, 3830–3831; (c) R. O. Ayinla, L. L. Schafer, *Dalton Trans.* **2011**, *40*, 7769–7776.

[9] L. L. Schafer, J. C. H. Yim, N. Yonson, in *Metal-Catalyzed Cross-Coupling Reactions and More* (eds A. de Meijere, S. Bräse, M. Oestreich), Wiley-VCH Verlag GmbH, Weinheim, **2014**, pp. 1135–1258.

[10] (a) G. R. Giesbrecht, A. Shafir, J. Arnold, *Inorg. Chem.* **2001**, *40*, 6069–6072; (b) S. Liu, W.-H. Sun, Y. Zeng, D. Wang, W. Zhang, Y. Li, *Organometallics* **2010**, *29*, 2459–2464; (c) M. B. Harkness, E. Alvarado, A. C. Badaj, B. C. Skrela, L. Fan, G. G. Lavoie, *Organometallics* **2013**, *32*, 3309–3321.

[11] L. J. E. Stanlake, J. D. Beard, L. L. Schafer, *Inorg. Chem.* **2008**, *47*, 8062–8068.

[12] J. C. H. Yim, L. L. Schafer, *Eur. J. Org. Chem.* **2014**, 6825–6840.

[13] E. Chong, P. Garcia, L. L. Schafer, *Synthesis* **2014**, *46*, 2884–2896.

[14] R. H. Platel, L. L. Schafer, *Chem. Commun.* **2012**, *48*, 10609–10611.

[15] F. T. Edelmann, in *Advances Organometallic Chemistry, Vol. 61* (eds F. H. Anthony, J. F. Mark), Academic Press, New York, **2013**, pp. 55–374.

[16] (a) E. Chong, S. Qayyum, L. L. Schafer, R. Kempe, *Organometallics* **2013**, *32*, 1858–1865; (b) E. Smolensky, M. Kapon, M. S. Eisen, *Organometallics* **2007**, *26*, 4510–4527; (c) G. Zi, F. Zhang, X. Liu, L. Ai, H. Song, *J. Organomet. Chem.* **2010**, *695*, 730–739; (d) J. Dörfler, S. Doye, *Angew. Chem. Int. Ed.* **2013**, *52*, 1806–1809.

[17] R. K. Thomson, F. E. Zahariev, Z. Zhang, B. O. Patrick, Y. A. Wang, L. L. Schafer, *Inorg. Chem.* **2005**, *44*, 8680–8689.

[18] R. K. Thomson, B. O. Patrick, L. L. Schafer, *Can. J. Chem.* **2005**, *83*, 1037–1042.

[19] C. Li, R. K. Thomson, B. Gillon, B. O. Patrick, L. L. Schafer, *Chem. Commun.* **2003**, 2462–2463.
[20] A. V. Lee, L. L. Schafer, *Eur. J. Inorg. Chem.* **2007**, 2243–2255.
[21] (a) D. C. Leitch, P. R. Payne, C. R. Dunbar, L. L. Schafer, *J. Am. Chem. Soc.* **2009**, *131*, 18246–18247; (b) D. C. Leitch, J. D. Beard, R. K. Thomson, V. A. Wright, B. O. Patrick, L. L. Schafer, *Eur. J. Inorg. Chem.* **2009**, 2691–2701; (c) J. M. P. Lauzon, L. L. Schafer, *Z. Anorg. Allg. Chem.* **2015**, *641*, 128–135.
[22] (a) J. M. Rawson, R. E. P. Winpenny, *Coord. Chem. Rev.* **1995**, *139*, 313–374; (b) R. L. Webster, N. Noroozi, S. G. Hatzikiriakos, J. A. Thomson, L. L. Schafer, *Chem. Commun.* **2013**, *49*, 57–59; (c) E. Chong, L. L. Schafer, *Org. Lett.* **2013**, *15*, 6002–6005; (d) E. Chong, J. W. Brandt, L. L. Schafer, *J. Am. Chem. Soc.* **2014**, *136*, 10898–10901.
[23] P. Garcia, Y. Y. Lau, M. R. Perry, L. L. Schafer, *Angew. Chem. Int. Ed.* **2013**, *52*, 9144–9148.
[24] (a) P. J. Walsh, *Acc. Chem. Res.* **2003**, *36*, 739–749; (b) F. Zhang, H. Song, G. Zi, *J. Organomet. Chem.* **2010**, *695*, 1993–1999.
[25] G. Chandra, A. D. Jenkins, M. F. Lappert, R. C. Srivastava, *J. Chem. Soc. A* **1970**, 2550–2558.
[26] P. Braunstein, D. Nobel, *Chem. Rev.* **1989**, *89*, 1927–1945.
[27] H. Wang, H. S. Chan, Z. W. Xie, *Organometallics* **2005**, *24*, 3772–3779.
[28] D. C. Leitch, PhD thesis, University of British Columbia (Vancouver), **2010** DOI: 10.14288/1.0060312.
[29] L. J. E. Stanlake, PhD thesis, University of British Columbia (Vancouver), **2008** DOI: 10.14288/1.0061568.
[30] (a) D. C. Bradley, I. M. Thomas, *Can. J. Chem.* **1962**, *40*, 449–454; (b) R. R. Schrock, *J. Organomet. Chem.* **1976**, *122*, 209–225.
[31] P. Eisenberger, R. O. Ayinla, J. M. P. Lauzon, L. L. Schafer, *Angew. Chem. Int. Ed.* **2009**, *48*, 8361–8365.
[32] (a) J. C. Jeffrey, T. B. Rauchfuss, *Inorg. Chem.* **1979**, *18*, 2658–2666; (b) P. Braunstein, F. Naud, *Angew. Chem. Int. Ed.* **2001**, *40*, 680–699; (c) M. Bassetti, *Eur. J. Inorg. Chem.* **2006**, *2006*, 4473–4482; (d) C. S. Slone, D. A. Weinberger, C. A. Mirkin, *Prog. Inorg. Chem.* **1999**, *48*, 233–350.
[33] (a) A. Bader, E. Lindner, *Coord. Chem. Rev.* **1991**, *108*, 27–110; (b) P. Espinet, K. Soulantica, *Coord. Chem. Rev.* **1999**, *193–195*, 499–556; (c) Z. Weng, S. Teo, T. S. A. Hor, *Acc. Chem. Res.* **2007**, *40*, 676–684.
[34] P. R. Payne, R. K. Thomson, D. M. Medeiros, G. Wan, L. L. Schafer, *Dalton Trans* **2013**, *42*, 15670–15677.
[35] J. M. P. Lauzon, PhD thesis, University of British Columbia (Vancouver), **2013**, DOI: 10.14288/1.0073860.
[36] D. C. Leitch, L. L. Schafer, *Organometallics* **2010**, *29*, 5162–5172.
[37] D. C. Leitch, R. H. Platel, L. L. Schafer, *J. Am. Chem. Soc.* **2011**, *133*, 15453–15463.
[38] A. Antiñolo, F. Carrillo-Hermosilla, J. Fernández-Baeza, A. M. Fernández de Toro, S. García-Yuste, A. Otero, J. C. Pérez-Flores, A. M. Rodríguez, *Inorg. Chim. Acta* **2000**, *300–302*, 131–137.
[39] (a) R. Fandos, C. Hernandez, A. Otero, A. Rodriguez, M. J. Ruiz, P. Terreros, *Eur. J. Inorg. Chem.* **2003**, 493–498; (b) A. Antiñolo, F. Carrillo-Hermosilla, A. E. Corrochano, R. Fandos, J. Fernández-Baeza, A. M. Rodríguez, M. J. Ruiz, A. Otero, *Organometallics* **1999**, *18*, 5219–5224; (c) F. A. Cotton, G. E. Lewis, G. N. Mott, *Inorg. Chem.* **1983**, *22*, 378–382.
[40] D. J. Gilmour, R. L. Webster, M. R. Perry, L. L. Schafer, *Dalton Trans.* **2015**, *44*, 12411–12419.

[41] E. Chong, PhD thesis, University of British Columbia (Vancouver), **2014** DOI: 10.14288/1.0167040.
[42] (a) H. Takahashi, T. Kawakita, M. Yoshioka, S. Kobayashi, M. Ohno, *Tetrahedron Lett.* **1989**, *30*, 7095–7098; (b) H. Takahashi, T. Kawakita, M. Ohno, M. Yoshioka, S. Kobayashi, *Tetrahedron* **1992**, *48*, 5691–5700; (c) C. Lutz, P. Knochel, *J. Org. Chem.* **1997**, *62*, 7895–7898; (d) S. Vettel, C. Lutz, A. Diefenbach, G. Haderlein, S. Hammerschmidt, K. Kühling, M. R. Mofid, T. Zimmermann, P. Knochel, *Tetrahedron: Asymmetry* **1997**, *8*, 779–800.
[43] (a) L. T. Armistead, P. S. White, M. R. Gagné, *Organometallics* **1998**, *17*, 216–220; (b) S. Hamura, T. Oda, Y. Shimizu, K. Matsubara, H. Nagashima, *J. Chem. Soc., Dalton Trans.* **2002**, 1521–1527.
[44] F. Zhang, H. Song, G. Zi, *Dalton Trans.* **2011**, *40*, 1547–1566.
[45] F. G. Bordwell, *Acc. Chem. Res.* **1988**, *21*, 456–463.
[46] P. Galli, G. Vecellio, *Prog. Polym. Sci.* **2001**, *26*, 1287–1336.
[47] K. N. Ziegler, G. Natta, *Nobel Lectures, Chemistry 1963–1970*, Elsevier, Amsterdam, **1972**.
[48] E. Zurek, T. Ziegler, *Prog. Polym. Sci.* **2004**, *29*, 107–148.
[49] C. Vasile (ed.), *Handbook of Polyolefins*, 2nd edn, CRC Press, New York, NY, **2000**.
[50] S. Dutta, W.-C. Hung, B.-H. Huang, C.-C. Lin, in *Synthetic Biodegradable Polymers, Vol. 245* (eds B. Rieger, A. Künkel, G. W. Coates, R. Reichardt, E. Dinjus, T. A. Zevaco), Springer Berlin, **2012**, pp. 219–283.
[51] A. Sauer, A. Kapelski, C. Fliedel, S. Dagorne, M. Kol, J. Okuda, *Dalton Trans.* **2013**, *42*, 9007–9023.
[52] (a) N. Noroozi, J. Thomson, N. Noroozi, L. Schafer, S. Hatzikiriakos, *Rheol Acta* **2012**, *51*, 179–192; (b) F. Zhang, J. Zhang, H. Song, G. Zi, *Inorg. Chem. Commun.* **2011**, *14*, 72–74; (c) Q. Wang, F. Zhang, H. Song, G. Zi, *J. Organomet. Chem.* **2011**, *696*, 2186–2192; (d) W. Zhang, S. Liu, W.-H. Sun, X. Hao, C. Redshaw, *New J. Chem.* **2012**, *36*, 2392–2396; (e) X. Hu, C. Lu, B. Wu, H. Ding, B. Zhao, Y. Yao, Q. Shen, *J. Organomet. Chem.* **2013**, *732*, 92–101; (f) A. D. Schwarz, A. L. Thompson, P. Mountford, *Inorg. Chem.* **2009**, *48*, 10442–10454; (g) A. D. Schwarz, K. R. Herbert, C. Paniagua, P. Mountford, *Organometallics* **2010**, *29*, 4171–4188.
[53] T. Matsugi, T. Fujita, *Chem. Soc. Rev.* **2008**, *37*, 1264–1277.
[54] (a) D. V. Gribkov, K. C. Hultzsch, *Angew. Chem. Int. Ed.* **2004**, *43*, 5542–5546; (b) L. J. E. Stanlake, L. L. Schafer, *Organometallics* **2009**, *28*, 3990–3998; (c) F. Pohlki, S. Doye, *Chem. Soc. Rev.* **2003**, *32*, 104–114; (d) R. Severin, S. Doye, *Chem. Soc. Rev.* **2007**, *36*, 1407–1420.
[55] (a) G. Zi, F. Zhang, H. Song, *Chem. Commun.* **2010**, *46*, 6296–6298; (b) P. W. Roesky, *Angew. Chem. Int. Ed.* **2009**, *48*, 4892–4894.
[56] (a) A. L. Reznichenko, K. C. Hultzsch, in *Science of Synthesis Reference Library: Stereoselective Synthesis* (ed. J. G. de Vries), Thieme, Stuttgart, **2010**, pp. 689–729; (b) T. E. Müller, K. C. Hultzsch, M. Yus, F. Foubelo, M. Tada, *Chem. Rev.* **2008**, *108*, 3795–3892.
[57] T. E. Müller, M. Beller, *Chem. Rev.* **1998**, *98*, 675–704.
[58] I. Bytschkov, S. Doye, *Eur. J. Org. Chem.* **2003**, 935–946.
[59] (a) Z. Zhang, L. L. Schafer, *Org. Lett.* **2003**, *5*, 4733–4736; (b) Z. Zhang, D. C. Leitch, M. Lu, B. O. Patrick, L. L. Schafer, *Chem. Eur. J.* **2007**, *13*, 2012–2022.
[60] J. C. Yim, J. A. Bexrud, R. O. Ayinla, D. C. Leitch, L. L. Schafer, *J. Org. Chem.* **2014**, *79*, 2015–2028.

[61] A. V. Lee, L. L. Schafer, *Synlett* **2006**, 2973–2976.

[62] H. Zhai, A. Borzenko, Y. Y. Lau, S. H. Ahn, L. L. Schafer, *Angew. Chem. Int. Ed.* **2012**, *51*, 12219–12223.

[63] Sigma-Aldrich. Schafer's Ti-Amidate Catalyst - L511749, http://www.sigmaaldrich.com/catalog/product/aldrich/l511749 (accessed April 13, 2016).

[64] (a) P. J. Walsh, A. M. Baranger, R. G. Bergman, *J. Am. Chem. Soc.* **1992**, *114*, 1708–1719; (b) F. Pohlki, S. Doye, *Angew. Chem. Int. Ed.* **2001**, *40*, 2305–2308; (c) J. S. Johnson, R. G. Bergman, *J. Am. Chem. Soc.* **2001**, *123*, 2923–2924; (d) Y. H. Li, Y. H. Shi, A. L. Odom, *J. Am. Chem. Soc.* **2004**, *126*, 1794–1803; (e) R. K. Thomson, J. A. Bexrud, L. L. Schafer, *Organometallics* **2006**, *25*, 4069–4071; (f) A. L. Gott, A. J. Clarke, G. J. Clarkson, P. Scott, *Chem. Commun.* **2008**, 1422–1424; (g) L. E. N. Allan, G. J. Clarkson, D. J. Fox, A. L. Gott, P. Scott, *J. Am. Chem. Soc.* **2010**, *132*, 15308–15320.

[65] (a) M. C. Wood, D. C. Leitch, C. S. Yeung, J. A. Kozak, L. L. Schafer, *Angew. Chem. Int. Ed.* **2007**, *46*, 354–358; (b) M. C. Wood, D. C. Leitch, C. S. Yeung, J. A. Kozak, L. L. Schafer, *Angew. Chem. Int. Ed.* **2009**, *48*, 6937.

[66] (a) L. Ackermann, R. G. Bergman, *Org. Lett.* **2002**, *4*, 1475–1478; (b) L. Ackermann, R. G. Bergman, R. N. Loy, *J. Am. Chem. Soc.* **2003**, *125*, 11956–11963.

[67] J. A. Bexrud, J. D. Beard, D. C. Leitch, L. L. Schafer, *Org. Lett.* **2005**, *7*, 1959–1962.

[68] D. A. Watson, M. Chiu, R. G. Bergman, *Organometallics* **2006**, *25*, 4731–4733.

[69] R. O. Ayinla, T. Gibson, L. L. Schafer, *J. Organomet. Chem.* **2011**, *696*, 50–60.

[70] (a) G. Zi, *J. Organomet. Chem.* **2011**, *696*, 68–75; (b) G. Zi, X. Liu, L. Xiang, H. Song, *Organometallics* **2009**, *28*, 1127–1137; (c) A. L. Gott, A. J. Clarke, G. J. Clarkson, P. Scott, *Organometallics* **2007**, *26*, 1729–1737.

[71] S. Tobisch, *Inorg. Chem.* **2012**, *51*, 3786–3795.

[72] J. A. Bexrud, P. Eisenberger, D. C. Leitch, P. R. Payne, L. L. Schafer, *J. Am. Chem. Soc.* **2009**, *131*, 2116–2118.

[73] (a) I. Prochnow, R. Kubiak, O. N. Frey, R. Beckhaus, S. Doye, *ChemCatChem* **2009**, *1*, 162–172; (b) R. Kubiak, I. Prochnow, S. Doye, *Angew. Chem. Int. Ed.* **2009**, *48*, 1153–1156; (c) R. Kubiak, I. Prochnow, S. Doye, *Angew. Chem. Int. Ed.* **2010**, *49*, 2626–2629; (d) T. Preuß, W. Saak, S. Doye, *Chem. Eur. J.* **2013**, *19*, 3833–3837.

[74] (a) W. A. Nugent, D. W. Ovenall, S. J. Holmes, *Organometallics* **1983**, *2*, 161–162; (b) M. G. Clerici, F. Maspero, *Synthesis* **1980**, *1980*, 305–306.

[75] S. B. Herzon, J. F. Hartwig, *J. Am. Chem. Soc.* **2007**, *129*, 6690–6691.

[76] A. L. Reznichenko, K. C. Hultzsch, *J. Am. Chem. Soc.* **2012**, *134*, 3300–3311.

[77] J. M. Lauzon, L. L. Schafer, *Dalton Trans* **2012**, *41*, 11539–11550.

[78] S. B. Herzon, J. F. Hartwig, *J. Am. Chem. Soc.* **2008**, *130*, 14940–14941.

[79] P. R. Payne, P. Garcia, P. Eisenberger, J. C. H. Yim, L. L. Schafer, *Org. Lett.* **2013**, *15*, 2182–2185.

[80] P. Garcia, P. R. Payne, E. Chong, R. L. Webster, B. J. Barron, A. C. Behrle, J. A. R. Schmidt, L. L. Schafer, *Tetrahedron* **2013**, *69*, 5737–5743.

[81] D. Jaspers, W. Saak, S. Doye, *Synlett* **2012**, *23*, 2098–2102.

[82] (a) C. Müller, W. Saak, S. Doye, *Eur. J. Org. Chem.* **2008**, *2008*, 2731–2739; (b) J. A. Bexrud, L. L. Schafer, *Dalton Trans.* **2010**, *39*, 361–363.

[83] A. L. Reznichenko, T. J. Emge, S. Audörsch, E. G. Klauber, K. C. Hultzsch, B. Schmidt, *Organometallics* **2011**, *30*, 921–924.

[84] (a) S. Ge, A. Meetsma, B. Hessen, *Organometallics* **2009**, *28*, 719–726; (b) M. Nishiura, Z. Hou, Y. Wakatsuki, T. Yamaki, T. Miyamoto, *J. Am. Chem. Soc.* **2003**, *125*, 1184–1185; (c) K. Komeyama, T. Kawabata, K. Takehira, K. Takaki, *J. Org. Chem.* **2005**, *70*,

7260–7266; (d) A. Haskel, T. Straub, A. K. Dash, M. S. Eisen, *J. Am. Chem. Soc.* **1999**, *121*, 3014–3024; (e) K. Mach, R. Gyepes, M. Horáček, L. Petrusová, J. Kubišta, *Collect. Czech. Chem. Commun.* **2003**, *68*, 1877–1896; (f) M. Horáček, I. Císařová, J. Čejka, J. Karban, L. Petrusová, K. Mach, *J. Organomet. Chem.* **1999**, *577*, 103–112; (g) A. D. Horton, *J. Chem. Soc., Chem. Commun.* **1992**, 185–187; (h) M. Akita, H. Yasuda, A. Nakamura, *Bull. Chem. Soc. Jpn.* **1984**, *57*, 480–487; (i) M. Yoshida, R. F. Jordan, *Organometallics* **1997**, *16*, 4508–4510; (j) G. C. Midya, S. Paladhi, K. Dhara, J. Dash, *Chem. Commun.* **2011**, *47*, 6698–6700; (k) H. Katayama, H. Yari, M. Tanaka, F. Ozawa, *Chem. Commun.* **2005**, 4336–4338.

[85] Y. Liu, M. Nishiura, Y. Wang, Z. Hou, *J. Am. Chem. Soc.* **2006**, *128*, 5592–5593.

Index

σ-acceptor 243–244
acceptor sites, ambiphilic ligands 238–239
acetonitrile 258
acetophenone 209
  catalyst activity 224
  derivatives 211
achiral catalysts, aluminum 287
acid base concept
  Lewis, *see* Lewis …
  Pearson's 8
acid esters, β-arylamino 97
acid-labile groups 58
acidity, metal hydride complexes 208
acrylonitrile, CM 28
activation
  cooperative 258–261
  N_2 and CO 262–264
  reversible 261
active transmetalating species 166–168
activity
  ATH 224–226
  structure–activity relationship 26, 144
"actor ligands" 176–179
acyclic diaminocarbene-based catalysts 31
acyclic diene metathesis (ADMET) 18
adamantyl-activated catalysts 37, 39
adamantyl-based chelates 41
adamantylazide 197
addition
  anti-Markovnikov 324
  ATRA 324–325
  carbenes 312–314
  nitrenes 319–321
  oxidative, *see* oxidative addition
aerobic oxidation, benzyl alcohol 188
Ag, *see* silver
air-stable ligands 109–112, 115–116, 253
alcohols
  benzyl 188
  functionalization 314
  primary 194
  secondary 193
aldehydes
  epoxidation 323
  Pd-catalyzed umpolung allylation 87
  "protected" phosphine 220
  racemic 2-substituted 81–82
alkanes
  dehydrogenation 324
  transfer hydrogenation 258–261
alkenes
  carboamination 98
  catalytic hydrogenation 261
  hydroamination 384, 387
  polymerization 377–379
alkoxide bis(pyridonate) complexes 374
alkoxide initiator 273

Ligand Design in Metal Chemistry: Reactivity and Catalysis, First Edition. Edited by Mark Stradiotto and Rylan J. Lundgren.

alkyl–alkyl Negishi coupling 165–167
alkyl–alkyl Suzuki–Miyaura coupling 156
alkyl complexes 333
alkyl heterocycles 170
N'-alkyl NHCs 28
alkyl precatalyst, tantalum 394
(alkyl)(amino)carbene-based catalysts, cyclic 30
alkylation
diastereoselective C-H 393
1,4-dibromobenzene 165
direct C-H 393
alkylic C-H bonds 322
alkylidene ligands 38
alkylidenes 350
alkyls, homoleptic Ln(II) 334
alkylzinc reagents 168
alkynes
catalytic hydrogenation 261
[2 + 2] cycloaddition 386
cyclopropenation 314
enyne formation 399
hydroamination 384
reductive coupling 88
allenes 97–98
allylation 87
allylbenzene 24, 28–29, 31
allylboronic acid 158
allylic amination 95–97
allylic cyclization 97–98
alternative NHC ligands 29–31
aluminum, achiral catalysts 287
aluminum complexes 277–278, 285–286
salen 285–286
Alzheimer's disease treatment, rivastigmine 81
ambiphilic ligands 237–269
coordination 242–251
general structure 238–239
amidate complexes
tantalum 372
titanium 383
Y 381
amidate ligands 369–372
amidates 363–404
amide proligands 372
amido complexes 209
bis(pyridonate) 374
amidophenolato dianion 199
amination
BHA 104–105, 108–112
Pd-catalyzed allylic 95–97
sp^3 C-H 200
amine-aldehyde components 215
amine derivatives 135
amines
aromatic 92
α-chiral 388
functionalization 314
secondary 391
aminoalkenes 387–388, 396
aminobiphenyl palladacycles 112
aminophosphine ligands 77
ammonia, monoarylation selectivity 108–117
amorphous PLA 271
ancillary ligands 4
C(sp^2)-bonding 105–107
imidazole-based 120
in situ arylation 128
modification 125–127
N,O-chelates 366
"repurposing" 110
Andersson's quadrant model 60
anilines
as synthons 117
diphenyl hydrazine disproportionation 189
pharmaceutical 108
anionic ligands, variation 31–32, 38–40
ansa-metallocenes 338
anti-bacterial pharmaceuticals 123
anti-depressants 117
anti geometry 22
anti-Markovnikov addition 324
anti-metallacycles 42
apical position 220
arenes, zerovalent 342
aromatic amines 92
aromatic C-H bonds 322
direct oxidation 324
aryl bromides 148
aryl groups
variation 40–41
fluorinated 123
aryl halides 104, 123–129
KTC coupling 147–148

β-aryl β-ketoesters 80
N-aryl NHCs 28
aryl pseudohalides 104
  nucleophilic fluorination 123–129
aryl transfer 125
aryl triflates 124–125
β-arylamino acid esters 97
arylation
  in situ 128
  Rh-catalyzed 86–87
  Suzuki–Miyaura 156
2-arylaziridines 156
arylboranes 261–262
asymmetric catalysis 66–103
asymmetric cyclopropanation, styrene 314
asymmetric direct hydrogenation (ADH) 205, 212
  catalytic properties 230
asymmetric hydrogenation 205–236
  Ir-catalyzed 46–65
  rhodium-catalyzed 47
  spiro ligands 73–84
asymmetric Suzuki–Miyaura coupling 153–155
asymmetric transfer hydrogenation (ATH) 206–207, 212
  activity 224–226
  catalytic properties 229–230
atom transfer radical reactions 324–326
atorvastatin 123
Au, *see* gold
auto-aggregation prevention 77
axial chiral biarylphosphite moiety 54
2-azabiciclo[3.3.1] nonanes 326
azides
  organic 191
  unactivated 200
aziridination 319–320
  furans 321
azlactones 89
azobenzene 189

B-H bond insertion 95
backbones
  Binap 299
  cyclohexyl 298
  fused hetero spiro 68
  modification 139–140
  spiro[4.4]nona-1,6-diene 72
base-sensitive functionalities 114
bases, Lewis, *see* Lewis bases
bent geometries, lanthanides 334
benzene 313, 340–342
benzo spiro rings 71–72
benzoquinone 192
benzyl alcohol, aerobic oxidation 188
BHA (Buchwald–Hartwig amination) 104–105, 108–112
biaryl ligands 69
biaryl monophosphine ligands 112–113
biaryl monophosphines, XPhos-type 118
biaryl phosphines, Suzuki–Miyaura coupling 253–255
biaryl products, *ortho*-substituted, *see* *ortho*-substituted biaryl products
biarylphosphite moiety 54
bicyclic pyridine-phosphinite ligands 61–62
bidentate *N,O*-chelates 364
Binap backbones 299
binaphthyl compounds 153–155
binuclear Ru catalysts 182
biodegradable polyesters 270
biodegradable polymers 379–383
biodegradation, organic compounds 186
bioxazoline-derived NHC ligands 149
$\{BIPM^{TMS}\}^{2-}$ ligand 351
BippyPhos 110–111, 130
  hydroxylation of (hetero)aryl halides 121–122
bis-ene-amido complexes 226
bis-iminophenolate ligands, non-innocent 187
bis-iminopyridine Co(II) complex 183
bis(amidate) complexes, isomers 371
bis(amidate) tantalaziridines 392
bis(amidate) titanium complex 377
bis(amidate)dichloro complexes 369
bis(diphenylphosphino)methane, *see* dppm
bisphosphine-borane Rh complexes 245
bisphosphine ligands 9
bis(pyrazolyl)methane ligands 275
bis(pyridonate) complexes 374
bis(sulfonamidate) titanium species 375
bis(ureate) complexes 373
bond insertion 91–95
bonding
  alkylic/aromatic C-H bonds 322
  C-C 85–91, 140–141, 184, 190

carbon–heteroatom 91–98
covalent bond classification 2–4
C(sp^2)- 104–133
C(sp^3)-C(sp^3) 390
elementary concepts 2–4
ionic 331
metal–ligand 1
multiple 350–356
Nalewajski–Mrozek bond index 353
O-O 183
polar 211, 352
proton-assisted C-N 389
saturated bonds 315
sp^3-sp^3/sp^2-sp^3 bonding 156–158
transannular 246
boranes 255–262, 244
boron-free Buchwald-type ligand 254
boronic acid 149
ortho-boryl phosphines 253–255
"bottom-bound" pathway, metallacycles 19–21
bridged metallaziridine species 396–397
bridging coordination 243–248
M–X bonds 248–250
bromides
aryl 114
cyclohexylmagnesium 147
(hetero)aryl 127–128
isopropenylmagnesium 148
Pd-catalyzed cross-coupling 119
see also halides
3-bromo-5-phenylpyridine 120
2-bromoarylacetylenes, functionalized 110
bromomesitylene 121
Büchner reaction 313
Buchwald–Hartwig amination (BHA) 104–105, 108–112
Buchwald palladacycles 112–114
Buchwald-type ligand, boron-free 254
buried volume (V_{bur}) 11–12, 137, 152

C-C bonding
cross-coupling 140–141
redox non-innocent ligands 184
reductive elimination 190
spiro ligands 85–91
C-donor ligands 333–344
C-H alkylation 393
C-H bonds, functionalization 322
C-N bonding, proton-assisted 389
C_2-symmetric ligands 47
CAACs (cyclic (alkyl)(amino)carbenes) 30
cage complexes 248
ε-caprolactone 379–383
carbenes
addition 312–314
Ce complexes 353
Fischer 196
insertion 314–319
Ln 350
NHC, *see* *N*-heterocyclic carbenes (NHC)
Pd-*N*-heterocyclic complexes 134–175
scandium complexes 354
TEP values 7
transfer 196
transition metal 15
carboamination 98
carbodiimides 191
carbon dioxide, supercritical 317
carbon–heteroatom bonding 91–98
carbonyl compounds 86–87
carbonyl groups 114
carboxylic acids
heterocyclic 76
unsaturated 57–58, 75–78
carboxylic esters 63
carboxylic ketones 63
Carnegine 74–75
catalysis
asymmetric 66–103
coinage metal 308–329
N,O-chelates 376–399
redox non-innocent ligands 176–204
catalysts
achiral 287
acyclic diaminocarbene-based 31
adamantyl-activated 37, 39
(alkyl)(amino)carbene-based 30
asymmetric hydrogenation 205–236
binuclear Ru 182
catechol-based *Z*-selective 35
CGC 338–339
chiral, *see* chiral catalysts
constrained geometry 10, 338–340
Crabtree 60
cyclometallated *Z*-selective metathesis 36–42
deactivation pathway 51

catalysts (*cont'd*)
dithiolate-based Z-selective 34–36
Fe, *see* iron … 213–216
first-generation 25–26
fluorinated 317
hydrosilylation 261
ill-defined 16
lactide ROP 282–283
mesityl-activated 37, 39
monophosphate metathesis 32
monopivalate Z-selective 36
nitrate-substituted 40
Noyori-type Ru 207
oxazoline-based 53–54
phosphanyl oxazoline-derived 55
phosphonium alkylidene 21
pivalate-substituted 40
reactivity of Z-selective 42
Rh-, *see* rhodium …
ROP 272
Ru-, *see* ruthenium …
Sc, *see* scandium …
Schrock's olefin metathesis 8
second-generation 25–26
Sharpless' epoxidation 8
SIPHOX-based 55–57, 69–70
structure 15–45
thiazolium-based 29
thiophenolate-based Z-selective 33
Ziegler–Natta 376
Zr, *see* zirconium … 399
"catalytic pocket" 310
catechol-based Z-selective catalysts 35
cationic iridium complexes 75
cationic iron complexes 215
cerium complexes
carbene 353
oxidation 348
oxo 356
tethered NHC 343
CGC (constrained geometry catalysts) 10, 338–340
chalcogen bridges 278
chelates
adamantyl-based 41
iridium complexes 55
N,O- 363–404
optimal ring size 215
chemoselectivity 115–117, 397
α-chiral amines 388
chiral auxiliaries 67
chiral biarylphosphite moiety 54
chiral binaphthyl compounds 153–155
chiral catalysts, "privileged" 66, 71
chiral diaminophenolate ligands 292–297
chiral indium salen complexes 297
chiral inducing models 89
chiral ligand elements 207–208
"chiral pocket" 70
chiral proligands 293
chiral spiro ligands 66–72
chloramine-T 319
chlorides
aryl 114
Pd-catalyzed cross-coupling 119
see also halides
cholesterol, treatment of high cholesterol 123
ciprofloxacin 123
cis-cyclopropanation olefins 312
cis-1,4-diacetoxy-2-butene (CDAB) 24, 29, 31
cyclometallated metathesis 36–37
cis–trans selectivity 15–45
CM, *see* cross metathesis
cobalt complexes 183, 186–187
coinage metal catalysis 308–329
complexes
Ag 309
alkoxide 374
aluminum 277–278, 285–286
amido 209, 374
Au 246–247, 251
bis-ene-amido 226
bis-iminopyridine Co(II) 183
bis(amidate) 371
bis(amidate)dichloro 369
bis(pyridonate) 374
bis(ureate) 373
cage 248
cationic iridium 75
cationic iron 215
Ce 343, 348
Ce carbene 353
chelate, *see* chelates
chiral indium salen 297

cobalt 183, 186–187
coordination 176
copper 246–247, 309
diamagnetic iron 213–214
dianion 341
dimeric 398
dinuclear Pd 146
dinuclear titanium-sulfonamide 395
ethoxide 294
Ga 278
hafnium alkoxide 289
indium 294–297
Ir 61, 63
iron, *see* iron complexes
Ln-arene 340–341
Ln benzene 340–342
Ln bis(arene) 342
Ln-imido 354–355
magnesium 275
metal, *see* metal complexes
metal hydride 208
naphthoquinone 142
neodymium-gallyl 344
Ni 183, 250
open-shell porphyrin 195
oxidation states 177
oxo 187, 356
paramagnetic iron 214
Pd, *see* palladium ...
pendant Lewis acids 242
pentadentate ferraaziridine 218–219
pincer ligands 10, 191, 197
Re 187–188
reactivity 251–264, 256
SalBinap 298
salt metathesis 376
Sc carbene 354
Sm methanediide 351
square planar 6
tantalum amidate 372
tantalum pyridonate 374
tethered bis(amidate) titanium 377
Ti amidate 383
Ti imido 386
Ti pyridonate 383
Ti sulfonamidate 381
Ti ureate 378
transition metal 309
triphosphine-borane 246
v 336–338
Y amidate 381
Y methanediide 352
Zn 275–277, 292–294
Zr, *see* zirconium ...
see also ligands
condensation, Schiff base 210
cone angle
ligands 11, 136
Tolman 136, 224
conformation, “open-book” 351
conjugate ketones 79
constrained geometry catalyst (CGC) 10
Cp 338–340
cooperative activation 258–261
cooperative hydrogenation, styrene 260
cooperative ligand-centered reactivity 192–195
cooperative non-innocent ligands 12
cooperative substrate-centered radical-type reactivity 195–200
coordination chemistry 308
ambiphilic ligands 237–269, 242–251
bridging 243–248
coordination complexes, reactivity 176
coordination geometry 5
irregular 331
coordination–insertion mechanism 273
coordination sphere 60, 216
lanthanides 333
Ru 258
second 262
copper-catalyzed N-H bond insertion 91–93
copper-catalyzed O-H bond insertion 93–95
copper complexes
phosphine-borane 246–247
Tp^x ligands 309
“core-like” valence orbitals 330
coupling
cross-, *see* cross-coupling
dehydrogenative 252–253
“head-to-tail” 338
KTC 141–148
Negishi 163–170
Ni-catalyzed three-component 87–89
reductive 88, 263
Suzuki–Miyaura, *see* Suzuki–Miyaura coupling

covalent bonds, classification 2–4
Cp, *see* cyclopentadienyl
Crabtree catalyst 60
crispine A 74–75
cross-coupling
  C-C bonds 140–141
  (hetero)aryl triflates 124
  Pd-catalyzed 104–133
  Pd-*N*-heterocyclic carbene complexes 134–175
  selective 168–170
cross metathesis (CM) 17
  productive 20
crystalline PLA 271, 287
C(sp^2)-bonding 104–133
C(sp^3)-C(sp^3) bonding 390
Cu, *see* copper
cyano groups 114
cyclic (alkyl)(amino)carbenes (CAACs) 30
cyclic esters 270
  polymerization 295–296
  ROP 379–384
cyclization
  atom transfer radical 324
  intramolecular 388
  Pd-catalyzed allylic 97–98
  primary aminoalkenes 396
  Rh-catalyzed 89–90
cycloaddition, Au-catalyzed 90–91
[2 + 2] cycloaddition 386
  ring-closure reactions 186
[2π + 2π] cycloaddition 184
cyclohexene 181
cyclohexyl backbones 298
cyclohexylmagnesium bromide 147
cyclometallated NHCs 37
  substituents 41
cyclometallated Z-selective metathesis catalysts 36–42
1,5-cyclooctadiene 121, 127
cyclopentadienyl (Cp)
  constrained geometry 338–340
  N,O-chelates 363
  substituted 336–338
cyclopropanation 314, 91
cis-cyclopropanation, olefins 312
cyclopropenation, alkynes 314
CyPF-tBu 109
DalPhos ligands 115, 130
deactivation pathway, catalysts 51
dearomatization 125
dehydrogenation, alkanes 324
dehydrogenative coupling 252–253
density functional theory (DFT) calculations
  asymmetric hydrogenation 57–59
  ATH 226
  metathesis catalysts 33–34
  Negishi coupling 167
  NHC ligands 137
  Suzuki–Miyaura coupling 151
deprotonation, CM 367
design, ligand, *see* ligand design
desvenlafaxine 117
cis-1,4-diacetoxy-2-butene (CDAB)
  CM 24, 29, 31
  cyclometallated metathesis 36–37
diamagnetic iron complexes 213–214
diamidoamino ligands 276–277
diamidoether ligands 278–279
diaminocarbene-based catalysts, acyclic 31
(*R,R*)-diaminocyclohexane 225
diaminophenolate ligands 292–297
dianion complexes 341
diarylmethane products 159
diarylphosphinoaldehyde diethylacetals 222
diastereoselectivity
  alkene carboamination 98
  C-H alkylation 393
  cyclopropanation 312
α-diazocarbonyl compounds 91
α-diazoesters 92
dibenzodiazepines 112
1,4-dibromobenzene 165
2,6-dichloro,3-nitropyridine 254
Diels–Alder reaction, spiro ligands 67
dienes 18, 184
diethylacetals 221–222
1,2-dihydropyridines 321
N′-2,6-diisopropylphenyl (DIPP) 21, 28–29
  N-aryl group variation 40–41
dimeric complexes 398
dimerization
  prevention 107
  secondary alcohols 193
dimethyl itaconate 48
dinitrogen ligands, chiral 68

dinuclear In complexes 296–301
dinuclear Pd complex 146
dinuclear titanium-sulfonamide complex 395
diphenyl hydrazine 189
direct C-H alkylation 393
direct oxidation, aromatic C-H bonds 324
dispiro[2.0.2.1]heptane 67
disproportionation, diphenyl hydrazine 189
1,*w*-dithiaalkanediyl-bridged bis(phenolate) (OSSO) 291
dithiolate-based *Z*-selective catalysts 34–36
donicity, ligand 7
donor sites, ambiphilic ligands 238
double bonding, metal–amide 3
dppm 9
drug-like heterocycles 143
dynamic kinetic resolution (DKR) 81–83

"E" transfer 311
early transition metals 363–365
  polymerizations 376–384
EDA, *see* ethyl diazoacetate
electron shuttle 12
electronic configuration 4
electronic differentiation 50
electronic factors, stabilizing 38
electronic parameters, NHC ligands 138–140
electrophilic metal center 394
elimination
  β-hydride 212, 332
  reductive 106, 190
enamides, hydrogenation 73–75
enantiopure ligands 228
enantioselectivity
  asymmetric hydrogenation 57–63
  homogeneous hydrogenation 46
  O-H insertion 94
  ring-opening polymerization 270–307
  Suzuki–Miyaura coupling 153–156
enynes 398–399
  [2π + 2π] cycloaddition 184
enzymes 177, 192
epilepsy treatment 76
epoxidation
  Sharpless' catalyst 8
  styrene 323
esters
  β-arylamino acid 97
  carboxylic 63
  cyclic 270, 295–296, 379–384
  Ir-catalyzed hydrogenation 80
  pinacol 158
ethane, functionalization 317
ethene, purification 194
ethers, substituted 122
ethoxide complexes 294
ethyl diazoacetate (EDA) 312–316
  epoxidation 323
  intermolecular modification 318
ethylenes
  capture 195
  ethylene-derived metallacycles 23
  polymerization 377
E/Z selectivity 24–32

faces, *Re* and *Si* 78
Fe, *see* iron
ferraaziridine complexes, pentadentate 218–219
first-generation catalysts 25–26
Fischer carbenes 196
"flexible steric bulk" 149–150, 165, 170
fluorinated catalysts 317
fluorination, nucleophilic 123–129
N'-fluorophenyl NHCs 27
functionalization
  2-bromoarylacetylenes 110
  by "E" transfer 311
  C-H bonds 322
  furans 314
  hexane 316
  N_2 and CO 262–264
  polyolefin 316
furans 321, 314
fused hetero spiro backbone 68

galactose oxidase 192
gallium complexes 278
geometry
  anti 22
  bent 334
  constrained 10, 338–340
  coordination 5, 331
  irregular 331
  metallacycles 19–22
  syn 22

gold-catalyzed [2 + 2] cycloaddition 90–91
gold-catalyzed cyclopropanation 91
gold-catalyzed Mannich reactions, azlactones 89
gold complexes 246–247, 251
Grignard reagents 134
  organomagnesium 141

hafnium alkoxide complexes 289
halides
  (hetero) aryl 104, 114, 123–129
  KTC coupling 147–148
  organic 134
  selective hydroxylation 117–122
haloalkanes 318
halopyridines 161–162
"head-to-tail" coupling 338
Heck reactions, Pd-catalyzed 91
hemilabile ligands 368
hemilability 8–10
hetero aryl pseudohalide 104, 114
  nucleophilic fluorination 123–129
  selective hydroxylation 117–122
heteroanilines 108
heteroaryl compounds 158–163
heteroaryl triflates 124–125
heterobiaryls 158, 254
heterobidentate ligands 49–57, 115–117
heterocycles
  drug-like 143
  Negishi coupling 170
  synthesis 110–112
N-heterocyclic carbenes (NHC) 16
  alternative 29
  as catalysts 135–136
  bioxazoline-derived ligands 149
  bridges 278
  cyclometallated 37, 41
  electronic parameters 138–140
  imidazoline-based ligands 136
  orientation 22
  Pd-catalyzed cross-coupling
  steric parameters 136–137
  tethered 343–344
  unsymmetrical 26–29
heterocyclic carboxylic acids 76
heteroscorpionate ligands 274–276
hexamethyldisilazide 344–347
hexane, functionalization 316
homodimerization 40
homodimers 18–19
homogeneous hydrogenation, enantioselective 46
homogeneous ligands, iron-based 205–236
homoleptic Ln(II) alkyls 334
homoscorpionate ligands 274–276
hydrazine, diphenyl 189
hydride cluster, iridium 50–51
β-hydride elimination 212
  lanthanides 332
hydride–metal–nitrogen–proton motif 207–208
hydride relay 257
hydroacylation, Rh-catalyzed 85–86
hydroalkynylation 398–399
hydroamination 110
  intramolecular 387
  N,O-chelates 384–390
  regioselective 385
hydroaminoalkylation 390–398
  intramolecular 391, 395
hydrofunctionalization, *N,O*-chelates 384–399
hydrogenation 255–262
  ADH 205, 212
  asymmetric, *see* asymmetric hydrogenation
  ATH 206–207, 212
  cooperative 258–261, 260
  enamides 73–75
  enantioselective homogeneous 46
  Ir-catalyzed, *see* iridium-catalyzed hydrogenation
  Rh-catalyzed 73
  Ru-catalyzed 81–83
  selectivity 53
hydrosilylation 214, 255–262
  ketones 261
  Rh-catalyzed 89–90
hydrovinylation 85
hydroxylation, selective 117–122
hydroxyphthioceranic acid precursor 61

ill-defined catalysts 16
imidazole-based ancillary ligands 120
imidazole-derived monophosphines 112–114

imidazole-2-ylidine NHCs 138
imidazoles 136
imidazoline-based NHC ligands 136
imido complexes 354–355, 386
imines 72
asymmetric hydrogenation 205–236
catalytic hydrogenation 261–262
Ir-catalyzed hydrogenation 84
Pd-catalyzed umpolung allylation 87
prochiral 206
reductive coupling 88
Rh-catalyzed arylation 86–87
iminophosphine ligands 209–212
in situ arylation, ancillary ligands 128
indium complexes 294–297
indoles 111
inducing models, chiral 89
interatomic distance, covalent bonds 3
intramolecular cyclization 388
intramolecular hydroacylation 85
intramolecular hydroamination 387
intramolecular hydroaminoalkylation 391, 395
iodides, (hetero)aryl 127–128
ionic bonding 331
ionization, M–X bonds 250–251
irbesartan 160
iridium
complexes 75
dihydride species 50
oxidation of primary alcohols 194
proline-based complexes 63
pyridine-based complexes 61
complexes 256
iridium-catalyzed hydrogenation 73–81
asymmetric 46–65
esters 80
imines 84
ketones 79–81
olefins 78–79
iron catalysts
ligand design 213–216
O-H bond insertion 93–95
symmetrical ligands 216–227
synthetic routes 217–218
iron complexes 185–186
cationic 215
diamagnetic 213–214
paramagnetic 214
template-assisted synthesis 228–229
triphosphine-borane 247
iron ligands 205–216
iron-based homogeneous ligands 205–236
irregular coordination geometry 331
isocyanate insertion 367
isocyanides 191
isomerization, olefin-migration 38
isomers
bis(amidate) complexes 371
trisubstituted 53
isopropenylmagnesium bromide 148
itaconate 48

JosiPhos 109, 130

ketenes 196
ketiminate ligands 279–281
ketoesters 80
ketones
asymmetric hydrogenation 205–236
carboxylic 63
conjugate 79
hydrosilylation 261
Ir-catalyzed hydrogenation 79–81
Pd-catalyzed umpolung allylation 87
racemic 2-substituted 82–83
transfer hydrogenation 258–261
Kharasch reaction 325
kinetic resolution, dynamic 81–83
kinetics, lanthanide chemistry 331
Kumada–Tamao–Corriu (KTC) coupling 141–148

L ligands 2
lactides 270–307
racemic, *see* racemic lactides
ROP 379–383
ROP catalysts 282–283
stereoisomers 271
lanthanides, ligand design 330–362
lanthanum alkylidenes 350
lanthanum arene complexes 340–341
lanthanum bis(arene) complexes 342
lanthanum carbenes 350
lanthanum imido complexes 354–355
levothyroxine thyroid 117

Lewis acid–base properties, metals 178–181
Lewis acid moieties 237–269
Lewis acids
 enhancement effect 251–255
 pendant 242–243
Lewis bases 179–181
 adduct 252
ligand bite angle 8–10
ligand-centered reactivity 192–195
ligand design
 ambiphilic ligands 238–241
 enantioselective ring-opening polymerization 270–307
 iminophosphine 209–212
 key concepts 1–14
 lanthanide chemistry 330–362
 "rational" 126
 ruthenium ligands 205–212
ligands
 "actor" 176–179
 air-stable 109–112, 115–116, 253
 alkylidene 38
 ambiphilic, *see* ambiphilic ligands
 amidate 369–372
 amine derivatives 135
 aminophosphine 77
 ancillary, *see* ancillary ligands
 anionic 31–32, 38–40
 biaryl 69
 biaryl monophosphine 112–113, 118–119
 $\{BIPM^{TMS}\}^{2-}$ ligand 351
 bis-iminophenolate 187
 bisphosphine 9
 bis(pyrazolyl)methane 275
 boron-free Buchwald-type 254
 C-donor 333–344
 C_2-symmetric 47
 chiral 68, 207–208, 292–297
 cone angle 11, 136
 DalPhos 115, 130
 donicity 7
 enantiopure 228
 Fe-based homogeneous 205–236
 hemilabile 368
 heterobidentate 49–57
 homo-/heteroscorpionate 274–276
 iminophosphine 209–212
 Ir-catalyzed asymmetric hydrogenation 46–65
 JosiPhos 109
 ketiminate 279–281
 linker 240
 macrocyclic 284
 modifications 296–297
 N-donor 344–349
 NacNac 349
 NHC, *see* *N*-heterocyclic carbenes
 nitrogen donor 282
 non-innocent, *see* non-innocent ligands
 P-donor 349–350
 participation in substrate activation 13
 phosphane derivatives 135
 phosphine, *see* phosphine ligands
 phosphinomethyl-oxazoline 53
 phosphorous 67
 PHOX 49–52
 pincer 191, 197
 privileged polydentate 214–215
 pyridonate 374
 quinone/semiquinone 182
 Re 193–195
 reactive 4
 salen-type 285
 scrambling 332
 SerPHOX 52
 "spectator" 178–179
 spiro 66–103
 steric properties 10–12
 sterically demanding 104–133, 389
 strong-field 4–6
 strong field 213–214
 sulfonamidate 375–376
 sulfonamide 277
 symmetrical 216–227
 tetradentate 211, 282–284
 thiolphan 292
 thiophosphine 193
 ThrePHOX 52
 "throw-away" 144
 Tomioka's 61
 Tp^x, *see* trispyrazolylborate
 tricyclohexylphosphine 16
 tridentate 276–279
 tripodal 274–276, 282–284
 unsymmetrical 227–230

ureate 373–374
weak-field 4–6
Z-type 243–244
see also complexes
κ^2-P,N ligands, heterobidentate 115–117
linker, propylene 287
linker ligands 240

M–X bonding, bridging coordination 248–250
macrocyclic ligands 284
macromolecule modification 313
magnesium complexes 275
Mannich reactions 89
Markovnikov addition, anti- 324
MeCN 95–96, 184–185, 222–223, 324
mesityl-activated catalysts 39
targeted 37
mesityl-based chelates 41
N-mesityl NHCs 27
metal–amide bonding 3
metal complexes 366–376
acidity 208
electronic configuration 4
octahedral 5
preparation 366–376
reactivity 251–264
metal–ligand bonding 1
metallacycles
ethylene-derived 23
geometry 19–22
N,O-chelates 364
syn/anti 22–24, 35, 42
thiophenolate-based catalysts 34
metallacyclobutane 21, 37
metallaziridine, bridged species 396–397
metal—ligand cooperativity 13
metallocene 337
metals
coinage 309
electrophilic center 394
hydride–metal–nitrogen–proton motif 207–208
Lewis acid–base properties 179–181
oxophilic centers 363, 369
ROP catalysts 272
trivalent 288–289
metathesis
ADMET 18
CM, *see* cross metathesis
cyclometallated *Z*-selective catalysts 36–42
monosulfonate catalysts 32
olefins, *see* olefin metathesis
RCM 17
reaction types 17–19
ROMP 17, 35
salt, *see* salt metathesis
secondary 24
methane, functionalization 317
methanediide complexes 351–352
methyl stilbene, *trans*- 51
methyl(3-trifluoromethylphenyl)ketone 207
(*S*)-metolachlor 48
microwave-assisted Suzuki–Miyaura coupling 161–162
migratory insertion 59
milrinone 160
"molecular flask" 197–198
monoanionic *N,O*-chelates 364
monoarylation selectivity 107–117
monophosphate metathesis catalysts 32
monophosphine ligands, biaryl 112–113, 118–119
monophosphines 112–114
monopivalate *Z*-selective catalysts, targeted 36
monosulfonate metathesis catalysts 32
Mor-DalPhos 115–116, 130
multidenticity 8–10
multielectron reactions 181
multiple bonds, lanthanide chemistry 350–356

N-donor ligands 344–349
N-H bond insertion, Cu-catalyzed 91–93
NacNac ligand 349
Nalewajski–Mrozek bond index 353
naphthalenes, 1-phosphino 2-boryl 241
naphthoquinone complexes 142
α-naphthylmethyl groups 97
Negishi coupling 163–170
alkyl–alkyl 165–167
one-pot synthesis 166
neodymium-gallyl complex 344

NH effect 207
NHC, *see N*-heterocyclic carbenes
nickel-catalyzed hydrovinylation, olefins 85
nickel-catalyzed three-component coupling 87–89
nickel complexes 183
  zwitterionic 250
nitrate-substituted catalysts 40
nitrenes
  addition 319–321
  insertion 199, 321–322
  transfer 196
nitrile 258
nitrogen donor ligands 282
N,O-chelates 363–404
  bidentate 364
  hydrofunctionalization 384–399
N,O-proligands 365–366
non-innocent ligands 12
  redox 176–204
nonanes, 2-azabiciclo[3.3.1] 326
Noyori-type Ru catalysts 207
nucleophilic fluorination 123–129

O-H bond insertion 93–95
O-O bonding 183
occlusion, lanthanide chemistry 332
octahedral metal complexes 5
olefin metathesis
  cis–trans selectivity 15–45
  Ru-based 15–45, 19, 30, 33–36
  Schrock's catalyst 8
olefins
  aziridination 319–320
  cis-cyclopropanation 312
  Ir-catalyzed hydrogenation 78–79
  migration isomerization 38
  Ni-catalyzed hydrovinylation 85
  nitrene transfer 319
  oligomerization 339
  polar 72
  Sc polymerization catalysts 339
  unactivated 393
α-olefins, polymerization 377
oligomerization, regiospecific 338
one-pot synthesis
  Negishi coupling 166
  substituted ethers 122
ONNO 297
"open-book" conformation 351
open-shell porphyrin complexes 195
organic amide proligands 372
organic azides 191
organic compounds, biodegradation 186
organic halides 134
organic substrates, functionalization by "E" transfer 311
organoboron reagent 148–149
organomagnesium reagent 141
organomercury reagents 335
organometallic chemistry 308
  lanthanides 335–336
  transmetalators 141
organometallic complexes, salt metathesis 376
organozinc reagents 168–170
ORTEP diagram 370–371, 392
ortho-boryl phosphines 253–255
ortho-quinone/semiquinone ligands 182
ortho-substituted biaryl products 144–146
  Suzuki–Miyaura coupling 149–153
OSSO (1,*w*-dithiaalkanediyl-bridged bis(phenolate)) 291
oxazoline-based catalysts 53–54
oxidation
  aerobic 188
  cerium complexes 348
  direct 324
  primary alcohols 194
  water 181
oxidation states, complexes 177
oxidative addition 11
  C(sp^2)-bonding 106
oxo Ce complexes 356
oxo complexes 187
oxo transfer reactions 322–324
oxone 323
oxophilic metal centers 363, 369
oxophilicity, lanthanides 332
oxygen donors 284

P-donor ligands 349–350
palladacycles 112–114, 146
palladium catalysts
  alkene carboamination 98
  allylic amination 95–97

allylic cyclization 97–98
cross-coupling 104–133
Heck reactions 91
hydroxylation 119
O-H bond insertion 93–95
Suzuki–Miyaura coupling 253–255
umpolung allylation 87
palladium complexes
dinuclear 146
NHC 134–175
Pd-PEPPSI complexes 143–145
pendant Lewis acids 242–243
paramagnetic iron complexes 214
PCET (proton-coupled electron transfer) 183–184
$[Pd(cinnamyl)Cl]_2$ 111, 116
nucleophilic fluorinations 124
Pearson acid base concept 8
pendant Lewis acids 242
pentadentate ferraaziridine complex 218–219
PEPPSI, Negishi coupling 164, 169
percent buried volume (V_{bur}) 11–12, 137, 152
pharmaceuticals 1, 69
anti-bacterial 123
anti-depressant 117
drug-like heterocycles 143
fluorinated aryl groups 123
(hetero)anilines 108
irbesartan 160
milrinone 160
(*R*)-tiagabine 76
rivastigmine 81
synthons 117
phenols 117
phenylalkylketone 206
phosphane derivatives 135
phosphanyl oxazoline-derived catalysts 55
phosphazene 119
phosphine aldehydes 215
"protected" 220
phosphine boranes 244
phosphine ligands
cone angle 11
mono-, *see* monophosphine ligands
steric parameters 137
sterically demanding 104–133
phosphines
biaryl 253–255
ortho-boryl 253–255
TEP values 7
vinyl 239
1-phosphino 2-boryl naphthalenes 241
α-phosphinoacetaldehyde diethylacetals 221
phosphinomethyl-oxazoline ligands 53
phosphinooxazoline (PHOX) 49–51
phospholides 349–350
phosphonium alkylidene catalyst 21
phosphonium dimers 220–221
phosphoramidates 366, 376, 394
phosphorous ligands, chiral spiro 67
pinacol ester derivatives 158
pincer complexes 10
pincer ligands 191
redox non-innocent 197
pivalate-substituted catalysts 40
PLA, *see* poly(lactic acid)
platinum complexes, pendant Lewis acids 242
"pocket"
catalytic 310
chiral 70
polar bonds 211
polar olefins 72
polarized bonding 352
polyarylbenzenes 152
polydispersity index 380
polyesters, biodegradable 270
polyheteroaryl compounds 158–163
poly(lactic acid) (PLA)
salen-type ligands 286
tacticity 270–272
polylactide synthesis 381
polymerization
alkenes 377–379
atom transfer radical 324–326
cyclic esters 295–296
early transition metals 376–384
enantioselective ROP 270–307
ethylenes 377
olefins 339
α-olefins 377
ring-opening, *see* ring-opening polymerization
polymers, biodegradable 379–383

polyolefin functionalization 316
porphyrin complexes 195
potassium graphite columns 337
precatalysts
  hydroamination 385
  phosphoramidate tantalum alkyl 394
  tantalum 391
primary alcohols, oxidation 194
primary aminoalkenes 387
  cyclization 396
"privileged chiral catalysts" 66, 71
privileged polydentate ligands 214–215
prochiral imine 206
prochiral phenylalkylketone 206
productive CM 20
proligands
  chiral 293
  deprotonation 367
  N,O- 365–366
  organic amide 372
proline-based iridium complexes 63
propylene linker 287
"protected" phosphine aldehydes 220–221
proton-assisted C-N bonding 389
proton-coupled electron transfer (PCET) 183–184
proton exchange 23
proton shuttle 12
protonolysis 367, 369
pseudohalides
  (hetero) aryl 104, 114, 117–129
purification, ethene 194
pyrazolyl ring 324
pyrazolylborates 317
pyridine-based iridium complexes 61
pyridine-phosphinite ligands 61–62
2-pyridonate 366
pyridonate complexes, titanium 383
pyridonate ligands 374
  sterically demanding 389

quadrant model, Andersson's 60
ortho-quinone/semiquinone ligands 182

racemic lactide 274–292
  polymerization 299–300
  ROP 379–383
racemic 2-substituted aldehydes 81–82
racemic 2-substituted ketones 82–83
radical reactions, atom transfer 324–326
radical-type reactivity 195–200
"rational" ligand design 126
RCM (ring-closing metathesis) 17
Re face 78
reactive ligands 4
reactivity
  ambiphilic ligands 237–269
  C(sp^2)-bonding 107–108
  cooperative ligand-centered 192–195
  coordination complexes 176
  cyclometallated *Z*-selective catalysts 42
  (hetero)aryl bromides and iodides 127–128
  Ir complexes 256
  metallic complexes 251–264
  Negishi coupling 164–165, 168
  N,O-chelates 376–399
  palladacycles 146
  radical-type 195–200
  redox non-innocent ligands 176–204
  Rh complexes 256
  structure–reactivity relationship 136–140
  substrate-centered 195–200
  Suzuki–Miyaura coupling 153, 155, 157, 161
  Wittig-type 353
redox non-innocent ligands 12, 176–204
redox transmetalation 335–336
reduction
  nitrile 258
  styrene 259
reductive coupling 88
  CO 263
reductive elimination
  C(sp^2)-bonding 106
  C-C bonding 190
regioselective hydroamination 385
regiospecific oligomerization 338
"repurposing", ancillary ligands 110
respective control concept 47
reversible activation 261
rhenium complexes 187–188
rhenium ligands 193–195

rhodium-catalyzed arylation 86–87
rhodium-catalyzed asymmetric hydrogenation 47
rhodium-catalyzed hydroacylation 85–86
rhodium-catalyzed hydrogenation, enamides 73
rhodium-catalyzed hydrosilylation/cyclization 89–90
rhodium complexes
  bisphosphine-borane 245
  pendant Lewis acids 242
  reactivity 256
rigidity, spirobiindane ligands 69–70
ring-arylated RockPhos 126
ring-closing metathesis (RCM) 17
ring-closure reactions, [2 + 2] cycloaddition 186
ring dearomatization 125
ring-opening metathesis polymerization (ROMP) 17
  Z-selectivity 35
ring-opening polymerization (ROP)
  catalysts 282–283
  cyclic esters 379–384
  enantioselective 270–307
  metal catalysts 272
ring-opening/cross metathesis (ROCM) 35
rivastigmine 81
RockPhos 130
  ring-arylated 126
room temperature reactions 115–117
  Suzuki–Miyaura coupling 151
ruthenium-based olefin metathesis 15–45
  cyclic catalysts 30
  mechanisms 19
  Z-selective 33–36
ruthenium catalysts
  binuclear 182
  hydrogenation 81–83
  Noyori-type 207
ruthenium complexes 242
ruthenium ligands 205–216

S-H bond insertion 95
SalBinap complexes 298
salen complexes, chiral indium 297
salen-type ligands 285
salt metathesis 367–369
  diaminophenolates 294
  organometallic complexes 376
  silylalkyls 333–335
salts, transition metal 134
samarium methanediide complex 351
saturated bonds 315
scandium catalysts, olefin polymerization 339
scandium complexes, carbene 354
Schiff base condensation 210
Schrock's olefin metathesis catalyst 8
scrambling, ligands 332
second coordination sphere 262
second-generation catalysts 25–26
secondary alcohols, dimerization 193
secondary amines, hydroaminoalkylation 391
secondary metathesis 24
secondary organozinc reagents 168–170
selective hydroxylation 117–122
selectivity
  aziridination 320
  diastereo-, *see* diastereoselectivity
  chemo-, *see* chemoselectivity
  cis–trans 15–45
  cross-coupling 168–170
  enantio-, *see* enantioselectivity
  E/Z 24–32
  hydrogenation 53
  monoarylation 107–117
  regio-, *see* regioselectivity
  Z-selective ruthenium-based olefin metathesis 33–36
semiquinonate 192
SerPHOX ligands 52
Sharpless' epoxidation catalyst 8
Si face 78
Si-H bond insertion 95
"side-bound" pathway, metallacycles 19–21
silanes 258–261, 318
silver complexes, Tp^x ligands 309
silylalkyls 333–335
silylamides 344
single bonding, metal–amide 3
single molecule magnets (SMMs) 346
SIPHOX-based catalysts 55–57
  spiro ligands 69–70

solution chemistry, lanthanides 330
sp^3 C-H amination 200
sp^3-sp^3/sp^2-sp^3 bonding 156–158
"spectator ligands" 178–179
spin state, transition metal complexes 4–5
spiro-2,2′-bischroman scaffold 72
spiro ligands 66–103
1,1′-spirobiindane scaffold 69–70
spiro[4.4]nona-1,6-diene backbones 72
spiro[4,4]nonane-1,6-diol 67
SpiroNP 68
spiro[2.2]pentane 67
square planar complexes 6
stabilizing electronic factors 38
stereo-recognition model
  Ir-catalysts 77–78
  Rh-catalyst 87
stereoisomers 59, 271
steric differentiation 50
steric hindrance
  phosphine ligands 104–133
  pyridonate ligands 389
  Tp^x ligands 310
steric parameters
  ligands 10–12
  NHC ligands 136–137
  phosphine ligands 137
stilbene, *trans*-methyl 51
strong-field ligands 4–6, 213–214
structure, catalysts 15–45
structure–activity relationship 26
  KTC coupling 144
structure–reactivity relationship, NHC 136–140
styrene
  asymmetric cyclopropanation 314
  epoxidation 323
  reduction 259
substituted Cp 336–338
substituted ethers, one-pot synthesis 122
substituted indoles 111
substituted trispyrazolylborates 347–348
substrate activation 13
substrate-centered reactivity 195–200
sulfonamidate 366
  ligands 375–376
  Zr complexes 382
sulfonamidate , Ti complexes 381
sulfonamide 277, 395
supercritical CO_2 317
Suzuki–Miyaura arylation 156
Suzuki–Miyaura coupling 148–163
  alkyl–alkyl 156
  asymmetric 153–155
  enantioselective 153–156
  microwave-assisted 161–162
  ortho-boryl phosphines 253–255
  Pd-catalyzed 253–255
symmetrical ligands, iron catalysts 216–227
syn–anti preference, metallacycles 22–24
syn geometry 22
synthons 117

tacticity 270–272
  biodegradable polymers 379
  racemic lactide 300
tantalaziridines 392
tantalum alkyl precatalyst 394
tantalum amidate complexes 372
tantalum precatalysts 391
tantalum pyridonate complexes 374
targeted mesityl-activated catalysts 37
targeted monopivalate Z-selective catalyst 36
Tebbe's reagent 350–351
temperature, room, *see* room temperature reactions
template-assisted synthesis, iron complexes 228–229
TEP, *see* Tolman electronic parameter
terphenyls 335–336
tethered bis(amidate) titanium complex 377
tethered NHCs 343–344
tetradentate ligands 211
  tripodal 282–284
thermodynamics, lanthanide chemistry 331
thiazolium-based catalysts 29
thiolphan ligands 292
thiophenolate-based Z-selective catalysts 33
thiophosphine ligands 193
three-component coupling, Ni-catalyzed 87–89
ThrePHOX ligands 52
"throw-away" ligands 144
(*R*)-tiagabine 76
tight bite angle *N,O*-chelates 363–404
titanatranes 282

titanium amidate complexes 383
titanium bis(sulfonamidate) species 375
titanium imido complex 386
titanium pyridonate complexes 383
titanium sulfonamidate complex 381
titanium sulfonamide complexes,
    dinuclear 395
titanium ureate complexes 378
TM, *see* transition metals
γ-tocotrienyl acetate 60
Tolman cone angle 136, 224
Tolman electronic parameter (TEP) 6–8
  NHC ligands 138–139
Tomioka's ligand 61
Tp^x, *see* trispyrazolylborate
trans effect 6
trans-methyl stilbene 51
transannular bonding 246
transfer hydrogenation 255–262
transition metals
  carbenes 15
  complexes 4, 309
  early 363–365, 376–384
  hydride shuttles 255
  salts 134
  second coordination sphere 262
transmetalation 106
  active species 166–168
  cross-coupling reactions 140–141
  redox 335
"trapped" reactive fragment 351
tricyclohexylphosphine ligands 16
tridentate diamidoamino ligands 276–277
tridentate diamidoether ligands 278–279
triflates, (hetero)aryl 124–125
2,4,6-trimethylphenol 120
trinuclear iridium hydride cluster 51
triphosphine-borane complexes 246
tripodal ligands 274–276
  tetradentate 282–284
trispyrazolylborate (Tp^x) 308–329
  steric hindrance 310
  substituted ligands 347–348
trisubstituted isomers 53
trivalent metals 288–289

umpolung allylation 87
unactivated olefins 393
unsaturated carboxylic acids 57–58
  Ir-catalyzed hydrogenation 75–78
unsymmetrical ligands 227–230
unsymmetrical NHCs 26–29
ureates 363–404
  ligands 373–374
  structure 366
  titanium complexes 378
  Zr catalysts 399

valence orbitals, "core-like" 330
vinyl phosphines 239
vinyl radical intermediates 196
vitamine E precursor 61
volume, percent buried 12

water, oxidation 181
weak-field ligands 4–6
White's solid angle 11
Wittig-type reactivity 353

X ligands 2
Xantphos 9
XPhos 130
  biaryl monophosphines 118
Xyliphos 207

ylidine 138
ytterbium terphenyls 336
yttrium amidate complexes 381
yttrium methanediide complexes 352

Z ligands 2
Z-selective catalysts, reactivity 42
Z-selective ruthenium-based olefin
    metathesis 33–36
Z-type ligands 243–244
zero-field relaxation pathways 346
zerovalent arenes 342
Ziegler–Natta catalysts 376
zinc complexes 292–294
  bis(pyrazolyl)methane ligands 275–277
zincates 167
zirconium complexes 188–191
  salen ligands 289
  sulfonamidate 382
zirconium ureate catalyst 399
zwitterionic complexes 250–251